D1332706

Hadlow College L	
519.5 KAP	Feb 2010
00028294	£35.00

Biostatistics for Animal Science, 2nd Edition

Miroslav Kaps

University of Zagreb, Croatia

and

William R. Lamberson

University of Missouri, USA

www.cabi.org

CABI is a trading name of CAB International

CABI Head Office	CABI North American Office
Nosworthy Way	875 Massachusetts Avenue
Wallingford	7th Floor
Oxfordshire OX10 8DE	Cambridge, MA 02139
UK	USA
Tel: +44 (0)1491 832111	Tel: +1 617 395 4056
Fax: +44 (0)1491 833508	Fax: +1 617 354 6875
E-mail: cabi@cabi.org	E-mail: cabi-nao@cabi.org
Website: www.cabi.org	

© M. Kaps and W. Lamberson 2009. All rights reserved. No part of this publication may be reproduced in any form or by any means, electronically, mechanically, by photocopying, recording or otherwise, without the prior permission of the copyright owners.

A catalogue record for this book is available from the British Library, London, UK.

Library of Congress Cataloging-in-Publication Data

Kaps, Miroslav.
   Biostatistics for animal science / Miroslav Kaps and William R. Lamberson. -- 2nd ed.
      p. cm.
   Includes bibliographical references and index.
   ISBN 978-1-84593-540-5 (alk. paper)
   1. Livestock--Statistical methods. 2. Biometry. I. Lamberson, William R. II. Title.

   SF140.S72K37 2009
   636.0072'7--dc22

                                 2009012455

First edition published in 2004 with ISBN 978 0 85199 820 6

ISBN: 978 1 84593 540 5

Printed and bound in the UK from copy supplied by the authors by MPG Books Group.

The paper used for the text pages in this book is FSC certified. The FSC (Forest Stewardship Council) is an international network to promote responsible management of the world's forests.

List of Chapters

Table of Contents

Preface

We greatly appreciate those who have read and used the first edition of the book and then wrote a review or made a suggestion in personal communications. In response to these suggestions we have extended some topics, added new sections, and tried to improve clarity of the text and examples.

The purpose and main structure of the book have not been changed. Its primary purpose is to help students and researchers to learn about and apply appropriate experimental designs and statistical methods. It has been written mainly for students and researchers of the animal sciences; however, we believe that many sections of the book may be of interest for those studying other agricultural, or biological and veterinary sciences.

Statistical methods applied to biological sciences are known as biostatistics or biometrics, and they have their origins in agricultural research. The characteristic of biological measurements is that they are variable, not only because of measurement error, but also from their natural variability from genetic and environmental sources. Accounting for these sources of variation in making inferences has led to the development of experimental designs that incorporate blocking, covariates and repeated measures.

In this second edition, readers are again presented basic principles of statistics so they will be able to follow subsequent applications with familiarity and understanding, and without having to switch to another book of introductory statistics. Later chapters cover statistical methods most frequently used in the animal sciences for analysis of continuous and categorical variables. Each chapter begins by introducing a problem with practical questions, followed with a brief theoretical background and short proofs. The text is augmented with examples, mostly from animal sciences and related fields, with the purpose of making applications of the statistical methods familiar. Some examples are very simple and are presented in order to provide basic understanding and the logic behind calculations. These examples can be solved using a pocket calculator. Some examples are more complex, especially those in the later chapters. Most examples are also solved using SAS statistical software. Both sample SAS programs and SAS listings are given with brief explanations. Further, the solutions are often given with sufficient decimal digits, more than is practically necessary, so that readers can compare results to verify calculation technique.

The first five chapters of the book are: 1) Presenting and Summarizing Data; 2) Probability; 3) Random Variables and Their Distributions; 4) Population and Sample; and 5) Estimation of Parameters. These chapters provide a basic introduction to biostatistics including definitions of terms, coverage of descriptive statistics and graphical presentation of data, the basic rules of probability, methods of parameter estimation, and descriptions of distributions including the Bernoulli, binomial, hypergeometric, Poisson, multinomial, uniform, normal, *chi*-square, *t*, and *F* distributions. Chapter 6 describes hypothesis testing and includes explanations of the null and alternate hypotheses, use of probability or density functions, critical values, critical region and *P* values. Hypothesis tests for many specific cases are shown such as population means and proportions, expected and empirical frequency, and test of homogeneity of variances. The difference between statistical and practical significance, types of errors in making conclusions, power of test, and sample size are discussed. New topics include an equivalence test for concluding that there is no difference between treatments. There are updates of power analyses incorporating the new procedure POWER in SAS.

Chapters 7 to 10 present the topics of correlation and regression. The coverage begins with simple linear regression and describes the model, its parameters and assumptions. Least squares and maximum likelihood methods of parameter estimation are shown. The concept of partitioning the total variance to explained and unexplained sources is introduced. In chapter 8 the general meaning and definition of the correlation coefficient, and the estimation of the correlation coefficient from samples and testing of hypothesis are shown. Partial correlation is explained. In chapters 9 and 10 multiple and curvilinear regressions are described. Important facts are explained using matrices in the same order of argument as for simple regression. Model building is introduced including the definitions of partial and sequential sum of squares, test of model adequacy using a likelihood function, and Conceptual Predictive and Akaike criteria. Some common problems of regression analysis including outliers and multicollinearity are described, and their detection and possible remedies are explained, including ridge and robust regression. Polynomial, nonlinear and segmented regressions are introduced. Segmented regression has been updated with cubic splines. Some examples are shown including estimating growth curves and functions with a plateau such as for determining nutrient requirements.

One-way analysis of variance is introduced in chapters 11 and 12. In chapter 11, a fixed effect one-way analysis of variance model is used to define hypotheses, partition sums of squares in order to use an F test, and estimate means and effects. Post-test comparison of means, including least significant difference, Tukey test and contrasts are shown. Characteristics of a random effect one-way model are covered in chapter 12, and mixed models which include fixed and random effects is introduced in chapter 13.

Chapters 14 to 23 focus on specific experimental designs and their analyses. Specific topics include: general concepts of design, blocking, change-over designs, factorials, nested designs, double blocking, split-plots, analysis of covariance, repeated measures and analysis of numerical treatment levels. Examples with sample SAS programs, updated with new approaches of computing appropriate degrees of freedom if necessary, are provided for each topic.

The final chapter covers the special topic of discrete dependent variables. Logit and probit models for binary and binomial dependent variables and log-linear models for count data are explained. Also, diagnostics test and ROC curve analysis are included.

The updates from the first edition also include a nonparametric approach to one- and two-way analysis of variance, useful when the distribution of data is unknown, discussions and notes about incomplete data and missing values, SAS programs updates with some new useful options and graph capabilities. Appropriate new examples are added.

We wish to express our gratitude to everyone who helped us produce this book. We extend our special acknowledgement for their kindly help for the first edition to Matt Lucy, Duane Keisler, Henry Mesa, Kristi Cammack, Marijan Posavi and Vesna Luzar-Stiffler for their reviews, and Cyndi Jennings, Cinda Hudlow and Dragan Tupajic for their assistance with editing. Thank you to Steven Lukefahr, Denise McNamara, Catherine Selby, Jackie Atkins and Laura School, in addition to those listed above, for providing assistance with this second edition.

Zagreb, Croatia   Miroslav Kaps
Columbia, Missouri   William R. Lamberson
March 2009

Chapter 1

Presenting and Summarizing Data

1.1   Data and Variables

Data are the material with which statisticians work. They are records of measurement, counts or observations. Examples of data are records of weights of calves, milk yield in lactation of a group of cows, male or female sex, and blue or green color of eyes. A set of observations on a particular character is termed a variable. For example, variables denoting the data listed above are weight, milk yield, sex, and eye color. Data are the values of a variable, for example, a weight of 200 kg, a daily milk yield of 20 kg, male, or blue eyes. The expression variable depicts that measurements or observations can be different, i.e., they show variability. Variables can be defined as quantitative (numerical) and qualitative (attributive, categorical, or classification).

Quantitative variables have values expressed as numbers and the differences between values have numerical meaning. Examples of quantitative variables are weight of animals, litter size, temperature or time. They also can include ratios of two numerical variables, count data, and proportions. A quantitative variable can be continuous or discrete. A continuous variable can take on an infinite number of values over a given interval. Its values are real numbers. A discrete variable is a variable that has countable values, and the number of those values can either be finite or infinite. Its values are natural numbers or integers. Examples of continuous variables are milk yield or weight, and examples of discrete variables are litter size or number of laid eggs per month.

Qualitative variables have values expressed in categories. Examples of qualitative variables are eye color or whether or not an animal is ill. A qualitative variable can be an ordinal or nominal. An ordinal variable has categories that can be ranked. A nominal variable has categories that cannot be ranked. No category of a nominal variable is more valuable than another. Examples of nominal variables are identification number, color or gender, and an example of an ordinal variable is calving ease score. For example, calving ease can be described in 5 categories, but those categories can be enumerated: 1. normal calving, 2. calving with little intervention, 3. calving with considerable intervention, 4. very difficult calving, and 5. Caesarean section. We can assign numbers (scores) to ordinal categories; however, the differences among those numbers do not have numerical meaning. For example, for calving ease, the difference between scores 1 and 2 (normal calving and calving with little intervention) does not have the same meaning as the difference between scores 4 and 5 (very difficult calving and Caesarean section). As a rule those scores depict ordered categories, but not a numerical scale. On the basis of the definition of a qualitative variable it may be possible to assign some quantitative variables, for example, the number of animals that belong to a category, or the proportion of animals in one category out of the total number of animals.

1.2   Graphical Presentation of Data

1.2.1   Graphical Presentation of Qualitative Data

When describing qualitative data each observation is assigned to a specific category. Data are then described by the number of observations in each category or by the proportion of the total number of observations. The frequency for a certain category is the number of observations in that category. The relative frequency for a certain category is the proportion of the total number of observations. Graphical presentations of qualitative variables can include bar, column or pie charts.

Example: The numbers of cows in Croatia under milk recording by breed are listed in the following table:

Breed	Number of cows	Percentage
Simmental	62672	76.7%
Holstein-Friesian	15195	18.6%
Brown	3855	4.7%
Total	81722	100%

The number of cows can be presented using bars with each bar representing a breed (Figure 1.1). The proportions or percentage of cows by breed can also be shown using a pie chart (Figure 1.2).

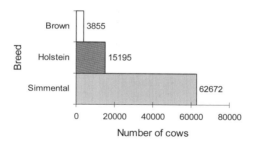

Figure 1.1 Number of cows under milk recording by breed

Figure 1.2 Percentage of cows under milk recording by breed

1.2.2   Graphical Presentation of Quantitative Data

The purpose of graphical presentation of quantitative data is to give information about the distribution of the data. The most widely used graphs include the histogram, stem and leaf and box plot. The histogram and stem and leaf graph will be introduced here. The box plot is connected with numerical presentation of data and will be explained in section 1.3.5.

Histogram

A histogram is a frequency distribution of a set of data. In order to present a distribution, the quantitative data are partitioned into classes and the histogram shows the number or relative frequency of observations for each class. Instructions for drawing a histogram are as follows:

1. Calculate the range (range = maximum – minimum value).
2. Divide the range into five to 20 classes, depending on the number of observations. The class width is obtained by rounding the result up to an integer number. The lowest class boundary must be defined below the minimum value, and the highest class boundary must be defined above the maximum value.
3. For each class, count the number of observations belonging to that class. This is the true frequency.
4. The relative frequency is calculated by dividing the true frequency by the total number of observations (relative frequency = true frequency / total number of observations).
5. The histogram is a column (or bar) graph with class boundaries defined on one axis and frequencies on the other axis.

Example: Construct a histogram for the 7-month weights (kg) of 100 calves:

233	208	306	300	271	304	207	254	262	231
279	228	287	223	247	292	209	303	194	268
263	262	234	277	291	277	256	271	255	299
278	290	259	251	265	316	318	252	316	221
249	304	241	249	289	211	273	241	215	264
216	271	296	196	269	231	272	236	219	312
320	245	263	244	239	227	275	255	292	246
245	255	329	240	262	291	275	272	218	317
251	257	327	222	266	227	255	251	298	255
266	255	214	304	272	230	224	250	255	284

Minimum = 194
Maximum = 329
Range = 329 – 194 = 135

For a total 15 classes, the with of a class is:

$$135 / 15 = 9$$

The class width can be rounded to 10 and the following table constructed:

Class limits	Class midrange	Number of calves	Relative Frequency (%)	Cumulative number of calves
185 - 194	190	1	1	1
195 - 204	200	1	1	2
205 - 214	210	5	5	7
215 - 224	220	8	8	15
225 - 234	230	8	8	23
235 - 244	240	6	6	29
245 - 254	250	12	12	41
255 - 264	260	16	16	57
265 - 274	270	12	12	69
275 - 284	280	7	7	76
285 - 294	290	7	7	83
295 - 304	300	8	8	91
305 - 314	310	2	2	93
315 - 324	320	5	5	98
325 - 334	330	2	2	100

Figure 1.3 presents the histogram of weights of calves. The classes are on the horizontal axis and the numbers of animals are on the vertical axis. Class values are expressed as the class midranges (midpoint between the limits), but could alternatively be expressed as class limits.

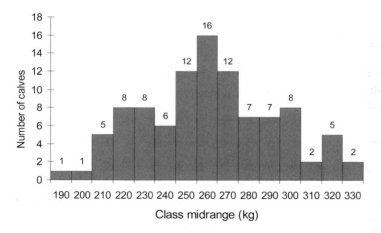

Figure 1.3 Histogram of weights of calves at 7 months of age (*n* = 100)

Stem and Leaf Graph

Another well-known way of presenting quantitative data is by the use of a 'stem and leaf' graph. The construction of a stem and leaf can be shown in three steps:

1. Each value is divided into two parts, 'stem' and 'leaf'. 'Stem' corresponds to higher decimal places, and 'leaf' corresponds to lower decimal places. For the example of calf weights, the first two digits of each weight would represent the stem and the third digit the leaf.
2. 'Stems' are sorted in ascending order in the first column.
3. The appropriate 'leaf' for each observation is recorded in the row with the appropriate 'stem'.

A 'stem and leaf' plot of the weights of calves is shown below.

```
Stem      Leaf
19 | 4 6
20 | 7 8 9
21 | 1 4 5 6 8 9
22 | 1 2 3 4 7 7 8
23 | 0 1 1 3 4 6 9
24 | 0 1 1 4 5 5 6 7 9 9
25 | 0 1 1 1 2 4 5 5 5 5 5 5 5 6 7 9
26 | 2 2 2 3 3 4 5 6 6 8 9
27 | 1 1 1 2 2 2 3 5 5 7 7 8 9
28 | 4 7 9
29 | 0 1 1 2 2 6 8 9
30 | 0 3 4 4 4 6
31 | 2 6 6 7 8
32 | 0 7 9
```

For example, in the next to last row the 'stem' is 31 and 'leaves' are 2, 6, 6, 7 and 8. This indicates that the category includes the measurements 312, 316, 316, 317 and 318. When the data are suited to a stem and leaf plot it shows a distribution similar to the histogram and also shows each value of the data.

1.3   Numerical Methods for Presenting Data

Numerical methods for presenting data are often called descriptive statistics. They include: a) measures of central tendency; b) measures of variability; c) measures of the shape of a distribution; and d) measures of relative position.

Descriptive statistics			
a) measures of central tendency	b) measures of variability	c) measures of the shape of a distribution	d) measures of relative position
- arithmetic mean	- range	- skewness	- percentiles
- median	- variance	- kurtosis	- z values
- mode	- standard deviation		
	- coefficient of variation		

Before descriptive statistics are explained in detail, it is useful to explain a system of symbolic notation that is used not only in descriptive statistics, but in statistics in general. This includes the symbols for the sum, sum of squares and sum of products.

1.3.1 Symbolic Notation

The Greek letter Σ (sigma) is used as a symbol for summation, and y_i for the value for observation i.

The sum of n numbers $y_1, y_2, \ldots, y_n$ can be expressed:

$$\Sigma_i y_i = y_1 + y_2 + \ldots + y_n$$

The sum of squares of n numbers $y_1, y_2, \ldots, y_n$ is:

$$\Sigma_i y^2_i = y^2_1 + y^2_2 + \ldots + y^2_n$$

The sum of products of two sets of n pairs of numbers $(x_1, x_2, \ldots, x_n)$ and $(y_1, y_2, \ldots, y_n)$ is:

$$\Sigma_i x_i y_i = x_1 y_1 + x_2 y_2 + \ldots + x_n y_n$$

Example: Consider a set of three numbers: 1, 3 and 6. The numbers are symbolized by: $y_1 = 1$, $y_2 = 3$ and $y_3 = 6$.

The sum and sum of squares of those numbers are:

$$\Sigma_i y_i = 1 + 3 + 6 = 10$$

$$\Sigma_i y^2_i = 1^2 + 3^2 + 6^2 = 46$$

Consider another set of numbers: $x_1 = 2$, $x_2 = 4$ and $x_3 = 5$.

The sum of products of x and y is:

$$\Sigma_i x_i y_i = (1)(2) + (3)(4) + (6)(5) = 44$$

Three main rules of summation are:

1. The sum of addition of two sets of numbers is equal to the addition of the sums:

$$\Sigma_i (x_i + y_i) = \Sigma_i x_i + \Sigma_i y_i$$

2. The sum of products of a constant k and a variable y is equal to the product of the constant and the sum of the values of the variable:

$$\Sigma_i k \, y_i = k \, \Sigma_i y_i$$

3. The sum of n constants with value k is equal to the product $n\,k$:

$$\Sigma_i k = n \, k$$

1.3.2   Measures of Central Tendency

We will describe the measures of central tendency and variability as being calculated on a sample of values from some larger population. The measures on the sample can be used to describe the population. A more complete description of samples and populations is given in chapter 4. Commonly used measures of central tendency are the arithmetic mean, median and mode.

The *arithmetic mean* of a sample of n numbers $y_1, y_2,..., y_n$ drawn from some population of observations is:

$$\bar{y} = \frac{\sum_i y_i}{n}$$

The arithmetic mean for grouped data is:

$$\bar{y} = \frac{\sum_i f_i y_i}{n}$$

with f_i being the frequency or proportion of observations having value y_i. If f_i is a proportion then $n = 1$.

Important properties of the arithmetic mean are:

1. $\sum_i (y_i - \bar{y}) = 0$

    The sum of deviations from the arithmetic mean is equal to zero. This means that only $(n - 1)$ observations are independent and the nth can be expressed as:

    $$y_n = n\bar{y} - y_1 - ... - y_{n-1}$$

2. $\sum_i (y_i - \bar{y})^2 = \text{minimum}$

    The sum of squared deviations from the arithmetic mean is smaller than the sum of squared deviations from any other value.

The *median* of a sample of n observations $y_1, y_2,..., y_n$ is the value of the observation that is in the middle when observations are sorted from smallest to the largest. It is the value of the observation located such that one half of the area of a histogram is on the left and the other half is on the right. If n is an odd number the median is the value of the ${(n+1)}/{2}$-th observation. If n is an even number the median is the average of ${(n)}/{2}$-th and ${(n+2)}/{2}$-th observations.

The *mode* of a sample of n observations $y_1, y_2,..., y_n$ is the value among the observations that has the highest frequency.

Figure 1.4 presents frequency distributions illustrating the mean, median and mode. When the frequency distribution is symmetric, the measures are equal. Although the mean is the measure that is most commonly used, when distributions are asymmetric, the median and mode can give better information about the set of data. Unusually extreme values in a sample will affect the arithmetic mean more than the median. In that case the median is a more representative measure of central tendency than the arithmetic mean. For extremely asymmetric distributions the mode is the best measure.

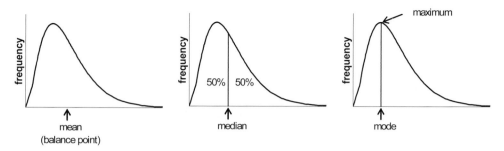

Figure 1.4 Interpretation of mean, median and mode

1.3.3 Measures of Variability

Commonly used measures of variability are the range, variance, standard deviation and coefficient of variation.

The *range* is defined as the difference between the maximum and minimum values in a set of observations.

The *sample variance* (s^2) of n observations (measurements) y_1, y_2,..., y_n drawn from a population of measurements is:

$$s^2 = \frac{\sum_i (y_i - \bar{y})^2}{n-1}$$

This formula is valid if $\bar{y}$ is calculated from the same sample, i.e., the mean of a population is not known. If the mean of a population (μ) is known then the variance is:

$$s^2 = \frac{\sum_i (y_i - \mu)^2}{n}$$

The variance is the average squared deviation about the mean.

The sum of squared deviations about the arithmetic mean is often called the corrected sum of squares or just sum of squares, and it is denoted by SS_{yy}. The corrected sum of squares can be calculated:

$$SS_{yy} = \sum_i (y_i - \bar{y})^2 = \sum_i y_i^2 - \frac{\left(\sum_i y_i\right)^2}{n}$$

Further, the sample variance is often called the mean square, denoted by MS_{yy}, because:

$$s^2 = MS_{yy} = \frac{SS_{yy}}{n-1}$$

For grouped data, the variance of a sample of values drawn from a population with an unknown mean is:

$$s^2 = \frac{\sum_i f_i (y_i - \bar{y})^2}{n-1}$$

where f_i is the frequency of observation y_i, and the total number of observations is $n = \sum_i f_i$.

The *sample standard deviation* (*s*) is equal to square root of the variance. It is the average absolute deviation from the mean:

$$s = \sqrt{s^2}$$

The *coefficient of variation* (*CV*) is defined as:

$$CV = \frac{s}{\bar{y}} 100\%$$

The coefficient of variation is a relative measure of variability expressed as a percentage. It is often easier to understand the importance of variability if it is expressed as a percentage. This is especially true when variability is compared among sets of data that have different units. For example if *CV* for weight and height are 40% and 20%, respectively, we can conclude that weight is more variable than height.

1.3.4 Measures of the Shape of a Distribution

The measures of the shape of a distribution are the coefficients of skewness and kurtosis.

Skewness (*sk*) is a measure of asymmetry of a frequency distribution. It shows if deviations from the mean are larger on one side than the other side of the distribution. If the population mean (μ) is known, then skewness is:

$$sk = \frac{1}{(n-1)(n-2)} \sum_i \left(\frac{y_i - \mu}{s} \right)^3$$

If the population mean is unknown, the sample mean ($\bar{y}$) is substituted for μ and skewness is:

$$sk = \frac{n}{(n-1)(n-2)} \sum_i \left(\frac{y_i - \bar{y}}{s} \right)^3$$

For a symmetric distribution skewness is equal to zero. It is positive when the right tail is longer, and negative when left tail is longer (Figure 1.5).

a)   b)

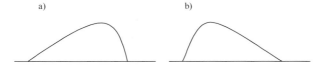

Figure 1.5 Illustrations of skewness: a) negative, b) positive

Kurtosis (*kt*) is a measure of flatness or steepness of a distribution, or a measure of the heaviness of the tails of a distribution. If the population mean (μ) is known, kurtosis is:

$$kt = \frac{1}{n}\sum_i \left(\frac{y_i - \mu}{s}\right)^4 - 3$$

If the population mean is unknown, the sample mean ($\bar{y}$) is used instead, and kurtosis is:

$$kt = \frac{n(n+1)}{(n-1)(n-2)(n-3)}\sum_i \left(\frac{y_i - \bar{y}}{s}\right)^4 - \frac{3(n-1)^2}{(n-2)(n-3)}$$

For variables such as weight, height or milk yield, frequency distributions are expected to be symmetric about the mean and bell-shaped. These are normal distributions. If observations follow a normal distribution then kurtosis is equal to zero. A distribution with positive kurtosis has a large frequency of observations close to the mean and thin tails. A distribution with a negative kurtosis has thicker tails and a lower frequency of observations close to the mean than does the normal distribution (Figure 1.6).

a)   b)

Figure 1.6 Illustrations of kurtosis: a) positive, b) negative

1.3.5   Measures of Relative Position

Measures of relative position include percentiles and *z* values.
The *percentile value* (*p*) of an observation y_i, in a data set has 100*p*% of observations smaller than y_i and has 100(1 – *p*)% of observations greater than y_i. A lower quartile is the 25th percentile, an upper quartile is 75th percentile, and the median is the 50th percentile.

The *z value* is the deviation of an observation from the mean in standard deviation units:

$$z_i = \frac{y_i - \bar{y}}{s}$$

Box plot Percentiles can be used to construct a box plot. The purpose of a box plot is to show the amount of data within particular percentiles, especially the median, and lower and upper quartiles, thus showing whether the distribution of data is symmetric or skewed. Extreme observations, so called outliers, are also presented.

A box plot can be constructed by following these steps:
1. Calculate the median, and lower and upper quartiles. The range between lower and upper quartile is called interquartile.
2. Calculate (lower quartile – 1.5 interquartile) and (upper quartile + 1.5 interquartile)
3. The interquartile is shown as a box, with the median as a line within that box.
4. Limits of the distances (lower quartile – 1.5 interquartile) and (upper quartile + 1.5 interquartile) are shown as lines.

5. Observations beyond (lower quartile − 1.5 interquartile) and (upper quartile + 1.5 interquartile) are defined as extreme observations (outliers) and can be shown by using special symbols.

Example: Calculate the arithmetic mean, variance, standard deviation, coefficient of variation, median and mode, and draw a box plot of the following weights of calves (kg):

260 260 230 280 290 280 260 270 260 300

280 290 260 250 270 320 320 250 320 220

Arithmetic mean:

$$\bar{y} = \frac{\sum_i y_i}{n}$$

$$\sum_i y_i = 260 + 260 + \ldots + 220 = 5470 \text{ kg}$$

$$\bar{y} = \frac{5470}{20} = 273.5 \text{ kg}$$

Sample variance:

$$s^2 = \frac{\sum_i (y_i - \bar{y})^2}{n-1} = \frac{\sum_i y_i^2 - \frac{\left(\sum_i y_i\right)^2}{n}}{n-1}$$

$$\sum_i y_i^2 = (260^2 + 260^2 + \ldots + 220^2) = 1510700 \text{ kg}^2$$

$$s^2 = \frac{1510700 - \frac{(5470)^2}{20}}{19} = 771.3158 \text{ kg}^2$$

Sample standard deviation:

$$s = \sqrt{s^2} = \sqrt{771.3158} = 27.77 \text{ kg}$$

Coefficient of variation:

$$CV = \frac{s}{\bar{y}} 100\% = \frac{27.77}{273.5} 100\% = 10.15\%$$

To find the median, the observations are sorted from smallest to the largest:

220 230 250 250 260 260 260 260 260 $\boxed{270}$ $\boxed{270}$ 280 280 280 290 290 300 320 320 320

Since $n = 20$ is an even number, the median is the average of $n/2 = 10^{\text{th}}$ and $(n+2)/2 = 11^{\text{th}}$ observations when the data are sorted. The values of those observations are 270 and 270, respectively, and their average is 270, thus, the median is 270 kg. The mode is 260 kg because this is the value occurring with highest frequency.

For the box plot we need:

Median = 270; lower quartile = 260; upper quartile = 290; interquartile = 30

(lower quartile – 1.5 interquartile) = 260 – 45 = 215, the smallest observation within that range is 220

(upper quartile + 1.5 interquartile) = 290 + 45 = 335, the largest observation within that range is 320

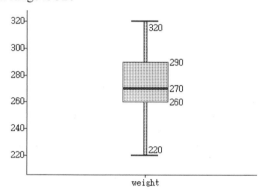

Note that all observations are within the range 215 to 335, thus there are no outliers.

1.3.6 SAS Example

Descriptive statistics for the example set of weights of calves are calculated using SAS software. For a more detailed explanation how to use the computing and graphical capability of SAS, we recommend the exhaustive SAS literature, part of which is included in the list of literature at the end of this book. This SAS program consists of two parts: 1) the DATA step, which is used for entry and transformation of data; and 2) the PROC step, which defines the procedure(s) for data analysis. SAS has three basic windows: a Program or Editor window (PGM) in which the program is written, an Output window (OUT) in which the user can see the results, and LOG window in which the user can view details regarding program execution or error messages. Returning to the example of weights of 20 calves:

SAS program:

```
DATA calves;
INPUT weight @@;
DATALINES;
260 260 230 280 290 280 260 270 260 300
280 290 260 250 270 320 320 250 320 220
;
PROC MEANS DATA = calves N MEAN MIN MAX VAR STD CV ;
VAR weight;
RUN;
```

Explanation: The SAS statements will be written with capital letters for emphasis, although it is not generally mandatory, i.e. the program does not distinguish between lower case letters and capitals. Names that the user assigns to variables, data files, etc., will be written with lower case letters. In this program, the DATA statement defines the name of the file that contains data. Here, *calves* is the name of the file. The INPUT statement defines the name(s) of the variable, and the DATALINES statement indicates that data are on the following lines. Here, the name of the variable is *weight*. SAS needs data in columns, for example,

```
INPUT weight;
DATALINES;
260
260
...
220
;
```

reads values of the variable *weight*. Data can be written in rows if the symbols @@ are used with the INPUT statement. SAS reads observations one by one and stores them into a column named *weight*. This program uses the procedure (PROC) MEANS. The option *DATA = calves* defines the data file that will be used in the calculation of statistics, followed by the list of statistics to be calculated: N = the number of observations, MEAN = arithmetic mean, MIN = minimum, MAX = maximum, VAR = sample variance, STD = sample standard deviation, and CV = coefficient of variation. The VAR statement defines the variable (*weight*) to be analyzed.

SAS output:

```
     Analysis Variable: WEIGHT

N   Mean   Minimum   Maximum   Variance   Std Dev      CV
---------------------------------------------------------------
20  273.5    220       320    771.31579  27.77257   10.1545
---------------------------------------------------------------
```

The SAS output lists the variable that was analyzed (*Analysis Variable: WEIGHT*). The descriptive statistics are then listed.

Alternatively, the UNIVARIATE procedure could be used. An example of SAS program is shown:

SAS program:

```
PROC UNIVARIATE DATA = calves modes  PLOT PLOTSIZE= 50;
VAR weight;
HISTOGRAM / MIDPOINTS = 240 260 280 300 320;
RUN;
```

Explanation: The program computes descriptive (basic) statistics, lists extreme observations and quantiles and modes (if there are more than one). The options PLOT and PLOTSIZE direct construction of stem and leaf and box plots. The HISTOGRAM statement directs construction of histogram with 5 classes defined at 240 260 280 300 320.

Exercises

1.1. The numbers of eggs laid per month in a sample of 40 hens are shown below:

30	23	26	27
29	25	27	24
28	26	26	26
30	26	25	29
26	23	26	30
25	28	24	26
27	25	25	28
27	28	26	30
26	25	28	28
24	27	27	29

Calculate descriptive statistics and present a frequency distribution.

1.2. Calculate the sample variance given the following sums:

$\sum_i y_i = 600$ (sum of observations); $\sum_i y_i^2 = 12656$ (sum of squared observations); $n = 30$ (number of observations)

1.3. Draw the histogram of the values of a variable y and its frequencies f:

y	12	14	16	18	20	22	24	26	28
f	1	3	4	9	11	9	6	1	2

Calculate descriptive statistics for this sample.

1.4. The following are milk fat yields (kg) per month from 17 Holstein cows:

27 17 31 20 29 22 40 28 26 28 34 32 32 32 30 23 25

Calculate descriptive statistics. Show that if 3 kg are added to each observation, the mean will increase by three and the sample variance will stay the same. Show that if each observation is divided by two, the mean will be two times smaller and the sample variance will be four times smaller. How will the standard deviation be changed?

Chapter 2

Probability

The word probability is used to indicate the likelihood that some event will happen. For example, 'there is high probability that it will rain tonight'. We can conclude this according to some signs, observations or measurements. If we can count or make a conclusion about the number of favorable events, we can express the probability of occurrence of an event by using a proportion or percentage of all events. Probability is important in drawing inferences about a population. Statistics deals with drawing inferences by using observations and measurements, and applying the rules of mathematical probability.

A probability can be a-priori or a-posteriori. An a-priori probability comes from a logical deduction on the basis of previous experiences. Our experience tells us that if it is cloudy, we can expect with high probability that it will rain. If an animal has particular symptoms, there is high probability that it has or will have a particular disease. An a-posteriori probability is established by using a planned experiment. For example, assume that changing a ration will increase milk yield of dairy cows. Only after an experiment was conducted in which numerical differences were measured, it can be concluded with some probability or uncertainty, that a positive response can be expected for other cows as well. Generally, each process of collecting data is an experiment. For example, throwing a die and observing the number is an experiment.

Mathematically, probability is:

$$P = \frac{m}{n}$$

where m is the number of favorable trials and n is the total number of trials.

An observation of an experiment that cannot be partitioned to simpler events is called an elementary event or simple event. For example, we throw a die once and observe the result. This is a simple event. The set of all possible simple events is called the sample space. All the possible simple events in an experiment consisting of throwing a die are 1, 2, 3, 4, 5 and 6. The probability of a simple event is a probability that this specific event occurs. If we denote a simple event by E_i, such as throwing a 4, then $P(E_i)$ is the probability of that event.

2.1 Rules about Probabilities of Simple Events

Let E_1, E_2,..., E_k be the set of all simple events in some sample space of simple events. Then we have:

1. The probability of any simple event occurring must be between 0 and 1 inclusively:

$$0 \le P(E_i) \le 1, \qquad i = 1,...,k$$

2. The sum of the probabilities of all simple events is equal to 1:

$$\sum_i P(E_i) = 1$$

Example: Assume an experiment consists of one throw of a die. Possible results are 1, 2, 3, 4, 5 and 6. Each of those possible results is a simple event. The probability of each of those events is $^1/_6$, i.e., $P(E_1) = P(E_2) = P(E_3) = P(E_4) = P(E_5) = P(E_6)$. This can be shown in a table:

Observation	Event (E_i)	$P(E_i)$
1	E_1	$P(E_1) = {}^1/_6$
2	E_2	$P(E_2) = {}^1/_6$
3	E_3	$P(E_3) = {}^1/_6$
4	E_4	$P(E_4) = {}^1/_6$
5	E_5	$P(E_5) = {}^1/_6$
6	E_6	$P(E_6) = {}^1/_6$

Both rules about probabilities are satisfied. The probability of each event is ($^1/_6$), which is less than one. Further, the sum of probabilities, $\sum_i P(E_i)$ is equal to one. In other words, the probability is equal to one that any number between one and six will result from the throw of a die.

Generally, an event A is a specific set of simple events; that is, an event consists of one or more simple events. The probability of an event A is equal to the sum of probabilities of the simple events in the event A. This probability is denoted with $P(A)$. For example, assume the event that is defined as a number less than 3 in one throw of a die. The simple events are 1 and 2 each with the probability ($^1/_6$). The probability of A is then ($^1/_3$).

2.2   Counting Rules

Recall that probability is:

P = number of favorable trials / total number of trials

Or, if we are able to count the number of simple events in an event A and the total number of simple events:

P = number of favorable simple events / total number of simple events

A logical way of estimating or calculating probability is to count the number of favorable trials or simple events and divide by the total number of trials. However, practically this can often be very cumbersome, and we can use counting rules instead.

2.2.1 Multiplicative Rule

Consider k sets of elements of size n_1, n_2,..., n_k. If one element is randomly chosen from each set, then the total number of different results is:

$$n_1\ n_2\ n_3 ...\ n_k$$

Example: Consider three pens with animals marked as listed:

Pen 1: 1, 2, 3
Pen 2: *A*, *B*, *C*
Pen 3: *x*, *y*

The number of animals per pen are $n_1 = 3$, $n_2 = 3$, $n_3 = 2$.

The possible triplets with one animal taken from each pen are:

1*Ax*, 1*Ay*, 1*Bx*, 1*By*, 1*Cx*, 1*Cy*
2*Ax*, 2*Ay*, 2*Bx*, 2*By*, 2*Cx*, 2*Cy*
3*Ax*, 3*Ay*, 3*Bx*, 3*By*, 3*Cx*, 3*Cy*

The number of possible triplets is: (3) (3) (2) = 18

2.2.2 Permutations

From a set of n elements, the number of ways those n elements can be rearranged, i.e., put in different orders, is the *permutations* of n elements:

$$P_n = n!$$

The symbol $n!$ (factorial of n) denotes the product of all natural numbers from 1 to n:

$$n! = (1)\ (2)\ (3)\ ...\ (n)$$

Also, by definition $0! = 1$.

Example: In how many ways can three animals, x, y and z, be arranged in triplets?

The number of permutations of $n = 3$ elements: $P(3) = 3! = (1)\ (2)\ (3) = 6$

The six possible triplets: *xyz xzy yxz yzx zxy zyx*

More generally, we can define permutations of n elements taken k at a time in particular order as:

$$P_{n,k} = \frac{n!}{(n-k)!}$$

Example: In how many ways can three animals, x, y and z, be arranged in pairs such that the order in the pairs is important (*xz* is different than *zx*)?

$$P_{n,k} = \frac{3!}{(3-2)!} = 6$$

The six possible pairs are: *xy xz yx yz zx zy*

2.2.3 Combinations

From a set of *n* elements, the number of ways those *n* elements can be taken *k* at a time regardless of order (*xz* is not different than *zx*) is:

$$\binom{n}{k} = \frac{n!}{k!(n-k)!} = \frac{n(n-1)...(n-k+1)}{k!}$$

Example: In how many ways can three animals, *x*, *y* and *z*, be arranged in pairs when the order in the pairs is not important?

$$\binom{n}{k} = \binom{3}{2} = \frac{3!}{2!(3-2)!} = 3$$

There are three possible pairs: *xy xz yz*

2.2.4 Partition Rule

From a set of *n* elements to be assigned to *j* groups of size n_1, n_2, n_3,..., n_j, the number of ways in which those elements can be assigned is:

$$\frac{n!}{n_1!n_2!...n_j!}$$

where $n = n_1 + n_2 + ... + n_j$

Example: In how many ways can a set of five animals be assigned to *j* = 3 stalls with $n_1 = 2$ animals in the first, $n_2 = 2$ animals in the second and $n_3 = 1$ animal in the third?

$$\frac{5!}{2!\,2!\,1!} = 30$$

Note that the previous rule for combinations is a special case of partitioning a set of size *n* into two groups of size *k* and *n* – *k*.

2.2.5 Tree Diagram

The tree diagram illustrates counting, the representation of all possible outcomes of an experiment. This diagram can be used to present and check the probabilities of a particular event. As an example, a tree diagram of possible triplets, one animal taken from each of three pens, is shown below:

     Pen 1: 1, 2, 3
     Pen 2: *x, y*
     Pen 3: *A, B, C*

The number of all possible triplets is:

     (3)(3)(2) = 18

The tree diagram is:

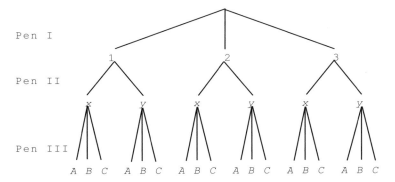

The first triplet has animal 1 from pen 1, animal x from pen 2, and animal A from pen 3. If we assign the probabilities to each of the events then that tree diagram is called a probability tree.

2.3   Compound Events

A compound event is an event composed of two or more events. Consider two events A and B. The compound event such that both events A and B occur is called the *intersection of the events,* denoted by $A \cap B$. The compound event such that either event A or event B occurs is called the *union of events*, denoted by $A \cup B$. The probability of an intersection is $P(A \cap B)$ and the probability of union is $P(A \cup B)$. Also:

$$P(A \cup B) = P(A) + P(B) - P(A \cap B)$$

The *complement* of an event A is the event that A does not occur, and it is denoted by A^c. The probability of a complement is:

$$P(A^c) = 1 - P(A)$$

Example: Let the event A be such that the result of a throw of a die is an even number. Let the event B be such that the number is greater than 3.
The event A is the set: $\{2,4,6\}$
The event B is the set: $\{4,5,6\}$

The intersection A and B is an event such that the result is an even number and a number greater than 3 at the same time. This is the set:

$(A \cap B) = \{4,6\}$

with the probability:

$P(A \cap B) = P(4) + P(6) = {}^2/_6$, because the probability of an event is the sum of probabilities of the simple events that make up the set.

The union of the events A and B is an event such that the result is an even number or a number greater than 3. This is the set:

$(A \cup B) = \{2,4,5,6\}$

with the probability

$P(A \cup B) = P(2) + P(4) + P(5) + P(6) = {}^4/_6$

Figure 2.1 presents the intersection and union of the events A and B.

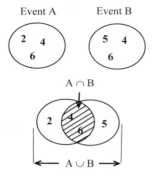

Figure 2.1 Intersection and union of two events

A conditional probability is the probability that an event will occur if some assumptions are satisfied. In other words a conditional probability is a probability that an event B will occur if it is known that an event A has already occurred. The conditional probability of B given A is calculated by using the formula:

$$P(B \mid A) = \frac{P(A \cap B)}{P(A)}$$

Events can be dependent or independent. If events A and B are independent then:

$$P(B \mid A) = P(B) \quad \text{and} \quad P(A \mid B) = P(A)$$

If independent the probability of B does not depend on the probability of A. Also, the probability that both events occur is equal to the product of each probability:

$$P(A \cap B) = P(A) \, P(B)$$

If the two events are dependent, for example, the probability of the occurrence of event B depends on the occurrence of event A, then:

$$P(B \mid A) = \frac{P(A \cap B)}{P(A)}$$

and consequently the probability that both events occur is:

$$P(A \cap B) = P(A) \, P(B \mid A)$$

An example of independent events: We throw a die two times. What is the probability of obtaining two sixes?

Consider the first throw as event A and the second as event B. The probability that both events occur is $P(A \cap B)$. The probability of each event is: $P(A) = 1/6$, and $P(B) = 1/6$. The events are independent which means:

$$P(A \cap B) = P(A)\,P(B) = (1/6)(1/6) = (1/36).$$

The probability that in two throws we get two sixes is $(1/36)$.

An example of dependent events: From a deck of 52 playing cards we draw two cards. What is the probability that both cards drawn are aces?

The first draw is event A and the second is event B. Recall that in a deck there are four aces. The probability that both are aces is $P(A \cap B)$. The events are obviously dependent, namely the drawing of the second card depends on which card has been drawn first.

$$P(A = \text{Ace}) = (4/52) = (1/13).$$

$P(B = \text{Ace} \mid A = \text{Ace}) = (3/51)$, that is, if the first card was an ace, only 51 cards were left and only 3 aces. Thus,

$$P(A \cap B) = P(A)\,P(B|A) = (4/52)(3/51) = (1/221).$$

The probability of drawing two aces is $(1/221)$.

Example: In a pen there are 10 calves: 2 black, 3 red and 5 spotted. They are let out one at the time in completely random order. The probabilities of the first calf being of a particular color are in the following table:

	A_i	$P(A_i)$
2 black	A_1	$P(\text{black}) = 2/10$
3 red	A_2	$P(\text{red}) = 3/10$
5 spotted	A_3	$P(\text{spotted}) = 5/10$

Here, the probability $P(A_i)$ is the relative number of animals of a particular color. We can see that:

$$\sum_i P(A_i) = 1$$

Find the following probabilities:

    a) the first calf is spotted,
    b) the first calf is either black or red,
    c) the second calf is black if the first was spotted,
    d) the first calf is spotted and the second black,
    e) the first two calves are spotted and black, regardless of order.

Solutions:

a) There is a total of 10 calves, and 5 are spotted. The number of favorable outcomes is $m = 5$ and the total number of outcomes is $n = 10$. Thus, the probability that a calf is spotted is:

$$P(\text{spotted}) = {}^5/_{10} = {}^1/_2$$

b) The probability that the first calf is either black or red is an example of union. P(black or red) = P(black) + P(red) = ${}^2/_{10} + {}^3/_{10} = {}^5/_{10} = \frac{1}{2}$. Also, this is equal to the probability that the first calf is not spotted, the complement of the event described in a):

$$P(\text{black} \cup \text{red}) = 1 - P(\text{spotted}) = 1 - {}^1/_2 = {}^1/_2$$

c) This is an example of conditional probability.
The probability that the second calf is black if we know that the first one was spotted is the number of black calves (2) divided by the number of calves remaining after removing a spotted one from the pen (9):

$$P(\text{black} \mid \text{spotted}) = {}^2/_9$$

d) This is an example of the probability of an intersection of events. The probability that the first calf is spotted is $P(\text{spotted}) = 0.5$. The probability that the second calf is black when the first was spotted is:

$$P(\text{black} \mid \text{spotted}) = {}^2/_9$$

The probability that the first calf is spotted and the second is black is the intersection:
$$P[\text{spotted} \cap (\text{black} \mid \text{spotted})] = ({}^5/_{10})\,({}^2/_9) = {}^1/_9$$

e) We have already seen that the probability that the first calf is spotted and the second is black is:

$$P[\text{spotted} \cap (\text{black} \mid \text{spotted})] = {}^1/_9.$$

Similarly, the probability that the first is black and the second is spotted is:

$$P[\text{black} \cap (\text{spotted} \mid \text{black})] = ({}^2/_{10})\,({}^5/_9) = {}^1/_9$$

Since we are looking for a pair (black, spotted) regardless of the order, then we have either a (spotted, black) or a (black, spotted) event. This is an example of union, so the probability is:

$$P\{[\text{spotted} \cap (\text{black} \mid \text{spotted})] \cup [\text{black} \cap (\text{spotted} \mid \text{black})]\} = ({}^1/_9) + ({}^1/_9) = {}^2/_9$$

We can illustrate the previous examples using a tree diagram:

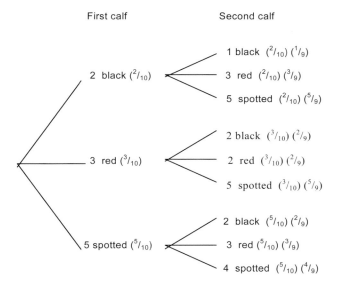

First calf   Second calf

2 black ($^2/_{10}$)
   1 black ($^2/_{10}$) ($^1/_9$)
   3 red ($^2/_{10}$) ($^3/_9$)
   5 spotted ($^2/_{10}$) ($^5/_9$)

3 red ($^3/_{10}$)
   2 black ($^3/_{10}$) ($^2/_9$)
   2 red ($^3/_{10}$) ($^2/_9$)
   5 spotted ($^3/_{10}$) ($^5/_9$)

5 spotted ($^5/_{10}$)
   2 black ($^5/_{10}$) ($^2/_9$)
   3 red ($^5/_{10}$) ($^3/_9$)
   4 spotted ($^5/_{10}$) ($^4/_9$)

2.4   Bayes Theorem

Bayes theorem is useful for stating the probability of some event A if there is information about the probability of some event E that happened after the event A. Bayes theorem is applied to an experiment that occurs in two or more steps. Consider two cages K_1 and K_2, in the first cage there are three mice, two brown and one white, and in the second there are two brown and two white mice. Each brown mouse is designated with the letter B, and each white mouse with the letter W.

Cage K_1 Cage K_2
B, B, W B, B, W, W

A cage is randomly chosen and then a mouse is randomly chosen from that cage. If the chosen mouse is brown, what is the probability that it is from the first cage?

The first step of the experiment is choosing a cage. Since it is chosen randomly, the probability of choosing the first cage is $P(K_1) = (^1/_2)$. The second step is choosing a mouse from the cage. The probability of choosing a brown mouse from the first cage is $P(B \mid K_1) = (^2/_3)$, and of choosing a brown mouse from the second cage is $P(B \mid K_2) = (^2/_4)$. The probability that the first cage was chosen if it is known that the mouse is brown is an example of conditional probability:

$$P(K_1 \mid B) = \frac{P(K_1 \cap B)}{P(B)}$$

The probability that the mouse is from the first cage and that it is brown is:

$$P(K_1 \cap B) = P(K_1)\, P(B \mid K_1) = (^1/_2)\,(^2/_3) = (^1/_3)$$

The probability that the mouse is brown regardless from which cage it is chosen is $P(B)$, which is the probability that the brown mouse is either from the first cage and brown, or from the second cage and brown:

$$P(B) = P(K_1)\,P(B \mid K_1) + P(K_2)\,P(B \mid K_2) = (^1/_2)\,(^2/_3) + (^1/_2)\,(^2/_4) = \,^7/_{12}$$

Those probabilities assigned to the proposed formula:

$$P(K_1 \mid B) = (^1/_3) / (^7/_{12}) = \,^4/_7$$

Thus, the probability that a mouse is from the first cage if it is known that it is brown is $(^4/_7)$.

This problem can be presented using Bayes theorem:

$$P(K_1 \mid B) = P(K_1 \cap B) / P(B) = \frac{P(K_1)P(B \mid K_1)}{P(K_1)P(B \mid K_1) + P(K_2)P(B \mid K_2)}$$

Generally, there is an event A with k possible outcomes $A_1, A_2,...,A_k$, that are independent and the sum of their probabilities is 1, $(\sum_i P(A_i) = 1)$. Also, there is an event E, that occurs after event A. Then:

$$P(A_i \mid E) = \frac{P(A_i \cap E)}{P(E)} = \frac{P(A_i)P(E \mid A_i)}{P(A_1)P(E \mid A_1) + P(A_2)P(E \mid A_2) + ... + P(A_k)P(E \mid A_k)}$$

To find a solution to some Bayes problems one can use a tree diagram. The example with two cages and mice can be presented as follows:

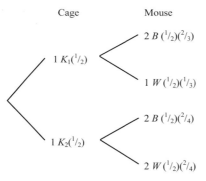

From the diagram we can easily read the probability of interest. For example, the probability that the mouse is brown and from the first cage is $(^1/_2)\,(^2/_3) = (^1/_3)$, and the probability that it is brown and from the second cage is $(^1/_2)\,(^2/_4) = (^1/_4)$.

Second example: For artificial insemination of a large dairy herd, semen from two bulls is utilized. Bull 1 has been used on 60% of the cows, and bull 2 on 40%. We know that the percentages of successful inseminations for bull 1 and bull 2 are 65% and 82%, respectively. For a certain calf the information about its father has been lost. What is the probability that the father of that calf is bull 2?

We define:

$P(A_1) = 0.6$ is the probability of having used bull 1
$P(A_2) = 0.4$ is the probability of having used bull 2
E = the event that a calf is born (because of successful insemination)

$P(E | A_1) = 0.65$ = the probability of successful insemination if bull 1
$P(E | A_2) = 0.82$ = the probability of successful insemination if bull 2

$$P(A_2 | E) = \frac{P(A_2 \cap E)}{P(E)} = \frac{P(A_2)P(E | A_2)}{P(A_1)P(E | A_1) + P(A_2)P(E | A_2)} = \frac{(.4)(.82)}{(.6)(.65) + (.4)(.82)} = 0.457$$

Thus, the probability that the father of that calf is bull 2 is 0.457.

Exercise

2.1. In a barn there are 9 cows. Their previous lactation milk records are:

Cow	1	2	3	4	5	6	7	8	9
Milk (kg)	3700	4200	4500	5300	5400	5700	6100	6200	6900

If we randomly choose a cow, what is the probability: a) that it produced more than 5000 kg; b) that it produced less than 5000 kg? If we randomly choose two cows, what is the probability: c) that both cows produced more than 5000 kg; d) that at least one cow produced more than 5000 kg; e) that one cow produced more than 4000 kg, and the other produced more than 5000 kg?

Chapter 3

Random Variables and their Distributions

A random variable is a rule or function that assigns numerical values to observations or measurements. It is called a random variable because the number that is assigned to the observation is a numerical event which varies randomly. It can take different values for different observations or measurements of an experiment. A random variable takes a numerical value with some probability.

Random variables can be discrete or continuous. A continuous variable can take on all values in an interval of real numbers. For example, calf weight at the age of 6 months might take any possible value in an interval from 160 to 260 kg, say the value of 180.0 kg or 191.23456 kg; however, precision of scales or practical use determines the number of decimal places to which the values will be reported. A discrete variable can take only particular values (often integers) and not all values in some interval. For example, the number of eggs laid in a month, litter size, etc.

The value of a variable y is a numerical event and thus it has some probability. A table, graph or formula that shows that probability is called the probability distribution for the random variable y. For the set of observations that is finite and countable, the probability distribution corresponds to a frequency distribution. Often, in presenting the probability distribution we use a mathematical function as a model of empirical frequency. Functions that present a theoretical probability distribution of discrete variables are called probability functions. Functions that present a theoretical probability distribution of continuous variables are called probability density functions.

3.1 Expectations and Variances of Random Variables

Important parameters describing a random variable are the mean (expectation) and variance. The term expectation is interchangeable with mean, because the expected value of the typical member is the mean. The expectation of a variable y is denoted with:

$$E(y) = \mu_y$$

The variance of y is:

$$Var(y) = \sigma^2_y = E[(y - \mu_y)^2] = E(y^2) - \mu_y^2$$

which is the mean square deviation from the mean. Recall that the standard deviation is the square root of the variance:

$$\sigma = \sqrt{\sigma^2}$$

There are certain rules that apply when a constant is multiplied or added to a variable, or two variables are added to each other.

1) The expectation of a constant c is the value of the constant itself:

$$E(c) = c$$

2) The expectation of the sum of a constant c and a variable y is the sum of the constant and expectation of the variable y:

$$E(c + y) = c + E(y)$$

This indicates that when the same number is added to each value of a variable the mean increases by that number.

3) The expectation of the product of a constant c and a variable y is equal to the product of the constant and the expectation of the variable y:

$$E(cy) = cE(y)$$

This indicates that if each value of the variable is multiplied by the same number, then the expectation is multiplied by that number.

4) The expectation of the sum of two variables x and y is the sum of the expectations of the two variables:

$$E(x + y) = E(x) + E(y)$$

5) The variance of a constant c is equal to zero:

$$Var(c) = 0$$

6) The variance of the product of a constant c and a variable y is the product of the squared constant multiplied by the variance of the variable y:

$$Var(cy) = c^2\, Var(y)$$

7) The covariance of two variables x and y:

$$Cov(x,y) = E[(x - \mu_x)(y - \mu_y)]$$
$$= E(xy) - E(x)E(y)$$
$$= E(xy) - \mu_x\mu_y$$

The covariance is a measure of simultaneous variability of two variables.

8) The variance of the sum of two variables is equal to the sum of the individual variances plus two times the covariance:

$$Var(x + y) = Var(x) + Var(y) + 2Cov(x,y)$$

3.2   Probability Distributions for Discrete Random Variables

The probability distribution for a discrete random variable y is the table, graph or formula that assigns the probability $P(y)$ for each possible value of the variable y. The probability distribution $P(y)$ must satisfy the following two assumptions:

1) $0 \leq P(y) \leq 1$

The probability of each value must be between 0 and 1, inclusively.

2) $\sum_{(all\ y)} P(y) = 1$

The sum of probabilities of all possible values of a variable y is equal to 1.

Example: An experiment consists of tossing two coins. Let H and T denote head and tail, respectively. A random variable y is defined as the number of heads in one tossing of two coins. Possible outcomes are 0, 1 and 2. What is the probability distribution for the variable y?

The events and associated probabilities are shown in the following table. The simple events are denoted with E_1, E_2, E_3 and E_4. There are four possible simple events HH, HT, TH and TT.

Simple event	Description	y	$P(y)$
E_1	HH	2	$1/4$
E_2	HT	1	$1/4$
E_3	TH	1	$1/4$
E_4	TT	0	$1/4$

From the table we can see that:

The probability that $y = 2$ is $P(y = 2) = P(E_1) = 1/4$.
The probability that $y = 1$ is $P(y = 1) = P(E_2) + P(E_3) = 1/4 + 1/4 = 1/2$.
The probability that $y = 0$ is $P(y = 0) = P(E_4) = 1/4$.

Thus, the probability distribution of the variable y is:

y	$P(y)$
0	$1/4$
1	$1/2$
2	$1/4$

Checking the previously stated assumptions:

1. $0 \leq P(y) \leq 1$
2. $\sum_{(all\ y)} P(y) = P(y = 0) + P(y = 1) + P(y = 2) = \frac{1}{4} + \frac{1}{2} + \frac{1}{4} = 1$

A cumulative probability distribution $F(y_i)$ describes the probability that a variable y has values less than or equal to some value y_i:

$$F(y_i) = P(y \leq y_i)$$

Example: For the example of tossing two coins, what is the cumulative distribution?

We have:

y	$P(y)$	$F(y)$
0	$^1/_4$	$^1/_4$
1	$^1/_2$	$^3/_4$
2	$^1/_4$	$^4/_4$

For example, the probability $F(1) = {}^3/_4$ denotes the probability that y, the number of heads, is 0 or 1, that is, in tossing two coins that we have at least one tail (or we do not have two heads).

3.2.1   Expectation and Variance of a Discrete Random Variable

The expectation or mean of a discrete variable y is defined:

$$E(y) = \mu = \sum_i P(y_i)\, y_i \qquad\qquad i = 1,\ldots, n$$

The variance of a discrete random variable y is defined:

$$Var(y) = \sigma^2 = E\{[y - E(y)]^2\} = \sum_i P(y_i)\, [y_i - E(y)]^2 \quad i = 1,\ldots, n$$

Example: Calculate the expectation and variance of the number of heads resulting from tossing two coins.

Expectation:

$$E(y) = \mu = \sum_i P(y_i)\, y_i = (^1/_4)\,(0) + (^1/_2)\,(1) + (^1/_4)\,(2) = 1$$

The expected value is one head and one tail when tossing two coins.

Variance:

$$Var(y) = \sigma^2 = \sum_i P(y_i)\, [y_i - E(y)]^2 = (^1/_4)\,(0 - 1)^2 + (^1/_2)\,(1 - 1)^2 + (^1/_4)\,(2 - 1)^2 = {}^1/_2$$

Example: Let y be a discrete random variable with values 1 to 5 with the following probability distribution:

y	1	2	3	4	5
Frequency	1	2	4	2	1
$P(y)$	$^1/_{10}$	$^2/_{10}$	$^4/_{10}$	$^2/_{10}$	$^1/_{10}$

Check if the table shows a correct probability distribution. What is the probability that y is greater than three, $P(y > 3)$?

$$1)\ 0 \leq P(y) \leq 1 \Rightarrow \text{OK}$$
$$2)\ \Sigma_i\, P(y_i) = 1 \Rightarrow \text{OK}$$

The cumulative frequency of $y = 3$ is 7.

$$F(3) = P(y \leq 3) = P(1) + P(2) + P(3) = (^1/_{10}) + (^2/_{10}) + (^4/_{10}) = {}^7/_{10}$$
$$P(y > 3) = P(4) + P(5) = (^2/_{10}) + (^1/_{10}) = {}^3/_{10}$$
$$P(y > 3) = 1 - P(y \leq 3) = 1 - (^7/_{10}) = {}^3/_{10}$$

Expectation:

$$E(y) = \mu = \Sigma_i\, y_i\, P(y_i) = (1)\,(^1/_{10}) + (2)\,(^2/_{10}) + (3)\,(^4/_{10}) + (4)\,(^2/_{10}) + (5)\,(^1/_{10}) = (^{30}/_{10}) = 3$$

Variance:

$$Var(y) = \sigma^2 = E\{[y - E(y)]^2\} = \Sigma_i\, P(y_i)\, [y_i - E(y)]^2 =$$
$$(^1/_{10})\,(1 - 3)^2 + (^2/_{10})\,(2 - 3)^2 + (^4/_{10})\,(3 - 3)^2 + (^2/_{10})\,(4 - 3)^2 + (^1/_{10})\,(5 - 3)^2 = 1.2$$

3.2.2 Bernoulli Distribution

Consider a random variable that can take only two values, for example Yes and No, or 0 and 1. Such a variable is called a binary or Bernoulli variable. For example, let a variable y be the incidence of some illness. Then the variable takes the values:

$y_i = 1$ if an animal is ill
$y_i = 0$ if an animal is not ill

The probability distribution of y has the Bernoulli distribution:

$$p(y) = p^y q^{1-y} \quad \text{for } y = 0, 1$$

Here, $q = 1 - p$

Thus,

$$P(y_i = 1) = p$$
$$P(y_i = 0) = q$$

The expectation and variance of a Bernoulli variable are:

$$E(y) = \mu = p \quad \text{and} \quad \sigma^2 = Var(y) = \sigma^2 = pq$$

3.2.3 Binomial Distribution

Assume a single trial that can take only two outcomes, for example, Yes and No, success and failure, or 1 and 0. Such a variable is called a binary or Bernoulli variable. Now assume that such single trial is repeated n times. A binomial variable y is the number of successes in those n trials. It is the sum of n binary variables. The binomial probability distribution describes the distribution of different values of the variable y {0, 1, 2, …, n} in a total of n trials. Characteristics of a binomial experiment are:

1. The experiment consists of n equivalent trials, independent of each other
2. There are only two possible outcomes of a single trial, denoted with Y (yes) and N (no) or equivalently 1 and 0
3. The probability of obtaining Y is the same from trial to trial, denoted with p. The probability of N is denoted with q, so $p + q = 1$
4. The random variable y is the number of successes (Y) in the total of n trials.

The probability distribution of a random variable y is determined by the parameter p and the number of trials n:

$$P(y) = \binom{n}{y} p^y q^{n-y} \qquad\qquad y = 0, 1, 2,..., n$$

where:

p = the probability of success in a single trial
$q = 1 - p$ = the probability of failure in a single trial

The expectation and variance of a binomial variable are:

$$E(y) = \mu = np \qquad \text{and} \qquad Var(y) = \sigma^2 = npq$$

The shape of the distribution depends on the parameter p. The binomial distribution is symmetric only when $p = 0.5$, and asymmetric in all other cases. Figure 3.1 presents two binomial distributions for $p = 0.5$ and $p = 0.2$ with $n = 8$.

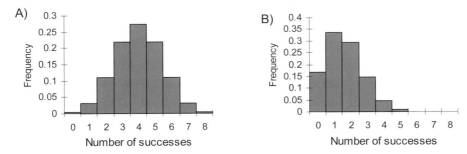

Figure 3.1 Binomial distribution ($n = 8$): A) $p = 0.5$ and B) $p = 0.2$

The binomial distribution is used extensively in research on and selection of animals, including questions such as whether an animal will meet some standard, whether a cow is pregnant or open, etc.

Example: Determine the probability distribution of the number of female calves in three consecutive calvings. Assume that only a single calf is possible at each calving, and that the probability of having a female in a single calving is $p = 0.5$.

The random variable y is defined as the number of female calves in three consecutive calvings. Possible outcomes are 0, 1, 2 and 3. The distribution is binomial with $p = 0.5$ and $n = 3$:

$$P(y) = \binom{3}{y}(0.5)^y (0.5)^{3-y} \qquad y = 0, 1, 2, 3$$

Possible values with corresponding probabilities are presented in the following table:

y	$p(y)$
0	$\binom{3}{0}(0.5)^0(0.5)^3 = 0.125$
1	$\binom{3}{1}(0.5)^1(0.5)^2 = 0.375$
2	$\binom{3}{2}(0.5)^2(0.5)^1 = 0.375$
3	$\binom{3}{3}(0.5)^3(0.5)^0 = 0.125$

The sum of the probabilities of all possible values is:

$$\sum_i p(y_i) = 1$$

The expectation and variance are:

$$\mu = E(y) = np = (3)(0.5) = 1.5$$
$$\sigma^2 = var(y) = npq = (3)(0.5)(0.5) = 0.75$$

Second example: In a swine population susceptibility to a disease is genetically determined at a single locus. This gene has two alleles: B and b. The disease is associated with the recessive allele b; animals with the genotype bb will have the disease, while animals with Bb are only carriers. The frequency of the b allele is equal to 0.5. If a boar and sow both with Bb genotypes are mated and produce a litter of 10 piglets: a) how many piglets are expected to have the disease; b) what is the probability that none of the piglets has the disease; c) what is the probability that at least one piglet has the disease; d) what is the probability that exactly a half of the litter has the disease?

The frequency of the b allele is 0.5. The probability that a piglet has the disease (has the bb genotype) is equal to $(0.5)(0.5) = 0.25$. Further, the probability that a piglet is healthy is $1 - 0.25 = 0.75$. Thus, a binomial distribution with $p = 0.25$ and $n = 10$ can be used.

a) Expectation $= np = 2.5$, that is, between two and three piglets can be expected to have the disease.

b) $P(y=0) = \binom{10}{0} p^0 q^{10} = 1(0.25)^0 (0.75)^{10} = 0.056$

c) $P(y \geq 1) = 1 - P(y=0) = 1 - 0.056 = 0.944$

d) $P(y=5) = \binom{10}{5} p^5 q^5 = \frac{10!}{5!5!}(0.25)^5(0.75)^5 = 0.058$

Third example: A farmer buys an expensive cow with hopes that she will produce a future elite bull. How many calves must that cow produce such that the probability of having at least one male calf is greater than 0.99?

Solution: Assume that the probability of having a male calf in a single calving is 0.5. For at least one male calf the probability must be greater than 0.99:

$$P(y \geq 1) > 0.99$$

Using a binomial distribution, the probability that at least one calf is male is equal to one minus the probability that n calves are female:

$$P(y \geq 1) = 1 - P(y < 1) = 1 - P(y=0) = 1 - \binom{n}{0}\left(\tfrac{1}{2}\right)^0\left(\tfrac{1}{2}\right)^n$$

Thus:

$$1 - \binom{n}{0}\left(\tfrac{1}{2}\right)^0\left(\tfrac{1}{2}\right)^n > 0.99$$

$$\left(\tfrac{1}{2}\right)^n < 0.01$$

Solving for n in this inequality:

$$n > 6.64$$

Or rounded to an integer:

$$n = 7$$

3.2.4 Hyper-geometric Distribution

Assume a set of size N with R successes and $N - R$ failures. A single trial has only two outcomes, but the set is finite, and each trial depends on the outcomes of previous trials. The random variable y is the number of successes in a sample of size n drawn from the source set of size N. Such a variable has a hyper-geometric probability distribution:

$$P(y) = \frac{\binom{R}{y}\binom{N-R}{n-y}}{\binom{N}{n}}$$

where:

y = random variable, the number of successful trials in the sample

n = size of the sample

$n - y$ = the number of failures in the sample

N = size of the source set
R = the number of successful trials in the source set
$N - R$ = the number of failures in the source set

Properties of a hyper-geometric distribution are:

1) $n < N$
2) $0 < y < min(R,n)$

The expectation and variance are:

$$E(y) = \mu = \frac{nR}{N} \qquad Var(y) = \sigma^2 = \frac{nR(N-R)}{N^2}\left(1 - \frac{n-1}{N-1}\right)$$

Example: In a box, there are 12 male and 6 female piglets. If 6 piglets are chosen at random, what is the probability of getting 5 males and 1 female?

$$P(y) = \frac{\binom{R}{y}\binom{N-R}{n-y}}{\binom{N}{n}} = \frac{\binom{12}{5}\binom{6}{1}}{\binom{18}{6}} = 0.2559$$

Thus, the probability of choosing five male and one female piglets is 0.2559.

3.2.5 Poisson Distribution

The Poisson distribution is a model for the relative frequency of rare events and data defined as counts. It is often used for determining the probability that some event will happen in a specific time, volume or area. For example, the number of microorganisms within a microscope field, or the number of mutations in a sample or distribution of animals in a plot may have a Poisson distribution. A Poisson random variable y is defined as how many times some event occurs in specific time, given volume, or area. If we know that each single event occurs with the same probability, that is, the probability that some event will occur is equal for any part of time, volume or area, and the expected number of events is λ, then the probability function is defined as:

$$P(y) = \frac{e^{-\lambda}\lambda^y}{y!}$$

where λ is the average number of successes in a given time, volume or area, and e is the base of the natural logarithm ($e = 2.71828$).

Often, instead of the expected number, the proportion of successes is known, which is an estimate of the probability of success in a single trial (p). When p is small and the total number of trials (n) large, the binomial distribution can be approximated with a Poisson distribution, $\lambda = np$.

A characteristic of the Poisson variable is that both the expectation and variance are equal to the parameter λ:

$$E(y) = \mu = \lambda \quad \text{and} \quad Var(y) = \sigma^2 = \lambda$$

Example: In a population of mice 2% have cancer. In a sample of 100 mice, what is the probability that more than one mouse has cancer?

$\mu = \lambda = 100 \, (0.02) = 2$ (expectation, the mean is 2% of 100)

$$P(y) = \frac{e^{-2} 2^y}{y!}$$

$$P(y > 1) = 1 - P(y = 0) - P(y = 1) = 1 - 0.1353 - 0.2706 = 0.5941$$

The probability that in the sample of 100 mice more than one mouse has cancer is 0.5941.

3.2.6 Multinomial Distribution

The multinomial probability distribution is a generalization of the binomial distribution. The outcome of a single trial can be not only Yes or No, or 1 or 0, but there can be more than two outcomes. Each outcome has a probability. Therefore, there are k possible outcomes of a single trial, each with its own probability: $p_1, p_2,..., p_k$. Single trials are independent. The numbers of particular outcomes in a total of n trials are random variables, that is, y_1 for outcome 1; y_2 for outcome 2; ..., y_k for outcome k. The probability function is:

$$p(y_1, y_2,..., y_k) = \frac{n!}{y_1! y_2!.....y_k!} p_1^{y_1} p_2^{y_2} ... p_k^{y_k}$$

Also,

$$n = y_1 + y_2 + ... + y_k$$
$$p_1 + p_2 + ... + p_k = 1$$

The number of occurrences y_i of an outcome i has its expectation and variance:

$$E(y_i) = \mu_i = np_i \qquad \text{and} \qquad Var(y_i) = \sigma^2_i = np_i(1 - p_i)$$

The covariance between the numbers of two outcomes i and j is:

$$Cov(y_i, y_j) = -np_i p_j$$

Example: Assume calving ease is defined in three categories labeled 1, 2 and 3. What is the probability that out of 10 cows, 8 cows are in the first category, 1 cow is in the second, and 1 cow is in the third, if the probabilities of a single calving being in categories 1, 2 and 3 are 0.6, 0.3 and 0.1, respectively? What is the expected number of cows in each category?

$$p_1 = 0.6, \ p_2 = 0.3, \ p_3 = 0.1$$

$$n = 10, \ y_1 = 8, \ y_2 = 1, \ y_3 = 1$$

$$p(y_1, y_2, y_3) = \frac{n!}{y_1! y_2! y_3!} p_1^{y_1} p_2^{y_2} p_3^{y_3} = p(8,1,1) = \frac{10!}{8! \ 1! \ 1!} (0.6)^8 \, (0.3)^1 (0.1)^1 = 0.045$$

The probability that out of 10 cows exactly 8 are in category 1, one is in category 2, and 1 is in category 3 is 0.045.

The expected number in each category is:

$$\mu_1 = np_1 = 10 \ (0.6) = 6, \quad \mu_2 = np_2 = 10 \ (0.3) = 3, \quad \mu_3 = np_3 = 10 \ (0.1) = 1$$

For 10 cows, the expected numbers of cows in categories 1, 2 and 3 are 6, 3 and 1, respectively.

3.3   Probability Distributions for Continuous Random Variables

A continuous random variable can take on an uncountable and infinite possible number of values, and because of that it is impossible to define the probability of occurrence of any single numerical event. The value of a single event is a point, a point does not have a dimension, and consequently the probability that a random variable has a specific value is equal to zero. Although it is not possible to define the probability of a particular value, the probability that a variable y takes values in some interval is defined. A probability is defined to the numerical event that is applicable to that interval. For example, take weight of calves as a random variable. Numbers defining a particular interval depend on the precision of the measuring device or practical usefulness. If the precision is 1 kg, a measurement of 220 kg indicates a value in the interval from 219.5 to 220.5 kg. Such a numerical event has a probability. A function used to model the probability distribution of a continuous random variable is called a probability *density* function.

A cumulative distribution function $F(y_0)$ for a random variable y, which yields values y_0 is:

$$F(y_0) = P(y \le y_0)$$

From the previous example, $F(220)$ represents the probability of all measurements less than 220 kg. A property of a continuous random variable is that its cumulative distribution function is continuous.

If a random variable y contains values between y_0 and $y_0 + \Delta y$, a density function is defined:

$$f(y_0) = \lim_{\Delta y \to 0} \frac{P(y_0 \le y \le y_0 + \Delta y)}{\Delta y}$$

It follows that:

$$f(y) = dF(y) / dy$$

The density function is the first derivative of the cumulative distribution function. The cumulative distribution function is:

$$F(y_0) = \int_{-\infty}^{y_0} f(y)dy,$$

an integral of the function representing the area under the density function in the interval $(-\infty, y)$.

A function is a density function if it has the following properties:

1) $f(y_i) \ge 0$

2) $\int_{-\infty}^{\infty} f(y)dy = 1$

   or written differently $P(-\infty \le y \le +\infty) = 1$, that is, the probability that any value of y occurs is equal to 1.

The probability that y is any value between y_1 and y_2 is:

$$P(y_1 \leq y \leq y_2) = \int_{y_1}^{y_2} f(y)dy$$

which is the area under $f(y)$ bounded by y_1 and y_2.

The expected value of a continuous random variable y is:

$$E(y) = \mu_y = \int_{-\infty}^{\infty} y\, f(y)dy$$

The variance of a continuous variable y is:

$$Var(y) = \sigma_y^2 = E\left[\left(y - \mu_y\right)^2\right] = \int_{-\infty}^{\infty} \left(y - \mu_y\right)^2 f(y)dy$$

Again the properties of a continuous variable are:

1. The cumulative distribution $F(y)$ is continuous;
2. The random variable y has infinite number of values;
3. The probability that y has a particular value is equal to zero.

3.3.1 Uniform Distribution

The uniform variable y is a variable that has the same probability for any value y_i in an interval ($a \leq y \leq b$). The density function is:

$$f(y) = \begin{cases} \frac{1}{b-a} & \text{if } a \leq y \leq b \\ 0 & \text{for all other } y \end{cases}$$

The expectation and variance are:

$$E(y) = \mu = \frac{a+b}{2} \qquad\qquad Var(y) = \sigma^2 = \frac{(b-a)^2}{12}$$

3.3.2 Normal Distribution

The normal curve models the frequency distributions of many biological events. In addition, many statistics utilized in making inferences follow the normal distribution. Often, the normal curve is called a Gauss curve, because it was introduced by C. F. Gauss as a model for relative frequency of measurement error. The normal curve has the shape of a bell, and its location and form are determined by two parameters, the mean μ and variance σ^2. The density function of normal distribution is:

$$f(y) = \frac{1}{\sqrt{2\pi\sigma^2}} e^{-(y-\mu)^2/2\sigma^2} \qquad\qquad -\infty < y < +\infty$$

where μ and σ^2 are parameters, e is the base of the natural logarithm ($e = 2.71828...$) and $\pi = 3.14...$. The following describes a variable y as a normal random variable:

$$y \sim N(\mu, \sigma^2)$$

The parameters μ and σ^2 are the mean and variance of the distribution. Recall, that the standard deviation is:

$$\sigma = \sqrt{\sigma^2}$$

and represents the mean deviation of values from the mean.

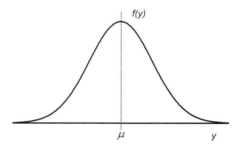

Figure 3.2 Normal or Gauss curve

The normal curve is symmetric about its mean, and the maximum value of its ordinate occurs at the mean of y, i.e. ($f(\mu)$ = maximum). That indicates that the mode and median are equal to the mean. In addition, the coefficient of skewness is equal to zero:

$$sk = E\left[\left(\frac{y-\mu}{\sigma}\right)^3\right] = 0$$

The coefficient of kurtosis is also equal to zero:

$$sk = E\left[\left(\frac{y-\mu}{\sigma}\right)^4\right] - 3 = 0$$

The inflection points of the curve are at $(\mu - \sigma)$ and $(\mu + \sigma)$, the distance of ± 1 standard deviation from the mean. The interval $\mu \pm 1.96\sigma$ is expected to contain 95% of the observations (Figure 3.3):

$$P(\mu \pm 1.96\sigma \leq y \leq \mu \pm 1.96\sigma) = 0.95$$

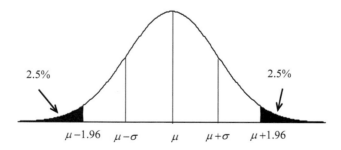

Figure 3.3 Some characteristics of the normal curve

The height and dispersion of the normal curve depend on the variance σ^2, (or the standard deviation σ). A higher σ leads to decreased height of the curve and increased dispersion. Figure 3.4 shows two curves with $\sigma = 1$ and $\sigma = 1.5$. Both curves have the same central location, $\mu = 0$.

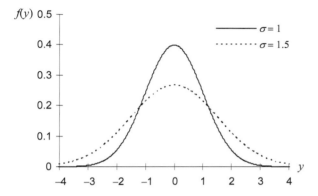

Figure 3.4 Normal curves with standard deviations $\sigma = 1$ and $\sigma = 1.5$

As for all density functions, the properties of the normal density function are:

1. $f(y_i) \geq 0$,

2. $\displaystyle\int_{-\infty}^{\infty} f(y)dy = 1$

The probability that the value of a normal random variable is in an interval (y_1, y_2) is:

$$P(y_1 < y < y_2) = \int_{y_1}^{y_2} \frac{1}{\sqrt{2\pi\sigma^2}} e^{-\frac{1}{2}\left(\frac{y-\mu}{\sigma}\right)^2}$$

This corresponds to the area under a normal curve bounded by the values y_1 and y_2, when the total area under the curve is defined as 1 or 100% (Figure 3.5). The area bounded by the values y_1 and y_2 is the proportion of values between y_1 and y_2 with respect to all possible values.

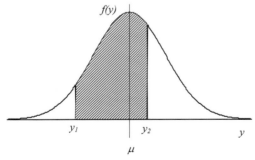

Figure 3.5 Area under the normal curve bounded with values y_1 and y_2

The value of a cumulative distribution for some value y_0, $F(y_0) = P(y \leq y_0)$, is explained by the area under the curve from $-\infty$ to y_0 (Figure 3.6).

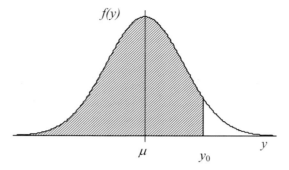

Figure 3.6 Value of the cumulative distribution for y_0 corresponds to the shaded area under the curve

The value of the cumulative distribution for the mean μ is equal to 0.5, because the curve is symmetric:

$$F(\mu) = P(y \leq \mu) = 0.5$$

The shape of the normal curve depends only on the standard deviation σ, thus all normal curves can be standardized and transformed to a standard normal curve with $\mu = 0$ and $\sigma = 1$. The standardization of a random normal variable y, symbolized by z, implies that its values are expressed as deviations from the mean in standard deviation units:

$$z = \frac{y - \mu}{\sigma}$$

The values of a standard normal variable z tell us by how many standard deviations the values of y deviate from the mean. Original values of y can be expressed as:

$$y = \mu + z\sigma$$

A density function of the standard normal variable is:

$$f(z) = \frac{e^{-\frac{1}{2}z^2}}{\sqrt{2\pi}} \qquad -\infty < z < +\infty$$

where e is the base of the natural logarithm ($e = 2.71828...$) and $\pi = 3.14...$ A standard normal variable z is usually written as:

$$z \sim Z \text{ or } z \sim N(0, 1)$$

This transformation is important to create a single curve used to determine the area under the curve bounded with some interval. Recall that the area under the curve over some interval (y_1, y_2) is equal to the probability that a random variable y takes values in that interval. The area under the curve is equal to the integral of a density function. Since an explicit formula for that integral does not exist, a table is used (either from a book or computer software). The standardization allows use of one table for any mean and variance (see the table of areas under the normal curve, Appendix B). The probability that a variable

y takes values between y_1 and y_2 is equal to the probability that the standard normal variable z takes values between the corresponding values z_1 and z_2:

$$P(y_1 \leq y \leq y_2) = P(z_1 \leq z \leq z_2)$$

where $z_1 = \dfrac{y_1 - \mu}{\sigma}$ and $z_2 = \dfrac{y_2 - \mu}{\sigma}$

For example, for a normal curve $P(-1.96\sigma \leq y \leq 1.96\sigma) = 0.95$. For the standard normal curve $P(-1.96 \leq z \leq 1.96) = 0.95$. The probability is 0.95 that the standard normal variable z is in the interval -1.96 to $+1.96$ (Figure 3.7).

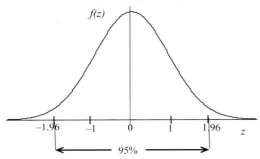

Figure 3.7 Standard normal curve ($\mu = 0$ and $\sigma^2 = 1$)

A related question is, what is the mean of selected values? The standard normal curve can also be utilized in finding the values of a variable y determined with a given probability. Figure 3.8 shows that concept. Here, z_S is the mean of z values greater than z_0, $z > z_0$. For the standard normal curve, the mean of selected animals is:

$$z_S = \frac{f(z_0)}{P}$$

where P is the area under the standard normal curve for $z > z_0$, and $f(z_0)$ is the ordinate for the value z_0. Recall that $f(z_0) = \dfrac{e^{-\frac{1}{2}z_0^2}}{\sqrt{2\pi}}$.

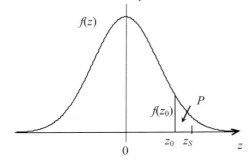

Figure 3.8 The mean of selected z values. $f(z_0)$ = the curve ordinate for $z = z_0$, P = area under the curve, i.e., the probability $P(z>z_0)$, and z_S is the mean of $z > z_0$

Example: Assume a theoretical normal distribution of calf weights at age 6 months defined with $\mu = 200$ kg and $\sigma = 20$ kg. Determine theoretical proportions of calves: a) more than 230 kg; b) less than 230 kg; and c) less than 210 and more than 170 kg; d) what is the theoretical lowest value for an animal to be included among the heaviest 20%; e) what is the theoretical mean of animals with weights greater than 230 kg?

a) The proportion of calves weighing more than 230 kg also denotes the probability that a randomly chosen calf weighs more than 230 kg. This can be shown by calculating the area under the normal curve for an interval $y > y_0 = 230$, that is $P(y > 230)$ (Figure 3.9).

First, determine the value of the standard normal variable, z_0, which corresponds to the value $y_0 = 230$ (Figure 3.9).

$$z_0 = \frac{230 - 200}{20} = 1.5$$

This indicates that 230 is 1.5 standard deviations above the mean.

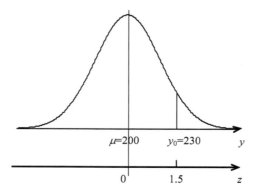

Figure 3.9 Normal curve with the original scale y and standard normal scale z. The value $y_0 = 230$ corresponds to the value $z_0 = 1.5$

The probability that y is greater than y_0 is equal to the probability that z is greater than z_0.

$$P(y > y_0) = P(z > z_0) = P(z > 1.5) = 0.0668$$

The value of the probability for $z_0 = 1.5$ can be read from the table (Appendix B: Area under the standard normal curve). The percentage of calves expected to be heavier than 230 kg is 6.68%.

b) Since the total area under the curve is equal to 1, then the probability that y has a value less than $y_0 = 230$ kg is:

$$P(y < y_0) = P(z < z_0) = 1 - P(z > 1.5) = 1 - 0.0668 = 0.9332$$

This is the value of the cumulative distribution for $y_0 = 230$ kg:

$$F(y_0) = F(230) = P(y \le y_0) = P(y \le 230)$$

Thus, 93.32% of calves are expected to weigh less than 230 kg.

Note that $P(y \le y_0) = P(y < y_0)$ because $P(y = y_0) = 0$.

c) $y_1 = 170$ kg, $y_2 = 210$ kg

The corresponding standardized values, z_1 and z_2, are:

$$z_1 = \frac{170 - 200}{20} = -1.5$$

$$z_2 = \frac{210 - 200}{20} = 0.5$$

Find the probability that the variable takes values between −1.5 and 0.5 standard deviations from the mean (Figure 3.10).

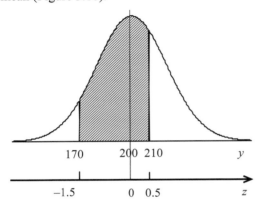

Figure 3.10 Area under the normal curve between 170 and 210 kg

The probability that the values of y are between 170 and 210 is:

$$P(y_1 \le y \le y_2) = P(170 \le y \le 210) = P(z_1 \le z \le z_2) = P(-1.5 \le z \le 0.5)$$

Recall that the curve is symmetric, which means that:

$$P(z \le -z_0) = P(z \ge z_0) \quad \text{or for this example:}$$

$$P(z \le -1.5) = P(z \ge 1.5)$$

The following values are from the table Area under the standard normal curve (Appendix B):

$$P(z > 1.5) = 0.0668$$

$$P(z > 0.5) = 0.3085$$

Now:

$$P(170 \le y \le 210) = P(-1.5 \le z \le 0.5) = 1 - [P(z > 1.5) + P(z > 0.5)] =$$

$$1 - (0.0668 + 0.3085) = 0.6247$$

Thus, 62.47% of calves are expected to weigh between 170 and 210 kg.

d) The heaviest 20% corresponds to the area under the standard normal curve for values greater than some value z_0:

$$P(z_0 \le z \le +\infty) = 0.20$$

First z_0 must be determined. From the table 'Area under the standard normal curve' (Appendix B), the value of z_0 is 0.84. Now, z_0 must be transformed to y_0, on the original scale using the formula:

$$z_0 = \frac{y_0 - \mu}{\sigma}$$

that is:

$$y_0 = \mu + z_0\,\sigma$$

$$y_0 = 200 + (0.84)(20) = 216.8 \text{ kg}$$

Animals greater than or equal to 216.8 kg are expected to be among the heaviest 20%.

e) The corresponding z value for 230 kg is:

$$z_0 = \frac{230 - 200}{20} = 1.5$$

From the table of areas under the normal curve:

$$P(z > z_0) = 1 - P(z \leq z_0) = 0.0668$$

The ordinate for $z_0 = 1.5$ is:

$$f(z_0) = \frac{e^{-\frac{1}{2}z_0^2}}{\sqrt{2\pi}} = \frac{e^{-\frac{1}{2}(1.5)^2}}{\sqrt{2\pi}} = 0.129518$$

The mean of standardized values greater than 1.5 is:

$$z_S = \frac{f(z_0)}{P} = \frac{0.129518}{0.0668} = 1.94$$

Transformed to the original scale:

$$y_S = \mu + z_S\,\sigma = 200 + (1.94)(20) = 238.8 \text{ kg}$$

Thus, the mean of the selected animals is expected to be 238.8 kg.

3.3.3 Multivariate Normal Distribution

Consider a set of n random variables $y_1, y_2, \ldots, y_n$ with means $\mu_1, \mu_2, \ldots, \mu_n$, variances σ_1^2, $\sigma_2^2, \ldots, \sigma_n^2$, and covariances among them $\sigma_{12}, \sigma_{13}, \ldots, \sigma_{(n-1)n}$. These can be expressed as vectors and a matrix as follows:

$$\mathbf{y} = \begin{bmatrix} y_1 \\ y_2 \\ \ldots \\ y_n \end{bmatrix}_{n \times 1} \quad \boldsymbol{\mu} = \begin{bmatrix} \mu_1 \\ \mu_2 \\ \ldots \\ \mu_n \end{bmatrix}_{n \times 1} \quad \text{and} \quad \mathbf{V} = \begin{bmatrix} \sigma_1^2 & \sigma_{12} & & \sigma_{1n} \\ \sigma_{12} & \sigma_2^2 & & \sigma_{2n} \\ & & & \\ \sigma_{1n} & \sigma_{2n} & & \sigma_n^2 \end{bmatrix}_{n \times n}$$

where $\mathbf{V}$ denotes the variance-covariance matrix of the vector $\mathbf{y}$.

The vector **y** has a multivariate normal distribution **y** ~ *N*(**μ**, **V**) if its probability density function is:

$$f(\mathbf{y}) = \frac{e^{-\frac{1}{2}(\mathbf{y}-\boldsymbol{\mu})'\mathbf{V}^{-1}(\mathbf{y}-\boldsymbol{\mu})}}{\sqrt{(2\pi)^n|\mathbf{V}|}}$$

where |**V**| denotes the determinant of **V**.

Some useful properties of the multivariate normal distribution include:

1. $E(\mathbf{y}) = \boldsymbol{\mu}$ and $Var(\mathbf{y}) = \mathbf{V}$
2. The marginal distribution of y_i is $N(\mu_i, \sigma_i^2)$
3. The conditional distribution of $y_i \mid y_j$ is $N\left(\mu_i + \frac{\sigma_{ij}}{\sigma_i^2}(y_j - \mu_j), \sigma_i^2 - \frac{\sigma_{ij}\sigma_{ij}}{\sigma_j^2}\right)$.

Generally, expressing the vector **y** as two subvectors $\mathbf{y} = \begin{bmatrix} \mathbf{y}_1 \\ \mathbf{y}_2 \end{bmatrix}$ and its distribution

$N\left(\begin{bmatrix} \boldsymbol{\mu}_1 \\ \boldsymbol{\mu}_2 \end{bmatrix}, \begin{bmatrix} \mathbf{V}_{11} & \mathbf{V}_{12} \\ \mathbf{V}_{21} & \mathbf{V}_{22} \end{bmatrix}\right)$, the conditional distribution of $\mathbf{y}_1 \mid \mathbf{y}_2$ is:

$$N\left(\boldsymbol{\mu}_1 + \mathbf{V}_{12}\mathbf{V}_{22}^{-1}(\mathbf{y}_2 - \boldsymbol{\mu}_2), \mathbf{V}_{11} - \mathbf{V}_{12}\mathbf{V}_{22}^{-1}\mathbf{V}_{21}\right)$$

Example: For weight (y_1) and heart girth (y_2) of cows, the following parameters are known: $\mu_1 = 660$ kg and $\mu_2 = 220$ cm; $\sigma_1^2 = 17400$ and $\sigma_2^2 = 4200$; and $\sigma_{12} = 5900$.

These can be expressed as:

$$\boldsymbol{\mu} = \begin{bmatrix} 660 \\ 220 \end{bmatrix} \text{ and } \mathbf{V} = \begin{bmatrix} 17400 & 5900 \\ 5900 & 4200 \end{bmatrix}$$

The bivariate normal probability density function of weight and heart girth is:

$$f(\mathbf{y}) = \frac{e^{-\frac{1}{2}\left(\mathbf{y}-\begin{bmatrix}660\\220\end{bmatrix}\right)'\begin{bmatrix}17400&5900\\5900&4200\end{bmatrix}^{-1}\left(\mathbf{y}-\begin{bmatrix}660\\220\end{bmatrix}\right)}}{\sqrt{(2\pi)^n\begin{vmatrix}17400&5900\\5900&4200\end{vmatrix}}}$$

The conditional mean of weight given the value of heart girth, for example $y_2 = 230$, is:

$$E(y_1 \mid y_2 = 230) = \mu_1 + \frac{\sigma_{12}}{\sigma_2^2}(y_2 - \mu_2) = 660 + \frac{5900}{4200}(230 - 220) = 674.0$$

Note that this is also the regression of weight on heart girth. The conditional variance of weight given the value of heart girth is:

$$Var(y_1 \mid y_2) = 17400 - \frac{(5900)(5900)}{4200} = 9111.9$$

The conditional distribution of weight given the value of heart girth $y_2 = 230$ is:

$$N(674.0, 9111.9)$$

Example: Consider a vector of data **y** such that the elements of the vector are independent and identically distributed, all have the same mean μ and variance σ^2, and the covariance among them is zero.

Assume that **y** has a multivariate normal distribution with mean $E(\mathbf{y}) = \mu$ and variance $Var(\mathbf{y}) = \mathbf{I}\sigma^2$.

$$f(\mathbf{y}) = \frac{e^{-\frac{1}{2}(\mathbf{y}-\boldsymbol{\mu})'\mathbf{V}^{-1}(\mathbf{y}-\boldsymbol{\mu})}}{\sqrt{(2\pi)^n|\mathbf{V}|}} = \frac{e^{-\frac{1}{2}(\mathbf{y}-\boldsymbol{\mu})'(\mathbf{I}\sigma^2)^{-1}(\mathbf{y}-\boldsymbol{\mu})}}{\sqrt{(2\pi)^n|\mathbf{I}\sigma^2|}}$$

Here **I** is an identity matrix.

Then $|\mathbf{I}\sigma^2| = (\sigma^2)^n$ and $(\mathbf{y}-\boldsymbol{\mu})'(\mathbf{I}\sigma^2)^{-1}(\mathbf{y}-\boldsymbol{\mu}) = \frac{1}{\sigma^2}\sum_i(y_i-\mu)^2$ and knowing that values of y are independent, the density function is:

$$f(y) = \prod_i f(y_i) = \frac{1}{\left(\sqrt{2\pi\sigma^2}\right)^n} e^{-\sum_i(y-\mu)^2/2\sigma^2}$$

Here, $\prod_i$ is the product symbol.

3.3.4 *Chi*-square Distribution

Consider a set of standard normal random variables z_i, ($i = 1,\ldots, v$), that are identical and independently distributed with the mean $\mu = 0$ and standard deviation $\sigma = 1$. Define a random variable:

$$\chi^2 = \sum_i z^2_i \qquad i = 1,\ldots, v$$

The variable χ^2 (*chi*-square variable) has a *chi*-square distribution with v degrees of freedom. The shape of the *chi*-square distribution depends on degrees of freedom which is the number of independently distributed simple variables used to build a composite variable; here, standard normal variables used to build a *chi*-square variable. Figure 3.11 shows *chi*-square density functions with 2, 6 and 10 degrees of freedom.

The expectation and variance of a χ^2 variable are:

$$E[\chi^2] = v \qquad \text{and} \qquad Var[\chi^2] = 2v$$

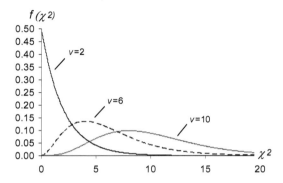

Figure 3.11 The density functions of χ^2 variables with $v = 2$, $v = 6$ and $v = 10$ degrees of freedom

Because the mean of the standard normal variable is equal to zero, this *chi*-square distribution is called a central *chi*-square distribution. A noncentral *chi*-square distribution is:

$$\chi^2 = \sum_i y^2_i \qquad i = 1,\dots, v;\ v \text{ is degrees of freedom}$$

where y_i is a normal variable with mean μ_i and variance $\sigma^2 = 1$. This distribution is defined by degrees of freedom and the noncentrality parameter $\lambda = \sum_i \mu^2_i$, $(i = 1,\dots, v)$.

The expectation of the noncentral χ^2_v variable is:

$$E[\chi^2] = v + \lambda$$

Comparing the noncentral to the central distribution, the mean is shifted to the right for the parameter λ. Figure 3.12 presents a comparison of central and noncentral *chi*-square distributions for different λ.

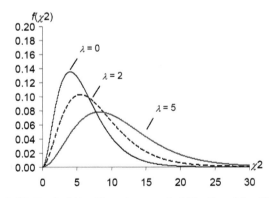

Figure 3.12 Central ($\lambda = 0$) and noncentral ($\lambda = 2$ and $\lambda = 5$) *chi*-square distributions with $v = 6$ degrees of freedom

3.3.5 Student *t* Distribution

Let z be a standard normal random variable with $\mu = 0$ and $\sigma = 1$, and let χ^2 be a *chi*-square random variable with v degrees of freedom. Then:

$$t = \frac{z}{\sqrt{\chi^2/v}}$$

is a random variable with a Student *t* distribution with v degrees of freedom.

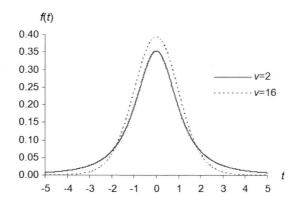

Figure 3.13 The density functions of *t* variables with degrees of freedom $v = 2$ and $v = 16$

The shape of the Student *t* distribution is similar to that of the normal distribution when degrees of freedom is large. With decreasing degrees of freedom, the curve flattens in the middle and it is more expanded ('fatter') toward the tails (Figure 3.13). The expectation and variance of the *t* variable are:

$$E\,[t] = 0 \quad \text{and} \quad Var[t] = \frac{v}{v-2}$$

Because the numerator of a *t* variable is a standard normal variable (centered around zero), this *t* distribution is often called a *central t* distribution. A noncentral *t* distribution is a distribution of:

$$t = \frac{y}{\sqrt{\chi^2/v}}$$

where y is a normal variable with the mean μ and variance $\sigma^2 = 1$. This distribution is defined by degrees of freedom and the noncentrality parameter λ. Figure 3.14 presents a comparison of central and noncentral *t* distribution with 20 degrees of freedom.

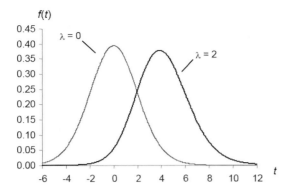

Figure 3.14 Central ($\lambda = 0$) and noncentral ($\lambda = 2$) *t* distributions with $v = 20$ degrees of freedom

3.3.6 *F* Distribution

Let χ^2_1 and χ^2_2 be two independent *chi*-square random variables with v_1 and v_2 degrees of freedom, respectively. Then:

$$F = \frac{\chi^2_1 / v_1}{\chi^2_2 / v_2}$$

is a random variable with an *F* distribution with degrees of freedom v_1 and v_2. The shape of the *F* distribution depends on the degrees of freedom (Figure 3.15).

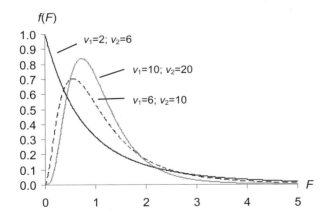

Figure 3.15 The density function of *F* variables with degrees of freedom: a) $v_1 = 2$, $v_2 = 6$; b) $v_1 = 6$, $v_2 = 10$; c) $v_1 = 10$, $v_2 = 20$

The expectation and variance of the *F* variable are:

$$E(F) = \frac{v_2}{v_2 - 2} \quad \text{and} \quad Var(F) = \frac{2v_2^2 (v_1 + v_2 - 2)}{v_1 (v_2 - 2)^2 (v_2 - 4)}$$

If the χ^2_1 variable in the numerator has a noncentral *chi*-square distribution with a noncentrality parameter λ, then the corresponding F variable has a noncentral F distribution with the noncentrality parameter λ.

The expectation of a noncentral F variable is:

$$E(F) = \frac{v_2}{v_2 - 2}\left(1 + \frac{2\lambda}{v_1}\right)$$

The mean of a noncentral F variable is shifted to the right compared to the central distribution. Figure 3.16 presents a comparison of central and noncentral F distributions with different parameters λ.

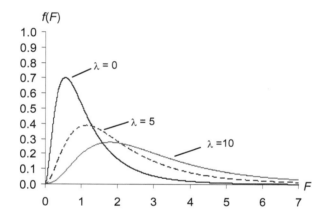

Figure 3.16 Central ($\lambda = 0$) and noncentral ($\lambda = 5$ and $\lambda = 10$) F distributions with $v_1 = 6$ and $v_2 = 10$ degrees of freedom

Exercises

3.1. The expected proportion of cows with more than 4000 kg milk in the standard lactation is 30%. If a farmer buys 10 cows, knowing nothing about their previous records, what is the probability: a) that exactly 5 of them have more than 4000 kg milk yield; b) that at least two have more than 4000 kg?

3.2. What is the ordinate of the standard normal curve for $z = -1.05$?

3.3. Consider a population of dairy cows with mean milk fat yield in a lactation of 180 kg, and standard deviation of 36 kg. What are the theoretical proportion of cows: a) with less than 180 kg fat; b) with more than 250 kg fat; and c) with less than 200 and more than 190 kg of fat? d) If the best 45% of cows are selected, what is the theoretical minimal fat yield an animal would have to have to be selected? e) What is the expected mean of the best 45% of animals?

3.4. Let the expected value of a variable y be $E(y) = \mu = 50$. Let the variance be $Var(y) = \sigma^2 = 10$. Calculate the following expectations and variances:

    a) $E(2 + y) =$ b) $Var(2 + y) =$

    c) $E(2 + 1.3y) =$ d) $Var(2 + 1.3y) =$

    e) $E(4y + 2y) =$ f) $Var(4y + 2y) =$

3.5. Consider a population of dairy cows with mean fat percentage of 4.1%, and standard deviation of 0.3%. What are the theoretical proportions of cows: a) with less than 4.0% fat; b) with more than 4.0% fat; and c) with more than 3.5% and less than 4.5%? d) If the best 25% of cows are selected, what is the theoretical minimum value for an animal to be included in the best 20%? e) What is the mean of the best 25% of cows?

Chapter 4

Population and Sample

A population is a set of all units that share some characteristics of interest. Usually a population is defined in order to make an inference about it. For example a population could be all Simmental cattle in Croatia, but it could also be a set of steers 1 year of age fed a particular diet. A population can be finite or infinite. An example of a finite population is a set of fattening steers on some farm in the year 2001. Such a population is defined as the number of steers on that farm, and the exact number and particular steers that belong to the population are known. On the contrary, an infinite population is a population for which the exact number of units is not known. This is for example the population of pigs in Croatia. The exact number of pigs is not known, if for nothing else because at any minute the population changes.

In order to draw a conclusion about a specified population, measures of location and variability must be determined. The ideal situation would be that the frequency distribution is known, but very often that is impossible. An alternative is to use a mathematical model of frequency distribution. The mathematical model is described and defined by parameters. The parameters are constant values that connect random variables with their frequency. They are usually denoted with Greek letters. For example, μ is the mean, and σ^2 is the variance of a population. Most often the true values of parameters are unknown, but they can be estimated from a sample. The sample is a set of observations drawn from a population. The way a sample is chosen will determine if it is a good representation of the population. Randomly drawn samples are usually considered most representative of a population. A sample of n units is a random sample if it is drawn in a way such that every set of n units has the same probability of being chosen. Numerical descriptions of a sample are called statistics. The arithmetic mean ($\bar{y}$) and sample variance (s^2) are examples of statistics. Statistics are functions of the random variable, and consequently they are random variables themselves. Generally, statistics are used in parameter estimation, but some statistics are used in making inferences about the population, although they themselves are not estimators of parameters.

4.1   Functions of Random Variables and Sampling Distributions

The frequency distribution of a sample can be presented by using graphs or tables. If a sample is large enough and representative, the frequency distribution of the sample is a good representation of the frequency distribution of the population. Although the sample may not be large in most cases, it can still give enough information to make a good inference about the population. The sample can be used to calculate values of functions of the random variable (statistics), which can be used in drawing conclusions about the population. The statistics are themselves random variables, that is, their values vary from sample to sample, and as such they have characteristic theoretical distributions called sampling distributions. If the sampling distribution is known, it is easy to estimate the probability of the particular value of a statistic such as the arithmetic mean or sample variance.

Inferences about a specified population can be made in two ways: by estimating parameters and by testing hypotheses. A conclusion based on a sample rests on probability. It is essential to use probability, because conclusions are based on just one part of a population (the sample) and consequently there is always some uncertainty that such conclusions based on a sample are true for the whole population.

4.2   Central Limit Theorem

One of the most important theorems in statistics describes the distribution of arithmetic means of samples. The theorem is as follows: if random samples of size n are drawn from some population with mean μ and variance σ^2, and n is large enough, the distribution of sample means can be represented with a normal density function with mean $\mu_{\bar{y}} = \mu$ and

standard deviation $\sigma_{\bar{y}} = \dfrac{\sigma}{\sqrt{n}}$. This standard deviation is often called the standard error of an

estimator of the population mean, or shortly, the standard error.

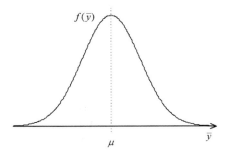

Figure 4.1 Distribution of sample means

If the population standard deviation σ is unknown, then the standard error $\sigma_{\bar{y}}$ can be estimated by a standard error of the sample:

$$s_{\bar{y}} = \frac{s}{\sqrt{n}}$$

4.3   Statistics with Distributions Other than Normal

Some statistics, for example the arithmetic mean, have normal distributions. However, from a sample we can calculate values of some other statistics that will not be normally distributed, but those statistics can also be useful in making inferences. The distributions of those statistics are known if it is assumed that the samples are drawn from a population with a normal distribution. For example the ratio:

$$\frac{(n-1)s^2}{\sigma^2} = \frac{\sum_i (y - \bar{y})^2}{\sigma^2}$$

has a *chi*-square distribution with $(n-1)$ degrees of freedom. Also, the statistic $\dfrac{\bar{y} - \mu}{\sqrt{s^2/n}}$

follows the Student t distribution with $(n-1)$ degrees of freedom. It will be shown later that some statistics have F distributions.

4.4   Degrees of Freedom

In the discussion about theoretical distributions, the term degrees of freedom has been mentioned. Although the mathematical and geometrical explanation is beyond the scope of this book, a practical meaning will be described. Degrees of freedom are the number of independent observations connected with variance estimation, or more generally, with the calculation of mean squares.

In calculating the sample variance from a sample of n observations using the formula $s^2 = \dfrac{\sum_i (y_i - \bar{y})^2}{n-1}$, the degrees of freedom are $(n-1)$. To calculate the sample variance, the arithmetic mean must first be calculated as an estimate of the population mean. Thus only $(n-1)$ observations used in calculating the variance are independent because there is a restriction concerning the arithmetic average, which is:

$$\sum_i (y_i - \bar{y}) = 0$$

Only $(n-1)$ of the observations are independent and the nth observation can be represented using the arithmetic average and the other observations:

$$y_n = (n-1)\bar{y} - y_1 - ... - y_{n-1}$$

Practically speaking, the degrees of freedom are equal to the total number of observations minus the number of estimated parameters used in the calculation of the variance.

Degrees of freedom are of importance when using statistics for estimation or making inference from samples. These procedures use the *chi*-square, t and F distributions. The shapes of these distributions affect the resulting estimates and inferences and the shapes depends on the degrees of freedom.

Chapter 5

Estimation of Parameters

Inferences can be made about a population either by parameter estimation or by hypothesis testing. Parameter estimation includes point estimation and interval estimation. A rule or a formula that describes how to calculate a single estimate using observations from a sample is called a point estimator. The number calculated by that rule is called a point estimate. Interval estimation is a procedure that is used to calculate an interval that estimates a population parameter.

5.1   Point Estimation

A point estimator is a function of a random variable, and as such is itself a random variable and a statistic. This means that the values of a point estimator vary from sample to sample, and have a distribution called a sampling distribution. For example, according to the central limit theorem, the distribution of sample means for large samples is approximately normal, with a mean μ and standard deviation $\sigma / \sqrt{n}$. Since the distribution is normal, all rules generally valid for a normal distribution apply here as well. The probability that the sample mean $\bar{y}$ is less than μ is 0.50. Further, the probability that $\bar{y}$ will not deviate from μ by more than $1.96 \sigma / \sqrt{n}$ is 0.95.

The distribution of an estimator is centralized about the parameter. If $\hat{\theta}$ denotes an estimator of a parameter θ and it is true that $E(\hat{\theta}) = \theta$, then the estimator is unbiased. Another property of a good estimator is that its variance should be as small as possible. The best estimator is the estimator with the minimal variance, that is, a minimal dispersion about θ compared to all other estimators. Estimation of the variability of $\hat{\theta}$ about θ can be expressed with the mean square for $\hat{\theta}$:

$$MS_{\hat{\theta}} = E\left[\left(\hat{\theta} - \theta\right)^2\right]$$

There are many methods for finding a point estimator. Most often used are methods of moments and the maximum likelihood method. Here, the maximum likelihood method will be described.

5.2   Maximum Likelihood Estimation

Consider a random variable y with a probability distribution $p(y|\theta)$, where θ denotes parameters. This function is thus the function of a variable y for given parameters θ. Assume now a function with the same algebraic form as the probability function, but defined as a function of the parameters θ for a given set of values of the variable y. That function is called a likelihood function and is denoted with $L(\theta\,|\,y)$ or shortly L. Briefly, the difference between probability and likelihood is that a probability refers to the occurrence of future events, while a likelihood refers to past events with known outcomes.

For example, the probability function for a binomial variable is:

$$p(y\,|\,p) = \binom{n}{y} p^{y}(1-p)^{n-y}$$

The likelihood function for given y_1 positive responses out of n trials is:

$$L(p\,|\,y_1) = \binom{n}{y_1} p^{y_1}(1-p)^{n-y_1}$$

The likelihood function can be used to estimate parameters for a given set of observations of some variable y. The desired value of an estimator will maximize the likelihood function. Such an estimate is called a maximum likelihood estimate and can be obtained by finding the solution of the first derivative of the likelihood function equated to zero. Often it is much easier to find the maximum of the log likelihood function, which has its maximum at the same value of the estimator as the likelihood function itself. This function is denoted with $logL(\theta\,|y)$ or shortly $logL$. For example, the log likelihood function for a y_1 value of a binomial variable is:

$$logL(p\,|\,y_1) = log\binom{n}{y_1} + y_1\,log(p) + (n-y_1)\,log(1-p)$$

Example: Consider the response of 10 cows given some treatment. A positive response is noted in 4 cows. Assume a binomial distribution.

The likelihood function for a binomial distribution is:

$$logL(p\,|\,y_1) = log\binom{n}{y_1} + y_1\,log(p) + (n-y_1)\,log(1-p)$$

In this example $n = 10$ and $y_1 = 4$. We seek an estimate of the parameter p which will maximize the log likelihood function. Taking the first derivative with respect to p:

$$\frac{\partial\,logL}{\partial p} = \frac{y_1}{p} - \frac{n-y_1}{1-p}$$

To obtain the maximum likelihood estimator this expression is equated to zero:

$$\frac{y_1}{\hat{p}} - \frac{n - y_1}{1 - \hat{p}} = 0$$

The solution is:

$$\hat{p} = \frac{y_1}{n}$$

For $n = 10$ and $y_1 = 4$ the estimate is:

$$\hat{p} = \frac{4}{10} = 0.4$$

Figure 5.1 presents the likelihood function for this example. The solution for p is at the peak of the function L.

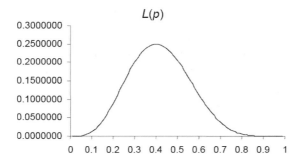

Figure 5.1 Likelihood function of binomial variable

5.3   Interval Estimation

Recall that a point estimator is a random variable with some probability distribution. If that distribution is known, it is possible to determine an interval estimator for a given probability. For example, let $\hat{\theta}$ denote an estimator of some parameter θ. Assume that $\hat{\theta}$ has a normal distribution with mean $E(\hat{\theta}) = \theta$ and standard error $\sigma_{\hat{\theta}}$. Define a standard normal variable:

$$z = \frac{\hat{\theta} - \theta}{\sigma_{\hat{\theta}}}$$

The probability is $(1 - \alpha)$ that the values of the standard normal variable are in the interval $\pm z_{\alpha/2}$ (Figure 5.2):

$$P(-z_{\alpha/2} \leq z \leq z_{\alpha/2}) = 1 - \alpha$$

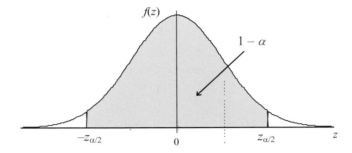

Figure 5.2 Interval of the standard normal variable defined with $(1 - \alpha)$ probability

Replacing z with $\dfrac{\hat{\theta} - \theta}{\sigma_{\hat{\theta}}}$ yields:

$$P(-z_{\alpha/2} \leq \frac{\hat{\theta} - \theta}{\sigma_{\hat{\theta}}} \leq z_{\alpha/2}) = 1 - \alpha$$

Further,

$$P(-z_{\alpha/2}\sigma_{\hat{\theta}} \leq \hat{\theta} - \theta \leq z_{\alpha/2}\sigma_{\hat{\theta}}) = 1 - \alpha$$

$$P(\hat{\theta} - z_{\alpha/2}\sigma_{\hat{\theta}} \leq \theta \leq \hat{\theta} + z_{\alpha/2}\sigma_{\hat{\theta}}) = 1 - \alpha$$

The expression $(\hat{\theta} - z_{\alpha/2}\sigma_{\hat{\theta}} \leq \theta \leq \hat{\theta} + z_{\alpha/2}\sigma_{\hat{\theta}})$ is called an interval estimator. Generally, the interval estimator is:

$$\left(\hat{\theta} - Error \leq \theta \leq \hat{\theta} + Error\right)$$

The error describes interval limits and depends on the probability distribution of the estimator. However, when the value for $\hat{\theta}$ is calculated from a given sample, the calculated interval does not include the probability of the random variable since the parameter θ is unknown and the exact position of the value for $\hat{\theta}$ in the distribution is unknown. This interval based on a single value of the random variable is called a confidence interval. A confidence interval includes a range of values on either side of the parameter estimated from a sample, such that the probability that the true value of the parameter θ lies within the given interval is equal to $1 - \alpha$. This probability is known as the confidence level. The upper and lower limits of the interval are known as confidence limits. A confidence interval at confidence level $1 - \alpha$ contains the true value of the parameter θ with probability $1 - \alpha$, regardless of the calculated value for $\hat{\theta}$. A confidence interval is interpreted as follows: if a large number of samples of size n are drawn from a population and for each sample a 0.95 (or 95%) confidence interval is calculated, then 95% of these intervals are expected to contain the true parameter θ. For example, if a 95% confidence interval for cow height based on the arithmetic mean and sample variance is 130 to 140 cm, we can say there is 95% confidence that the mean cow height for the population is between 130 and 140 cm.

Thus, if an estimator has a normal distribution, the confidence interval is:

$$\hat{\theta} \pm z_{\alpha/2}\sigma_{\hat{\theta}}$$

Here, $\hat{\theta}$ is the point estimate of the parameter θ calculated from a given sample. If the estimator has normal or Student distribution, then the general expression for the confidence interval is:

(Estimate) $\pm$ (standard error) (value of standard normal or Student variable for $\alpha/2$)

The calculation of a confidence interval can be accomplished in four steps:

1. Determine the point estimator and corresponding statistic with a known distribution,
2. Choose a confidence level $(1 - \alpha)$,
3. Calculate the estimate and standard error from the sample,
4. Calculate interval limits using the limit values for α, the estimate and its standard error.

5.4 Estimation of Parameters of a Normal Population

5.4.1 Maximum Likelihood Estimation

Recall the normal density function for a normal variable y is:

$$f(y \mid \mu, \sigma^2) = \frac{1}{\sqrt{2\pi\sigma^2}} e^{-(y-\mu)^2/2\sigma^2}$$

The likelihood function of n values of a normal variable y is:

$$L(\mu, \sigma \mid y_1, y_2, ..., y_n) = \prod_i \frac{1}{\sqrt{2\pi\sigma^2}} e^{-(y_i-\mu)^2/2\sigma^2}$$

or

$$L(\mu, \sigma \mid y_1, y_2, ..., y_n) = \frac{1}{\left(\sqrt{2\pi\sigma^2}\right)^n} e^{-\sum_i (y_i-\mu)^2/2\sigma^2}$$

The log likelihood function for n values of a normal variable is:

$$logL(\mu, \sigma \mid y_1, y_2 ... y_n) = -\frac{n}{2} log(\sigma^2) - \frac{n}{2} log(2\pi) - \frac{\sum_i (y_i - \mu)^2}{2\sigma^2}$$

The maximum likelihood estimators are obtained by taking the first derivative of the log likelihood function with respect to σ^2 and μ:

$$\frac{\partial\, logL(\mu, \sigma^2 \mid y)}{\partial \mu} = -\frac{1}{2\sigma^2} \sum_i 2(y_i - \mu)(-1)$$

$$\frac{\partial\, logL(\mu, \sigma^2 \mid y)}{\partial \sigma^2} = -\frac{n}{2\sigma^2} + \frac{1}{2\sigma^4} \sum_i (y_i - \mu)^2$$

By setting both terms to zero the estimators are:

$$\hat{\mu}_{ML} = \frac{\sum_i y_i}{n} = \overline{y}$$

$$\hat{\sigma}^2_{ML} = s^2_{ML} = \frac{\sum_i (y - \hat{\mu})^2}{n}$$

The expectation of $\overline{y}$ is $E(\overline{y}) = \mu$, thus, $\overline{y}$ is an unbiased estimator:

$$E(\overline{y}) = E\left(\frac{\sum_i y_i}{n}\right) = \frac{1}{n}\sum_i E(y_i) = \frac{n}{n} E(y_i) = \mu$$

However, the estimator of the variance is not unbiased. An unbiased estimator is obtained when the maximum likelihood estimator is multiplied by $n / (n-1)$:

$$s^2 = \frac{n}{n-1} s^2_{ML}$$

A variance estimator can be also obtained by using restricted maximum likelihood estimation (*REML*). The *REML* estimator is a maximum likelihood estimator adjusted for the degrees of freedom:

$$s^2_{REML} = \frac{\sum_i (y_i - \overline{y})^2}{n-1}$$

5.4.2 Interval Estimation of the Mean

A point estimator of a population mean μ is the sample arithmetic mean $\overline{y}$. The expectation of $\overline{y}$ is $E(\overline{y}) = \mu$, thus, $\overline{y}$ is an unbiased estimator. Also, it can be shown that $\overline{y}$ has the minimum variance of all possible estimators.

Recall that according to the central limit theorem, $\overline{y}$ has a normal distribution with mean μ and standard deviation $\sigma_{\overline{y}} = \sigma/\sqrt{n}$.

The statistic $z = \dfrac{\overline{y} - \mu}{\sigma_{\overline{y}}}$ is a standard normal variable. The interval estimator of the parameter μ is such that:

$$P(\overline{y} - z_{\alpha/2}\sigma_{\overline{y}} \le \mu \le \overline{y} + z_{\alpha/2}\sigma_{\overline{y}}) = 1 - \alpha$$

where $-z_{\alpha/2}$ and $z_{\alpha/2}$ are the values of the standard normal variable for $\alpha/2$ of the area under the standard normal curve at the tails of the distribution (Figure 5.2). Note that $\overline{y}$ is a random variable; however the interval does not include probability of a random variable since the population mean is unknown. The probability is $1 - \alpha$ that the interval includes the true population mean μ. The confidence interval around the estimate $\overline{y}$ is:

$$\overline{y} \pm z_{\alpha/2}\,\sigma_{\overline{y}}$$

If the population standard deviation (σ) is unknown, it can be replaced with the estimate from the sample. Then, the standard error is:

$$s_{\bar{y}} = \frac{s}{\sqrt{n}}$$

and the confidence interval is:

$$\bar{y} \pm z_{\alpha/2}\, s_{\bar{y}}$$

Example: Milk yields for one lactation for 50 cows sampled from a population have an arithmetic mean of 4000 kg and a sample standard deviation of 800 kg. Estimate the population mean with a 95% confidence interval.

$$\bar{y} = 4000 \ \text{kg}$$
$$s = 800 \ \text{kg}$$
$$n = 50 \ \text{cows}$$

For 95% confidence interval $\alpha = 0.05$, because $(1 - \alpha)$ 100% = 95%. The value $z_{\alpha/2} = z_{0.025} = 1.96$.

$$s_{\bar{y}} = \frac{s}{\sqrt{n}} = \frac{800}{\sqrt{50}} = 113.14$$

The confidence interval is:

$$\bar{y} \pm z_{\alpha/2}\, s_{\bar{y}}$$
$$4000 \pm (1.96)(113.14) = 4000 \pm 221.75$$

It can be stated with 95% confidence that the population mean μ is between 3778.2 and 4221.7 kg.

The central limit theorem is applicable only for large samples. For a small sample the distribution of $\bar{y}$ may not be approximately normal. However, assuming that the population from which the sample is drawn is normal, the t distribution can be used. A confidence interval is:

$$\bar{y} \pm t_{\alpha/2}\, s_{\bar{y}}$$

The value $t_{\alpha/2}$ can be found in the table Critical Values of the Student t distribution in Appendix B. Using $(n - 1)$ degrees of freedom, the procedure of estimation is the same as when using a z value.

5.4.3 Interval Estimation of the Variance

It can be shown that an unbiased estimator of the population variance σ^2 is equal to the sample variance:

$$s^2 = \frac{\sum_i (y - \bar{y})^2}{n - 1}$$

since $E(s^2) = \sigma^2$.

The sample variance has neither a normal nor a t distribution. If the sample is drawn from a normal population with mean μ and variance σ^2, then:

$$\chi^2 = \frac{(n-1)s^2}{\sigma^2}$$

is a random variable with a *chi*-square distribution with $(n-1)$ degrees of freedom. The interval estimator of the population variance is based on a *chi*-square distribution. With probability $(1-\alpha)$:

$$P(\chi^2_{1-\alpha/2} \leq \chi^2 \leq \chi^2_{\alpha/2}) = 1 - \alpha$$

that is

$$P(-\chi^2_{1-\alpha/2} \leq \frac{(n-1)s^2}{\sigma^2} \leq \chi^2_{1-\alpha/2}) = 1 - \alpha$$

where $\chi^2_{1-\alpha/2}$ and $\chi^2_{\alpha/2}$ are the values of the χ^2 variable that correspond to an area of $\alpha/2$ at each side of the *chi*-square distribution (Figure 5.3).

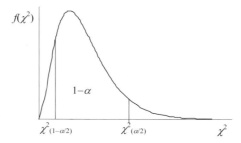

Figure 5.3 Interval of the χ^2 variable defined with $(1-\alpha)$ probability

Using numerical operations and the calculated sample variance from the expression above, the $(1-\alpha)100\%$ confidence interval is:

$$\frac{(n-1)s^2}{\chi^2_{\alpha/2}} \leq \sigma^2 \leq \frac{(n-1)s^2}{\chi^2_{(1-\alpha/2)}}$$

where s^2 is the sample variance.

Exercises

5.1. Using the sample from exercise 1.1, calculate the confidence interval for the population mean.

5.2. Using the sample from exercise 1.3, calculate the confidence interval for the population mean.

5.3. Using the sample from exercise 1.4, calculate the confidence interval for the population mean.

Chapter 6

Hypothesis Testing

The foundation of experimental research involves testing of hypotheses. There are two types of hypotheses: research and statistical hypotheses. The research hypothesis is postulated by the researcher on the basis of previous investigations, literature, or experience. For example, from experience and previous study, a researcher might hypothesize that in a certain region a new type of housing will be better than a traditional one. The statistical hypothesis, which usually follows the research hypothesis, formally describes the statistical alternatives that can result from the experimental evaluation of the research hypothesis.

There are two statistical hypotheses: the null hypothesis (H_0) and the alternative hypothesis (H_1). The null hypothesis is usually an assumption of unchanged state. For example, the H_0 states that there is no difference between some characteristics, for example means or variances of two populations. The alternative hypothesis, H_1, describes a changed state or existence of a difference. The research hypothesis can be postulated as two possibilities: there is a difference or there is no difference. Usually the statistical alternative hypothesis H_1 is identical to the research hypothesis, thus the null hypothesis is opposite to what a researcher expects. It is generally easier to prove a hypothesis is false than that it is true, thus a researcher usually attempts to reject H_0.

A statistical test based on a sample leads to one of two conclusions: 1) a decision to reject H_0 (because it is found to be false), or 2) a decision not to reject H_0, because there is insufficient proof for rejection. The null and alternative hypotheses, H_0 and H_1, always exclude each other. Thus, when H_0 is rejected H_1 is assumed to be true. On the other hand, it is difficult to prove that H_0 is true. Rather than accepting H_0, it is better to say that it is not rejected since there is not enough proof to conclude that H_0 is false. It could be that a larger amount of information would lead to rejecting H_0.

For example, a researcher suspects that ration A will give greater daily gains than ration B. The null hypothesis is defined that the two rations are equal, or will give the same daily gains. The alternative hypothesis is that rations A and B are not equal or that ration A will give larger daily gains. The alternative hypothesis is a research hypothesis. The researcher seeks to determine if ration A is better than B. An experiment is conducted and if the difference between sample means is large enough, he can conclude that generally the rations are different. If the difference between the sample means is small, he will fail to reject the null hypothesis. Failure to reject the null hypothesis does not show the rations to be the same. If a larger number of animals had been fed the two rations a difference might have been shown to exist, but the difference was not revealed in this experiment.

The rules of probability and characteristics of known theoretical distributions are used to test hypotheses. Probability is utilized to reject or fail to reject a hypothesis, because a sample is measured and not the whole population, and there cannot be 100% confidence that the conclusion from an experiment is the correct one.

6.1   Hypothesis Test of a Population Mean

One use of a hypothesis test is to determine if a sample mean is significantly different from a predetermined value. This example of hypothesis testing will be used to show the general principles of statistical hypotheses.

First, a researcher must define null and alternative hypotheses. To determine if a population mean is different from some value μ_0, the null and alternative hypotheses are:

$H_0: \mu = \mu_0$
$H_1: \mu \neq \mu_0$

The null hypothesis, H_0, states that the population mean is equal to μ_0, the alternative hypothesis, H_1, states that the population mean is different from μ_0.

The next step is to define an estimator of the population mean. This is the sample mean $\bar{y}$. Now, a test statistic with a known theoretical distribution is defined. For large samples the sample mean has a normal distribution, so a standard normal variable is defined:

$$z = \frac{\bar{y} - \mu_0}{\sigma_{\bar{y}}}$$

where $\sigma_{\bar{y}} = \dfrac{\sigma}{\sqrt{n}}$ is the standard error. This z statistic has a standard normal distribution if the population mean is $\mu = \mu_0$, that is, if H_0 is correct (Figure 6.1). Recall that generally the z statistic is of the form:

$$z = \frac{Estimator - Parameter}{Standard\ error\ of\ the\ estimator}$$

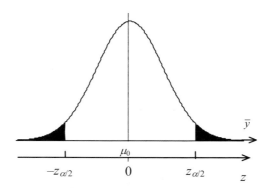

Figure 6.1 Distribution of $\bar{y}$. Lower scale is the standard scale $z = \dfrac{\bar{y} - \mu_0}{\sigma_{\bar{y}}}$

Recall if the population variance is unknown that the standard error $\sigma_{\bar{y}}$ can be estimated by a sample standard error $s_{\bar{y}} = s/\sqrt{n}$, and then:

$$z \approx \frac{\bar{y} - \mu_0}{s / \sqrt{n}}$$

From the sample the estimate (the arithmetic mean) is calculated. Next to be calculated is the value of the proposed test statistic for the sample. The question is, where is the position of the calculated value of the test statistic in the theoretical distribution? If the calculated value is unusually extreme, the calculated $\bar{y}$ is considerably distant from the μ_0, and there is doubt that $\bar{y}$ fits in the hypothetical population. If the calculated $\bar{y}$ does not fit in the hypothetical population the null hypothesis is rejected indicating that the sample belongs to a population with a mean different from μ_0. Therefore, it must be determined if the calculated value of the test statistic is sufficiently extreme to reject H_0. Here, sufficiently extreme implies that the calculated z is significantly different from zero in either a positive or negative direction, and consequently the calculated $\bar{y}$ is significantly smaller or greater than the hypothetical μ_0.

Most researchers initially determine a rule of decision against H_0. The rule is as follows: choose a probability α and determine the critical values of $z_{\alpha/2}$ and $-z_{\alpha/2}$ for the standard normal distribution for H_0 being correct. The critical values are the values of the z variable such that the probability of obtaining those or more extreme values is equal to α, $P(z > z_{\alpha/2}$ or $z < z_{\alpha/2}) = \alpha$, if H_0 is correct. The critical regions include the values of z that are greater than $z_{\alpha/2}$, or less than $-z_{\alpha/2}$, $(z > z_{\alpha/2}$ or $z < -z_{\alpha/2})$. The probability α is called the level of significance (Figure 6.2). Usually, $\alpha = 0.05$ or 0.01 is used, sometimes 0.10.

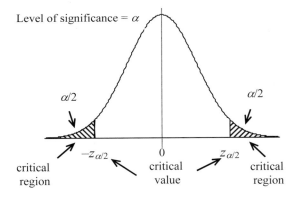

Figure 6.2 Illustration of significance level, critical value and critical region

The value of the test statistic calculated from the sample is compared with the critical value. If the calculated value of z is more extreme than one or the other of the critical values, thus is positioned in a critical region, H_0 is rejected. When H_0 is rejected, the value for which z was calculated does not belong to the distribution assumed given H_0 was correct (Figure 6.3). The probability that the conclusion to reject H_0 is incorrect and the calculated value belongs to the distribution of H_0 is less than α. If the calculated value z is not more extreme than $z_{\alpha/2}$ or $-z_{\alpha/2}$, H_0 cannot be rejected (Figure 6.4).

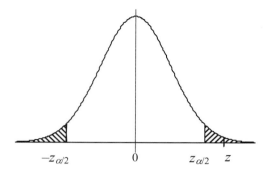

Figure 6.3 The calculated *z* is in the critical region and H_0 is rejected with an α level of significance. The probability that the calculated *z* belongs to the H_0 population is less than α

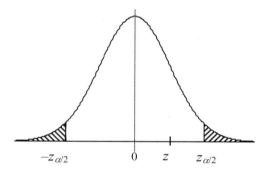

Figure 6.4 The calculated *z* is not in the critical region and H_0 is not rejected with α level of significance. The probability that calculated *z* belongs to the H_0 population is greater than α

Any hypothesis test can be performed by following these steps:

1. Define H_0 and H_1
2. Determine α
3. Calculate an estimate of the parameter
4. Determine a test statistic and its distribution when H_0 is correct and calculate its value from a sample
5. Determine the critical value and critical region
6. Compare the calculated value of the test statistic with the critical values and make a conclusion.

Example: Given a sample of 50 cows with an arithmetic mean for lactation milk yield of 4000 kg, does this herd belong to a population with a mean $\mu_0 = 3600$ kg and a standard deviation $\sigma = 1000$ kg?

The hypothetical population mean is $\mu_0 = 3600$ and the hypotheses are:

H_0: $\mu = 3600$
H_1: $\mu \neq 3600$

Known values are:

$$\bar{y} = 4000 \text{ kg}$$
$$\sigma = 1000 \text{ kg}$$
$$n = 50 \text{ cows}$$

The calculated value of the standard normal variable is:

$$z = \frac{4000 - 3600}{1000 / \sqrt{50}} = 2.828$$

A significance level of $\alpha = 0.05$ is chosen. The critical value corresponding to $\alpha = 0.05$ is $z_{\alpha/2} = 1.96$. The calculated z is 2.828. The sample mean (4000 kg) is 2.828 standard errors distant from the hypothetical population mean (3600 kg) if H_0 is correct. The question is if the calculated $z = 2.828$ is sufficiently extreme that the sample does belong to the population with a mean of 3600 kg. The calculated $|z| > z_{\alpha/2}$, numerically, $|2.828| > 1.96$, which means that the calculated z is in the critical region for rejection of H_0 with $\alpha = 0.05$ level of significance (Figure 6.5). The probability is less than 0.05 that the sample belongs to the population with the mean of 3600 kg and standard deviation of 1000.

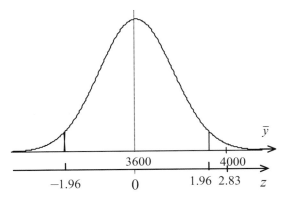

Figure 6.5 A distribution of sample means of milk yield with the mean $\mu = 3600$ and the standard deviation $\sigma = 1000$. The lower line presents the standard normal scale

6.1.1 *P* Value

Another way to conclude whether or not to reject H_0 is to determine the probability that the calculated value of a test statistic belongs to the distribution when H_0 is correct. This probability is denoted as the P value. The P value is the observed level of significance. Many computer software packages give P values and leave to the researcher the decision about rejecting H_0. The researcher can reject H_0 with a probability equal to the P value of being in error. The P value can also be used when a significance level is determined beforehand. For a given level of significance α, if a P value is less than α, H_0 is rejected with the α level of significance.

6.1.2   A Hypothesis Test can be One- or Two-sided

In the discussion about testing hypotheses given above, the question was whether the sample mean $\bar{y}$ was different than some value μ_0. That is a two-sided test. That test has two critical values, and H_0 is rejected if the calculated value of the test statistic is more extreme than either of the two critical values. A test can also be one-sided. In a one-sided test there is only one critical value and the rule is to reject H_0 if the calculated value of the test statistic is more extreme than that critical value.

If the question is to determine if $\mu > \mu_0$ then:

H_0: $\mu \leq \mu_0$
H_1: $\mu > \mu_0$

For testing these hypotheses the critical value and the critical region are defined in the right side of the distribution (Figure 6.6).

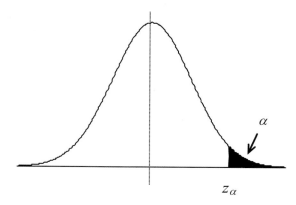

Figure 6.6 The critical value and critical region for $z > z_\alpha$

The critical value is z_α. The critical region consists of all z values greater than z_α. Thus, the probability that the random variable z has values in the interval (z_α, ∞) is equal to α, $P(z > z_\alpha) = \alpha$. If the calculated z is in the critical region, or greater than z_α, H_0 is rejected with α level of significance.

Alternatively, the question can be to determine if $\mu < \mu_0$ then:

H_0: $\mu \geq \mu_0$
H_1: $\mu < \mu_0$

For testing these hypotheses the critical value and the critical region are defined in the left side of the distribution (Figure 6.7).

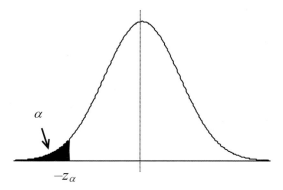

α

$-z_\alpha$

Figure 6.7 The critical value and critical region for $z < -z_\alpha$

The critical value is $-z_\alpha$. The critical region consists of all z values less than $-z_\alpha$. Thus, the probability that the random variable z has values in the interval $(-\infty, -z_\alpha)$ is equal to α, $P(z < -z_\alpha) = \alpha$. If the calculated z is in the critical region or is less than z_α, H_0 is rejected with α level of significance.

6.1.3 Hypothesis Test of a Population Mean for a Small Sample

The Student t distribution is used for testing hypotheses about the population mean for a small sample (say $n < 30$) drawn from a normal population. The test statistic is a t random variable:

$$t = \frac{\bar{y} - \mu_0}{s / \sqrt{n}}$$

The approach to reaching a conclusion is similar to that for a large sample. The calculated value of the t statistic is tested to determine if it is more extreme than the critical value t_α or $t_{\alpha/2}$ with α level of significance. For a two-sided test the null hypothesis H_0: $\mu = \mu_0$ is rejected if $|t| > t_{\alpha/2}$, where $t_{\alpha/2}$ is a critical value such that $P(t > t_{\alpha/2}) = \alpha/2$. For a one-sided test the null hypothesis H_0: $\mu \leq \mu_0$ is rejected if $t > t_\alpha$ or H_0: $\mu \geq \mu_0$ is rejected if $t < -t_\alpha$, depending on whether it is a right- or left-sided test. Critical values can be found in the table Critical Values of Student t Distributions in Appendix B. The shape of the distribution and the value of the critical point depend on degrees of freedom. The degrees of freedom are $(n - 1)$, where n is the number of observations in a sample.

Example: The data are lactation milk yields of 10 cows. Is the arithmetic mean of the sample, 3800 kg, significantly different from 4000 kg? The sample standard deviation is 500 kg.

The hypothetical mean is $\mu_0 = 4000$ kg and the hypotheses are as follows:

H_0: $\mu = 4000$ kg
H_1: $\mu \neq 4000$ kg

The sample mean is $\bar{y} = 3800$ kg.

The sample standard deviation is $s = 500$ kg.

The standard error is:

$$s/\sqrt{n} = 500/\sqrt{10}$$

The calculated value of the t statistic is:

$$t = \frac{\bar{y} - \mu_0}{s/\sqrt{n}} = \frac{3800 - 4000}{500/\sqrt{10}} = -1.26$$

For $\alpha = 0.05$ and degrees of freedom $(n - 1) = 9$, the critical value is $-t_{\alpha/2} = -2.262$. Since the calculated $t = -1.26$ is not more extreme than the critical value $-t_{\alpha/2} = -2.262$, H_0 is not rejected with an $\alpha = 0.05$ level of significance. The sample mean is not significantly different from 4000 kg.

6.1.4 SAS Example

Example: The data are lactation milk yields of 10 cows. Is the arithmetic mean of the sample given in the SAS program significantly different from 4000 kg?

SAS program:

```
DATA cows;
INPUT milk @@;
DATALINES;
3000 3270 3400 3720 3780
3800 3940 4090 4300 4700
;

PROC TTEST DATA = cows H0 = 4000;
VAR milk;
RUN;
```

Explanation: The TTEST procedure is used. The VAR statement defines the variable that is to be analyzed. The hypothetical population mean is defined by H0.

SAS output:

Variable	N	Lower CL Mean	Mean	Upper CL Mean	Std Dev	Std Err	Minimum	Maximum
milk	10	3442.2	3800	4157.8	500.16	158.16	3000	4700

T-Tests

Variable	DF	t Value	Pr > |t|
milk	9	-1.26	0.2378

Explanation: The program gives descriptive statistics and confidence limits for the variable *milk*. Values for *N, Lower CL Mean, Mean, Upper CL, Mean, Std Dev, Std Err, minimum and maximum* are sample size, lower confidence limit, mean, upper confidence limit, standard deviation of values, standard error of the mean, minimum and maximum, respectively. The program calculates t tests, together with corresponding degrees of freedom and P values (*Prob>|T|*). The t test is valid if observations are drawn from a normal distribution. The P value is 0. 2378 and thus H_0 cannot be rejected.

6.1.5   A Simple Check if a Sample Comes from a Normal Distribution using SAS

Many statistics test are based on the assumption that a sample is drawn from a normal distribution. A simple check if there is serious and systematic deviation from the normal distribution can be done by using ranks. Then ranks are plotted against the original values, and if a sample is a random sample from a normal distribution that plot should be approximately a straight line.

The rank of observations is determined in a way such that observations are sorted in ascending order and ranks are assigned to them. If some observations have the same value, then the mean of their ranks is assigned to them. For example, if the 10th and 11th observations have the same value, say 20, their ranks are $(10 + 11)/2 = 10.5$. Such ranks are also known as Wilcoxon scores.

Example: Check if a sample of 7-month weights (kg) of 100 calves are drawn from a normal distribution:

233	208	306	300	271	304	207	254	262	231
279	228	287	223	247	292	209	303	194	268
263	262	234	277	291	277	256	271	255	299
278	290	259	251	265	316	318	252	316	221
249	304	241	249	289	211	273	241	215	264
216	271	296	196	269	231	272	236	219	312
320	245	263	244	239	227	275	255	292	246
245	255	329	240	262	291	275	272	218	317
251	257	327	222	266	227	255	251	298	255
266	255	214	304	272	230	224	250	255	284

SAS program:

```
PROC RANK DATA = calves OUT = rankings;
VAR wt;
RANKS wtrank;
RUN;

PROC GPLOT DATA = rankings;
PLOT wtrank*wt = "x";
RUN;
```

Explanation: The RANK procedure is used to compute ranks. The input file was defined with *DATA=calves* and the output file with ranks with *OUT = ranks*. The VAR statement defines the variable *wt* values which will be used to compute ranks. The RANKS statement defines a new variable *wtrank*, values of which are the ranks. The GPLOT procedure is used to plot the data.

SAS output:

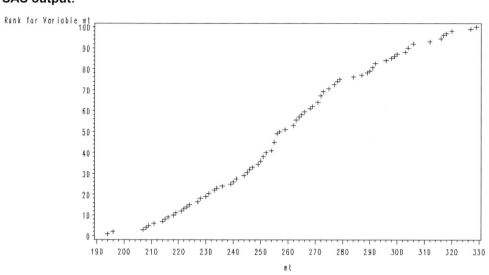

Explanation: The plot shows a slight deviation of ranks from a straight line in the middle section, indicating inconsequentially higher values around the mean than expected.

6.2 Hypothesis Test of the Difference between Two Population Means

Assume that samples are drawn from two populations with means μ_1 and μ_2. The samples can be used to test if the two means are different. A z or t statistic will be used depending on the sample size. The form of the test also depends on whether the two samples are dependent or independent of each other, and whether the variances are equal or unequal. Further, the hypotheses can be stated as one- or two-sided. The hypotheses for the two-sided test are:

$$H_0: \mu_1 - \mu_2 = 0$$
$$H_1: \mu_1 - \mu_2 \neq 0$$

The null hypothesis H_0 states that the population means are equal, and the alternative hypothesis H_1 states that they are different.

6.2.1 Large Samples

Let $\bar{y}_1$ and $\bar{y}_2$ denote arithmetic means and let n_1 and n_2 denote numbers of observations of samples drawn from two corresponding populations. The problem is to determine if the two populations differ. If the arithmetic means of those two samples are significantly different, it implies that the population means are different. The difference between the sample means is an estimator of the difference between the means of the populations. The z statistic is defined as:

$$z = \frac{(\bar{y}_1 - \bar{y}_2) - 0}{\sigma_{(\bar{y}_1 - \bar{y}_2)}}$$

where $\sigma_{(\bar{y}_1 - \bar{y}_2)} = \sqrt{\dfrac{\sigma_1^2}{n_1} + \dfrac{\sigma_2^2}{n_2}}$ is the standard error of the difference between the two means

and σ_1^2 and σ_2^2 are the variances of the two populations. If the variances are unknown, they can be estimated from the samples and the standard error of the difference is:

$$ s_{(\bar{y}_1 - \bar{y}_2)} = \sqrt{\frac{s_1^2}{n_1} + \frac{s_2^2}{n_2}} $$

where s_1^2 and s_2^2 are estimates of the variances of the samples.

The z statistic is:

$$ z \approx \frac{\bar{y}_1 - \bar{y}_2}{s_{(\bar{y}_1 - \bar{y}_2)}} $$

For a two-sided test H_0 is rejected if the calculated value $|z| > z_{\alpha/2}$, for which $z_{\alpha/2}$ is the critical value for the significance level α. In order to reject H_0, the calculated value of the test statistic z must be more extreme than the critical value $z_{\alpha/2}$.

Example: Two groups of 40 cows were fed two different rations (A and B) to determine which of those two rations will cause cows to produce more milk in lactation. At the end of the experiment the following sample means and variances (in thousand kg) were calculated:

	Ration *A*	Ration *B*
Mean ($\bar{y}$)	5.20	6.50
Variance (s^2)	0.25	0.36
Size (n)	40	40

The hypotheses for a two-sided test are:

$$ H_0: \mu_1 - \mu_2 = 0 $$
$$ H_1: \mu_1 - \mu_2 \neq 0 $$

The standard error of difference is:

$$ s_{(\bar{y}_1 - \bar{y}_2)} = \sqrt{\frac{s_1^2}{n_1} + \frac{s_2^2}{n_2}} = \sqrt{\frac{0.25}{40} + \frac{0.36}{40}} = 0.1235 $$

The calculated value of z statistic is:

$$ z \approx \frac{\bar{y}_1 - \bar{y}_2}{s_{(\bar{y}_1 - \bar{y}_2)}} = \frac{5.20 - 6.50}{0.1235} = -10.527 $$

Since the calculated value $z = -10.527$ is more extreme than the critical value $-z_{\alpha/2} = -z_{0.025} = -1.96$, the null hypothesis is rejected with $\alpha = 0.05$ level of significance, suggesting that feeding cows ration B will result in greater milk yield than feeding ration A.

6.2.2   Small Samples and Equal Variances

For comparisons involving small samples a t statistic is used. The t statistic is defined as:

$$t = \frac{(\bar{y}_1 - \bar{y}_2) - 0}{s_{\bar{y}_1 - \bar{y}_2}}$$

where $\bar{y}_1$ and $\bar{y}_2$ are the sample means, and $s_{\bar{y}_1 - \bar{y}_2}$ is the standard error of the difference of the sample means. The definition of the standard error, $s_{\bar{y}_1 - \bar{y}_2}$, depends on whether variances are equal or unequal. The standard error for small samples with equal variances is:

$$s_{\bar{y}_1 - \bar{y}_2} = \sqrt{s_p^2 \left(\frac{1}{n_1} + \frac{1}{n_2} \right)}$$

in which n_1 and n_2 are sample sizes, and s_p^2 is the pooled variance:

$$s_p^2 = \frac{(n_1 - 1)s_1^2 + (n_2 - 1)s_2^2}{n_1 + n_2 - 2}$$

or

$$s_p^2 = \frac{\sum_i (y_{1i} - \bar{y}_1)^2 + \sum_j (y_{2j} - \bar{y}_2)^2}{n_1 + n_2 - 2} =$$

$$\frac{\sum_i (y_{1i})^2 + \sum_j (y_{2j})^2 - \dfrac{\left(\sum_i y_{1i}\right)^2}{n_1} - \dfrac{\left(\sum_j y_{2j}\right)^2}{n_2}}{n_1 + n_2 - 2}$$

Here $i = 1$ to n_1, $j = 1$ to n_2. Since the variances are assumed to be equal, the estimate of the pooled variance s_p^2 is calculated from the observations from both samples. Degrees of freedom $= n_1 + n_2 - 2$.

When the number of observations is equal in two samples, that is $n_1 = n_2 = n$, the expression for the t statistic simplifies to:

$$t = \frac{(\bar{y}_1 - \bar{y}_2) - 0}{\sqrt{\dfrac{s_1^2 + s_2^2}{n}}}$$

The H_0 is rejected if the calculated value $|t| > t_{\alpha/2}$, for which $t_{\alpha/2}$ is the critical value of t with significance level α.

Example: Consider the same experiment as in the previous example with large samples, except that only 20 cows were fed each ration. From the first group 2 cows were culled because of illness. Thus, groups of 18 and 20 cows were fed two rations A and B, respectively. Again, the question is to determine which ration causes cows to produce more milk in a lactation. The sample means and variances (in thousand kg) at the end of the experiment were:

	Ration A	Ration B
Mean ($\bar{y}$)	5.50	6.80
$\sum_i y_i =$	99	136
$\sum_i y^2_i =$	548	932
Variance (s^2)	0.206	0.379
Size (n)	18	20

The estimated pooled variance is:

$$s^2_p = \frac{\sum_i (y_{1i})^2 + \sum_i (y_{2i})^2 - \frac{\left(\sum_i y_{1i}\right)^2}{n_1} - \frac{\left(\sum_i y_{2i}\right)^2}{n_2}}{n_1 + n_2 - 2} = \frac{548 + 932 - \frac{(99)^2}{18} - \frac{(136)^2}{20}}{18 + 20 - 2} = 0.297$$

The estimated variance can also be calculated from:

$$s^2_p = \frac{(n_1 - 1)s^2_1 + (n_2 - 1)s^2_2}{n_1 + n_2 - 2} = \frac{(18 - 1)(0.206) + (20 - 1)(0.379)}{18 + 20 - 2} = 0.297$$

The calculated value of the t statistic is:

$$t = \frac{(\bar{y}_1 - \bar{y}_2) - 0}{\sqrt{s^2_p \left(\frac{1}{n_1} + \frac{1}{n_2}\right)}} = \frac{(5.50 - 6.80) - 0}{\sqrt{0.297\left(\frac{1}{18} + \frac{1}{20}\right)}} = -7.342$$

The critical value is $-t_{\alpha/2} = -t_{0.025} = -2.03$. Since the calculated value of $t = -7.342$ is more extreme than the critical value $-t_{0.025} = -2.03$, the null hypothesis is rejected with $\alpha = 0.05$ level of significance, which implies that feeding cows ration B will cause them to give more milk than feeding ration A.

6.2.3   Small Samples and Unequal Variances

A statistic for testing the difference between two population means with unequal variances is also a t statistic:

$$t = \frac{(\bar{y}_1 - \bar{y}_2) - 0}{s_{\bar{y}_1 - \bar{y}_2}}$$

The standard error is defined as:

$$s_{\bar{y}_1 - \bar{y}_2} = \sqrt{\left(\frac{s^2_1}{n_1} + \frac{s^2_2}{n_2}\right)}$$

We can no longer use pooled variance, because the standard deviations are not assumed to be equal. For unequal variances, degrees of freedom, denoted v, are:

$$v = \frac{\left(s_1^2/n_1 + s_2^2/n_2\right)^2}{\dfrac{\left(s_1^2/n_1\right)^2}{n_1 - 1} + \dfrac{\left(s_2^2/n_2\right)^2}{n_2 - 1}}$$

6.2.4 Dependent Samples

Under some circumstances two samples are not independent of each other. A typical example is taking measurements on the same animal before and after applying a treatment. The effect of the treatment can be thought of as the average difference between the two measurements. The value of the second measurement is related to or depends on the value of the first measurement. In that case the difference between measurements before and after the treatment for each animal is calculated and the mean of those differences is tested to determine if it is significantly different from zero. Let d_i denote the difference for an animal i. The test statistic for dependent samples is:

$$t = \frac{\bar{d} - 0}{s_d/\sqrt{n}}$$

where $\bar{d}$ and s_d are the mean and standard deviation of the differences, respectively, and n is the number of animals. The testing procedure and definition of critical values is as before, except that degrees of freedom are $(n - 1)$. For this test to be valid the distribution of observations must be approximately normal.

Example: The effect of a treatment is tested on milk production of dairy cows. The cows were in the same parity and stage of lactation. The milk yields were measured before and after administration of the treatment:

Measurement	Cow 1	Cow 2	Cow 3	Cow 4	Cow 5	Cow 6	Cow 7	Cow 8	Cow 9
1	27	45	38	20	22	50	40	33	18
2	31	54	43	28	21	49	41	34	20
Difference (*d*)	4	9	5	8	−1	−1	1	1	2

$$n = 9$$

$$\bar{d} = \frac{\sum_i d_i}{n} = \frac{4 + 9 + \ldots + 2}{9} = 3.11$$

$$s_d = \sqrt{\frac{\sum_i (y_i - \mu)^2}{n - 1}} = \sqrt{\frac{\sum_i y_i^2 - \dfrac{\left(\sum_i y_i\right)^2}{n}}{n - 1}} = 3.655$$

$$t = \frac{\bar{d} - 0}{s_d/\sqrt{n}} = \frac{3.11 - 0}{3.655/\sqrt{9}} = 2.553$$

The critical value for $(n-1) = 8$ degrees of freedom is $t_{0.05} = 2.306$. Since the calculated value $t = 2.553$ is more extreme than the critical value 2.306, the null hypothesis is rejected with $\alpha = 0.05$ level of significance. The treatment thus increases milk yield.

Pairing measurements before and after treatment results in removal of variation due to differences among animals. When this design can be appropriately used there is greater power of test or an increased chance of finding a treatment effect to be significant when compared to a design involving two separate samples of animals.

6.2.5 Nonparametric Test

When samples are drawn from populations with unknown sampling distributions, it is not appropriate to use the previously shown z or t tests. Indications of such distributions are when the mode is near an end of the range or when some observations are more extreme than one would expect. Nonparametric tests are appropriate for such samples because no particular theoretical distribution is assumed to exist. Many nonparametric tests compare populations according to some central point such as the median or mode. Rank transformations are also utilized. The use of ranks diminishes the importance of the distribution and the influence of extreme values in samples. One such test is the simple rank test. The null hypothesis is that no effect of groups exists. It is assumed that the distributions of the groups are equal, but not necessarily known. This test uses an estimator of the ranks of observations. The estimator of ranks in a group is:

$T =$ the sum of ranks in a group

The simple test involves determining if the sum of ranks in one group is significantly different from the expected sum of ranks calculated on the basis of ranks of observations for both groups. The expected sum of ranks for a group if the groups are not different is:

$$E(T) = n_1 \, \overline{R}$$

where n_1 is the number of observations in group 1, and $\overline{R}$ is the mean rank using both groups together. The standard deviation in the combined groups is:

$$SD(T) = s_R \sqrt{\frac{n_1 n_2}{(n_1 + n_2)}}$$

where s_R is the standard deviation of the ranks using both groups together, and n_1 and n_2 are the numbers of observations in groups 1 and 2, respectively. If the standard deviations of ranks are approximately equal for both groups, then the distribution of T can be approximated by a standard normal distribution. The statistic:

$$z = \frac{T - E(T)}{SD(T)}$$

has a standard normal distribution. A practical rule is that the sample size must be greater than 5 and the values with same value should be distributed approximately equally to both groups. The rank of observations is determined in the following manner:

The observations of the combined groups are sorted in ascending order and ranks are assigned to them. If some observations have the same value, then the mean of their ranks is assigned to them. For example, if the 10th and 11th observations have the same value, say 20, their ranks are $(10 + 11)/2 = 10.5$.

Example: Groups of sows were injected with gonadotropin or saline. The aim of the experiment was to determine if the gonadotropin would result in higher ovulation rate. The following ovulation rates were measured:

Gonadotropin	14	14	7	45	18	36	15
Saline	12	11	12	12	14	13	9

The observations were sorted regardless of the treatment:

Treatment	Ovulation rate	Rank
Gonadotropin	7	1
Saline	9	2
Saline	11	3
Saline	12	5
Saline	12	5
Saline	12	5
Saline	13	7
Gonadotropin	14	9
Gonadotropin	14	9
Saline	14	9
Gonadotropin	15	11
Gonadotropin	18	12
Gonadotropin	36	13
Gonadotropin	45	14
	$\overline{R}$	7.5
	s_R	4.146

$n_1 = 7$

$n_2 = 7$

$T = 2 + 3 + 5 + 5 + 5 + 7 + 9 = 36$

$E(T) = n_1 \overline{R} = (7)(7.5) = 52.5$

$$SD(T) = s_R \sqrt{\frac{n_1 n_2}{(n_1 + n_2)}} = 4.146 \sqrt{\frac{(7)(7)}{(7+7)}} = 7.756$$

$$z = \frac{T - E(T)}{SD(T)} = \frac{36 - 52.5}{7.756} = -2.127$$

Since the calculated value of $z = -2.127$ is more extreme than the critical value, -1.96, the null hypothesis is rejected with $\alpha = 0.05$ significance level. It can be concluded that gonadotropin treatment increased ovulation rate. Note that the extreme values, 7, 36 and 45, did not have undue influence on the test.

Again test the difference between treatments for the same example, but now using a t test with unequal variances. The following values have been calculated from the samples:

	Gonadotropin	Saline
Mean ($\bar{y}$)	21.286	11.857
Variance (s^2)	189.905	2.476
Size (n)	7	7

The calculated value of the t statistic is:

$$t = \frac{(\bar{y}_1 - \bar{y}_2) - 0}{\sqrt{\left(\dfrac{s_1^2}{n_1} + \dfrac{s_2^2}{n_2}\right)}} = \frac{(11.857 - 21.286) - 0}{\sqrt{\left(\dfrac{189.905}{7} + \dfrac{2.426}{7}\right)}} = -1.799$$

Here, the degrees of freedom are $v = 6.16$ (because of the unequal variance) and the critical value of the t distribution is -2.365. Since the calculated value $t = -1.799$ is not more extreme than the critical value (-2.365) the null hypothesis is not rejected. Here, the extreme observations, 7, 36 and 45, have influenced the variance estimation, and consequently the test of difference.

6.2.6   SAS Examples for Hypotheses Tests of Two Population Means

The SAS program for the evaluation of superovulation of sows is as follows.

SAS program:

```
DATA superov;
INPUT trmt $ OR @@;
DATALINES;
G 14   G 14   G  7   G 45   G 18   G 36   G 15
S 12   S 11   S 12   S 12   S 14   S 13   S  9
;
PROC TTEST DATA = superov;
CLASS trmt;
VAR OR;
RUN;
```

Explanation: The TTEST procedure is used. The file with observations must have a categorical variable that determines allocation of each observation to a group (*trmt*). The CLASS statement defines which variable determines *trmt*. The VAR statement defines the variable that is to be analyzed.

SAS output:

```
                              Statistics
                   Lower CL          Upper CL Lower CL          Upper CL
Variable  trmt   N   Mean     Mean    Mean    Std Dev  Std Dev  Std Dev  Std Err
OR        G      7  8.5408   21.286  34.031   8.8801   13.781   30.346   5.2086
OR        S      7 10.402    11.857  13.312   1.014    1.5736   3.4652   0.5948
OR        Diff (1-2) -1.994   9.428  20.851   7.0329   9.8077   16.19    5.2424
```

```
                               T-Tests
Variable      Method        Variances     DF     t Value    Pr > |t|
OR            Pooled          Equal        12      1.80       0.0973
OR         Satterthwaite     Unequal      6.16     1.80       0.1209

                     Equality of Variances
Variable      Method      Num DF    Den DF    F Value    Pr > F
OR            Folded F       6         6        76.69    <.0001
```

Explanation: The program gives descriptive statistics and confidence limits for both treatments and their difference. *N, Lower CL Mean, Mean, Upper CL, Mean, Lower CL Std Dev, Std Dev, Upper CL Std Dev* and *Std Err* are sample size, lower confidence limit, mean, upper confidence limit, lower confidence limit of standard deviation, standard deviation of values, upper confidence limit of standard deviation and standard error of the mean, respectively. The program calculates *t* tests, for *Unequal* and *Equal* variances, together with corresponding degrees of freedom and *P* values (*Prob>|T|*). The Satterthwaite correction is applied for unequal variances. The *t* test is valid if observations are drawn from a normal distribution. Since in the test for equality of variances, $F = 76.69$ is greater than the critical value and the *P* value is <0.0001, the variances are different and it is appropriate to apply the *t* test for unequal variances. The *P* value is 0.1209 and thus H_0 cannot be rejected.

This alternative program uses the Wilcoxon test (the simple rank test):

```
DATA superov;
INPUT trmt $ OR @@;
DATALINES;
G 14   G 14   G 7   G 45   G 18   G 36   G 15
S 12   S 11   S 12   S 12   S 14   S 13   S 9
;
PROC NPAR1WAY DATA=superov
WILCOXON;
CLASS trmt;
VAR OR;
EXACT;
RUN;
```

Explanation: The program uses the NPAR1WAY procedure with the WILCOXON option for a Wilcoxon or simple rank test. The CLASS statement defines the variable that classifies observations to a particular treatment. The VAR statement defines the variable with observations.

SAS output:

```
            Wilcoxon Scores (Rank Sums) for Variable OR
                    Classified by Variable trmt


              Sum of        Expected       Std Dev        Mean
   trmt    N  Scores      Under H0        Under H0        Score
   ---------------------------------------------------------------
   G       7    69.0          52.50       7.757131      9.857143
   S       7    36.0          52.50       7.757131      5.142857

          Average scores were used for ties.


          Wilcoxon Two-Sample Test

      Statistic (S)                     69.0000

        Normal Approximation
        Z                                2.0626
        One-Sided Pr >   Z               0.0196
        Two-Sided Pr >  |Z|              0.0391


        t Approximation
        One-Sided Pr >   Z               0.0299
        Two-Sided Pr >  |Z|              0.0597


        Exact Test
        One-Sided Pr >=  S               0.0192
        Two-Sided Pr >=  |S - Mean|      0.0385


        Z includes a continuity correction of 0.5.


      Kruskal-Wallis Test
      Chi-Square           4.5244
      DF                        1
      Pr > Chi-Square      0.0334
```

Explanation: The sum of ranks (*Sum of scores*) = 69.0. The expected sum of ranks (*Expected Under H_0*) = 52.5. The P values for *One-Sided* and *Two-Sided Exact Tests* are 0.0192 and 0.0385, respectively. This suggests that H_0 should be rejected and that there is an effect of the superovulation treatments. Also, the output presents a z value with the correction (0.5) for a small sample. Again, it is appropriate to conclude that the populations are different since the P value for the two-sided test (*Prob* > $|z|$) = 0.0391. The same conclusion can be obtained from *Kruskal-Wallis Test* which uses *chi*-square distribution.

6.3   Hypothesis Test of a Population Proportion

Recall that a proportion is the probability of a successful trial in a binomial experiment. For a sample of size n and a number of successes y, the proportion is equal to:

$$p = \frac{y}{n}$$

Thus, the test of a proportion can utilize a binomial distribution for sample size n; however, for a large sample a normal approximation can be used. The distribution of an estimated proportion from a sample, $\hat{p}$, is approximately normal if the sample is large enough. A sample is assumed to be large enough if the interval $\hat{p} \pm \sqrt{\hat{p}\hat{q}/n}$ holds neither 0 nor 1. Here, n is the sample size and $\hat{q} = 1 - \hat{p}$.

The hypothesis test indicates whether the proportion calculated from a sample is significantly different from a hypothetical value. In other words, does the sample belong to a population with a predetermined proportion? The test can be one- or two-sided. The two-sided test for a large sample has the following hypotheses:

H_0: $p = p_0$
H_1: $p \neq p_0$

A z random variable is used as a test statistic:

$$z = \frac{\hat{p} - p_0}{\sqrt{p_0 q_0 / n}}$$

Example: There is a suspicion that due to ecological pollution in a region, the sex ratio in a population of field mice is not 1:1, but there are more males. An experiment was conducted to catch a sample of 200 mice and determine their sex. There were 90 females and 110 males captured.

The hypotheses are:

H_0: $p = 0.50$
H_1: $p > 0.50$

Let y = 110 be the number of males, n = 200 the total number of captured mice, $\hat{p}$ = 110/200 = 0.55 the proportion of captured mice that were males, and $\hat{q}$ = 0.45 the proportion of captured mice that were females. The hypothetical proportion that are males is $p_0 = 0.5$, and the hypothetical proportion that are females is $q_0 = 0.5$.
The calculated value of the test statistic is:

$$z = \frac{\hat{p} - p_0}{\sqrt{p_0 q_0 / n}} = \frac{0.55 - 0.50}{\sqrt{(0.50)(0.50)/200}} = 1.41$$

For a significance level of $\alpha = 0.05$, the critical value is $z_\alpha = 1.65$. Since the calculated value $z = 1.41$ is not more extreme than 1.65, we cannot conclude that the sex ratio is different than 1:1.

The z value can also be calculated using the number of individuals:

$$z = \frac{y - \mu_0}{\sqrt{np_0q_0}} = \frac{110 - 100}{\sqrt{200(0.5)(0.5)}} = 1.41$$

The z value is the same. Here μ_0 is the expected number of males if H_0 holds.

6.4 Hypothesis Test of the Difference between Proportions from Two Populations

Let y_1 and y_2 be the number of successes in two binomial experiments with sample sizes n_1 and n_2, respectively. For the estimation of $p_1 - p_2$, where p_1 and p_2 are the proportions of successes in two populations, the proportions $\hat{p}_1$ and $\hat{p}_2$ from two samples can be used:

$$\hat{p}_1 = \frac{y_1}{n_1} \quad \text{and} \quad \hat{p}_2 = \frac{y_2}{n_2}$$

The problem is to determine if the proportions from the two populations are different. An estimator of the difference between proportions is:

$$\hat{p}_1 - \hat{p}_2$$

The estimator has variance:

$$s^2_{\hat{p}_1 - \hat{p}_2} = \frac{\hat{p}_1\hat{q}_1}{n_1} + \frac{\hat{p}_2\hat{q}_2}{n_2}$$

For which $\hat{q}_1 = 1 - \hat{p}_1$ and $\hat{q}_2 = 1 - \hat{p}_2$. The hypotheses for a two-sided test are:

$$H_0: p_1 - p_2 = 0$$
$$H_1: p_1 - p_2 \neq 0$$

The test statistic is the standard normal variable:

$$z = \frac{(\hat{p}_1 - \hat{p}_2) - 0}{s_{\hat{p}_1 - \hat{p}_2}}$$

In which $s_{\hat{p}_1 - \hat{p}_2}$ is the standard error of the estimated difference between proportions ($\hat{p}_1 - \hat{p}_2$). Since the null hypothesis is that the proportions are equal, then:

$$s_{\hat{p}_1 - \hat{p}_2} = \sqrt{\frac{\hat{p}\hat{q}}{n_1} + \frac{\hat{p}\hat{q}}{n_2}}$$

that is:

$$s_{\hat{p}_1 - \hat{p}_2} = \sqrt{\hat{p}\hat{q}\left(\frac{1}{n_1} + \frac{1}{n_2}\right)}$$

where $\hat{q} = 1 - \hat{p}$

The proportion $\hat{p}$ is an estimator of the total proportion based on both samples:

$$\hat{p} = \frac{y_1 + y_2}{n_1 + n_2}$$

The proportion $\hat{p}$ can also be calculated from the given sample proportions:

$$\hat{p} = \frac{\hat{p}_1 n_1 + \hat{p}_2 n_2}{n_1 + n_2}$$

The normal approximation and use of a z statistic is appropriate if the intervals

$\hat{p}_1 \pm 2\sqrt{\dfrac{\hat{p}_1 \hat{q}_1}{n_1}}$ and $\hat{p}_2 \pm 2\sqrt{\dfrac{\hat{p}_2 \hat{q}_2}{n_2}}$ hold neither 0 nor 1.

The null hypothesis H_0 is rejected if the calculated $|z| > z_{\alpha/2}$, with $z_{\alpha/2}$ being the critical value for the significance level α.

Example: Test the difference between proportions of cows that returned to estrus after first insemination on two farms. Data are in the following table:

Farm 1	Farm 2
$y_1 = 40$	$y_2 = 30$
$n_1 = 100$	$n_2 = 100$
$p_1 = 0.4$	$p_2 = 0.3$

Here y_1 and y_2 are the number of cows that returned to estrus, and n_1 and n_2 are the total numbers of cows on farms 1 and 2, respectively.

$$\hat{p} = \frac{y_1 + y_2}{n_1 + n_2} = \frac{40 + 30}{100 + 100} = \frac{70}{200} = 0.35$$

$$\hat{q} = 1 - 0.35 = 0.65$$

$$s_{\hat{p}_1 - \hat{p}_2} = \sqrt{\hat{p}\hat{q}\left(\frac{1}{n_1} + \frac{1}{n_2}\right)} = \sqrt{(0.35)(0.65)\left(\frac{1}{100} + \frac{1}{100}\right)} = 0.067454$$

$$z = \frac{(\hat{p}_1 - \hat{p}_2) - 0}{s_{\hat{p}_1 - \hat{p}_2}} = \frac{(0.40 - 0.30) - 0}{0.067454} = 1.48$$

For the level of significance $\alpha = 0.05$, the critical value is 1.96. Since 1.48 is less than 1.96, there is not sufficient evidence to conclude that the proportion of cows that returned to estrus differs between the two farms.

6.5   *Chi*-square Test of the Difference between Observed and Expected Frequencies

Assume for some categorical characteristic the number of individuals in each of k categories has been counted. A common problem is to determine if the numbers in the categories are significantly different from hypothetical numbers defined by the theoretical proportions in populations:

$$H_0: p_1 = p_{1,0}, p_2 = p_{2,0}, ..., p_k = p_{k,0}$$
$$\text{(that is } H_0: p_i = p_{i,0} \text{ for each } i)$$
$$H_1: p_i \neq p_{i,0} \text{ for at least one } i$$

where $p_i = \dfrac{y_i}{n}$ is the proportion in any category i, $p_{i,0}$ is the expected proportion, and n is the total number of observations, $n = \sum_i y_i$, $i = 1,..., k$.

A test statistic:

$$\chi^2 = \sum_i \frac{[y_i - E(y_i)]^2}{E(y_i)}$$

has a *chi*-square distribution with $(k-1)$ degrees of freedom. Here, k is the number of categories, and $E(y_i) = n\, p_{i,0}$ is the expected number of observations in category i.

The null hypothesis, H_0, is rejected if the calculated $\chi^2 > \chi^2_a$, for which χ^2_a is the critical value for significance level α, that is a value of χ^2 such that $P(\chi^2 > \chi^2_a) = \alpha$. This holds when the samples are large enough, usually defined as when the expected number of observations in each category is greater than five.

Example: The expected proportions of white, brown and pied rabbits in a population are 0.36, 0.48 and 0.16, respectively. In a sample of 400 rabbits there were 140 white, 240 brown and 20 pied. Are the proportions in that sample of rabbits different than expected?

The observed and expected frequencies are presented in the following table:

Color	Observed (y_i)	Expected ($E[y_i]$)
White	140	(0.36)(400) = 144
Brown	240	(0.48)(400) = 192
Pied	20	(0.16)(400) = 64

$$\chi^2 = \sum_i \frac{[y_i - E(y_i)]^2}{E(y_i)} = \frac{[140 - 144]^2}{144} + \frac{[240 - 192]^2}{192} + \frac{[20 - 64]^2}{64} = 42.361$$

The critical value of the *chi*-square distribution for $k - 1 = 2$ degrees of freedom and significance level of $\alpha = 0.05$ is 5.991. Since the calculated χ^2 is greater than the critical value it can be concluded that the sample is different from the population with 0.05 level of significance.

6.5.1 SAS Example for Testing the Difference between Observed and Expected Frequencies

The SAS program for the example of white, brown and pied rabbits is as follows. Recall that the expected proportions of white, brown and pied rabbits are 0.36, 0.48 and 0.16, respectively. In a sample of 400 rabbits, there were 140 white, 240 brown and 20 pied. Are the proportions in that sample of rabbits different than expected?

SAS program:

```
DATA color;
INPUT color$ number;
DATALINES;
white 140
brown 240
pied 20
;
PROC FREQ DATA = color;
   WEIGHT number;
   TABLES color / TESTP = (36 48 16);
RUN;
```

Explanation: The FREQ procedure is used. The WEIGHT statement denotes a variable that defines the numbers in each category. The TABLES statement defines the category variable. The TESTP option defines the expected percentages.

SAS output:

Color	Frequency	Percent	Test Percent	Cumulative Frequency	Cumulative Percent
white	140	35.00	36.00	140	35.00
brown	240	60.00	48.00	380	95.00
pied	20	5.00	16.00	400	100.00

```
           Chi-Square Test
       for Specified Proportion
       ------------------------
       Chi-Square        42.3611
       DF                      2
       Pr > ChiSq        <.0001

       Sample Size = 400
```

Explanation: The first table presents categories (*Color*), the number and percentage of observations in each category (*Frequency* and *Percent*), the expected percentage (*Test Percent*), and the cumulative frequencies and percentages. In the second table the *chi*-square value (*Chi-Square*), degrees of freedom (*DF*) and P-value (*Pr > ChiSq*) are presented. The highly significant *Chi-Square* ($P < 0.0001$) indicates that color percentages differ from those expected.

6.6 Hypothesis Test of Differences among Proportions from Several Populations

For testing the difference between two proportions or two frequencies of successes the *chi-square* test can be used. Further, this test is not limited to only two samples, but can be used to compare the number of successes of more than two samples or categories. Each category or group represents a random sample. If there are no differences among proportions in the populations, the expected proportions will be the same in all groups. The expected proportion can be estimated by using the proportion of successes in all groups together. For k groups, the expected proportion of successes is:

$$p_0 = \frac{\sum_i y_i}{\sum_i n_i} \qquad i = 1,\ldots, k$$

The expected proportion of failures is:

$$q_0 = 1 - p_0$$

The expected number of successes in category i is:

$$E(y_i) = n_i\, p_0$$

In which n_i is the number of observations in category i.
The expected number of failures in category i is:

$$E(n_i - y_i) = n_i\, q_0$$

The hypotheses are:

$$H_0: p_1 = p_2 = \ldots = p_k = p_0 \quad (H_0: p_i = p_0 \text{ for every } i)$$
$$H_1: p_i \neq p_0 \text{ for at least one } i$$

The test statistic is:

$$\chi^2 = \sum_i \frac{[y_i - E(y_i)]^2}{E(y_i)} + \sum_i \frac{[(n_i - y_i) - E(n_i - y_i)]^2}{E(n_i - y_i)} \qquad i = 1,\ldots, k$$

with a *chi*-square distribution with $(k-1)$ degrees of freedom, k is the number of categories.

Example: Are the proportions of cows with mastitis significantly different among three farms? The numbers of cows on farms *A*, *B* and *C* are 96, 132 and 72, respectively. The numbers of cows with mastitis on farms *A*, *B* and *C* are 36, 29 and 10, respectively.

The numbers of cows are: $n_1 = 96$, $n_2 = 132$, and $n_3 = 72$
The numbers of with mastitis cows are: $y_1 = 36$, $y_2 = 29$, and $y_3 = 10$
The expected proportion of cows with mastitis is:

$$p_0 = \frac{\sum_i y_i}{\sum_i n_i} = \frac{36 + 29 + 10}{96 + 132 + 72} = 0.25$$

The expected proportion of healthy cows is:

$$q_0 = 1 - p_0 = 1 - 0.25 = 0.75$$

The expected numbers of cows with mastitis and healthy cows on farm A are:

$$E(y_1) = (96)(0.25) = 24$$
$$E(n_1 - y_1) = (96)(0.75) = 72$$

The expected numbers of cows with mastitis and healthy cows on farm B are:

$$E(y_2) = (132)(0.25) = 33$$
$$E(n_2 - y_2) = (132)(0.75) = 99$$

The expected numbers of cows with mastitis and healthy cows on farm C are:

$$E(y_3) = (72)(0.25) = 18$$
$$E(n_3 - y_3) = (72)(0.75) = 54$$

The example is summarized below as a 'Contingency Table'.

	Number of cows			Expected number of cows	
Farm	Mastitis	No mastitis	Total	Mastitis	No mastitis
A	36	60	96	(0.25)(96) = 24	(0.75)(96) = 72
B	29	103	132	(0.25)(132) = 33	(0.75)(132) = 99
C	10	62	72	(0.25)(72) = 18	(0.75)(72) = 54
Total	75	225	300	75	225

The calculated value of the *chi*-square statistic is:

$$\chi^2 = \sum_i \frac{[y_i - E(y_i)]^2}{E(y_i)} + \sum_i \frac{[(n_i - y_i) - E(n_i - y_i)]^2}{E(n_i - y_i)} =$$

$$\chi^2 = \frac{(36-24)^2}{24} + \frac{(29-33)^2}{33} + \frac{(10-18)^2}{18} + \frac{(60-72)^2}{72} + \frac{(103-99)^2}{99} + \frac{(62-54)^2}{54} = 13.387$$

For the significance level $\alpha = 0.05$ and degrees of freedom $(3-1) = 2$, the critical value $\chi^2_{0.05} = 5.991$. The calculated value (13.387) is greater than the critical value, thus there is sufficient evidence to conclude that the incidence of mastitis differs among these farms.

6.6.1 SAS Example for Testing Differences among Proportions from Several Populations

The SAS program for the example of mastitis in cows on three farms is as follows:

SAS program:

```
DATA a;
INPUT farm $ mastitis $ number;
DATALINES;
A  YES    36
A   NO    60
B  YES    29
B   NO   103
C  YES    10
C   NO    62
;
```

```
PROC FREQ DATA = a ORDER = DATA;
WEIGHT number;
TABLES farm*mastitis / CHISQ ;
RUN;
```

Explanation: The FREQ procedure is used. The ORDER option keeps the order of data as they are entered in the DATA step. The WEIGHT statement denotes a variable that defines the numbers in each category. The TABLES statement defines the categorical variables. The CHISQ option calculates a *chi*-square test.

SAS output:

```
                Table of farm by mastitis


        farm        mastitis


        Frequency|
        Percent  |
        Row Pct  |
        Col Pct  |YES      |NO       |    Total
        ---------|---------|---------|
        A        |      36 |      60 |       96
                 |   12.00 |   20.00 |    32.00
                 |   37.50 |   62.50 |
                 |   48.00 |   26.67 |
        ---------|---------|---------|
        B        |      29 |     103 |      132
                 |    9.67 |   34.33 |    44.00
                 |   21.97 |   78.03 |
                 |   38.67 |   45.78 |
        ---------|---------|---------|
        C        |      10 |      62 |       72
                 |    3.33 |   20.67 |    24.00
                 |   13.89 |   86.11 |
                 |   13.33 |   27.56 |
        ---------|---------|---------|
        Total          75       225       300
                     25.00     75.00    100.00


         Statistics for Table of farm by mastitis
```

Statistic	DF	Value	Prob
Chi-Square	2	13.3872	0.0012
Likelihood Ratio Chi-Square	2	13.3550	0.0013
Mantel-Haenszel Chi-Square	1	12.8024	0.0003
Phi Coefficient		0.2112	
Contingency Coefficient		0.2067	
Cramer's V		0.2112	

```
                Sample Size = 300
```

Explanation: The first table presents *farm by mastitis* categories, the number and percentage of observations in each category (*Frequency* and *Percent*), the percentage by farm (*Col Pct*) and the percentage by incidence of mastitis (*Row Pct*). In the second table *chi*-square value (*Chi-Square*), degrees of freedom (*DF*) and the *P* value (*Prob*) along with some other similar tests and coefficients, are presented. The *P* value is 0.0012 and thus H_0 is rejected.

6.7 Hypothesis Test of Population Variance

Populations can differ not only in their means, but also in the dispersion of observations. In other words, populations can have different variances. A test that the variance is different from a hypothetical value can be one- or two-sided. The two-sided hypotheses are:

$$H_0: \sigma^2 = \sigma^2_0$$
$$H_1: \sigma^2 \neq \sigma^2_0$$

The following test statistic can be used:

$$\chi^2 = \frac{(n-1)s^2}{\sigma^2_0}$$

The test statistic has a *chi*-square distribution with $(n-1)$ degrees of freedom. For the two-sided test H_0 is rejected if the calculated χ^2 is less than $\chi^2_{1-\alpha/2}$ or the calculated χ^2 is greater than $\chi^2_{\alpha/2}$. Here, $\chi^2_{\alpha/2}$ is a critical value such that $P(\chi^2 > \chi^2_{\alpha/2}) = \alpha/2$, and $\chi^2_{1-\alpha/2}$ is a critical value such that $P(\chi^2 < \chi^2_{1-\alpha/2}) = \alpha/2$.

6.8 Hypothesis Test of the Difference of Two Population Variances

To test whether the variances of two populations are different an *F* test can be used, providing that the observations are normally distributed. Namely, the ratio:

$$\frac{s^2_1}{\sigma^2_1} \div \frac{s^2_2}{\sigma^2_2}$$

has an *F* distribution with $(n_1 - 1)$ and $(n_2 - 1)$ degrees of freedom, where n_1 and n_2 are sample sizes. The test can be one- or two-sided. Hypotheses for the two-sided test can be written as:

$$H_0: \sigma^2_1 = \sigma^2_2$$
$$H_1: \sigma^2_1 \neq \sigma^2_2$$

As a test statistic the following quotient is used:

$$\frac{s^2_1}{s^2_2}$$

The quotient is always expressed with the larger estimated variance in the numerator. The H_0 is rejected if $\frac{s^2_1}{s^2_2} \geq F_{\alpha/2, n_1-1, n_2-1}$, where $F_{\alpha/2, n_1-1, n_2-1}$ is a critical value such that the probability $P(F > F_{\alpha/2, n_1-1, n_2-1}) = \alpha/2$.

An alternative to be used for populations in which observations are not necessary normally distributed is the *Levene* test. The *Levene* statistic is:

$$Le = \frac{(N-k)\sum_i n_i (\bar{u}_{i.} - \bar{u}..)^2}{(k-1)\sum_i \sum_j (u_{ij} - \bar{u}_{i.})^2} \qquad i = 1,..., k, \; j = 1,..., n_i$$

where:

N = the total number of observations
k = the number of groups
n_i = the number of observations in a group i
$u_{ij} = |y_{ij} - \bar{y}_{i.}|$
y_{ij} = observation j in group i
$\bar{y}_{i.}$ = the mean of group i
$\bar{u}_{i.}$ = the mean of group for u_{ij}
$\bar{u}..$ = the overall mean for u_{ij}

An F distribution is used to test the differences in variances. The variances are different if the calculated value Le is greater than $F_{\alpha, k-1, N-k}$.

6.9 Hypothesis Tests Using Confidence Intervals

Calculated confidence intervals can be used in hypothesis testing such that if the calculated interval contains a hypothetical parameter value, then the null hypothesis is not rejected. For example, for testing hypotheses about a population mean:

H_0: $\mu = \mu_0$
H_1: $\mu \neq \mu_0$

the following confidence interval is calculated:

$$\bar{y} \pm z_{\alpha/2} \, \sigma_{\bar{y}}$$

If that interval contains μ_0, the null hypothesis is not rejected.

Example: Assume that milk production has been measured on 50 cows sampled from a population and the mean lactation milk yield was 4000 kg. Does that sample belong to a population with a mean $\mu_0 = 3600$ kg and standard deviation $\sigma = 1000$ kg?

The hypothetical mean is $\mu_0 = 3600$ kg and the hypotheses are:

H_0: $\mu = 3600$ kg
H_1: $\mu \neq 3600$ kg

We have:

$\bar{y} = 4000$ kg
$n = 50$ cows
$\sigma = 1000$ kg

A calculated confidence interval is:

$$\bar{y} \pm z_{\alpha/2}\ \sigma_{\bar{y}}$$

For a 95% confidence interval, $\alpha = 0.05$ and $z_{\alpha/2} = z_{0.025} = 1.96$.

$$\sigma_{\bar{y}} = \frac{\sigma}{\sqrt{n}} = \frac{1000}{\sqrt{50}} = 141.4$$

The interval is (3722.9 to 4277.1 kg).

Since the interval does not contain $\mu_0 = 3600$, it can be concluded that the sample does not belong to the population with mean 3600, and that these cows have higher milk yield than those in the population.

The confidence interval approach can be used in a similar way to test other hypotheses, such as the difference between proportions or populations means, etc.

6.10 Statistical and Practical Significance

Statistical significance does not always indicate a practical significance. For example, consider the use of a feed additive that results in a true increase of 20 g of daily gain in cattle. This difference is relatively small and may be of neither practical nor economic importance, but if sufficiently large samples are tested the difference between them can be found to be statistically significant. Alternatively, the difference between the populations can be of practical importance, but if small samples are used for testing a statistically significant difference may not be detected.

The word 'significant' is often used improperly. The term significant is valid only for samples. The statement: 'There is a significant difference between sample means', denotes that their calculated difference leads to a P value small enough that H_0 is rejected. It is not appropriate to state that 'the population means are significantly different', because population means can be only practically different. Therefore, they are different or they are not different. Samples are taken from the populations and tested to determine if there is evidence that the population means are different.

6.11   Types of Errors in Inferences and Power of Test

A statistical test can have only two results: to reject or to fail to reject the null hypothesis H_0. Consequently, based on sample observations, there are two possible errors:

a) *type I error* = rejection of H_0, when H_0 is actually true
b) *type II error* = failure to reject H_0, when H_0 is actually false

The incorrect conclusions each have probabilities. The probability of a type I error is denoted as α, and the probability of a type II error is denoted as β. The probability of a type I error is the same as the P value if H_0 is rejected. The probability that H_1 is accepted and H_1 is actually true is called the power of test and is denoted as $(1 - \beta)$. The relationships of conclusions and true states and their probabilities are presented in the following table:

Decision of a statistical test	True situation	
	H_0 correct no true difference	H_0 not correct a true difference exists
H_0 not rejected	Correct acceptance $P = 1 - \alpha$	Type II error $P = \beta$
H_0 rejected	Type I error $P = \alpha$	Correct rejection $P = 1 - \beta$

The following have influence on making a correct conclusion:

1. sample size
2. level of significance α
3. effect size (desired difference considering variability)
4. power of test $(1 - \beta)$.

When planning an experiment at least three of those factors should be given, while the fourth can be determined from the others. To maximize the likelihood of reaching the correct conclusion, the type I error should be as small as possible, and the power of test as large as possible. To approach this, the sample size can be increased, the variance decreased, or the effect size increased. Thus, the level of significance and power of test must be taken into consideration when planning the experiment. When a sample has already been drawn, α and β cannot be simultaneously decreased. The probability of a type I error is usually either known or easily computed. It is established by the researcher as the level of significance, or is calculated as the P value. On the other hand, it is often difficult to calculate the probability of a type II error (β) or analogous power of test $(1 - \beta)$. In order to determine β, some distribution of H_1 must be assumed to be correct. The problem is that usually the distribution is unknown. Figure 6.8 shows the probability β for a given probability α with normal distributions assumed. If H_0 is correct, the mean is μ_0, and if H_1 is correct, the mean is μ_1. The case where $\mu_0 < \mu_1$ is shown. The value α can be used as the level of significance (for example 0.05), or the level of significance can be the observed P value. The critical value y_α, and the critical region is determined with the α or P value. The probability β is determined with the distribution for H_1, and corresponds to an area under the normal curve determined by the critical region:

$$\beta = P[y < y_\alpha = y_\beta]$$

using the H_1 distribution with mean μ_1, where $y_\alpha = y_\beta$ is the critical value.

The power of test is equal to $(1 - \beta)$ and this is an area under the curve H_1 determined by the critical region:

$$Power = (1 - \beta) = P[y > y_\alpha = y_\beta] \quad \text{using the } H_1 \text{ distribution with the mean } \mu_1.$$

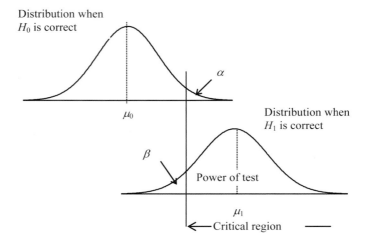

Figure 6.8 Probabilities of type I and type II errors

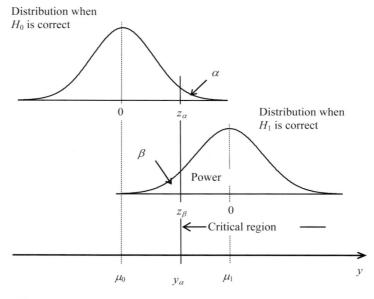

Figure 6.9 Standard normal distributions for H_0 and H_1. The power, type I error (α) and type II error (β) for a one-sided test are shown. The original scale of variable y is shown on the bottom

If the parameters of H_0 and H_1 are known, power can be determined by using the corresponding standard normal distributions. Let μ_0 and μ_1 be the means, and σ_{D0} and σ_{D1} the standard deviations of the H_0 and H_1 distributions, respectively. Using the standard normal distribution, and if for example $\mu_0 < \mu_1$, the power for a one-sided test is the probability $P(z > z_\beta)$ determined on the H_1 distribution (Figure 6.9).

The value z_β can be determined as usual, as a deviation from the mean divided by the corresponding standard deviation.

$$z_\beta = \frac{y_\alpha - \mu_1}{\sigma_{D1}}$$

The value y_α is the critical value, expressed in the original scale, which is determined by the value z_α:

$$y_\alpha = (\mu_0 + z_\alpha \sigma_{D0})$$

Recall that the value α is determined by the researcher as the significance level. It follows that:

$$z_\beta = \frac{(\mu_0 + z_\alpha \sigma_{D0}) - \mu_1}{\sigma_{D1}}$$

Therefore, if $\mu_0 < \mu_1$, the power is:

$$Power = (1 - \beta) = P[z > z_\beta] \quad \text{using the } H_1 \text{ distribution}$$

If $\mu_0 > \mu_1$, the power is:

$$Power = (1 - \beta) = P[z < z_\beta] \quad \text{using the } H_1 \text{ distribution}$$

The appropriate probability can be determined from the table Area under the Standard Normal Curve (Appendix B).

For specific tests the appropriate standard deviations must be defined. For example, for the test of hypothesis $\mu_1 > \mu_0$:

$$z_\beta = \frac{\left(\mu_0 + z_\alpha \, \sigma_0 / \sqrt{n}\right) - \mu_1}{\sigma_1 / \sqrt{n}}$$

where $\sigma_0 / \sqrt{n}$ and $\sigma_1 / \sqrt{n}$ are the standard errors of sample means for H_0 and H_1, respectively.

Often, $\sigma_0 = \sigma_1 = \sigma$, and then:

$$z_\beta = \frac{(\mu_0 - \mu_1)}{\sigma / \sqrt{n}} + z_\alpha$$

In testing the hypotheses of the difference of two means:

$$z_\beta = \frac{(\mu_1 - \mu_2)}{\sqrt{\dfrac{\sigma_1^2}{n_1} + \dfrac{\sigma_2^2}{n_1}}} + z_\alpha$$

where n_1 and n_2 are sample sizes, and σ_1^2 and σ_2^2 are the population variances.

In testing hypotheses of a population proportion when a normal approximation is used:

$$z_\beta = \frac{\left(p_0 + z_\alpha \sqrt{p_0(1 - p_0)/n}\right) - p_1}{\sqrt{p_1(1 - p_1)/n}}$$

where p_0 and p_1 are the proportions for H_0 and H_1, respectively.

The power of test can be calculated by applying a similar approach based on other estimators and distributions.

For a two-sided test the power is determined on the basis of two critical values $-z_{\alpha/2}$ and $z_{\alpha/2}$ (Figure 6.10). Expressions for calculating $z_{\beta 1}$ and $z_{\beta 2}$ are similar to the one-sided test, except z_α is replaced by $-z_{\alpha/2}$ or $z_{\alpha/2}$:

$$z_{\beta 1} = \frac{\left(\mu_0 - z_{\alpha/2}\sigma_{D0}\right) - \mu_1}{\sigma_{D1}}$$

$$z_{\beta 2} = \frac{\left(\mu_0 + z_{\alpha/2}\sigma_{D0}\right) - \mu_1}{\sigma_{D1}}$$

The power is again $(1 - \beta)$, the area under the H_1 curve held by the critical region. Thus, for the two-sided test the power is the sum of probabilities:

$$Power = (1 - \beta) = P[z < z_{\beta 1}] + P[z > z_{\beta 2}] \quad \text{using the } H_1 \text{ distribution}$$

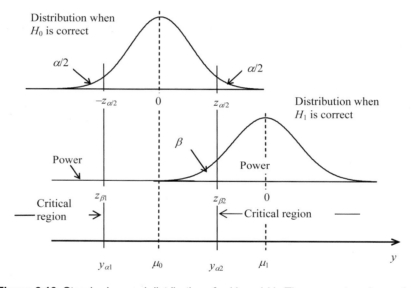

Figure 6.10 Standard normal distributions for H_0 and H_1. The power, type I error (α) and type II error (β) for the two-sided test are shown. On the bottom is the original scale of the variable y

One approach for estimation of power based on the sample is to set as an alternative hypothesis the estimated parameters or measured difference between samples. Using that difference, the theoretical distribution for H_1 is defined and the deviation from the assumed critical value is analyzed. Power of test is also important when H_0 is not rejected. If the hypothesis test has a considerable power and H_0 is not rejected, H_0 is likely correct. If the test has small power and H_0 is not rejected, there is a considerable chance of a type II error.

Example: The mean milk yield from a sample of 30 cows is 4100 kg. Is that value significantly greater than 4000 kg? The variance is 250000. Calculate the power of the test.

$\mu_0 = 4000$ kg (if H_0)
$\bar{y} = 4100$ kg ($= \mu_1$ if H_1)
$\sigma^2 = 250000$, and the standard deviation is $\sigma = 500$ kg

$$z = \frac{\bar{y} - \mu_0}{\sigma/\sqrt{n}} = \frac{4100 - 4000}{500/\sqrt{30}} = 1.095$$

For $\alpha = 0.05$, $z_\alpha = 1.65$, since the calculated value $z = 1.095$ is not more extreme than the critical value $z_\alpha = 1.65$, H_0 is not rejected with $\alpha = 0.05$. The sample mean is not significantly different than 4000 kg. The power of the test is:

$$z_\beta = \frac{(\mu_0 - \mu_1)}{\sigma/\sqrt{n}} + z_\alpha = \frac{4000 - 4100}{500/\sqrt{30}} + 1.65 = 0.55$$

Using the H_1 distribution, the power is $P(z > z_\beta) = P(z > 0.55) = 0.29$. The type II error, that is, the probability that H_0 is incorrectly accepted, is $1 - 0.29 = 0.71$. The high probability of error is because the difference between means for H_0 and H_1 is relatively small compared to the variability.

Example: Earlier in this chapter there was an example with mice and a test to determine if the sex ratio was different than 1:1. Out of a total of 200 captured mice, the number of males was $\mu_1 = 110$. Assume that this is the real number of males if H_1 is correct. If H_0 is correct, then the expected number of males is $\mu_0 = 100$.

The critical value for the significance level $\alpha = 0.05$ and the distribution if H_0 is correct is $z_\alpha = 1.65$. The proportions of males if H_1 holds is $p_1 = \frac{110}{200} = 0.55$. Then:

$$z_\beta = \frac{\left(0.5 + 1.65\sqrt{(0.5)(0.5)/200}\right) - 0.55}{\sqrt{(0.45)(0.55)/200}} = 0.24$$

The power, $(1 - \beta)$, is the probability $P(z > 0.24) = 0.41$. As the power is relatively low, the sample size must be increased in order to show that sex ratio has changed.

For samples from a normal population with unknown variance, the power of test can be calculated by using a Student t distribution. If H_0 holds, then the test statistic t has a central t distribution with df degrees of freedom. However, if H_1 holds, then the t statistic has a noncentral t distribution with the noncentrality parameter λ and df degrees of freedom. Let t_α be the critical value for the α level of significance. The power of test is calculated by using the value t_α and the probability (areas) from the noncentral t distribution.

For a one-sided test of hypotheses of the population mean, H_0: $\mu = \mu_0$, versus H_1: $\mu = \mu_1$, and for example, $\mu_1 > \mu_0$, the power is:

$$Power = (1 - \beta) = P[t > t_\alpha = t_\beta]$$

using the t distribution for H_1, with the noncentrality parameter $\lambda = \dfrac{|\mu_1 - \mu_0|}{s}\sqrt{n}$ and

degrees of freedom $df = n - 1$. Here, s = the sample standard deviation, and n is the sample size. The difference $\mu_1 - \mu_0$ is defined as a positive value, then the noncentral distribution for H_1 is situated on the right of the distribution for H_0 and the power is observed at the right side of the H_1 curve (Figure 6.11).

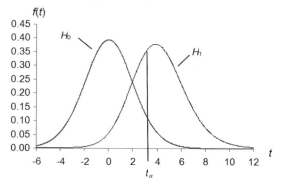

Figure 6.11 Significance and power of the one-sided t test. The t statistic has a central t distribution if H_0 is true, and a noncentral distribution if H_1 is true. The distributions with 20 degrees of freedom are shown. The critical value is t_α. The area under the H_0 curve on the right of the critical value is the level of significance (α). The area under the H_1 curve on the right of the critical value is the power ($1 - \beta$). The area under the H_1 curve on the left of the critical value is the type II error (β)

For a two-sided test of the population mean the power is:

$$Power = (1 - \beta) = P[t < -t_{\alpha/2} = t_{\beta1/2}] + P[t > t_{\alpha/2} = t_{\beta2/2}]$$

using a t distribution for H_1, with the noncentrality parameter $\lambda = \dfrac{|\mu_1 - \mu_0|}{s}\sqrt{n}$ and degrees of freedom $df = n - 1$ (Figure 6.12).

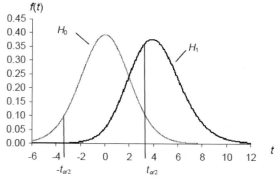

Figure 6.12 The significance and power of the two-sided t test. The t statistic has a central t distribution if H_0 is true, and a noncentral distribution if H_1 is true. The distributions with 20 degrees of freedom are shown. The critical values are $-t_{\alpha/2}$ and $t_{\alpha/2}$. The sum of areas under the H_0 curve on the left of $-t_{\alpha/2}$ and on the right of $t_{\alpha/2}$ is the level of significance (α). The sum of areas under the H_1 curve on the left of $-t_{\alpha/2}$ and on the right of $t_{\alpha/2}$ is the power ($1 - \beta$). The area under the H_1 curve between $-t_{\alpha/2}$ and $t_{\alpha/2}$ is the type II error (β)

For a one-sided test of the difference of two population means, H_0: $\mu_1 - \mu_2 = 0$, versus H_1: $\mu_1 - \mu_2 = \delta$, and for $\mu_1 > \mu_2$, the power is:

$$Power = (1 - \beta) = P[t > t_\alpha = t_\beta]$$

using a t distribution for H_1 with the noncentrality parameter $\lambda = \dfrac{|\mu_1 - \mu_2|}{s_p}\sqrt{\dfrac{n}{2}}$ and degrees

of freedom $df = 2n - 2$. Here, $s_p = \sqrt{\dfrac{(n_1 - 1)s_1^2 + (n_2 - 1)s_2^2}{n_1 + n_2 - 2}}$ denotes the pooled standard

deviation calculated from both samples, s_1 and s_2 are standard deviations, and n_1 and n_2 are the sample sizes drawn from populations 1 and 2.

For a two-sided test of the difference between two population means, the power is:

$$Power = (1 - \beta) = P[t < -t_{\alpha/2} = t_{\beta 1}] + P[t > t_{\alpha/2} = t_{\beta 2}]$$

using a t distribution for H_1 with the noncentrality parameter $\lambda = \dfrac{|\mu_1 - \mu_2|}{s_p}\sqrt{\dfrac{n}{2}}$ and degrees

of freedom $df = 2n - 1$.

6.11.1 SAS Examples for the Power of Test

Example: Test the hypothesis that the sample mean of milk yield of 4300 kg is different than the population mean of 4000 kg. The sample size is 9 dairy cows, and the sample standard deviation is 600 kg. Calculate the power of the test by defining H_1: $\mu = \bar{y} = 4300$ kg.

$\mu_0 = 4000$ kg (if H_0)
$\bar{y} = 4300$ kg ($= \mu_1$ if H_1)
$s = 600$ kg.
$$t = \frac{\bar{y} - \mu_0}{s/\sqrt{n}} = \frac{4300 - 4000}{600/\sqrt{9}} = 1.5$$

For $\alpha = 0.05$ and degrees of freedom $(n - 1) = 8$, the critical value for a one-sided test is $t_{0.05} = 1.86$. The calculated $t = 1.5$ is not more extreme than the critical value, and thus H_0 is not rejected. The sample mean is not significantly greater than 4000 kg.

The power of test is:

$$Power = (1 - \beta) = P[t > t_{0.05} = t_\beta]$$

Using a t distribution for H_1 with the noncentrality parameter

$$\lambda = \frac{|\mu_1 - \mu_0|}{s}\sqrt{n} = \frac{|4300 - 4000|}{600}\sqrt{9} = 1.5 \text{ and degrees of freedom } df = 8,$$

the power is:

$$Power = (1 - \beta) = P[t > 1.86] = 0.393$$

The power of test can be calculated by using a simple SAS program. One- and two-sided tests are given:

```
DATA a;
alpha=0.05;
n=9;
mi0=4000;
mi1=4300;
stdev=600;
df=n-1;
lambda=(ABS(mi1-mi0)/stdev)*SQRT(n);
tcrit_one_sided=TINV(1-alpha,df);
tcrit_low=TINV(alpha/2,df);
tcrit_up=TINV(1-alpha/2,df);
power_one_sided=1-CDF('t',tcrit_one_sided,df,lambda);
power_two_sided=CDF('t',tcrit_low,df,lambda)+ 1-CDF('t',tcrit_up,df,lambda);
PROC PRINT;
RUN;
```

Explanation: First are defined: *alpha* = significance level, n = sample size, *mi0* = μ_0 = the population mean if H_0 is true, *mi1* = $\bar{y}$ = μ_1 = the population mean if H_1 is true, *stdev* = the sample standard deviation, df = degrees of freedom. Then, the noncentrality parameter (*lambda*) and critical values, (*tcrit_one_sided*) for a one-sided test, and (*tcrit_low* and *tcrit_up*) for a two-sided test, are calculated. The critical value is computed by using the TINV function, which must have cumulative values of percentiles ($1 - \alpha = 0.95$, $\alpha/2 = 0.025$ and $1 - \alpha/2 = 0.975$) and degrees of freedom df. The *power* is calculated with the CDF function. This is a cumulative function of the t distribution, which requires definition of the critical value, degrees of freedom, and the noncentrality parameter *lambda*. As an alternative to the *CDF('t',tcrit,df,lambda)*, the function *PROBT(tcrit,df,lambda)* can also be used. The PRINT procedure gives the following SAS output:

SAS output:

alpha	n	mi0	mi1	stdev	df	lambda
0.05	9	4000	4300	600	8	1.5

tcrit_ one_sided	tcrit_ low	tcrit_up	power_ one_sided	power_ two_sided
1.85955	-2.30600	2.30600	0.39277	0.26275

Thus, the powers of test are 0.39277 for the one-sided and 0.26275 for the two-sided test.

The same analysis can be done by using the POWER procedure (SAS version 9.1 or higher).

SAS program:

```
PROC POWER;
   ONESAMPLEMEANS
      TEST = T
      ALPHA = 0.05
      NULLMEAN = 4000
      MEAN = 4300
```

```
        STDEV = 600
        SIDES = 1 2
        NTOTAL = 9
        POWER = .
    ;
RUN;
```

Explanation: The POWER procedure is used with ONESAMPLEMEANS and TEST = T statements indicating two sample test. Values are from the milk production example shown above, with sample mean of 4300 kg and sample standard deviation of 600 kg. Alpha of the test = 0.05, and number of observations is *NTOTAL* = 9. *SIDES* = 1 2 asks for one- and two-sided tests to be performed. *POWER* = . directs the power of test to be calculated by the program.

SAS output:

```
One-sample t Test for Mean


Distribution           Normal
Method                 Exact
Null Mean                4000              Computed Power
Alpha                    0.05
Mean                     4300        Index    Sides    Power
Standard Deviation        600          1        1      0.393
Total Sample Size           9          2        2      0.263
```

Second example: Two groups of eight cows were fed two different diets (*A* and *B*) in order to test the difference in milk yield. From the samples the following was calculated:

	Diet *A*	Diet *B*
Mean ($\bar{y}$), kg	21.8	26.4
Std. deviation (*s*)	4.1	5.9
Number of cows (*n*)	7	7

Test the null hypothesis, H_0: $\mu_2 - \mu_1 = 0$, and calculate the power of test by defining the alternative hypothesis H_1: $\mu_2 - \mu_1 = \bar{y}_2 - \bar{y}_1 = 4.6$ kg.

The test statistic is:

$$t = \frac{(\bar{y}_1 - \bar{y}_2) - 0}{s_p \sqrt{\left(\dfrac{1}{n_1} + \dfrac{1}{n_2}\right)}}$$

The standard deviation is:

$$s_p = \sqrt{\frac{(n_1 - 1)s_1^2 + (n_2 - 1)s_2^2}{n_1 + n_2 - 2}} = \sqrt{\frac{(7-1)(4.1^2) + (7-1)(5.9^2)}{7+7-2}} = 5.080$$

The calculated *t* value is:

$$t = \frac{(\bar{y}_1 - \bar{y}_2) - 0}{s_p\sqrt{\left(\dfrac{1}{n_1} + \dfrac{1}{n_2}\right)}} = \frac{(26.4 - 21.8) - 0}{5.080\sqrt{\left(\dfrac{1}{7} + \dfrac{1}{7}\right)}} = 1.694$$

For $\alpha = 0.05$ and degrees of freedom $(n_1 + n_2 - 2) = 12$, the critical value for a two-sided test is $t_{0.25} = 2.179$. The calculated $t = 1.694$ is not more extreme than the critical value and H_0 is not rejected.

The power for this test is:

$$Power = (1 - \beta) = P[t < -t_{\alpha/2} = t_{\beta1}] + 1 - P[t > t_{\alpha/2} = t_{\beta2}]$$

Using a *t* distribution for H_1 with the noncentrality parameter

$$\lambda = \frac{|\mu_1 - \mu_2|}{s_p}\sqrt{\frac{n}{2}} = \frac{|21.8 - 26.4|}{5.080}\sqrt{\frac{7}{2}} = 1.694 \text{ and degrees of freedom } df = 12, \text{ the}$$

power is:

$$Power = (1 - \beta) = P[t < -2.179] + P[t > 2.179] = 0.000207324 + 0.34429 = 0.34450$$

The SAS program for this example is as follows:

```
DATA aa;
alpha=0.05;
n1=7;
n2=7;
mi1=21.8;
mi2=26.4;
stdev1=4.1;
stdev2=5.9;
df=n1+n2-2;
sp = SQRT(((n1-1)*stdev1*stdev1+(n2-1)*stdev2*stdev2)/(n1+n2-2));
lambda=(ABS(mi2-mi1)/sp)/sqrt(1/n1+1/n2);
tcrit_low=TINV(alpha/2,df);
tcrit_up=TINV(1-alpha/2,df);
tcrit_one_sided=TINV(1-alpha,df);
power_one_sided=1-CDF('t',tcrit_one_sided,df,lambda);
power_two_sided=CDF('t',tcrit_low,df,lambda)+ 1-CDF('t',tcrit_up,df,lambda);
PROC PRINT;
RUN;
```

Explanation: First are defined: *alpha* = significance level, *n1 and n2* = sample sizes, *mi1* = $\mu_1 = \bar{y}_1$ = the mean of population 1, *mi2* = $\mu_2 = \bar{y}_2$ = the mean of population 2, *stdev1* and *stdev2* = standard deviation of samples 1 and 2, *df* = degrees of freedom, *sp* calculates the pooled standard deviation. The noncentrality parameter (*lambda*) and critical values (*tcrit_one_sided*) for a one-sided test, and *tcrit_low* and *tcrit_up* for a two-sided test) are calculated. The critical value is computed by using the TINV function, which must have cumulative values of percentiles $(1 - \alpha = 0.95, \alpha/2 = 0.025 \text{ and } 1 - \alpha/2 = 0.975)$ and degrees of freedom *df*. The *power* is calculated with the CDF function. This is a cumulative function of the *t* distribution, which needs the critical value, degrees of freedom and the noncentrality parameter *lambda* to be defined. As an alternative to *CDF('t',tcrit,df,lambda)* the function *PROBT(tcrit,df,lambda)* can also be used. The PRINT procedure gives the following:

alpha	n1	n2	mi1	mi2	stdev1	stdev2	df	sp	lambda
0.05	7	7	21.8	26.4	4.1	5.9	12	5.08035	1.69394

tcrit_low	tcrit_up	tcrit_one_sided	power_one_sided	power_two_sided
-2.17881	2.17881	1.78229	0.48118	0.34450

Thus, the powers are 0.48118 for the one-sided and 0.34450 for the two-sided tests.

The same analysis can be done by using the POWER procedure (SAS version 9.1 and higher).

SAS program:

```
PROC POWER;
   TWOSAMPLEMEANS
     TEST = DIFF
     ALPHA = 0.05
     GROUPMEANS = 21.8 | 26.4
     STDEV = 5.08
     NPERGROUP = 7
     SIDES = 1 2
     POWER = .
   ;
RUN;
```

Explanation: The POWER procedure is used with TWOSAMPLEMEANS and TEST = DIFF statements indicating a two sample test. Values are from the milk production example shown above, with means for A and B diets of 21.8 and 26.4 kg, respectively, the standard deviation for milk production of 5.08 kg, alpha of the test = 0.05, and number per group *NPERGROUP* = 7. *SIDES* = 1 2 asks for one- and two-sided tests to be performed. *POWER* = . directs the power of test to be calculated by the program.

SAS output:

```
Two-sample t Test for Mean
          Difference
```

Distribution	Normal
Method	Exact
Alpha	0.05
Group 1 Mean	21.8
Group 2 Mean	26.4
Standard Deviation	5.08
Sample Size Per Group	7
Null Difference	0

Computed Power

Index	Sides	Power
1	1	0.481
2	2	0.345

6.12   Sample Size

In many experiments the primary goal is to estimate the population mean from a normal distribution. What is the minimum sample size required to obtain a confidence interval of 2δ measurement units? To do so requires solving the inequality:

$$z_{\alpha/2}\frac{\sigma}{\sqrt{n}} \leq \delta$$

Rearranging:

$$n \geq \left(z_{\alpha/2}\frac{\sigma}{\delta}\right)^2$$

Here:

n = required sample size
$z_{\alpha/2}$ = the value of a standard normal variable determined with $\alpha/2$
δ = one-half the confidence interval
σ = the population standard deviation

With a similar approach, the sample size can be determined for the difference between population means, the population regression coefficient, etc.

In determining the sample size needed for rejection of a null hypothesis, type I and type II errors must be taken into consideration. An estimate of sample size depends on:

1. The minimum size of difference that is desired for detection
2. The variance
3. The power of test $(1 - \beta)$, or the certainty with which the difference is detected
4. The significance level, which is the probability of type I error
5. The type of statistical test.

Expressions to calculate the sample size needed to obtain a significant difference with a given probability of type I error and power can be derived from the formulas for calculation of power, as shown on the previous pages.

An expression for a one-sided test of a population mean is:

$$n = \frac{\left(z_\alpha - z_\beta\right)^2}{\delta^2}\sigma^2$$

An expression for a one-sided test of the difference of two population means is:

$$n = \frac{\left(z_\alpha - z_\beta\right)^2}{\delta^2}2\sigma^2$$

where:

n = required sample size
z_α = the value of a standard normal variable determined with α probability of type I error
z_β = the value of a standard normal variable determined with β probability of type II error
δ = the desired minimum difference which can be declared to be significant
σ^2 = the variance

For a two-sided test replace z_α with $z_{\alpha/2}$ in the expressions.

The variance σ^2 can be taken from the literature or similar previous research. Also, if the range of data is known, the variance can be estimated from:

$$\sigma^2 = [(\text{range}) / 4]^2$$

Example: What is the sample size required in order to show that a sample mean of 4100 kg for milk yield is significantly larger than 4000 kg? It is known that the standard deviation is 800 kg. The desired level of significance is 0.05 and power is 0.80.

$\mu_0 = 4000$ kg and $\bar{y} = 4100$ kg, thus the difference is $\delta = 100$ kg, and $\sigma^2 = 640000$

For $\alpha = 0.05$, $z_\alpha = 1.65$
For power $(1 - \beta) = 0.80$, $z_\beta = -0.84$
Then:

$$n = \frac{\left(z_\alpha - z_\beta\right)^2}{\delta^2}\sigma^2 = \frac{\left(1.65 + 0.84\right)^2}{100^2}640000 = 396.8$$

Thus, 397 cows are needed to have an 80% chance of proving that a difference as small as 100 kg is significant with $\alpha = 0.05$.

For observations drawn from a normal population and when the variance is unknown, the required sample size can be determined by using a noncentral t distribution. If the variance is unknown, the difference and variability are estimated.

6.12.1 SAS Examples for Sample Size

The required sample size for a t test can be determined by using SAS. The simple way to do that is to calculate powers for different sample sizes n. The smallest n resulting in power greater than that desired is the required sample size.

Example: Using the example of milk yield of dairy cows with a sample mean of 4300 kg and standard deviation of 600 kg, determine the sample size required to find the sample mean significantly different from 4000 kg, with the power of 0.80 and level of significance of 0.05.

SAS program:

```
DATA a;
DO n = 2 TO 100;
alpha=0.05;
mi0=4000;
mi1=4300;
stdev=600;
df=n-1;
lambda=(abs(mi1-mi0)/stdev)*sqrt(n);
tcrit_one_sided=TINV(1-alpha,df);
tcrit_low=TINV(alpha/2,df);
tcrit_up=TINV(1-alpha/2,df);
```

```
power_one_sided=1-CDF('t',tcrit_one_sided,df,lambda);
power_two_sided=CDF('t',tcrit_low,df,lambda)+ 1-CDF('t',tcrit_up,df,lambda);
OUTPUT;
END;
PROC PRINT data=a (obs=1 );
TITLE 'one-sided';
WHERE power_one_sided > .80;
VAR alpha n df power_one_sided;
RUN;
PROC PRINT data=a (obs=1 );
TITLE 'two-sided';
WHERE power_two_sided > .80;
VAR alpha n df power_two_sided;
RUN;
```

Explanation: The statement, *DO n = 2 to 100,* directs calculation of the power for sample sizes from 2 to 100. The following are defined: *alpha* = significance level, *n* = sample size, *mi0* = μ_0 = the population mean if H_0 is true, *mi1* = $\bar{y}$ = μ_1 = the population mean if H_1 is true, *stdev* = the sample standard deviation.

SAS output:

```
                    one-sided

    Obs    alpha     n     df    power_one_sided
    26     0.05     27     26       0.81183

                    two-sided

    Obs    alpha     n     df    power_two_sided
    33     0.05     34     33       0.80778
```

To find the the difference between these samples significant with 0.05 level of significance and the power of 0.80, the required sample sizes are at least 27 and 34 for one- and two-sided tests, respectively.

The same analysis can be done by using the POWER procedure (SAS version 9.1 or higher).

SAS program:

```
PROC POWER;
  ONESAMPLEMEANS
    TEST = T
    ALPHA = 0.05
    NULLMEAN = 4000
    MEAN = 4300
    STDEV = 600
    SIDES = 1 2
    NTOTAL = .
    POWER = 0.80
    ;
RUN;
```

Explanation: The POWER procedure is used with ONESAMPLEMEANS and TEST = T statements indicating a one sample *t* test. The sample mean is 4300 kg, and the sample standard deviation is 600. The hypothetical population mean is *NULLMEAN* = 4000 kg. Alpha of the test = 0.05, and *POWER* = 0.80. *SIDES* = 1 2 asks for one- and two-sided tests to be performed. *NTOTAL* = . directs number of observations required to be calculated by the program.

SAS output:

```
  One-sample t Test for Mean


  Distribution             Normal
  Method                    Exact            Computed N Total
  Null Mean                  4000
  Alpha                      0.05                       Actual      N
  Mean                       4300      Index   Sides    Power    Total
  Standard Deviation          600        1       1      0.812      27
  Nominal Power               0.8        2       2      0.808      34
```

Explanation: Input parameters of the program are listed first. For one- and two-sided tests the numbers of observations required are 27 and 34, respectively, and yield powers of 0.812 and 0.808.

Second example: Two groups of eight cows were fed two different diets (*A* and *B*) in order to test the difference in milk yield. From the samples the following was calculated:

	Diet *A*	Diet *B*
Mean ($\bar{y}$)	21.8 kg	26.4
Standard deviation (*s*)	4.1	5.9
Number of cows (*n*)	7	7

Determine the sample sizes required to find the samples means significantly different with power of 0.80 and level of significance $\alpha = 0.05$.

SAS program:

```
DATA aa;
DO n = 2 to 100;
alpha=0.05;
mi1=21.8;
mi2=26.4;
stdev1=4.1;
stdev2=5.9;
df=2*n-2;
sp = sqrt(((n-1)*stdev1*stdev1+(n-1)*stdev2*stdev2)/(n+n-2));
lambda=(abs(mi2-mi1)/sp)/sqrt(1/n+1/n);
tcrit_low=TINV(alpha/2,df);
tcrit_up=TINV(1-alpha/2,df);
tcrit_one_sided=TINV(1-alpha,df);
power_one_sided=1-CDF('t',tcrit_one_sided,df,lambda);
power_two_sided=CDF('t',tcrit_low,df,lambda)+ 1-CDF('t',tcrit_up,df,lambda);
OUTPUT;
END;
```

```
PROC PRINT DATA=aa (obs=1 );
TITLE 'one-sided';
WHERE power_one_sided > .80;
VAR alpha n df power_one_sided;
RUN;
PROC PRINT DATA=aa (obs=1 );
TITLE 'two-sided';
WHERE power_two_sided > .80;
VAR alpha n df power_two_sided;
RUN;
```

Explanation: The statement, *DO n = 2 to 100*; directs calculation of power for sample sizes from 2 to 100. The following are defined: *alpha* = significance level, *mi1* = μ_1 = $\bar{y}_1$ = the mean of population 1, *mi2* = μ_2 = $\bar{y}_2$ = the mean of population 2, *stdev1* = the standard deviation of sample 1 and *stdev2* = the standard deviation of sample 2, *df* = degrees of freedom. *sp* calculates the pooled standard deviation. Next, the noncentrality parameter (*lambda*) and critical values (*tcrit_one_sided* for a one-sided test, and *tcrit_low* and *tcrit_up* for a two-sided test) are calculated. The critical value is calculated by using the TINV function with cumulative values of percentiles ($1 - \alpha = 0.95$, $\alpha/2 = 0.025$ and $1 - \alpha/2 = 0.975$) and degrees of freedom *df*. The *power* is calculated with the CDF function. This is a cumulative function of the *t* distribution, which needs the critical value, degrees of freedom and the noncentrality parameter *lambda* to be defined. As an alternative to *CDF('t',tcrit,df,lambda)* the function *PROBT(tcrit,df,lambda)* can be used. The PRINT procedures give the following SAS output:

SAS output:

	one-sided				two-sided		
alpha	n	df	power_one_sided	alpha	n	df	power_two_sided
0.05	16	30	0.80447	0.05	21	40	0.81672

In order for the difference between the two sample means to be significant with $\alpha = 0.05$ level of significance and power of 0.80, the required sizes for each sample are at least 16 and 21 for one- and two-sided tests, respectively.

The same analysis can be done by using the POWER procedure (SAS version 9.1 or higher):

SAS program:

```
PROC POWER;
    TWOSAMPLEMEANS
      TEST = DIFF
      ALPHA = 0.05
      GROUPMEANS = 21.8 | 26.4
      STDEV = 5.08
      NPERGROUP = .
      SIDES = 1 2
      POWER = 0.80
      ;
RUN;
```

Explanation: The POWER procedure is used with TWOSAMPLEMEANS and TEST = DIFF statements indicating a two sample test. Values are from the milk production example shown above, with means on the old and new diets of 21.8 and 26.4 kg, respectively, the standard deviation for milk production of 5.08 kg, alpha of the test = 0.05, and *POWER* = 0.80. *SIDES* = 1 2 asks for one- and two-sided tests to be performed. *NPERGROUP* = . directs number of observations per group to be calculated by the program.

SAS output:

```
Two-sample t Test for Mean
            Difference


Distribution            Normal
Method                  Exact
Alpha                    0.05           Computed N Per Group
Group 1 Mean            21.8
Group 2 Mean            26.4                    Actual    N Per
Standard Deviation      5.08    Index   Sides   Power    Group
Nominal Power            0.8      1       1      0.805     16
Null Difference           0       2       2      0.817     21
```

Explanation: Input parameters of the program are listed first. For one- and two-sided tests the numbers of observations required are 16 and 21, respectively, and yield powers of 0.805 and 0.817.

6.13 Equivalence Tests

So far in this chapter the main goal has been to determine if evidence is present of a difference between populations. For example, in research comparing two populations for difference between their means, the null hypothesis has been that these two means were equal. We have been trying to find enough evidence that our conclusion regarding rejection of the null hypothesis has a small probability of being false. If the null hypothesis is not rejected, this means that there is not enough evidence to do so, not that we can conclude there is no difference.

We may be interested in concluding that there is no difference between means despite different treatments applied to animals. For example, if cows are fed a restrictive diet we might like to see no decrease in milk production, or at least not a practically significant decrease. Or, we might consider using a less expensive type of litter for chickens if it does not result in a practical decrease in daily gain.

The most common misconception is to conclude that there is no difference between populations because the null hypothesis, defined as no difference between treatments, was not rejected. It is not correct to conclude that two treatments (or in statistical terms two populations) are equal, when samples from them are not found to be significantly different. They may not be found to be significantly different because of unknown type II error. It may be that the sample was not large enough, and because of lack of evidence we could not conclude that there was a difference.

Recall that usually the statistical alternative hypothesis H_1 is identical to the research hypothesis, and thus the null hypothesis is opposite to what a researcher expects. It is generally easier to prove a hypothesis is false than that it is true, thus a researcher usually attempts to reject H_0. In the example with the litter and chickens, we hope to prove that use of a different litter would give the same or similar daily gain. In an equivalence test, a difference is defined as the hypothesis to refute. By disproving that hypothesis we can prove that there is no difference.

The procedure is to define H_0 as a difference between populations (treatments). The reasoning is as follows: by giving the animals different treatments they might be considered to be divergent in their means, μ_1 and μ_2, by at least some fixed value, say d. Then the null hypothesis is:

$$H_0: \mu_1 - \mu_2 \geq d \quad \text{or} \quad \mu_1 - \mu_2 \leq -d$$

The alternative hypothesis is what we hope to prove, that the means do not differ by more than the fixed value d. Then the alternative hypothesis is:

$$H_1: -d < \mu_1 - \mu_2 < d$$

This means that the difference between treatment means is within some interval ($-d$ to d). Note that we do not attempt to prove that the means are equal. We cannot expect that any two groups of animals are exactly equal. Because of unexplained effects on the two groups they can be unequal regardless of treatments. The question is what is the maximum difference that we can say is not of practical importance? That is defined as d. What value of d to use depends on the particular variable. For daily gain of pigs it may be that a difference of 50 g is of no noteworthy importance, but larger than that would be of economic importance. Then $d = 50$.

It can be difficult to define d, as the value considered as a maximum negligible difference can change from one experiment to another. Alternatively, the value of d can be defined using a percentage or proportion of the mean: d / μ_1.

For example, if the mean of daily gain is 600 g, and we define a difference of 5% of the mean as being of no practical difference, then we can use +/- 5% as limits of maximum negligible difference, thus the interval in grams is ($d = 600 \times 0.05 = 30$ g). Limits of 570 and 630 g would be considered not practically different than 600 g.

To test the hypothesis of equivalence a t test can be used. The hypotheses can be defined as two one-sided hypotheses for $-d$ and d:

$$H_0: \mu_1 - \mu_2 \leq -d \qquad \text{or} \qquad H_0: \mu_1 - \mu_2 \geq d$$
$$H_1: \mu_1 - \mu_2 > -d \qquad\qquad H_1: \mu_1 - \mu_2 < d$$

We conclude no difference, or equivalence, between the populations if both of the above null hypotheses are rejected.

Example: The following experiment was conducted to test if change to a new, cheaper diet will not substantially change daily milk yield in dairy cows. The following values were calculated:

	Old diet (y_1)	New diet (y_2)
Mean ($\bar{y}$, kg)	32	30
Standard deviation (s)	5	6
Size (n)	12	12

Considering the savings resulting from the reduced cost of the new diet, it was reasonable to conclude that a difference within ±10% of the mean of the old diet could be considered of no practical importance. This gives $d = (32)(0.1) = 3.2$ kg.

The two *t* statistics are:

$$t = \frac{d - (\bar{y}_1 - \bar{y}_2)}{\sqrt{s_p^2 \left(\frac{1}{n_1} + \frac{1}{n_2} \right)}} \quad \text{if } (\bar{y}_1 - \bar{y}_2) > 0 \quad \text{and}$$

$$t = \frac{-d - (\bar{y}_1 - \bar{y}_2)}{\sqrt{s_p^2 \left(\frac{1}{n_1} + \frac{1}{n_2} \right)}} \quad \text{if } (\bar{y}_1 - \bar{y}_2) < 0$$

where $\bar{y}_1$ and $\bar{y}_2$ are the sample means, n_1 and n_2 are sample sizes, and s_p^2 is the pooled variance:

$$s_p^2 = \frac{(n_1 - 1)s_1^2 + (n_2 - 1)s_2^2}{n_1 + n_2 - 2}$$

The estimated pooled variance is:

$$s_p^2 = \frac{(12 - 1)(5)^2 + (12 - 1)(6)^2}{12 + 12 - 2} = 30.5$$

The calculated value of the *t* statistics are:

$$t = \frac{3.2 - (32 - 30)}{\sqrt{30.5 \left(\frac{1}{12} + \frac{1}{12} \right)}} = 0.53 \quad \text{and } t = \frac{-3.2 - (32 - 30)}{\sqrt{30.5 \left(\frac{1}{12} + \frac{1}{12} \right)}} = -2.31$$

The critical value is $t_\alpha = t_{0.05} = 1.717$. Since the calculated value of $t = 0.53$ for the first test is not more extreme than the critical value, the null hypothesis that H_0: $\mu_1 - \mu_2 \geq d$ cannot be rejected with a 0.05 level of significance. For these two means to have been determined to be equivalent, the mean from the new diet must have been found to be significantly greater than the mean of the old diet – *d*, or 32 – 3.2, in the first test, and significantly less than the mean of the old diet + *d*, or 32 + 3.2, in the second test. Since we failed to reject H_0 in the first test, we cannot conclude that the two means are equivalent.

Recall the usual questions concerning evidence to reach a proper conclusion. What is the probability of type II error? What is the sample size? What is the variability?

Example: What is the power of the equivalence test shown above?

The power of the test or the required sample size to have a particular power of test can be calculated by using the POWER procedure of SAS. In the example shown below we calculate the number of observations required to find equivalence between the old and new diet means in the example of the effects of diet on milk production using power of 0.80 and significance level $\alpha = 0.05$.

SAS program:

```
PROC POWER;
    TWOSAMPLEMEANS
      TEST = EQUIV_DIFF
      GROUPMEANS = 32 | 30
      STDDEV = 5.52
      ALPHA = 0.05
      LOWER = -3.2
      UPPER = 3.2
      NPERGROUP = .
      POWER = 0.8
      ;
RUN;
```

Explanation: The POWER procedure is used with TWOSAMPLEMEANS and TEST = EQUIV_DIFF statements indicating a two sample equivalence test. Values are drawn from the milk production example shown above, with means on the old and new diets of 32 and 30 kg, respectively, the standard deviation for milk production of $\sqrt{30.5} = 5.52$ kg, $d = 10\%$ of the old mean = 3.2 kg, alpha of the test = 0.05, and *POWER* = 0.80. *NPERGROUP* = . directs number per group to be calculated by the program.

SAS output:

```
Equivalence Test for Mean Difference

Distribution                    Normal
Method                          Exact
Lower Equivalence Bound         -3.2
Upper Equivalence Bound          3.2
Alpha                           0.05
Reference Mean                    32          Computed N Per Group
Treatment Mean                    30
Standard Deviation              5.52        Actual      N Per
Nominal Power                    0.8        Power       Group
                                            0.801        263
```

Explanation: The distribution of the data is assumed to be normal. The calculated N per group to have yielded a significant equivalence between the old and new means for the milk production example with alpha = 0.05, power = 0.80 and parameters as shown above is 263 cows per diet.

Exercises

6.1. The mean of a sample is 24 and the standard deviation is 4. Sample size is $n = 50$. Is there sufficient evidence to conclude that this sample does not belong to a population with mean = 25?

6.2. For two groups, A and B, the following measurements have been recorded:

A	120	125	130	131	120	115	121	135	115
B	135	131	140	135	130	125	139	119	121

Is the difference between group means significant at the 5% level? State the appropriate hypotheses, test the hypotheses, and write a conclusion.

6.3. Is the difference between the means of two samples A and B statistically significant if the following values are known:

Group	A	B
Sample size	22	22
Arithmetic mean	20	25
Sample standard deviation	2	3

6.4. In an experiment 120 cows were treated five times and the number of positive responses is shown below. The expected proportion of positive responses is 0.4. Is it appropriate to conclude that this sample follows a binomial distribution with $p \neq 0.4$?

The number of positive responses	0	1	2	3	4	5
Number of cows	6	20	42	32	15	5

6.5. The progeny resulting from crossing two rabbit lines consist of 510 gray and 130 white rabbits. Is there evidence to conclude that the hypothetical ratio between gray and white rabbits is different than 3:1?

6.6. The expected proportion of cows with a defective udder is 0.2 (or 20%). In a sample of 60 cows, 20 have the udder defect. Is there sufficient evidence to conclude that the proportion in the sample is significantly different from the expected proportion?

6.7. Two groups of 60 sheep received different diets. During the experiment 18 and 5 sheep from the first and the second groups, respectively, experienced digestion problems. Is it appropriate to conclude that the number of sheep that were ill is the result of different treatments or the differences are accidental?

Chapter 7

Simple Linear Regression

It is often of interest to determine how changes of values of one variable influence the change of values of another variable, for example, how alteration of air temperature affects feed intake, or how increasing the protein level in a feed affects daily gain. In both examples, the relationship between variables can be described with a function, a function of temperature to describe feed intake, or a function of protein level to describe daily gain. A function that explains such relationships is called a regression function, and analysis of such problems and estimation of the regression function is called regression analysis. Regression includes a set of procedures designed to study statistical relationships among variables such that one variable is defined as dependent upon others defined as independent variables. By using regression the cause-consequence relationship between the independent and dependent variables can be determined. In the examples above, feed intake and daily gain are dependent variables, and temperature and protein level are independent variables. The dependent variable is usually denoted by y, and the independent variables by x. The dependent variable is often called the response variable, and the independent variables are called regressors or predictors. When the change of the dependent variable is described with just one independent variable and the relationship between them is linear, the appropriate procedures are called simple linear regression. Regression procedures used when the change of a dependent variable is explained by changes of two or more independent variables are called multiple regression.

Two main applications of regression analysis are:

1. Estimation of a function of dependency between variables
2. Prediction of future measurements or means of the dependent variable using new measurements of the independent variable(s).

7.1   The Simple Regression Model

A regression that explains linear change of a dependent variable based on changes of one independent variable is called a simple linear regression. For example, the weight of cows can be predicted by using measurements of heart girth. The aim is to determine a linear function that will explain changes in weight as heart girth changes. Hearth girth is the independent variable and weight is the dependent variable. To estimate the function it is necessary to choose a sample of cows and to measure the heart girth and weight of each cow. In other words, pairs of measurements of the dependent variable y and independent variable x are needed. Let the symbols y_i and x_i denote the measurements of weight and heart girth for animal i. For n animals in this example the measurements are:

Animal number	1	2	3	...	n
Heart girth (x)	x_1	x_2	x_3	...	x_n
Weight (y)	y_1	y_2	y_3	...	y_n

In this example it can be assumed that the relationship between the x and y variables is linear and that each value of variable y can be shown using the following model:

$$y = \beta_0 + \beta_1 x + \varepsilon$$

where:

  y = dependent variable
  x = independent variable
  β_0, β_1 = regression parameters
  ε = random error

Here, β_0 and β_1 are unknown constants called regression parameters. They describe the location and shape of the linear function. The parameter β_1 is often called the regression coefficient, because it explains the slope. The random error ε is included in the model because changes of the values of the dependent variable are usually not completely explained by changes of values of the independent variable, but there is also an unknown part of that change. The random error describes deviations from the model due to factors unaccounted for in the equation, for example, differences among animals, environments, measurement errors, etc. Generally, a mathematical model in which we allow existence of random error is called a statistical model. If a model exactly describes the dependent variable by using a mathematical function of the independent variable, the model is deterministic. For example, if the relationship is linear the model is:

$$y = \beta_0 + \beta_1 x$$

Note again, the existence of random deviations is the main difference between a deterministic and a statistical model. In a deterministic model the x variable exactly explains the y variable, and in a statistical model the x variable explains the y variable, but with random error.

A regression model uses pairs of measurements (x_1, y_1), (x_2, y_2),..., (x_n, y_n). According to the model each observation y_i can be shown as:

$$y_i = \beta_0 + \beta_1 x_i + \varepsilon_i \qquad i = 1,..., n$$

that is:

$$y_1 = \beta_0 + \beta_1 x_1 + \varepsilon_1$$
$$y_2 = \beta_0 + \beta_1 x_2 + \varepsilon_2$$
...
$$y_n = \beta_0 + \beta_1 x_n + \varepsilon_n$$

For example, in a population of cows it is assumed that weights can be described as a linear function of heart girth. If the variables' values are known, for example:

Weight (y):	641	633	651	...	...
Heart girth (x):	214	215	216	...	...

The measurements of the dependent variable y can be expressed as:

$$641 = \beta_0 + \beta_1 214 + \varepsilon_1$$
$$633 = \beta_0 + \beta_1 215 + \varepsilon_2$$
$$651 = \beta_0 + \beta_1 216 + \varepsilon_3$$
...

For a regression model, assumptions and properties also must be defined. The assumptions describe expectations and variance of random error.

Model assumptions:

1. $E(\varepsilon_i) = 0$, mean of errors is equal to zero
2. $Var(\varepsilon_i) = \sigma^2$, variance is constant for every ε_i, that is, variance is homogeneous
3. $Cov(\varepsilon_i, \varepsilon_{i'}) = 0$, $i \neq i'$, errors are independent, the covariance between them is zero
4. Usually, it is assumed that ε_i are normally distributed, $\varepsilon_i \sim N(0, \sigma^2)$. When that assumption is met the regression model is said to be normal.

The following model properties follow directly from these model assumptions.

Model properties:

1. $E(y_i|x_i) = \beta_0 + \beta_1 x_i$ for a given value x_i, the expected mean of y_i is $\beta_0 + \beta_1 x_i$
2. $Var(y_i) = \sigma^2$, the variance of any y_i is equal to the variance of ε_i and is homogeneous
3. $Cov(y_i, y_{i'}) = 0$, $i \neq i'$, y are independent, the covariance between them is zero.

The expectation (mean) of the dependent variable y for a given value of x, denoted by $E(y|x)$, is a straight line (Figure 7.1). Often, the mean of y for given x is also denoted by $\mu_{y|x}$.

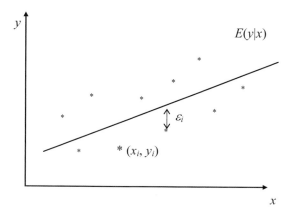

Figure 7.1 Linear regression. Dots represent observations (x_i, y_i). The line $E(y|x)$ shows the expected value of dependent variable. The errors ε_i are the deviation of the observations from their expected values

An interpretation of parameters is shown in Figure 7.2. The expectation or mean of y for a given x, $E(y_i|x_i) = \beta_0 + \beta_1 x_i$, represents a straight line; β_0 denotes the intercept, a value of

$E(y_i|x_i)$ when $x_i = 0$; β_1 describes the slope of the line, this is the change $\Delta E(y_i|x_i)$ when the value of x is increased by one unit. Also:

$$\beta_1 = \frac{Cov(x, y)}{Var(x)}$$

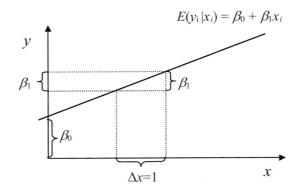

Figure 7.2 Interpretation of parameters for simple linear regression

A simple linear regression can be positive or negative (Figure 7.3). A positive regression, $\beta_1 > 0$, is represented by an upward sloping line and y increases as x is increased. A negative regression, $\beta_1 < 0$, is represented by a downward sloping line and y decreases as x is increased. A regression with slope $\beta_1 = 0$ indicates no linear relationship between the variables.

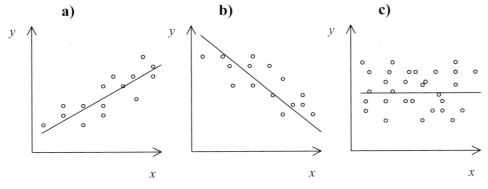

Figure 7.3 a) positive regression, $\beta_1 > 0$; b) negative regression, $\beta_1 < 0$, c) no linear relationship, $\beta_1 = 0$

7.2   Estimation of the Regression Parameters – Least Squares Estimation

Regression parameters of a population are usually unknown and estimated from data collected on a sample of the population. The aim is to determine a line that will best describe the linear relationship between the dependent and independent variables given data from the sample. Parameter estimators are usually denoted by $\hat{\beta}_0$ and $\hat{\beta}_1$ or b_0 and b_1. Therefore, the regression line $E(y|x)$ is unknown, but can be estimated by using a sample with:

$$\hat{y}_i = \hat{\beta}_0 + \hat{\beta}_1 x_i \qquad \text{or}$$

$$\hat{y}_i = b_0 + b_1 x_i$$

This line is called the estimated or fitted line, or more generally, the estimated regression line or estimated model. The best fitted line will have estimates of b_0 and b_1 such that it gives the least possible cumulative deviations from the given y_i values from the sample. In other words the line is as close to the dependent variables as possible.

The most widely applied method for estimation of parameters in linear regression is least squares estimation. The least squares estimators b_0 and b_1 for a given set of observations from a sample are such estimators that the expression

$$\sum_i (y_i - \hat{y}_i)^2 = \sum_i [y_i - (b_0 + b_1 x_i)]^2 \quad \text{is minimized.}$$

To determine such estimators assume a function in which the observations x_i and y_i from a sample are known, and β_0 and β_1 are unknown variables:

$$\sum_i (y_i - \beta_0 - \beta_1 x_i)^2$$

This function is the sum of the squared deviations of the measurements from the values predicted by the line. The estimators of parameters β_0 and β_1, denoted b_0 and b_1, are determined such that the function will have the minimum value. Calculus is used to determine such estimators by finding the first derivative of the function with respect to β_0 and β_1:

$$\frac{\partial}{\partial \beta_0} \left[\sum_i (y_i - \beta_0 - \beta_1 x_i)^2 \right] = -2 \sum_i (y_i - \beta_0 - \beta_1 x_i)$$

$$\frac{\partial}{\partial \beta_1} \left[\sum_i (y_i - \beta_0 - \beta_1 x_i)^2 \right] = -2 \sum_i x_i (y_i - \beta_0 - \beta_1 x_i)$$

The estimators, b_0 and b_1, are substituted for β_0 and β_1 such that:

$$-2 \sum_i (y_i - b_0 - b_1 x_i) = 0$$

$$-2 \sum_i x_i (y_i - b_0 - b_1 x_i) = 0$$

With simple arithmetic operations we can obtain a system of two equations with two unknowns, usually called normal equations:

$$nb_0 + b_1 \sum_i x_i = \sum_i y_i$$

$$b_0 \sum_i x_i + b_1 \sum_i x_i^2 = \sum_i x_i y_i$$

The estimators, b_1 and b_0, are obtained by solving the equations:

$$b_1 = \frac{SS_{xy}}{SS_{xx}}$$

$$b_0 = \bar{y} - b_1 \bar{x}$$

where:

$$SS_{xy} = \sum_i (x_i - \bar{x})(y_i - \bar{y}) = \sum_i x_i y_i - \frac{\left(\sum_i x_i\right)\left(\sum_i y_i\right)}{n} = \text{sum of products of } y \text{ and } x.$$

$$SS_{xx} = \sum_i (x_i - \bar{x})^2 = \sum_i x_i^2 - \frac{\left(\sum_i x_i\right)^2}{n} = \text{sum of squares of } x.$$

n = sample size
$\bar{y}$ and $\bar{x}$ = arithmetic means

Recall that $\hat{y}_i = b_0 + b_1 x_i$ describes the estimated line. The difference between the measurement y_i and the estimated value $\hat{y}_i$ is called the residual and is denoted with e_i (Figure 7.4):

$$e_i = y_i - \hat{y}_i = \left[y_i - (b_0 + b_1 x_i)\right]$$

Each observation in the sample can be written as:

$$y_i = b_0 + b_1 x_i + e_i \qquad\qquad i = 1,..., n$$

Again, the estimators, b_1 and b_0, are such that the sum of squared residuals $\sum_i (e_i)^2$ is minimal compared to any other set of estimators.

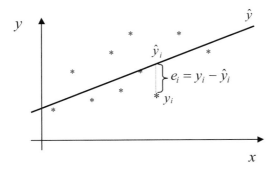

Figure 7.4 Estimated or fitted line of the simple linear regression

Example: Estimate the simple regression line of weight on heart girth of cows using the following sample:

Cow	1	2	3	4	5	6
Weight (y):	641	633	651	666	688	680
Heart girth (x):	214	215	216	217	219	221

Each measurement y_i in the sample can be expressed as:

$$641 = b_0 + b_1 214 + e_1$$
$$633 = b_0 + b_1 215 + e_2$$
$$651 = b_0 + b_1 216 + e_3$$
$$666 = b_0 + b_1 217 + e_4$$
$$688 = b_0 + b_1 219 + e_5$$
$$680 = b_0 + b_1 221 + e_6$$

To estimate the regression line the coefficients b_0 and b_1 must be calculated by using the sums $\sum_i x_i$ and $\sum_i y_i$, sum of squares $\sum_i x^2_i$ and sum of products $\sum_i x_i y_i$ as shown in the following table:

Weight (y)	Heart girth (x)	x^2	xy
641	214	45796	137174
633	215	46225	136095
651	216	46656	140616
666	217	47089	144522
688	219	47961	150672
680	221	48841	150280
Sums 3959	1302	282568	859359

$$n = 6$$
$$\sum_i x_i = 1302$$
$$\sum_i x^2_i = 282568$$
$$\sum_i y_i = 3959$$
$$\sum_i x_i y_i = 859359$$

$$SS_{xy} = \sum_i x_i y_i - \frac{\left(\sum_i x_i\right)\left(\sum_i y_i\right)}{n} = 859359 - \frac{(1302)(3959)}{6} = 256$$

$$SS_{xx} = \sum_i x^2_i - \frac{\left(\sum_i x_i\right)^2}{n} = 282568 - \frac{(1302)^2}{6} = 34$$

$$b_1 = \frac{SS_{xy}}{SS_{xx}} = \frac{254}{34} = 7.53$$

$$b_0 = \bar{y} - b_1\bar{x} = -974.05$$

The estimated line is:

$$\hat{y}_i = -974.05 + 7.53\, x_i$$

The observed and estimated values are shown in Figure 7.5. This figure provides information about the nature of the data, possible relationship between the variables, and about the adequacy of the model.

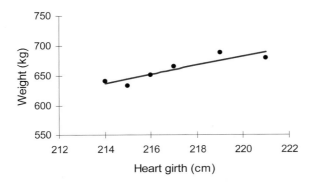

Figure 7.5 Regression of weight on heart girth of cows. Dots represent measured values

7.3 Residuals and Their Properties

Residual analysis can give useful information about the validity of a model. Residuals are values that can be thought of as 'errors' of the estimated model. Recall that an error of the true model is:

$$\varepsilon_i = y_i - E(y_i)$$

A residual is defined as:

$$e_i = y_i - \hat{y}_i$$

The residual sum of squares is:

$$SS_{RES} = \sum_i \left(y_i - \hat{y}_i\right)^2$$

Properties of residuals are:

1. $\sum_i \left(y_i - \hat{y}_i\right) = 0$
2. $\sum_i e_i^2 = \sum_i \left(y_i - \hat{y}_i\right)^2 = minimum$

The residual mean square is the residual sum of squares divided by its associated degrees of freedom, and is denoted by MS_{RES} or s^2:

$$MS_{RES} = s^2 = \frac{SS_{RES}}{n-2}$$

where $(n-2)$ is the degrees of freedom. The mean square $MS_{RES} = s^2$ is an estimator of the error variance in a population, $\sigma^2 = Var(\varepsilon)$. The square root of the mean square,

$s = \sqrt{s^2} = \sqrt{\dfrac{SS_{RES}}{n-2}}$, is called a standard deviation of the regression model.

A practical rule for determining the degrees of freedom is:

$n -$ (*number of parameters estimated for a particular sum of squares*), or
$n -$ (*number of restrictions associated with regression*)

In estimating a simple regression there are two restrictions:

1. $\sum_i (y_i - \hat{y}_i) = 0$
2. $\sum_i (y_i - \hat{y}_i) x_i = 0$

Also, two parameters, β_0 and β_1, are estimated, and consequently the residual degrees of freedom are $(n-2)$.

The expectation of a residual is:

$E(e_i) = 0$

The variance of residuals is not equal to the error variance, $Var(e_i) \neq \sigma^2$. The residual variance depends on x_i. For large n, $Var(e_i) \approx \sigma^2$, which is estimated by s^2, that is, $E(s^2) = \sigma^2$. Also, the covariance $Cov(e_i, e_{i'}) \neq 0$, but for large n, $Cov(e_i, e_{i'}) \approx 0$.

Example: For the example with weights and heart girths of cows, the residuals, squares of residuals, and sum of squares are shown in the following table:

	y	x	$\hat{y}$	e	e^2
	641	214	637.25	3.75	14.099
	633	215	644.77	−11.77	138.639
	651	216	652.30	−1.30	1.700
	666	217	659.83	6.17	38.028
	688	219	674.89	13.11	171.816
	680	221	689.95	−9.95	99.022
Sum	3959	1302	3959.0	0.0	463.304

The residual sum of squares is:

$$SS_{RES} = \sum_i (y_i - \hat{y}_i)^2 = 463.304$$

The estimate of the variance is:

$$s^2 = MS_{RES} = \frac{SS_{RES}}{n-2} = \frac{463.304}{4} = 115.826$$

7.4 Maximum Likelihood Estimation

Parameters of linear regression can also be estimated by using a likelihood function. Under the assumption of normality of the dependent variable, the likelihood function is a function of the parameters (β_0, β_1 and σ^2) for a given set of n observations of dependent and independent variables:

$$L(\beta_0, \beta_1, \sigma^2 \mid y) = \frac{1}{\left(\sqrt{2\pi\sigma^2}\right)^n} e^{-\sum_i (y_i - \beta_0 - \beta_1 x_i)^2 / 2\sigma^2}$$

The log likelihood is:

$$logL(\beta_0, \beta_1, \sigma^2 \mid y) = -\frac{n}{2} log(\sigma^2) - \frac{n}{2} log(2\pi) - \frac{\sum_i (y_i - \beta_0 - \beta_1 x_i)^2}{2\sigma^2}$$

A set of estimators is estimated that will maximize the log likelihood function. Such estimators are called maximum likelihood estimators. The maximum of the function can be determined by taking the partial derivatives of the log likelihood function with respect to the parameters:

$$\frac{\partial\, logL(\beta_0, \beta_1, \sigma^2 \mid y)}{\partial \beta_0} = \frac{1}{\sigma^2} \sum_i (y_i - \beta_0 - \beta_1 x_i)$$

$$\frac{\partial\, logL(\beta_0, \beta_1, \sigma^2 \mid y)}{\partial \beta_1} = \frac{1}{\sigma^2} \sum_i x_i (y_i - \beta_0 - \beta_1 x_i)$$

$$\frac{\partial\, logL(\beta_0, \beta_1, \sigma^2 \mid y)}{\partial \sigma^2} = -\frac{n}{2\sigma^2} + \frac{1}{2\sigma^4} \sum_i (y_i - \beta_0 - \beta_1 x_i)^2$$

These derivatives are set equal to zero in order to find the estimators b_0, b_1 and s^2_{ML}. Three equations are obtained:

$$\frac{1}{s^2_{ML}} \sum_i (y_i - b_0 - b_1 x_i) = 0$$

$$\frac{1}{s^2_{ML}} \sum_i x_i (y_i - b_0 - b_1 x_i) = 0$$

$$-\frac{n}{2s^2_{ML}} + \frac{1}{2s^4_{ML}} \sum_i (y_i - b_0 - b_1 x_i)^2 = 0$$

By simplifying the equations the following results are obtained:

$$nb_0 + b_1 \sum_i x_i = \sum_i y_i$$

$$b_0 \sum_i x_i + b_1 \sum_i x_i^2 = \sum_i x_i y_i$$

$$s^2_{ML} = \frac{1}{n} \sum_i (y_i - b_0 - b_1 x_i)^2$$

Solving the first two equations results in estimators identical to least squares estimation; however, the maximum likelihood estimator of the variance is not unbiased. This is why it is denoted with $s^2{}_{ML}$. An unbiased estimator is obtained when the maximum likelihood estimator is multiplied by $n / (n - 2)$:

$$s^2 = MS_{RES} = \frac{n}{n-2} s^2_{ML}$$

7.5 Expectations and Variances of the Parameter Estimators

In most cases inferences are based on the estimators b_0 and b_1. For that reason it is essential to know properties of the estimators, their expectations and variances. The expectations of b_0 and b_1 are:

$$E(b_1) = \beta_1$$
$$E(b_0) = \beta_0$$

The expectations of the estimators are equal to parameters, which implies that the estimators are unbiased. The variances of the estimators are:

$$Var(b_1) = \sigma^2_{b_1} = \frac{\sigma^2}{SS_{xx}}$$

$$Var(b_0) = \sigma^2_{b_0} = \sigma^2 \left(\frac{1}{n} + \frac{\overline{x}}{SS_{xx}} \right)$$

Assuming that the y_i are normal, then b_0 and b_1 are also normal, because they are linear functions of y_i. Since the estimator of the variance σ^2 is s^2, the variance of b_1 can be estimated by:

$$s^2_{b_1} = \frac{s^2}{SS_{xx}}$$

A standard error of b_1 is:

$$s_{b_1} = \sqrt{\frac{s^2}{SS_{xx}}}$$

7.6 Student t Test in Testing Hypotheses about the Parameters

If changing variable x effects change in variable y, then the regression line has a slope, the parameter β_1 is different from zero. To test if there is a regression in a population the following hypotheses about the parameter β_1 are stated:

$$H_0: \beta_1 = 0$$
$$H_1: \beta_1 \neq 0$$

The null hypothesis H_0 states that slope of the regression line is not different from zero and that there is no linear association between the variables. The alternative hypothesis H_1 states that the regression line is not horizontal and there is linear association between the variables. Assuming that the dependent variable y is normally distributed, the hypotheses about the parameter β_1 can be tested using a t distribution. It can be proved that the test statistic

$$t = \frac{b_1 - 0}{\sqrt{s^2 / SS_{xx}}}$$

has a t distribution with $(n - 2)$ degrees of freedom under H_0. Note the form of the t statistic:

$$t = \frac{Estimator - Paramater\ (under\ H_0)}{Standard\ error\ of\ estimator}$$

The null hypothesis H_0 is rejected if the computed value from a sample $|t|$ is 'large'. For a level of significance α, H_0 is rejected if $|t| \geq t_{\alpha/2,(n-2)}$, where $t_{\alpha/2,(n-2)}$ is a critical value (Figure 7.6).

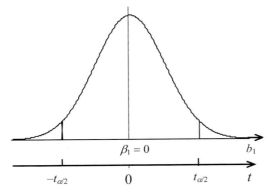

Figure 7.6 Theoretical distribution of the estimator b_1 and corresponding scale of the t statistic. Symbols $t_{\alpha/2}$ are the critical values

Example: Test the hypotheses about the regression of weights on heart girths of cows. The coefficient of regression was $b_1 = 7.53$, the residual sum of squares was $SS_{RES} = 463.304$, sum of squares of x was $SS_{xx} = 34$, and the estimated variance was:

$$s^2 = \frac{SS_{RES}}{n-2} = MS_{RES} = 115.826$$

The standard error of the estimator b_1 is:

$$s_{b_1} = \sqrt{\frac{s^2}{SS_{xx}}} = \sqrt{\frac{115.826}{34}} = 1.845$$

The calculated value of the t statistic from the sample is:

$$t = \frac{b_1 - 0}{\sqrt{s^2 / SS_{xx}}} = \frac{7.53}{1.845} = 4.079$$

The critical value is $t_{\alpha/2,(n-2)} = t_{0.025,4} = 2.776$ (See Appendix B: Critical values of Student t distributions).

The calculated $t = 4.079$ is more extreme than the critical value (2.776), thus the estimate of the regression slope $b_1 = 7.53$ is significantly different from zero and a linear relationship between weight and heart girth exists in the population.

7.7 Confidence Intervals of the Parameters

Recall that a confidence interval usually has the form:

Estimate ± standard error × value of standard normal or t variable for α/2

We have already stated that:

$$\frac{b_1 - \beta_1}{s_{b_1}}$$

has a t distribution. Here:

$$s_{b_1} = \sqrt{s^2/SS_{xx}}$$

is the standard error of the estimator b_1. It can be shown that:

$$P\left\{b_1 - t_{\alpha/2,n-2}s_{b_1} \leq \beta_1 \leq b_2 + t_{\alpha/2,n-2}s_{b_1}\right\} = 1 - \alpha$$

in which $t_{\alpha/2,n-2}$ is a critical value on the right side of the t distribution with $\alpha/2$. The $100(1 - \alpha)\%$ confidence interval is:

$$b_1 \pm t_{\alpha/2,n-2}s_{b_1}$$

It follows that a 95% confidence interval is:

$$b_1 \pm s_{b_1}t_{0.025,n-2}$$

Example: For the previous example of weight and heart girth of cows, construct a 95% confidence interval for β_1. The following parameters have been calculated, $\alpha = 0.05$, degrees of freedom = 4, $t_{0.025,4} = 2.776$, $s_{b_1} = 1.845$ and $b_1 = 7.529$.

The 95% confidence interval is:

$$b_1 \pm s_{b_1}t_{0.25,n-2}$$

$$7.529 \pm (1.845)(2.776) \qquad \text{or equal to}$$

$$(2.406,\ 12.654)$$

7.8   Mean and Prediction Confidence Intervals of the Response Variable

In regression analysis it is also important to estimate values of the dependent variable. Estimation of the dependent variable includes two approaches: a) estimation of the mean for a given value of the independent variable x_0; and b) prediction of future values of the variable y for a given value of the independent variable x_0.

The mean of variable y for a given value x_0 is $E[y|x_0] = \beta_0 + \beta_1 x_0$. Its estimator is $\hat{y}_0 = b_0 + b_1 x_0$. Assuming that the dependent variable y has a normal distribution, the estimator also has a normal distribution with mean $\beta_0 + \beta_1 x_0$ and variance:

$$Var(\hat{y}_0) = \sigma^2 \left[\frac{1}{n} + \frac{(x_0 - \bar{x})^2}{SS_{xx}} \right]$$

The standard error can be estimated from:

$$s_{\hat{y}_0} = \sqrt{s^2 \left[\frac{1}{n} + \frac{(x_0 - \bar{x})^2}{SS_{xx}} \right]}$$

Recall that SS_{xx} is the sum of squares of the independent variable x, and $s^2 = MS_{RES}$ is the residual mean square that estimates a population variance.

Prediction of the values of variable y given a value x_0 includes prediction of a random variable $y|x_0 = \beta_0 + \beta_1 x_0 + \varepsilon_0$. Note that $y|x_0$ is a random variable because it holds ε_0, compared to $E[y|x_0]$, which is a constant. An estimator of the new value is $\hat{y}_{0,NEW} = b_0 + b_1 x_0$. Assuming that the dependent variable y has a normal distribution, the estimator also has a normal distribution with mean $\beta_0 + \beta_1 x_0$ and variance:

$$Var(\hat{y}_{0,NEW}) = \sigma^2 \left[1 + \frac{1}{n} + \frac{(x_0 - \bar{x})^2}{SS_{xx}} \right]$$

Note that the estimators for both the mean and new values are the same. However, the variances are different. The standard error of the predicted values for a given value of x_0 is:

$$s_{\hat{y}_{0,NEW}} = \sqrt{s^2 \left[1 + \frac{1}{n} + \frac{(x_0 - \bar{x})^2}{SS_{xx}} \right]}$$

Confidence intervals follow the classical form:

$$Estimator \pm (standard\ error)\ (t_{\alpha/2, n-2})$$

A confidence interval for the population mean with a confidence level of $1 - \alpha$:

$$\hat{y}_0 \pm s_{\hat{y}_i} t_{\alpha/2, n-2}$$

A confidence interval for the prediction with a confidence level of $1 - \alpha$:

$$\hat{y}_{0,NEW} \pm s_{\hat{y}_{i,NEW}} t_{\alpha/2, n-2}$$

Example: For the previous example of weight and heart girth of cows calculate the mean and prediction confidence intervals. Recall that $n = 6$, $SS_{xx} = 34$, $MS_{RES} = 115.826$, $\bar{x} = 217$, $b_0 = -974.05$, and $b_1 = 7.53$.

For example, for a value $x_0 = 216$:

$$s_{\hat{y}_0} = \sqrt{s^2\left[\frac{1}{n} + \frac{(x_i - \bar{x})^2}{SS_{xx}}\right]} = \sqrt{115.826\left[\frac{1}{6} + \frac{(216 - 217)^2}{34}\right]} = 4.7656$$

$$s_{\hat{y}_{0,NEW}} = \sqrt{s^2\left[1 + \frac{1}{n} + \frac{(x_i - \bar{x})^2}{SS_{xx}}\right]} = \sqrt{115.826\left[1 + \frac{1}{6} + \frac{(216 - 217)^2}{34}\right]} = 11.7702$$

$$\hat{y}_{x=216} = -974.05 + 7.53(216) = 652.43$$

$$t_{0.025,4} = 2.776$$

The confidence interval for the population mean for a given value $x_0 = 216$ with a confidence level of $1 - \alpha = 0.95$ is:

$$652.43 \pm (4.7656)(2.776)$$

The confidence interval for the prediction for a given value $x_0 = 216$ with a confidence level of $1 - \alpha = 0.95$ is:

$$652.43 \pm (11.7702)(2.776)$$

Note that the interval for the new observation is wider than for the mean.

It could be of interest to estimate confidence intervals for several given values of the x variable. If a 95% confidence interval is calculated for each value of x, this implies that for each interval the probability that it is correct is 0.95. However, the probability that all intervals are correct is not 0.95. If all intervals were independent, the probability that all intervals are correct would be 0.95^k, where k is the number of intervals. The probability that at least one interval is not correct is not 0.05, but rather $(1 - 0.95^k)$. This means that, for example, for five intervals estimated together, the probability that at least one is incorrect is $(1 - 0.95^5) = 0.27$. Fortunately, as estimated values of the dependent variable also depend on the estimated regression, this probability is not enhanced so drastically. To estimate confidence intervals for means for several given values of x we can use the Working-Hotelling formula:

$$\hat{y}_i \pm s_{\hat{y}_i}\sqrt{pF_{\alpha,p,n-p}}$$

Similarly, for prediction confidence intervals:

$$\hat{y}_i \pm s_{\hat{y}_i,NEW}\sqrt{pF_{\alpha,p,n-p}}$$

where $F_{\alpha,p,n-p}$ is a critical value of the F distribution for p and $(n - p)$ degrees of freedom, p is the number of parameters, n is the number of observations, and α is the probability that at least one interval is incorrect.

These expressions are valid for any number of intervals. When intervals are estimated for all values of x, then we can define a confidence contour. A confidence contour for the example is shown in Figure 7.7. The mean and prediction confidence contours are shown. The prediction contour is wider than the mean contour and both intervals widen toward extreme values of variable *x*. The latter warns users to be cautious of using regression predictions beyond the observed values of the independent variables.

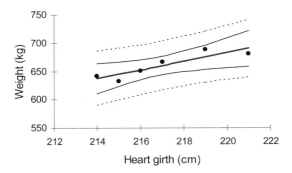

Figure 7.7 Confidence contours for the mean (___) and prediction (......) of the dependent variable for given values of *x*

7.9 Partitioning Total Variability

An intention of using a regression model is to explain as much variability of the dependent variable as possible. The variability accounted for by the model is called the explained variability. Unexplained variability is variability that remains unaccounted for by the model. In a sample, the total variability of the dependent variable is the variability of measurements y_i about the arithmetic mean $\bar{y}$, and is measured with the total sum of squares. The unexplained variability is variability of y_i about the estimated regression line ($\hat{y}$) and is measured with the residual sum of squares (Figure 7.8).

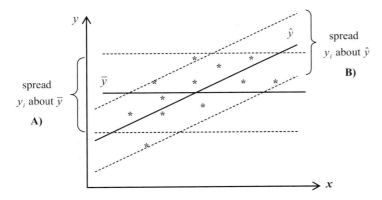

Figure 7.8 Distribution of variability about the arithmetic mean and estimated regression line:

(B) measured with the residual sum of squares, $SS_{RES} = \sum_i (y_i - \hat{y}_i)^2$

(A) measured with the total sum of squares, $SS_{TOT} = \sum_i (y_i - \bar{y})^2$

Comparison of the total and residual sums of squares measures the strength of association between the independent and dependent variables x and y (Figure 7.9). If SS_{RES} is considerably less than SS_{TOT}, it implies a linear association between x and y. If SS_{RES} is close to SS_{TOT} then the linear association between x and y is not clearly defined.

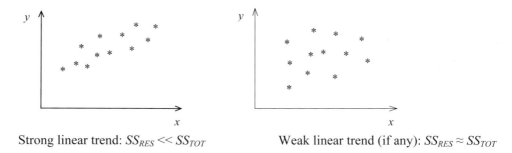

Strong linear trend: $SS_{RES} \ll SS_{TOT}$ Weak linear trend (if any): $SS_{RES} \approx SS_{TOT}$

Figure 7.9 Comparison of the sums of squares and linear trend

Recall that along with total and unexplained variability there is variability explained by the regression model, which is measured with the regression sum of squares $SS_{REG} = \sum_i (\hat{y}_i - \bar{y})^2$. Briefly, the three sources of variability are:

1. Total variability of the dependent variable
   - variability about $\bar{y}$, measured with the total sum of squares (SS_{TOT})
2. Variability accounted for by the model
   - explained variability, measured with the regression sum of squares (SS_{REG}).
3. Variability not accounted for by the model
   - unexplained variability, variability about $\hat{y}$, measured with the residual sum of squares (SS_{RES}).

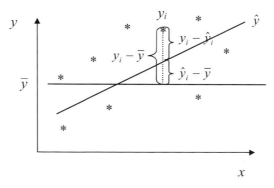

Figure 7.10 Measurement y_i expressed as deviations from the arithmetic mean and estimated regression line

7.9.1 Relationships among Sums of Squares

If measurements of a variable y are shown as deviations from the arithmetic mean and estimated regression line (Figure 7.10) then the following holds:

$$(y_i - \bar{y}) = (\hat{y}_i - \bar{y}) + (y_i - \hat{y}_i)$$

It can be shown that by taking squares of deviations for all y_i points and by summing those squares, the following also holds:

$$\sum_i (y_i - \bar{y}_i)^2 = \sum_i (\hat{y}_i - \bar{y}_i)^2 + \sum_i (y_i - \hat{y}_i)^2$$

This can be written as:

$$SS_{TOT} = SS_{REG} + SS_{RES}$$

Thus, total variability can be partitioned into variability explained by regression and unexplained variability.

The sums of squares can be calculated using shortcuts:

1. The total sum of squares is the sum of squares of the dependent variable:

$$SS_{TOT} = SS_{yy}$$

2. The regression sum of squares is:

$$SS_{REG} = \frac{(SS_{xy})^2}{SS_{xx}}$$

3. The residual sum of squares is the difference between the total sum of squares and the regression sum of squares:

$$SS_{RES} = SS_{yy} - \frac{(SS_{xy})^2}{SS_{xx}}$$

Example: Compute the total, regression and residual sums of squares for the example of weight and heart girth of cows. The sum of products is $SS_{xy} = 256$ and the sum of squares $SS_{xx} = 34$. The total sum of squares is the sum of squares for y.

The sums of squares are:

$$SS_{TOT} = SS_{yy} = \sum_i y_i^2 - \frac{\left(\sum_i y_i\right)^2}{n} = 2390.833$$

$$SS_{REG} = \frac{(SS_{xy})^2}{SS_{xx}} = \frac{(256)^2}{34} = 1927.529$$

$$SS_{RES} = SS_{TOT} - SS_{REG} = 2390.833 - 1927.529 = 463.304$$

7.9.2   Theoretical Distribution of Sum of Squares

Assuming a normal distribution of the residual, SS_{RES}/σ^2 has a *chi*-square distribution with $(n-2)$ degrees of freedom. Under the assumption that there is no regression, that is, $\beta_1 = 0$, SS_{REG}/σ^2 has *chi*-square distribution with 1 degree of freedom, and SS_{TOT}/σ^2 has a *chi*-square distribution with $(n-1)$ degrees of freedom. Recall that a χ^2 variable is equal to $\sum_i z_i^2$, where z_i are standard normal variables ($i = 1$ to v):

$$z_i = \frac{y_i - \overline{y}}{\sigma^2}$$

Thus the expression:

$$\frac{\sum_i (y_i - \overline{y})^2}{\sigma^2} = \frac{SS_{yy}}{\sigma^2}$$

is the sum of $(n-1)$ squared independent standard normal variables with a *chi*-square distribution. The same can be shown for other sums of squares. To compute the corresponding mean squares it is necessary to determine the degrees of freedom. Degrees of freedom can be partitioned similarly to sums of squares:

$$SS_{TOT} = SS_{REG} + SS_{RES} \qquad \text{(sums of squares)}$$

$$(n-1) = 1 + (n-2) \qquad \text{(degrees of freedom)}$$

where n is the number of pairs of observations. Degrees of freedom can be empirically determined as:

The total degrees of freedom: $(n-1) = 1$ degree of freedom is lost in estimating the arithmetic mean.
The residual degrees of freedom: $(n-2) = 2$ degrees of freedom are lost in estimating β_0 and β_1.
The regression degrees of freedom: $(1) = 1$ degree of freedom is used for estimating β_1.

Mean squares are obtained by dividing the sums of squares by their corresponding degrees of freedom:

$$\text{Regression mean square: } MS_{REG} = \frac{SS_{REG}}{1}$$

$$\text{Residual mean square: } MS_{RES} = \frac{SS_{RES}}{n-2}$$

These mean squares are used in hypotheses testing.

7.10   Test of Hypotheses – *F* test

The sums of squares and their distributions are needed for testing the statistical hypotheses: $H_0\colon \beta_1 = 0$ vs. $H_1\colon \beta_1 \neq 0$. It can be shown that the SS_{REG} and SS_{RES} are independent. That assumption allows an F test to be applied for testing the hypotheses. The F statistic is defined as:

$$F = \frac{\left(SS_{REG}/\sigma^2\right)/1}{\left(SS_{RES}/\sigma^2\right)/(n-2)} \leftarrow \frac{\chi_1^2/1}{\chi_{n-2}^2/(n-2)} \quad (\text{if } H_0)$$

The following mean squares have already been defined as:

$$\frac{SS_{REG}}{1} = MS_{REG} = \text{regression mean square}$$

$$\frac{SS_{RES}}{n-2} = MS_{RES} = \text{residual mean square}$$

The F statistic is:

$$F = \frac{MS_{REG}}{MS_{RES}}$$

The F statistic has an F distribution with 1 and $(n-2)$ degrees of freedom if H_0 is true.

The expectations of the sums of squares are:

$$E(SS_{RES}) = \sigma^2(n-2)$$
$$E(SS_{REG}) = \sigma^2 + \beta_1^2 SS_{xx}$$

The expectations of mean squares are:

$$E(MS_{RES}) = \sigma^2$$
$$E(MS_{REG}) = \sigma^2 + \beta_1^2 SS_{xx}$$

If H_0 is true, then $\beta_1 = 0$, $MS_{REG} \approx \sigma^2$, and $F \approx 1$. If H_1 is true, then $MS_{REG} > \sigma^2$ and $F > 1$. H_0 is rejected if the F statistic is 'large'. For the significance level α, H_0 is rejected if $F > F_{\alpha,1,n-2}$, where $F_{\alpha,1,n-2}$ is a critical value (Figure 7.11).

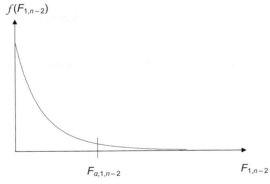

Figure 7.11 *F* distribution and the critical value for 1 and $(n-2)$ degrees of freedom. The $F_{\alpha,1,n-2}$ denotes a critical value of the *F* distribution

Note that for simple linear regression the F test is analogous to the t test for the parameter β_1. Further, it holds:

$$F = t^2$$

It is convenient to write sources of variability, sums of squares (*SS*), degrees of freedom (*df*) and mean squares (*MS*) in a table, which is called an analysis of variance or *ANOVA* table:

Source	SS	df	MS = SS / df	F
Regression	SS_{REG}	1	$MS_{REG} = SS_{REG} / 1$	$F = MS_{REG} / MS_{RES}$
Residual	SS_{RES}	$n - 2$	$MS_{RES} = SS_{RES} / (n - 2)$	
Total	SS_{TOT}	$n - 1$		

Note: in some literature Error as a source is used instead of Residual.

Example: Test the regression hypotheses using an *F* test for the example of weights and heart girths of cows. The following were previously computed:

$$SS_{TOT} = SS_{yy} = \sum_i y_i^2 - \frac{\left(\sum_i y_i\right)^2}{n} = 2390.833$$

$$SS_{REG} = \frac{(SS_{xy})^2}{SS_{xx}} = \frac{(256)^2}{34} = 1927.529$$

$$SS_{RES} = SS_{TOT} - SS_{REG} = 2390.833 - 1927.529 = 463.304$$

The degrees of freedom for total, regression and residual are $(n - 1) = 5$, 1 and $(n - 2) = 4$, respectively.

The mean squares are:

$$MS_{REG} = \frac{SS_{REG}}{1} = \frac{1927.529}{1} = 1927.529$$

$$MS_{RES} = \frac{SS_{RES}}{n - 2} = \frac{463.304}{4} = 115.826$$

The value of the *F* statistic is:

$$F = \frac{MS_{REG}}{MS_{RES}} = \frac{1927.529}{115.826} = 16.642$$

In the form of an *ANOVA* table:

Source	SS	df	MS	F
Regression	1927.529	1	1927.529	16.642
Residual	463.304	4	115.826	
Total	2390.833	5		

The critical value of the *F* distribution for $\alpha = 0.05$ and 1 and 4 degrees of freedom is $F_{0.05,1,4} = 7.71$ (See Appendix B: Critical values of *F* distributions). Since the calculated $F = 16.642$ is greater than the critical value, H_0 is rejected.

7.11 Likelihood Ratio Test

The hypotheses H_0: $\beta_1 = 0$ vs. H_1: $\beta_1 \neq 0$, can be tested using likelihood functions. The idea is to compare the values of likelihood functions using estimates for H_0 and H_1. Those values are maximums of the corresponding likelihood functions.

The likelihood function under H_0 is:

$$L(\mu, \sigma^2 \mid y) = \frac{1}{\left(\sqrt{2\pi\sigma^2}\right)^n} e^{-\sum_i (y_i - \mu)^2 / 2\sigma^2}$$

Note that $\mu = \beta_0$.

The corresponding maximum likelihood estimators are:

$$\hat{\mu}_{ML\_0} = \hat{\beta}_0 = \frac{\sum_i y_i}{n} = \bar{y}$$

$$\hat{\sigma}^2_{ML\_0} = s^2_{ML\_0} = \frac{\sum_i (y_i - \bar{y})^2}{n}$$

Using the estimators, the maximum of the likelihood function is:

$$L(\bar{y}, s^2_{ML\_0} \mid y) = \frac{1}{\left(\sqrt{2\pi s^2_{ML\_0}}\right)^n} e^{-\sum_i (y_i - \bar{y})^2 / 2 s^2_{ML\_0}}$$

The likelihood function when H_0 is not true is:

$$L(\beta_0, \beta_1, \sigma^2 \mid y) = \frac{1}{\left(\sqrt{2\pi\sigma^2}\right)^n} e^{-\sum_i (y_i - \beta_0 - \beta_1 x_i)^2 / 2\sigma^2}$$

and the corresponding maximum likelihood estimators are:

$$\hat{\beta}_1 = b_1 = \frac{SS_{xy}}{SS_{xx}}$$

$$\hat{\beta}_0 = b_0 = \bar{y} - b_1 \bar{x}$$

$$\hat{\sigma}^2_{ML\_1} = s^2_{ML} = \frac{1}{n} \sum_i (y_i - b_0 - b_1 x_i)^2$$

Using the estimators, the maximum of the likelihood function is:

$$L(b_0, b_1, s^2_{ML\_1} \mid y) = \frac{1}{\left(\sqrt{2\pi s^2_{ML\_1}}\right)^n} e^{-\sum_i (y_i - b_0 - b_1 x_i)^2 / 2 s^2_{ML\_1}}$$

The likelihood ratio statistic is:

$$\Lambda = \frac{L(\bar{y}, s^2_{ML\_0} \mid y)}{L(b_0, b_1, s^2_{ML\_1} \mid y)}$$

Further, a natural logarithm of this ratio multiplied by (–2) is distributed approximately as a *chi*-square with one degree of freedom:

$$-2log\Lambda = -2log\frac{L(\bar{y},s^2_{ML\_0}\,|\,y)}{L(b_0,b_1,s^2_{ML\_1}\,|\,y)} = -2\left[logL(\bar{y},s^2_{ML\_0}\,|\,y) - logL(b_0,b_1,s^2_{ML\_1}\,|\,y)\right]$$

For a significance level α, H_0 is rejected if $-2log\Lambda > \chi^2_\alpha$, where χ^2_α is a critical value.

For a regression model there is a relationship between the likelihood ratio test and the F test. The logarithms of likelihood expressions can be expressed as:

$$logL(\bar{y},s^2_{ML\_0}\,|\,y) = -\frac{n}{2}log(s^2_{ML\_0}) - \frac{n}{2}log(2\pi) - \frac{\sum_i(y_i-\bar{y})^2}{2s^2_{ML\_0}}$$

$$logL(b_0,b_1,s^2_{ML\_1}\,|\,y) = -\frac{n}{2}log(s^2_{ML\_1}) - \frac{n}{2}log(2\pi) - \frac{\sum_i(y_i-b_0-b_1x_i)^2}{2s^2_{ML\_1}}$$

Recall that:

$$s^2_{ML\_0} = \frac{\sum_i(y-\bar{y})^2}{n}$$

$$s^2_{ML} = \frac{1}{n}\sum_i(y_i-b_0-b_1x_i)^2$$

Thus:

$$-2log\Lambda = -2\left[-\frac{n}{2}log(s^2_{ML\_0}) - n\frac{\sum_i(y_i-\bar{y})^2}{2\sum_i(y_i-\bar{y})^2} + \frac{n}{2}log(s^2_{ML\_1}) + n\frac{\sum_i(y_i-b_0-b_1x_i)^2}{2\sum_i(y_i-b_0-b_1x_i)^2}\right]$$

$$= -n\left[-log(s^2_{ML\_0}) + log(s^2_{ML\_1})\right] = n\,log\left(\frac{s^2_{ML\_0}}{s^2_{ML\_1}}\right)$$

Assuming the variance σ^2 is known then:

$$-2log\Lambda = -2\left[logL(\bar{y}\,|\,\sigma^2,y) - logL(b_0,b_1\,|\,\sigma^2,y)\right]$$

$$= -2\left[-\frac{n}{2}log(\sigma^2) - \frac{\sum_i(y_i-\bar{y})^2}{2\sigma^2} + \frac{n}{2}log(\sigma^2) + \frac{\sum_i(y_i-b_0-b_1x_i)^2}{2\sigma^2}\right]$$

$$= \left[\frac{\sum_i(y_i-\bar{y})^2}{\sigma^2} - \frac{\sum_i(y_i-b_0-b_1x_i)^2}{\sigma^2}\right]$$

where:

$$\sum_i(y_i-\bar{y})^2 = SS_{TOT} = \text{the total sum of squares}$$

$$\sum_i(y_i-b_0-b_1x_i)^2 = SS_{RES} = \text{the residual sum of squares, and}$$

$$SS_{TOT} - SS_{RES} = SS_{REG} = \text{the regression sum of squares}$$

Thus:

$$-2log\Lambda = \left[\frac{SS_{REG}}{\sigma^2}\right]$$

Estimating σ^2 from the regression model as $s^2 = MS_{RES} = \dfrac{SS_{RES}}{n-2}$, and having

$MS_{REG} = \dfrac{SS_{REG}}{1}$, note that asymptotically $-2log\Lambda$ divided by the degrees of freedom (for linear regression they equal one) is equivalent to the F statistic as shown before.

7.12 Coefficient of Determination

The coefficient of determination is often used as a measure of the correctness of a model, that is, how well a regression model fits the data. A model is 'good' if the regression sum of squares is close to the total sum of squares, $SS_{REG} \approx SS_{TOT}$. A model is 'bad' if the residual sum of squares is close to the total sum of squares, $SS_{RES} \approx SS_{TOT}$. The coefficient of determination represents the proportion of the total variability explained by the model:

$$R^2 = \frac{SS_{REG}}{SS_{TOT}}$$

Since an increase of the regression sum of squares implies a decrease of the residual sum of squares, the coefficient of determination is also:

$$R^2 = 1 - \frac{SS_{RES}}{SS_{TOT}}$$

The coefficient of determination can have values $0 \le R^2 \le 1$. The 'good' model is indicated by an R^2 close to one.

7.12.1 Shortcut Calculation of Sums of Squares and the Coefficient of Determination

The regression and total sum of squares can be written as:

$$SS_{REG} = b^2_1\, SS_{xx}$$
$$SS_{TOT} = SS_{yy}$$

Since:

$$b_1 = \frac{SS_{xy}}{SS_{xx}}$$

The coefficient of determination is:

$$R^2 = \frac{SS_{REG}}{SS_{TOT}} = b_1^2 \frac{SS_{xx}}{SS_{yy}} = \frac{\dfrac{SS_{xy}^2}{SS_{xx}^2} SS_{xx}}{SS_{yy}} = \frac{SS_{xy}^2}{SS_{xx} SS_{yy}}$$

Example: Compute a coefficient of determination for the example of weights and heart girths of cows.

$$SS_{REG} = \frac{(SS_{xy})^2}{SS_{xx}} = \frac{(256)^2}{34} = 1927.529$$

or

$$SS_{REG} = b_1^2 SS_{xx} = (7.529)^2 (34) = 1927.529$$
$$SS_{TOT} = SS_{yy} = 2390.833$$
$$R^2 = \frac{SS_{REG}}{SS_{TOT}} = \frac{1927.529}{2390.833} = 0.81$$

Thus, the regression model explains 81% of the total variability, or variation in heart girth explains 81% of the variation in weight of these cows.

7.13 Matrix Approach to Simple Linear Regression

7.13.1 The Simple Regression Model

Since a regression model can be presented as a set of linear equations, those equations can be shown using matrices and vectors. Recall that the linear regression model is:

$$y_i = \beta_0 + \beta_1 x_i + \varepsilon_i \qquad i = 1,..., n$$

y_i = observation i of the dependent variable y
x_i = observation i of the independent variable x
β_0, β_1 = regression parameters
ε_i = random error

Thus:

$$y_1 = \beta_0 + \beta_1 x_1 + \varepsilon_1$$
$$y_2 = \beta_0 + \beta_1 x_2 + \varepsilon_2$$
$$...$$
$$y_n = \beta_0 + \beta_1 x_n + \varepsilon_n$$

The equivalently defined vectors and matrices are:

$$\mathbf{y} = \begin{bmatrix} y_1 \\ y_2 \\ ... \\ y_n \end{bmatrix} \quad \mathbf{X} = \begin{bmatrix} 1 & x_{11} \\ 1 & x_{21} \\ ... & ... \\ 1 & x_{n1} \end{bmatrix} \quad \boldsymbol{\beta} = \begin{bmatrix} \beta_0 \\ \beta_1 \end{bmatrix} \quad \boldsymbol{\varepsilon} = \begin{bmatrix} \varepsilon_1 \\ \varepsilon_2 \\ ... \\ \varepsilon_n \end{bmatrix}$$

where:

$\mathbf{y}$ = the vector of observations of a dependent variable
$\mathbf{X}$ = the matrix of observations of an independent variable
$\boldsymbol{\beta}$ = the vector of parameters
$\boldsymbol{\varepsilon}$ = the vector of random errors

Using those matrices and vectors the regression model is:

$$\mathbf{y} = \mathbf{X}\boldsymbol{\beta} + \boldsymbol{\varepsilon}$$

The expectation of $\mathbf{y}$ is:

$$E(\mathbf{y}) = \begin{bmatrix} E(y_1) \\ E(y_2) \\ \dots \\ E(y_n) \end{bmatrix} = \begin{bmatrix} \beta_0 + \beta_1 x_1 \\ \beta_0 + \beta_1 x_2 \\ \dots \\ \beta_0 + \beta_1 x_n \end{bmatrix} = \mathbf{X\beta}$$

The variance of $\mathbf{y}$ is:

$$Var(\mathbf{y}) = \sigma^2 \mathbf{I}$$

Also, $E(\boldsymbol{\varepsilon}) = \mathbf{0}$ and $Var(\boldsymbol{\varepsilon}) = \sigma^2 \mathbf{I}$, that is, the expectation of error is zero and the variance is constant. The $\mathbf{0}$ vector is a vector with all elements equal to zero, and $\mathbf{I}$ is an identity matrix.

Assuming a normal model the $\mathbf{y}$ vector includes random normal variables with multi-normal distribution with mean $\mathbf{X\beta}$ and variance $\mathbf{I}\sigma^2$. It is assumed that each observation y is drawn from a normal population and that all y are independent of each other, but with the same mean and variance.

The estimation model is defined as:

$$\hat{\mathbf{y}} = \mathbf{Xb}$$

where:

$\hat{\mathbf{y}} =$ the vector of estimated (fitted) values

$$\mathbf{b} = \begin{bmatrix} b_0 \\ b_1 \end{bmatrix} = \text{the vector of estimators}$$

The vector of residuals is the difference between the observed $\mathbf{y}$ vector and the vector of estimated values of the dependent variable:

$$\mathbf{e} = \begin{bmatrix} e_1 \\ e_2 \\ \dots \\ e_n \end{bmatrix} = \mathbf{y} - \hat{\mathbf{y}}$$

Thus, a vector of sample observations is:

$$\mathbf{y} = \mathbf{Xb} + \mathbf{e}$$

7.13.2 Estimation of Parameters

By using either least squares or maximum likelihood estimation, the following normal equations are obtained:

$$(\mathbf{X'X})\mathbf{b} = \mathbf{X'y}$$

Solving for $\mathbf{b}$ gives:

$$\mathbf{b} = (\mathbf{X'X})^{-1}\mathbf{X'y}$$

The $\mathbf{X'X}$, $\mathbf{X'y}$ and $(\mathbf{X'X})^{-1}$ matrices have the following elements:

$$\mathbf{X'X} = \begin{bmatrix} n & \sum_i x_i \\ \sum_i x_i & \sum_i x_i^2 \end{bmatrix}$$

$$\mathbf{X'y} = \begin{bmatrix} \sum_i y_i \\ \sum_i x_i y_i \end{bmatrix}$$

$$(\mathbf{X'X})^{-1} = \begin{bmatrix} \dfrac{1}{n} + \dfrac{\bar{x}^2}{SS_x} & -\dfrac{\bar{x}}{SS_{xx}} \\ -\dfrac{\bar{x}}{SS_{xx}} & \dfrac{1}{SS_{xx}} \end{bmatrix}$$

Properties of the estimators, the expectations and (co)variances are:

$$E(\mathbf{b}) = \boldsymbol{\beta}$$

$$\mathrm{Var}(\mathbf{b}) = \sigma^2 (\mathbf{X'X})^{-1} = \begin{bmatrix} Var(b_0) & Cov(b_0, b_1) \\ Cov(b_0, b_1) & Var(b_1) \end{bmatrix}$$

Using an estimate of the variance from a sample, s^2, the variance of the **b** vector equals:

$$\mathbf{s}^2(\mathbf{b}) = s^2 (\mathbf{X'X})^{-1}$$

The vector of estimated values of the dependent variable is:

$$\hat{\mathbf{y}} = \mathbf{Xb} = \mathbf{X}(\mathbf{X'X})^{-1}\mathbf{Xy}$$

The variance of estimated values is:

$$Var(\hat{\mathbf{y}}) = Var(\mathbf{Xb}) = \mathbf{X}Var(\mathbf{b})\mathbf{X'} = \mathbf{X}(\mathbf{X'X})^{-1}\mathbf{X'}\sigma^2$$

Using s^2 instead of σ^2 the estimator of the variance is:

$$\mathbf{s}_{\hat{y}}^2 = \mathbf{X}(\mathbf{X'X})^{-1}\mathbf{X'}s^2$$

The regression sum of squares (SS_{REG}), residual sum of squares (SS_{RES}) and total sum of squares (SS_{TOT}) are:

$$SS_{REG} = (\hat{\mathbf{y}} - \bar{\mathbf{y}})'(\hat{\mathbf{y}} - \bar{\mathbf{y}}) = \sum_i (\hat{y}_i - \bar{y})^2$$

$$SS_{RES} = (\mathbf{y} - \hat{\mathbf{y}})'(\mathbf{y} - \hat{\mathbf{y}}) = \sum_i (y_i - \hat{y}_i)^2$$

$$SS_{TOT} = (\mathbf{y} - \bar{\mathbf{y}})'(\mathbf{y} - \bar{\mathbf{y}}) = \sum_i (y_i - \bar{y})^2$$

or shortly using the **b** vector:

$$SS_{REG} = \mathbf{b'X'y} - n\bar{y}^2$$

$$SS_{RES} = \mathbf{y'y} - \mathbf{b'X'y}$$

$$SS_{TOT} = \mathbf{y'y} - n\bar{y}^2$$

Example: Estimate the regression for the example of weights and heart girths of cows. Measurements of six cows are given in the following table:

Cow	1	2	3	4	5	6
Weight (y):	641	633	651	666	688	680
Heart girth (x):	214	215	216	217	219	221

The **y** vector and **X** matrix are:

$$\mathbf{y} = \begin{bmatrix} 641 \\ 633 \\ 651 \\ 666 \\ 688 \\ 680 \end{bmatrix} \quad \text{and} \quad \mathbf{X} = \begin{bmatrix} 1 & 214 \\ 1 & 215 \\ 1 & 216 \\ 1 & 217 \\ 1 & 219 \\ 1 & 221 \end{bmatrix}$$

The first column of the **X** matrix consists of the number 1 to facilitate calculating the intercept b_0. Including **y** and **X** in the model gives:

$$\begin{bmatrix} 641 \\ 633 \\ 651 \\ 666 \\ 688 \\ 680 \end{bmatrix} = \begin{bmatrix} 1 & 214 \\ 1 & 215 \\ 1 & 216 \\ 1 & 217 \\ 1 & 219 \\ 1 & 2211 \end{bmatrix} \begin{bmatrix} b_0 \\ b_1 \end{bmatrix} + \begin{bmatrix} e_1 \\ e_2 \\ e_3 \\ e_4 \\ e_5 \\ e_6 \end{bmatrix} = \begin{bmatrix} b_0 + b_1 \cdot 214 + e_1 \\ b_0 + b_1 \cdot 215 + e_2 \\ b_0 + b_1 \cdot 216 + e_3 \\ b_0 + b_1 \cdot 217 + e_4 \\ b_0 + b_1 \cdot 219 + e_5 \\ b_0 + b_1 \cdot 221 + e_6 \end{bmatrix}$$

The $\mathbf{X'X}$, $\mathbf{X'y}$ and $(\mathbf{X'X})^{-1}$ matrices are:

$$\mathbf{X'X} = \begin{bmatrix} 1 & 1 & 1 & 1 & 1 & 1 \\ 214 & 215 & 216 & 217 & 219 & 221 \end{bmatrix} \begin{bmatrix} 1 & 214 \\ 1 & 215 \\ 1 & 216 \\ 1 & 217 \\ 1 & 219 \\ 1 & 2211 \end{bmatrix} = \begin{bmatrix} 6 & 1302 \\ 1302 & 282568 \end{bmatrix} = \begin{bmatrix} n & \sum_i x_i \\ \sum_i x_i & \sum_i x_i^2 \end{bmatrix}$$

$$\mathbf{X'y} = \begin{bmatrix} 1 & 1 & 1 & 1 & 1 & 1 \\ 214 & 215 & 216 & 217 & 219 & 221 \end{bmatrix} \begin{bmatrix} 641 \\ 633 \\ 651 \\ 666 \\ 688 \\ 680 \end{bmatrix} = \begin{bmatrix} 3959 \\ 859359 \end{bmatrix} = \begin{bmatrix} \sum_i y_i \\ \sum_i x_i y_i \end{bmatrix}$$

$$(\mathbf{X'X})^{-1} = \begin{bmatrix} 1385.137 & -6.38235 \\ -6.38235 & 0.02941 \end{bmatrix} = \begin{bmatrix} \dfrac{1}{n} + \dfrac{\bar{x}^2}{SS_x} & -\dfrac{\bar{x}}{SS_{xx}} \\ -\dfrac{\bar{x}}{SS_{xx}} & \dfrac{1}{SS_{xx}} \end{bmatrix}$$

The **b** vector is:

$$\mathbf{b} = (\mathbf{X'X})^{-1}\mathbf{X'y} = \begin{bmatrix} 1385.137 & -6.38235 \\ -6.38235 & 0.02941 \end{bmatrix} \begin{bmatrix} 3959 \\ 859359 \end{bmatrix} = \begin{bmatrix} -974.05 \\ 7.53 \end{bmatrix} = \begin{bmatrix} b_0 \\ b_1 \end{bmatrix} = \begin{bmatrix} \bar{y} - b_1\bar{x} \\ \dfrac{SS_{xy}}{SS_{xx}} \end{bmatrix}$$

Recall that:

$$SS_{xy} = \sum_i (x_i - \bar{x})(y_i - \bar{y})$$
$$SS_{xx} = \sum_i (x_i - \bar{x})^2$$

The estimated values are:

$$\hat{\mathbf{y}} = \mathbf{Xb} = \begin{bmatrix} 1 & 214 \\ 1 & 215 \\ 1 & 216 \\ 1 & 217 \\ 1 & 219 \\ 1 & 221 \end{bmatrix} \begin{bmatrix} -974.05 \\ 7.53 \end{bmatrix} = \begin{bmatrix} 637.25 \\ 644.77 \\ 652.30 \\ 659.83 \\ 674.89 \\ 689.95 \end{bmatrix}$$

$s^2 = 115.826$ is the residual mean square. The estimated variance of **b** is:

$$\mathbf{s^2(b)} = s^2(\mathbf{X'X})^{-1} = 115.826 \begin{bmatrix} 1385.137 & -6.38235 \\ -6.38235 & 0.02941 \end{bmatrix}$$
$$= \begin{bmatrix} 160434.9 & -739.242 \\ -739.242 & 3.407 \end{bmatrix}$$

For example, the estimated variance of b_1 is:

$$s^2(b_1) = 3.407$$

The tests of hypothesis are conducted as was previously shown in the scalar presentation.

7.13.3 Maximum Likelihood Estimation

Under the assumption of normality of the dependent variable, **y** has a multivariate normal distribution $\mathbf{y} \sim N(\mathbf{X\beta}, \sigma^2\mathbf{I})$. The likelihood function is a function with parameters $\boldsymbol{\beta}$ and σ^2 for a given set of n observations of dependent and independent variables:

$$L(\boldsymbol{\beta}, \sigma^2 \mid \mathbf{y}) = \dfrac{e^{-\frac{1}{2}(\mathbf{y}-\mathbf{X\beta})'(\sigma^2\mathbf{I})^{-1}(\mathbf{y}-\mathbf{X\beta})}}{\left(\sqrt{2\pi\sigma^2}\right)^n}$$

The log likelihood is:

$$logL(\boldsymbol{\beta}, \sigma^2 \mid \mathbf{y}) = -\frac{n}{2}log(\sigma^2) - \frac{n}{2}log(2\pi) - \frac{1}{2\sigma^2}(\mathbf{y} - \mathbf{X}\boldsymbol{\beta})'(\mathbf{y} - \mathbf{X}\boldsymbol{\beta})$$

A set of maximum likelihood estimators can be estimated that will maximize the log likelihood function. The maximum of the function can be determined by taking the partial derivatives of the log likelihood function with respect to the parameters. These derivatives are equated to zero in order to find the estimators $\mathbf{b}$ and s^2_{ML}. The following normal equations are obtained:

$$(\mathbf{X'X})\mathbf{b} = \mathbf{X'y}$$

Solving for $\mathbf{b}$ gives:

$$\mathbf{b} = (\mathbf{X'X})^{-1}\mathbf{X'y}$$

A maximum likelihood estimator of the variance is given by:

$$s^2_{ML} = \frac{1}{n}(\mathbf{y} - \mathbf{Xb})'(\mathbf{y} - \mathbf{Xb})$$

Note again that the maximum likelihood estimator of the variance is not unbiased. An unbiased estimator is obtained when the maximum likelihood estimator is multiplied by $n/(n-2)$, that is:

$$s^2 = MS_{RES} = \frac{n}{n-2}s^2_{ML}$$

7.14 SAS Example for Simple Linear Regression

The SAS program for the example of weights and heart girths of cows is as follows:

SAS program:

```
DATA cows;
INPUT wt h_girth;
DATALINES;
641 214
633 215
651 216
666 217
688 219
680 221
;
PROC REG;
MODEL wt = h_girth;
PLOT wt *h_girth / PRED CONF;
RUN;
QUIT;

*or;
```

```
PROC GLM;
MODEL wt =h_girth / ;
RUN;
QUIT;
```

Explanation: Either the GLM or REG procedures can be used. The MODEL statement, *wt = h_girth* denotes that the dependent variable is *wt*, and the independent variable is *h_girth*. The PLOT statement requests a plot of the original and fitted data with confidence bounds, for the mean (CONF) and predicted values (PRED).

SAS output:

Analysis of Variance

Source	DF	Sum of Squares	Mean Square	F Value	Prob>F
Model	1	1927.52941	1927.52941	16.642	0.0151
Error	4	463.30392	115.82598		
C Total	5	2390.83333			

Root MSE	10.76225	R-square	0.8062	
Dep Mean	659.83333	Adj R-sq	0.7578	
C.V.	1.63106			

Parameter Estimates

Variable	DF	Parameter Estimate	Standard Error	T for H0: Parameter=0	Prob > |T|
INTERCEP	1	-974.049020	400.54323178	-2.432	0.0718
H_GIRTH	1	7.529412	1.84571029	4.079	0.0151

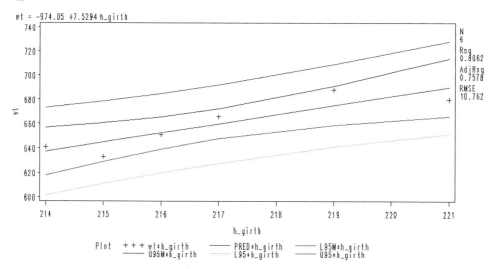

Explanation: The first table is an *ANOVA* table for the dependent variable *wt*. The sources of variability are *Model*, *Error* (residual) and *Corrected Total*. In the table are listed degrees of freedom (*DF*), *Sum of Squares*, *Mean Square*, calculated *F* (*F Value*) and *P* value (*Prob>F*) that the test of significance shows that $F = 16.642$ with a *P* value = 0.0151. This means that the sample regression coefficient is significantly different than zero. Below the *ANOVA* table, the standard deviation of the regression model (*Root MSE*) = 10.76225 and the coefficient of determination (*R-square*) = 0.8062 are given. Under the title *Parameter Estimates*, the parameter estimates are presented with standard errors and corresponding *t* tests indicating whether the estimates are significantly different than zero. Here, b_0 (*INTERCEP*) = –974.049020 and b_1 (*H_GIRTH*) = 7.529412. The *Standard Errors* are 400.54323178 and 1.84571029 for b_0 and b_1, respectively. The calculated *t* statistic is 4.079, with the *P* value (*Prob* > |*T*|) = 0.0151. This confirms that b_1 is significantly different to zero. The last item in the output is a plot of the original data and fitted line, and confidence (L95M and U95M, 95% lower and upper) and predicted (L95 and U95) bounds. On the plot are the regression equation and other statistics.

7.15 Power of Tests

The power of a linear regression based on a sample can be calculated by using either *t* or *F* central and noncentral distributions. Recall that the null and alternative hypotheses are H_0: $\beta_1 = 0$ and H_1: $\beta_1 \neq 0$. The power can be calculated by stating the alternative hypothesis as H_1: $\beta_1 = b_1$, where b_1 is the estimate from a sample. The *t* distribution is used in the following way. If H_0 holds, the test statistic *t* has a central *t* distribution with $n - 2$ degrees of freedom; however, if H_1 holds, the *t* statistic has a noncentral *t* distribution with a noncentrality parameter $\lambda = \dfrac{|b_1|}{s} \sqrt{SS_{xx}}$ and $n - 2$ degrees of freedom. Here, b_1 is the estimate of the regression coefficient, $s = \sqrt{MS_{RES}}$ is the estimated standard deviation of the regression model, MS_{RES} is the residual mean square, and SS_{xx} is the sum of squares for the independent variable *x*. For a two-sided test the power is a probability:

$$Power = 1 - \beta = P[t < -t_{\alpha/2} = t_{\beta 1}] + P[t > t_{\alpha/2} = t_{\beta 2}]$$

using the noncentral *t* distribution for H_1. Here, $t_{\alpha/2}$ is the critical value with α level of significance, and 1 and $n - 2$ degrees of freedom. Power for a linear regression with 20 degrees of freedom is shown in Figure 7.12.

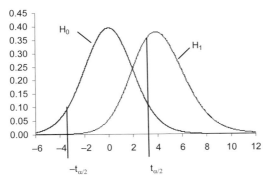

Figure 7.12 The significance and power of a two-sided t test for linear regression. The *t* statistic has a central *t* distribution if H_0 is true, and a noncentral distribution if H_1 is true. The distributions with 20 degrees of freedom are shown. The critical values are $-t_{\alpha/2}$ and $t_{\alpha/2}$. The sum of areas under the H_0 curve on the left of $-t_{\alpha/2}$ and on the right of $t_{\alpha/2}$ is the level of significance (α). The sum of areas under the H_1 curve on the left of $-t_{\alpha/2}$ and on the right of $t_{\alpha/2}$ is the power ($1 - \beta$). The area under the H_1 curve between $-t_{\alpha/2}$ and $t_{\alpha/2}$ is the type II error (β)

An F distribution is used to compute power as follows. If H_0 holds, then the test statistic F has a central F distribution with 1 and $n-2$ degrees of freedom. If H_1 is true the F statistic has a noncentral F distribution with the noncentrality parameter $\lambda = \dfrac{SS_{REG}}{MS_{RES}}$, and 1 and $n-2$ degrees of freedom. The power of test is a probability:

$$Power = 1 - \beta = P[F > F_{\alpha,1,(n-2)} = F_\beta]$$

using the noncentral F distribution for H_1. Here, $F_{\alpha,1,n-2}$ is the critical value with the α level of significance, and 1 and $n-2$ degrees of freedom. Power for a linear regression with 1 and 20 degrees of freedom is shown in Figure 7.13.

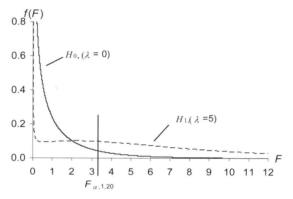

Figure 7.13 Significance and power of the F test for regression. The F statistic has a central F distribution if H_0 is true, and a noncentral F distribution if H_1 is true. The distributions with 1 and 20 degrees of freedom are shown. The critical value is $F_{\alpha,1,20}$. The area under the H_0 curve on the right of the $F_{\alpha,1,20}$ is the significance level (α). The area under the H_1 curve on the right of the $F_{\alpha,1,20}$ is the power ($1 - \beta$). The area under the H_1 curve on the left of the $F_{\alpha,1,20}$ is the type II error (β)

7.15.1 SAS Examples for Calculating the Power of Test

Example: Calculate the power of test for the example of weights and heart girths of cows by using a t distribution.

The following were previously calculated: $b_1 = 7.53$, the variance $s^2 = MS_{RES} = 115.826$, $SS_{xx} = 34$ and degrees of freedom $df = 4$.
The calculated value of the t statistic was:

$$t = \frac{b_1 - 0}{\sqrt{s^2/SS_{xx}}} = 4.079$$

The calculated $t = 4.079$ is more extreme than the critical value (2.776), thus the estimate of the regression slope $b_1 = 7.53$ is significantly different to zero and it can be concluded that linear relationship between weight and heart girth exists in the population.

For a two-sided test the power is:

$$Power = 1 - \beta = P[t > -t_{\alpha/2} = t_{\beta1}] + 1 - P[t > t_{\alpha/2} = t_{\beta2}]$$

using a noncentral t distribution for H_1 with the noncentrality parameter

$$\lambda = \frac{|b_1|}{\sqrt{MS_{RES}}}\sqrt{SS_{xx}} = \frac{7.53}{\sqrt{115.826}}\sqrt{34} = 4.079 \text{ and four degrees of freedom. The power is:}$$

$$Power = 1 - \beta = P[t > -2.776] + P[t > 2.776] = 0.000 + 0.856 = 0.856$$

Power can be calculated by using SAS:

```
DATA a;
alpha=0.05;
n=6;
b=7.52941;
msres=115.82598;
ssxx=34;
df=n-2;
lambda=abs(b)/sqrt(msres/ssxx);
tcrit_low=TINV(alpha/2,df);
tcrit_up=TINV(1-alpha/2,df);
power=CDF('t',tcrit_low,df,lambda)+ 1-CDF('t',tcrit_up,df,lambda);
PROC PRINT;
RUN;
```

Explanation: First we define: *alpha* = significance level, n = sample size, b = estimated regression coefficient, *msres* = residual (error) mean square (the estimated variance), *ssxx* = sum of squares for x, and *df* = degrees of freedom. The noncentrality parameter (*lambda*) and critical values (*tcrit_low* and *tcrit_up* for a two-sided test) are calculated. The critical value is computed by using the TINV function, which must have as input cumulative values of percentiles ($\alpha/2 = 0.025$ and $1 - \alpha/2 = 0.975$) and degrees of freedom *df*. The *power* is calculated with the CDF function. This is a cumulative function of the t distribution, which needs the critical value, degrees of freedom and the noncentrality parameter *lambda* to be defined. As an alternative to *CDF('t',tcrit,df,lambda)*, the function *PROBT(tcrit,df,lambda)* can be used. The PRINT procedure gives the following SAS output:

```
Alpha   n    b     msres    ssxx df lambda  tcrit_low   tcrit_up   power
0.05    6  7.529  115.826    34   4  4.079  -2.77645    2.77645    0.856
```

The power is 0.856.

Example: Calculate the power of test for the example of weights and heart girths of cows by using an F distribution.

The following were previously calculated: the regression sum of squares $SS_{REG} = 1927.529$ and the variance $s^2 = MS_{RES} = 115.826$. The regression and residual degrees of freedom were 1 and 4, respectively. The calculated value of the F statistic was:

$$F = \frac{MS_{REG}}{MS_{RES}} = \frac{1927.529}{115.826} = 16.642$$

The critical value for $\alpha = 0.05$ and 1 and 4 degrees of freedom is $F_{0.05,1,4} = 7.71$. Since the calculated $F = 16.642$ is greater than the critical value, H_0 is rejected.

The power of test is calculated using the critical value $F_{0.05,1,4} = 7.71$, and the noncentral F distribution for H_1 with the noncentrality parameter $\lambda = \dfrac{SS_{REG}}{MS_{RES}} = \dfrac{1927.529}{115.826} = 16.642$ and 1 and 4 degrees of freedom. The power is:

$$Power = 1 - \beta = P[F > 7.71] = 0.856$$

the same value obtained by using the t distribution.

The power can be calculated by using SAS:

```
DATA a;
alpha=0.05;
n=6;
ssreg=1927.52941;
msres=115.82598;
df=n-2;
lambda=ssreg/mse;
Fcrit=FINV(1-alpha,1,df);
power=1-PROBF(Fcrit,1,df,lambda);
PROC PRINT;
RUN;
```

Explanation: First we define: *alpha* = significance level, n = sample size, *ssreg* = regression sum of squares, *msres* = residual mean square, and *df* = degrees of freedom. Then, the noncentrality parameter (*lambda*) and critical value (*Fcrit*) are calculated. The critical value is computed with the FINV function, which must have cumulative values of percentiles ($1 - \alpha = 0.95$) and degrees of freedom 1 and *df*. The *power* is calculated with PROBF. This is a cumulative function of the F distribution which needs the critical value, degrees of freedom and the noncentrality parameter *lambda* to be defined. As an alternative to *PROBF(Fcrit,1,df,lambda)*, the *CDF('F',Fcrit,1,df,lambda)* function can be used. The PRINT procedure gives the following SAS output:

```
alpha    n    ssreg      msres     df    lambda    Fcrit      power
0.05     6    1927.53    115.826   4     16.6416   7.70865    0.856
```

The power is 0.856.

Exercises

7.1. Estimate the linear regression relating the influence of hen weights (x) on feed intake (y) in a year:

x	2.3	2.6	2.4	2.2	2.8	2.3	2.6	2.6	2.4	2.5
y	43	46	45	46	50	46	48	49	46	47

Test the null hypothesis that regression does not exist. Construct a confidence interval for the regression coefficient. Compute the coefficient of determination. Explain the results.

7.2. The aim of this study was to test effects of weight at slaughter on back-fat thickness. Eight pigs of the Poland China breed were measured. The measurements are shown in the following table:

Slaughter weight (kg)	100	130	140	110	105	95	130	120
Back fat (mm)	42	38	53	34	35	31	45	43

Test the H_0 that regression does not exist. Construct a confidence interval for the regression coefficient. Compute the coefficient of determination. Explain the results.

7.3. In the period from 1990 to 2001 on a horse farm there were the following numbers of horses:

Year	1990	1991	1992	1993	1994	1995	1996	1997	1998	1999	2000	2001
Number of horses	110	110	105	104	90	95	92	90	88	85	78	80

a) Describe the changes of the number of horses with a regression line.
b) Show the true and estimated numbers in a graph.
c) How many horses would be expected in the year 2002 if a linear trend is assumed?

Chapter 8

Correlation

The coefficient of correlation measures the strength of the linear relationship between two variables. Recall that the main goal of regression analysis is to determine the functional dependency of the dependent variable y on the independent variable x. The roles of variables y and x are clearly defined as dependent and independent. The values of y are expressed as a function of the values of x. The correlation is used when there is interest in determining the degree of association among variables, but when they cannot be easily defined as dependent or independent. For example, we may wish to determine the relationship between weight and height, but do not consider one to be dependent on the other, perhaps both are dependent on a third factor. The coefficient of correlation (ρ) is defined:

$$\rho = \frac{\sigma_{xy}}{\sqrt{\sigma_x^2 \sigma_y^2}}$$

where:

σ_y^2 = variance of y
σ_x^2 = variance of x
σ_{xy} = covariance between x and y

Variables x and y are assumed to be random normal variables jointly distributed with a bivariate normal distribution. Recall that the covariance is a measure of joint variability of two random variables. It is an absolute measure of association. If two variables are not associated then their covariance is equal to zero. The coefficient of correlation is a relative measure of association between two variables, and is equal to the covariance of the standardized variables x and y:

$$\rho = Cov\left(\frac{y - \mu_y}{\sigma_y}, \frac{x - \mu_x}{\sigma_x}\right)$$

where μ_y and μ_x are the means of y and x, respectively.

Values of the coefficient of correlation range between –1 and 1, inclusively ($-1 \le \rho \le 1$). For $\rho > 0$, the two variables have a positive correlation, and for $\rho < 0$, the two variables have a negative correlation. The positive correlation means that as values of one variable increase, increasing values of the other variable are observed. A negative correlation means that as values of one variable increase, decreasing values of the other variable are observed. The values $\rho = 1$ or $\rho = -1$ indicate an ideal or perfect linear relationship, and $\rho = 0$ indicates no linear association. The sign of the coefficient of correlation, ρ, is the same as that of the coefficient of linear regression β_1, and the numerical connection between those two coefficients is as follows:

$$\beta_1 = \frac{\sigma_{xy}}{\sigma_x^2}$$

Thus:

$$\rho = \frac{\sigma_{xy}}{\sigma_x \sigma_y} = \left(\frac{\sigma_x}{\sigma_x}\right)\frac{\sigma_{xy}}{\sigma_x \sigma_y} = \beta_1 \frac{\sigma_x}{\sigma_y}$$

In Figure 8.1, a positive correlation between x and y is evident in a), and a negative correlation in b). The lower figures illustrate two cases in which, by definition, there is no correlation; there is no clear association between x and y in c), and there is an association but it is not linear in d).

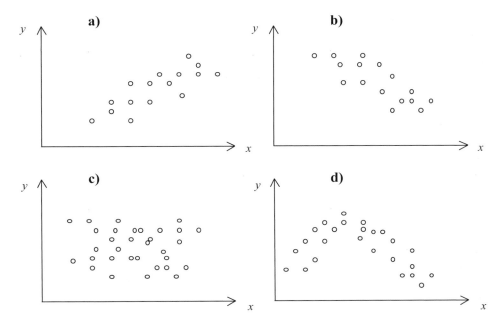

Figure 8.1 a) positive correlation, b) negative correlation, c) no association, and d) an association but it is not linear

8.1 Estimation of the Coefficient of Correlation and Tests of Hypotheses

The coefficient of correlation is estimated from a random sample by a sample coefficient of correlation (r):

$$r = \frac{SS_{xy}}{\sqrt{SS_{xx}SS_{yy}}}$$

where:

$$SS_{xy} = \sum_i (x_i - \bar{x})(y_i - \bar{y}) = \text{sum of products of } y \text{ and } x$$

$$SS_{xx} = \sum_i (x_i - \bar{x})^2 = \text{sum of squares of } x$$

$$SS_{yy} = \sum_i (y_i - \bar{y})^2 = \text{sum of squares of } y$$

n = sample size
$\bar{y}$ and $\bar{x}$ = arithmetic means of y and x

Values for r also range between –1 and 1, inclusively. The sample coefficient of correlation is equal to the mean product of the standardized values of variables from the samples. This is an estimator of the covariance of the standardized values of x and y in the population. Recall that the mean product is:

$$MS_{xy} = \frac{SS_{xy}}{n-1} = \frac{\sum_i (x_i - \bar{x})(y_i - \bar{y})}{n-1}$$

Let s_x and s_y denote the standard deviations of x and y calculated from a sample. Then the mean product of the standardized variables, $\frac{x_i - \bar{x}}{s_x}$ and $\frac{y_i - \bar{y}}{s_y}$, calculated from a sample is:

$$\frac{\sum_i \left(\frac{x_i - \bar{x}}{s_x} - 0\right)\left(\frac{y_i - \bar{y}}{s_y} - 0\right)}{n-1} = \frac{\sum_i (x_i - \bar{x})(y_i - \bar{y})}{(n-1)s_x s_y} = \frac{\sum_i (x_i - \bar{x})(y_i - \bar{y})}{\sqrt{\sum_i (x_i - \bar{x}) \sum_i (y_i - \bar{y})}}$$

The last term can be written:

$$\frac{SS_{xy}}{\sqrt{SS_{xx} SS_{yy}}} = r$$

which is the sample coefficient of correlation.

To test the significance of a correlation in a sample, the null and alternative hypotheses about the parameter ρ are:

$H_0: \rho = 0$
$H_1: \rho \neq 0$

The null hypothesis states that the coefficient of correlation in the population is not different from zero, that is, there is no linear association between variables in the population. The alternative hypothesis states that the correlation in the population differs from zero. In hypothesis testing, a t distribution can be used, because it can be shown that the t statistic:

$$t = \frac{r}{s_r}$$

has a t distribution with $(n-2)$ degrees of freedom assuming the following:
1. variables x and y have a joint bivariate normal distribution
2. the hypothesis $H_0: \rho = 0$ is true.

Here, $s_r = \sqrt{\dfrac{1-r^2}{n-2}}$ is the standard error of the coefficient of correlation. Further:

$$t = \frac{r-0}{\sqrt{\frac{1-r^2}{n-2}}}$$

or simplified:

$$t = \frac{r\sqrt{n-2}}{\sqrt{1-r^2}}$$

Example: Is there a linear association between weight and heart girth in this herd of cows? Weight was measured in kg and heart girth in cm on 10 cows:

Cow	1	2	3	4	5	6	7	8	9	10
Weight (y):	641	620	633	651	640	666	650	688	680	670
Heart girth (x):	205	212	213	216	216	217	218	219	221	226

The computed sums of squares and sum of products are: $SS_{xx} = 284.1$, $SS_{xy} = 738.3$, $SS_{yy} = 4218.9$. The sample coefficient of correlation is:

$$r = \frac{SS_{xy}}{\sqrt{SS_{xx}SS_{yy}}} \quad \frac{738.3}{\sqrt{(284.1)(4218.9)}} = 0.67$$

The calculated value of the t statistic is:

$$t = \frac{r\sqrt{n-2}}{\sqrt{1-r^2}} = \frac{0.67\sqrt{10-2}}{\sqrt{1-0.67^2}} = 2.58$$

The critical value with 5% significance level and 8 degrees of freedom is:

$$t_{\alpha/2,8} = t_{0.025,8} = 2.31$$

The calculated $t = 2.58$ is more extreme than 2.31, so H_0 is rejected. There is linear association between weight and heart girth in the population.

Note: We have seen earlier that the symbol of the coefficient of determination is R^2. The reason is that a numerical relationship exists between the R^2 and the sample coefficient of correlation r:

$$r^2 = R^2$$

This can be shown by the following:

$$r^2 = \frac{SS_{xy}^2}{SS_{xx}SS_{yy}} = \frac{SS_{REG}}{SS_{TOT}} = R^2$$

SS_{REG} and SS_{TOT} are the regression sum of squares and the total sum of squares, respectively. Note the conceptual difference between R^2 and r; R^2 determines the correctness of a linear model and r denotes linear association between variables.

8.1.1 SAS Example for Correlation

The SAS program for the example of weights and heart girths of cows is as follows:

SAS program:

```
DATA cows;
INPUT wt h_girth @@;
DATALINES;
641 205  620 212  633 213  651 216  640 216
666 217  650 218  688 219  680 221  670 226
;
PROC CORR DATA = cows ;
VAR wt h_girth;
RUN;
```

Explanation: The statement VAR defines the variables between which the correlation is computed.

SAS output:

```
                 Simple Statistics

Variable    N    Mean     Std Dev    Sum    Minimum    Maximum
wt         10   653.900   21.651    6539    620.000    688.000
h_girth    10   216.300    5.618    2163    205.000    226.000


          Pearson Correlation Coefficients, N = 10
                Prob > |r| under H0: Rho=0


                         wt         h_girth
          wt          1.00000       0.67437
                                    0.0325
          h_girth     0.67437       1.00000
                      0.0325
```

Explanation: First, the descriptive statistics are given. Next, the sample coefficient of correlation and its *P* value are shown (*Pearson Correlation Coefficients* and *Prob > |r| under H0: Rho=0*). The sample coefficient of correlation is 0.67437. The *P* value is 0.0325, which is less than 0.05. The conclusion is that correlation exists in the population.

8.2 Partial Correlation

In a correlation analysis the correlation between two variables can be influenced by a third variable. For example, we may be interested in estimating the relationship between percentages of milk fat and protein in dairy cows. Both variables depend on milk yield. The capacity of a cow to produce milk is greater than its capacity to produce dry matter, which leads to negative correlations between milk yield and both fat and protein percentages. In this situation we can call milk yield a controlling variable. So, the question is whether there is a 'true' relationship between fat and protein percentages ignoring (or adjusting for) milk yield. Another example is the relationship between height and weight of young bulls. Consider bulls that were not measured at the same age, but ranged from 6 to 12 months. Older animals will be heavier and taller, and we will assume that the age-weight and age-height relationships from 6 to 12 months are linear. For this example, age is a controlling variable. The question is whether age indirectly causes a strong relationship between weight and height, or if a relationship exists regardless of age.

To adjust on a control variable, we can estimate regressions of both variables of interest on the control variable, and then compute correlations between residuals obtained from those two regressions. If these two regressions are simple linear regressions, then the correlation after adjustment on the control variable is known as a partial correlation.

Formally, define the regression functions of two variables x and y on the third variable z:

$$x = \beta_{0x} + \beta_{1x} z + \varepsilon_x \quad \text{and} \quad y = \beta_{0y} + \beta_{1y} z + \varepsilon_y$$

where β_{0x}, β_{1x}, β_{0y}, and β_{1y} are regression parameters and ε_x and ε_y are random errors.

The partial correlation $\rho_{xy.z}$ is a correlation between ε_x and ε_y. The estimated partial correlation from a sample is the correlation between residuals of the variables x and y:

$$\left(x_i - \hat{x}_i\right) \text{ and } \left(y_i - \hat{y}_i\right)$$

where $\hat{x}_i$ and $\hat{y}_i$ are estimated values from sample regressions of x and y on z, respectively. The estimated partial correlation, often denoted as $r_{xy.z}$, can be computed as:

$$r_{xy.z} = \frac{r_{xy} - r_{xz} r_{yz}}{\sqrt{\left(1 - r_{xz}^2\right)\left(1 - r_{yz}^2\right)}}$$

where r_{xy}, r_{xz}, and r_{yz} are 'ordinary' correlations between variables x, y and z.

More generally, a set of variables can be used as control variables when we compute partial correlations. Residuals can come from multiple regression or any form of regression function.

Example: Twelve pigs of varying ages were weighed and had loin muscle area (lea) measured at slaughter. The data and simple correlations between all pairs of variables are shown below. The aim of the analysis is to determine the relationship between weight and loin muscle area independent of age at slaughter, which would be considered the control variable.

Age (days)	147	133	133	160	160	159	144	130	130	143	157	157
Lea (cm^2)	34.2	23.9	27.1	36.1	35.5	36.8	36.8	27.7	29.7	36.1	33.5	37.4
Weight (kg)	104	82	84	104	118	118	105	90	85	104	113	108

The simple correlations between age, loin muscle area (lea) and weight (wt) are:

$$r_{age\_lea} = 0.80214, \quad r_{age\_wt} = 0.91111, \text{ and } r_{lea\_wt} = 0.86630$$

The partial correlation coefficient between loin muscle area and weight adjusted on age is:

$$r_{lea\_wt.age} = \frac{r_{lea\_wt} - r_{lea\_age} r_{wt\_age}}{\sqrt{\left(1 - r_{lea\_age}^2\right)\left(1 - r_{wt\_age}^2\right)}} = 0.55041$$

The simple correlation between loin muscle area and slaughter weight was 0.86630. The partial correlation of loin muscle area with slaughter weight after adjustment for age at slaughter reduces to 0.55041. Calculating a partial correlation between loin muscle area and slaughter weight with age at slaughter as a control variable in effect adjusts the correlation for differences in age at slaughter.

To test the significance of a partial correlation in a sample, the null and alternative hypotheses about the parameter ρ are:

$$H_0: \rho_{xy.z} = 0$$
$$H_1: \rho_{xy.z} \neq 0$$

A t distribution can be used in the significance test, because it can be shown that the statistic:

$$t = \frac{r_{xy.z}\sqrt{n-3}}{\sqrt{1 - r_{xy.z}^2}}$$

has a t distribution with $(n-3)$ degrees of freedom.

For k control variables:

$$t = \frac{r_{xy.z}\sqrt{n-k-2}}{\sqrt{1 - r_{xy.z}^2}}$$

8.2.1 SAS Example for Partial Correlation

The SAS program for the example of weights (wt) and loin muscle area (lea) of pigs is as follows:

SAS program:

```
DATA pork;
INPUT age lea wt;
DATALINES;
147   5.3   104
133   3.7    82
133   4.2    84
160   5.6   104
160   5.5   118
159   5.7   118
144   5.7   105
130   4.3    90
130   4.6    85
143   5.6   104
157   5.2   113
157   5.8   108
;
```

```
PROC CORR DATA = pork;
VAR lea wt;
RUN;

PROC CORR DATA = pork;
VAR lea wt;
PARTIAL age;
RUN;
```

Explanation: Two programs are given. The first CORR procedure computes an ordinary correlation coefficient between *lea* and *wt*. The second computes partial correlation coefficient adjusted on *age*. The statement VAR defines the variables between which the correlation is computed, and the PARTIAL statement requests the partial correlation coefficient.

SAS output:

```
            2  Variables:     LEA        wt

    Pearson Correlation Coefficients, N = 12
            Prob > |r| under H0: Rho=0

                    lea              wt
        lea       1.00000         0.86630
                                  0.0003
        wt        0.86630         1.00000
                  0.0003
```

```
            1 Partial Variables:     age
            2 Variables:      lea     wt
```

```
                    Simple Statistics
                                                        Partial    Partial
Var   N     Mean      Std Dev    Sum    Minimum   Maximum  Variance    Std Dev
age  12   146.08333   12.31007   1753   130.00    160.00
lea  12    32.90       4.59130   394.80  23.90     37.40   8.26814     2.87544
wt   12   101.25000   12.93427   1215    82.00    118.00   31.26283    5.59132
```

```
    Pearson Partial Correlation Coefficients, N = 12
            Prob > |r| under H0: Partial Rho=0


                    lea              wt
        lea       1.00000         0.55041
                                  0.0793
        wt        0.55041         1.00000
                  0.0793
```

Explanation: The first part of the output shows the estimate of the ordinary correlation coefficient (0.86630). The second part is the output of the partial correlation analysis. First, descriptive statistics are given. Next, the partial coefficient of correlation and its *P* value are shown (*Pearson Partial Correlation Coefficients* and *Prob > |r| under H0: Rho=0*). The partial coefficient of correlation is 0.55041. The value was reduced by adjustment for age. Also, the *P* value is 0.0793, comparing to 0.0003 before the adjustment. We can conclude that part of the relationship between *lea* and *wt* is due to age.

8.3   Rank Correlation

In cases when variables are not normally distributed, but their values can be ranked, the nonparametric coefficient of rank correlation can be used as a measure of association. The rules are as follows. For each variable, values are sorted from lowest to highest and then ranks are assigned to them. For example, assume heights of four cows: 132, 130, 133 and 135. Assigned ranks are: 2, 1, 3 and 4. If there is a tie, then the average of their ranks is assigned. For heights 132, 130, 133, 130 and 130, assigned ranks are: 4, 2, 5, 2 and 2. Once the ranks are determined, the formula for the sample rank coefficient of correlation is the same as before:

$$r = \frac{SS_{xy}}{\sqrt{SS_{xx}SS_{yy}}} = \frac{\sum_i (x_i - \bar{x})(y_i - \bar{y})}{\sqrt{\sum_i (x_i - \bar{x}) \sum_i (y_i - \bar{y})}}$$

but now values x_i and y_i are the ranks of observation i for the respective variables.

Example: Is there a relationship between gene expression (RNA levels) and feed intake of lambs? The feed intake is kilograms consumed over a 1 week feeding period. The RNA measures expression of the leptin receptor gene.

Lamb	1	2	3	4	5	6	7	8	9	10	11	12
RNA	195	201	295	301	400	500	600	720	1020	3100	4100	6100
Rank	1	2	3	4	5	6	7	8	9	10	11	12
Feed intake	7.9	8.3	9.1	7.4	8.6	7.5	10.7	9.7	10.4	9.5	9.0	11.3
Rank	3	4	7	1	5	2	11	9	10	8	6	12

Using ranks as values we compute the following sums of squares and sum of products:

$$SS_{RNA} = 143$$
$$SS_{FeedIntake} = 143$$
$$SS_{RNA\_FeedIntake} = 95$$

Note that the sums of squares for both RNA and feed intake are the same as rank values go from 1 to 12. Using the usual formula the correlation is:

$$r = \frac{SS_{RNA\_FeedIntake}}{\sqrt{SS_{RNA}SS_{FeedIntake}}} = \frac{95}{\sqrt{(143)(143)}} = 0.664$$

8.3.1   SAS Example for Rank Correlation

The SAS program for the example of gene expression and feed intake of lambs is as follows:

SAS program:

```
DATA lambs;
INPUT rna intake @@;
DATALINES;
  195   7.9      201   8.3      295   9.1      301    7.4
  400   8.6      500   7.5      600  10.7      720    9.7
 1020  10.4     3100   9.5     4100   9.0     6100   11.3
;
```

```
PROC CORR DATA = lambs SPEARMAN;
VAR rna intake;
RUN;
```

Explanation: The SPEARMAN option computes the rank correlation. The VAR statement defines the variables between which the correlation is computed.

SAS output:

```
                  Simple Statistics
```

Variable	N	Mean	Std Dev	Median	Minimum	Maximum
rna	12	1461	1921	550.00000	195.00000	6100
intake	12	9.11667	1.25758	9.05000	7.40000	11.30000

```
     Spearman Correlation Coefficients, N = 12
              Prob > |r| under H0: Rho=0
```

	rna	intake
rna	1.00000	0.66434
		0.0185
intake	0.66434	1.00000
	0.0185	

Explanation: First, the descriptive statistics are given. Next, the sample coefficient of rank correlation and its P value are shown (*Spearman Correlation Coefficients* and *Prob > |r| under H0: Rho=0*). The sample coefficient of rank correlation is 0.66434. The P value is 0.0185, which is less than 0.05. The conclusion is that correlation between feed intake and leptin receptor gene expression exists in the population.

Exercises

8.1. Calculate the sample coefficient of correlation between estimated number of ovulated follicles and number of eggs laid by pheasants. Data of 11 pheasants were collected:

Number of eggs	39	29	46	28	31	25	49	57	51	21	42
Number of follicles	37	34	52	26	32	25	55	65	54	25	45

Test the null hypothesis that correlation is not different from zero.

8.2. An estimated coefficient of correlation is $r = 0.65$. Sample size is $n = 15$. Is this value significantly different from zero at the 5% level?

Chapter 9

Multiple Linear Regression

A simple linear regression explains the linear cause-consequence relationship between one independent variable x and a dependent variable y. Often, it is necessary to analyze effects of two or more independent variables on a dependent variable. For example, weight gain may be influenced by the protein level in feed, the amount of feed consumed, and the environmental temperature, etc. The variability of a dependent variable y can be explained by a function of several independent variables, x_1, x_2,..., x_p. A regression that has two or more independent variables is called a multiple regression.

Goals of multiple regression analysis can be:

1. To find a model (function) that best describes the dependent with the independent variables, that is, to estimate parameters of the model,
2. To predict values of the dependent variable based on new measurements of the independent variables,
3. To analyze the importance of particular independent variables, thus, to analyze whether all or just some independent variables are important in the model. This involves building an optimal model.

The multiple linear regression model is:

$$y = \beta_0 + \beta_1 x_1 + \beta_2 x_2 + ... + \beta_{p-1} x_{p-1} + \varepsilon$$

where:

y = dependent variable
x_1, x_2,..., x_{p-1} = independent variables
β_0, β_1, β_2,..., β_{p-1} = regression parameters
ε = random error

Data used in multiple regression have the general form:

y	x_1	x_2	...	x_{p-1}
y_1	x_{11}	x_{21}	...	$x_{(p-1)1}$
y_2	x_{12}	x_{22}	...	$x_{(p-1)2}$
...				
y_n	x_{1n}	x_{2n}	...	$x_{(p-1)n}$

Each observation y_i can be presented as:

$$y_i = \beta_0 + \beta_1 x_{1i} + \beta_2 x_{2i} + ... + \beta_{p-1} x_{(p-1)i} + \varepsilon_i \qquad i = 1,..., n$$

The assumptions of the model are:

1. $E(\varepsilon_i) = 0$
2. $Var(\varepsilon_i) = \sigma^2$, the variance is constant
3. $Cov\,(\varepsilon_i, \varepsilon_{i'}) = 0$, $i \neq i'$, different errors are independent
4. Usually, it is assumed that errors have a normal distribution

The following model properties follow directly from these model assumptions:

1. $E(y_i) = \beta_0 + \beta_1 x_i + \beta_2 x_{2i} + \ldots + \beta_{p-1} x_{(p-1)i}$
2. $Var(y_i) = Var(\varepsilon_i) = \sigma^2$
3. $Cov(y_i, y_{i'}) = 0$, $i \neq i'$

9.1 Two Independent Variables

Multiple linear regression will be explained by using a model with two independent variables. Estimating a model with three or more independent variables and testing hypotheses follows the same logic. The model for a linear regression with two independent variables and n observations is:

$$y_i = \beta_0 + \beta_1 x_{1i} + \beta_2 x_{2i} + \varepsilon_i \qquad\qquad i = 1, \ldots, n$$

where:

y_i = observation i of dependent variable y
x_{1i} and x_{2i} = observations i of independent variables x_1 and x_2
β_0, β_1, and β_2 = regression parameters
ε_i = random error

The regression model in matrix notation is:

$$\mathbf{y} = \mathbf{X}\boldsymbol{\beta} + \boldsymbol{\varepsilon}$$

where:

$\mathbf{y}$ = the vector of observations of a dependent variable
$\boldsymbol{\beta}$ = the vector of parameters
$\mathbf{X}$ = the matrix of observations of independent variables
$\boldsymbol{\varepsilon}$ = the vector of random errors with the mean $E(\boldsymbol{\varepsilon}) = \mathbf{0}$ and variance $Var(\boldsymbol{\varepsilon}) = \sigma^2 \mathbf{I}$

The matrices and vectors are defined as:

$$\mathbf{y} = \begin{bmatrix} y_1 \\ y_2 \\ \ldots \\ y_n \end{bmatrix} \quad \mathbf{X} = \begin{bmatrix} 1 & x_{11} & x_{21} \\ 1 & x_{12} & x_{22} \\ \ldots & \ldots & \ldots \\ 1 & x_{1n} & x_{2n} \end{bmatrix} \quad \boldsymbol{\beta} = \begin{bmatrix} \beta_0 \\ \beta_1 \\ \beta_2 \end{bmatrix} \quad \boldsymbol{\varepsilon} = \begin{bmatrix} \varepsilon_1 \\ \varepsilon_2 \\ \ldots \\ \varepsilon_n \end{bmatrix}$$

9.1.1 Estimation of Parameters

A vector of parameters β is estimated by a vector $\mathbf{b}$ from a sample of data assumed to be randomly chosen from a population. The estimation model for the sample is:

$$\hat{\mathbf{y}} = \mathbf{Xb}$$

where $\hat{\mathbf{y}} = \begin{bmatrix} \hat{y}_1 \\ \hat{y}_2 \\ \dots \\ \hat{y}_n \end{bmatrix} =$ the vector of estimated values, and $\mathbf{b} = \begin{bmatrix} b_0 \\ b_1 \\ b_2 \end{bmatrix} =$ the vector of estimators.

The vector of residuals is the difference between values from the $\mathbf{y}$ vector and corresponding estimated values from the $\hat{\mathbf{y}}$ vector:

$$\mathbf{e} = \mathbf{y} - \hat{\mathbf{y}} = \begin{bmatrix} e_1 \\ e_2 \\ \dots \\ e_n \end{bmatrix}$$

Using either least squares or maximum likelihood estimation, in the same way as shown for simple linear regression, the following normal equations are obtained:

$$\mathbf{X'Xb} = \mathbf{X'y}$$

Solving for $\mathbf{b}$ gives:

$$\mathbf{b} = (\mathbf{X'X})^{-1}\mathbf{X'y}$$

The elements of the $\mathbf{X'X}$ and $\mathbf{X'y}$ matrices are corresponding sums, sums of squares and sums of products:

$$\mathbf{X'X} = \begin{bmatrix} 1 & 1 & \dots & 1 \\ x_{11} & x_{12} & \dots & x_{1n} \\ x_{21} & x_{22} & \dots & x_{2n} \end{bmatrix} \begin{bmatrix} 1 & x_{11} & x_{21} \\ 1 & x_{12} & x_{22} \\ \dots & \dots & \dots \\ 1 & x_{1n} & x_{2n} \end{bmatrix} = \begin{bmatrix} n & \sum_i x_{1i} & \sum_i x_{2i} \\ \sum_i x_{1i} & \sum_i x_{1i}^2 & \sum_i x_{1i}x_{2i} \\ \sum_i x_{2i} & \sum_i x_{1i}x_{2i} & \sum_i x_{2i}^2 \end{bmatrix}$$

$$\mathbf{X'y} = \begin{bmatrix} 1 & 1 & \dots & 1 \\ x_{11} & x_{12} & \dots & x_{1n} \\ x_{21} & x_{22} & \dots & x_{2n} \end{bmatrix} \begin{bmatrix} y_1 \\ y_2 \\ \dots \\ y_n \end{bmatrix} = \begin{bmatrix} \sum_i y_i \\ \sum_i x_{1i}y_i \\ \sum_i x_{2i}y_i \end{bmatrix}$$

The vector of estimated values of a dependent variable can be expressed by using $\mathbf{X}$ and $\mathbf{y}$:

$$\hat{\mathbf{y}} = \mathbf{Xb} = \mathbf{X}(\mathbf{X'X})^{-1}\mathbf{Xy}$$

The variance σ^2 is estimated by:

$$s^2 = \frac{SS_{RES}}{n-p} = MS_{RES}$$

where:

$SS_{RES} = \mathbf{e'e}$ = the residual sum of squares

$(n - p)$ = the degrees of freedom

p = the number of parameters in the model, for two independent variables $p = 3$

MS_{RES} = the residual mean square

The square root of the variance estimator is the standard deviation of the regression model:

$$s = \sqrt{s^2}$$

Example: Estimate the regression of weight on heart girth and height, and its error variance, from the data of seven young bulls given in the following table:

Bull:	1	2	3	4	5	6	7
Weight, kg (y):	480	450	480	500	520	510	500
Heart girth, cm (x_1):	175	177	178	175	186	183	185
Height, cm (x_2):	128	122	124	128	131	130	124

The **y** vector and **X** matrix are:

$$\mathbf{y} = \begin{bmatrix} 480 \\ 450 \\ 480 \\ 500 \\ 520 \\ 510 \\ 500 \end{bmatrix} \qquad \mathbf{X} = \begin{bmatrix} 1 & 175 & 128 \\ 1 & 177 & 122 \\ 1 & 178 & 124 \\ 1 & 175 & 128 \\ 1 & 186 & 131 \\ 1 & 183 & 130 \\ 1 & 185 & 124 \end{bmatrix}$$

The matrices needed for parameter estimation are:

$$\mathbf{X'X} = \begin{bmatrix} 1 & 1 & 1 & 1 & 1 & 1 & 1 \\ 175 & 177 & 178 & 175 & 186 & 183 & 185 \\ 128 & 122 & 124 & 128 & 131 & 130 & 124 \end{bmatrix} \begin{bmatrix} 1 & 175 & 128 \\ 1 & 177 & 122 \\ 1 & 178 & 124 \\ 1 & 175 & 128 \\ 1 & 186 & 131 \\ 1 & 183 & 130 \\ 1 & 185 & 124 \end{bmatrix}$$

$$= \begin{bmatrix} 7 & 1259 & 887 \\ 1259 & 226573 & 159562 \\ 887 & 159562 & 112465 \end{bmatrix} = \begin{bmatrix} n & \sum_i x_{1i} & \sum_i x_{2i} \\ \sum_i x_{1i} & \sum_i x_{1i}^2 & \sum_i x_{1i} x_{2i} \\ \sum_i x_{2i} & \sum_i x_{1i} x_{2i} & \sum_i x_{2i}^2 \end{bmatrix}$$

$$\mathbf{X'y} = \begin{bmatrix} 1 & 1 & 1 & 1 & 1 & 1 & 1 \\ 175 & 177 & 178 & 175 & 186 & 183 & 185 \\ 128 & 122 & 124 & 128 & 131 & 130 & 124 \end{bmatrix} \begin{bmatrix} 480 \\ 450 \\ 480 \\ 500 \\ 520 \\ 510 \\ 500 \end{bmatrix} = \begin{bmatrix} 3440 \\ 619140 \\ 436280 \end{bmatrix} = \begin{bmatrix} \sum_i y_i \\ \sum_i x_{1i} y_i \\ \sum_i x_{2i} y_i \end{bmatrix}$$

$$(\mathbf{X'X})^{-1} = \begin{bmatrix} 365.65714 & -1.05347 & -1.38941 \\ -1.05347 & 0.00827 & -0.00342 \\ -1.38941 & -0.00342 & 0.01582 \end{bmatrix}$$

The **b** vector is:

$$\mathbf{b} = \begin{bmatrix} b_0 \\ b_1 \\ b_2 \end{bmatrix} = (\mathbf{X'X})^{-1}\mathbf{X'y} = \begin{bmatrix} 365.65714 & -1.05347 & -1.38941 \\ -1.05347 & 0.00827 & -0.00342 \\ -1.38941 & -0.00342 & 0.01582 \end{bmatrix} \begin{bmatrix} 3440 \\ 619140 \\ 436280 \end{bmatrix} = \begin{bmatrix} -495.014 \\ 2.257 \\ 4.581 \end{bmatrix}$$

Hence, the estimated regression is:

$$y = -495.014 + 2.257 x_1 + 4.581 x_2$$

The vector of estimated values is:

$$\hat{\mathbf{y}} = \mathbf{Xb} = \begin{bmatrix} 1 & 175 & 128 \\ 1 & 177 & 122 \\ 1 & 178 & 124 \\ 1 & 175 & 128 \\ 1 & 186 & 131 \\ 1 & 183 & 130 \\ 1 & 185 & 124 \end{bmatrix} \begin{bmatrix} -495.014 \\ 2.257 \\ 4.581 \end{bmatrix} = \begin{bmatrix} 486.35 \\ 463.38 \\ 474.80 \\ 486.35 \\ 524.93 \\ 513.57 \\ 490.60 \end{bmatrix}$$

The residual vector is:

$$\mathbf{e} = \mathbf{y} - \hat{\mathbf{y}} = \begin{bmatrix} -6.35 \\ -13.38 \\ 5.20 \\ 13.65 \\ -4.93 \\ -3.57 \\ 9.40 \end{bmatrix}$$

The residual sum of squares is:

$$SS_{RES} = \mathbf{e'e} = 558.059$$

The residual mean square, which is an estimate of the error variance, is:

$$MS_{RES} = s^2 = \frac{SS_{RES}}{n-p} = \frac{558.059}{7-3} = 139.515$$

9.1.2 Student *t* test in Testing Hypotheses

The expectation and variance of estimators are:

$$E(\mathbf{b}) = \boldsymbol{\beta} \quad \text{and} \quad Var(\mathbf{b}) = \sigma^2(\mathbf{X'X})^{-1}$$

If the variance σ^2 is unknown, it can be estimated from a sample. Then the variance of the **b** vector equals:

$$\mathbf{s}^2(\mathbf{b}) = s^2(\mathbf{X'X})^{-1}$$

The null and alternative hypotheses are:

$$H_0: \beta_1 = \beta_2 = 0$$
$$H_1: \text{at least one } \beta_i \neq 0, \, i = 1 \text{ and } 2$$

A test of the hypotheses, that is, the test that b_1 or b_2 are significantly different than zero, can be done by using a *t* test. The test statistic is:

$$t = \frac{b_i}{s(b_i)}$$

where $s(b_i) = \sqrt{s^2(b_i)}$. The critical value of the *t* distribution is determined by the level of significance α and degrees of freedom $(n-p)$, where p is the number of parameters.

Example: Recall the example of weights, heart girths and heights of young bulls. The following was previously calculated: the estimated variance $MS_{RES} = s^2 = 139.515$, and the parameter estimates $b_0 = -495.014$, $b_1 = 2.257$ and $b_2 = 4.581$, for the intercept, heart girth and height, respectively. What are the variances of the estimated parameters? Test H_0 that changes of height and heart girth do not influence changes in weight.

$$\mathbf{s}^2(\mathbf{b}) = s^2(\mathbf{X'X})^{-1} = 139.515 \begin{bmatrix} 365.65714 & -1.05347 & -1.38941 \\ -1.05347 & 0.00827 & -0.00342 \\ -1.38941 & -0.00342 & 0.01582 \end{bmatrix}$$

$$\mathbf{s}^2(\mathbf{b}) = \begin{bmatrix} 51017.083 & -146.975 & -193.843 \\ -146.975 & 1.153 & -0.477 \\ -193.843 & -0.477 & 2.207 \end{bmatrix}$$

Thus, the variance estimates for b_1 and b_2 are $s^2(b_1) = 1.153$ and $s^2(b_2) = 2.207$, respectively. The *t* statistics are:

$$t = \frac{b_i}{s(b_i)}$$

The calculated t for b_1 is:

$$t = \frac{2.257}{\sqrt{1.153}} = 2.10$$

The calculated t for b_2 is:

$$t = \frac{4.581}{\sqrt{2.207}} = 3.08$$

For the significance level $\alpha = 0.05$, the critical value of the t distribution is $t_{0.025,4} = 2.776$ (See Appendix B: Critical values of Student t distributions). Only the t for b_2 is greater than the critical value, H_0: $\beta_2 = 0$ is rejected at the 5% level of significance.

9.1.3 Partitioning Total Variability and Tests of Hypotheses

As for the simple regression model, the total sum of squares can be partitioned into regression and residual sums of squares:

$$SS_{TOT} = SS_{REG} + SS_{RES}$$

The regression sum of squares is: $SS_{REG} = (\hat{\mathbf{y}} - \overline{\mathbf{y}})'(\hat{\mathbf{y}} - \overline{\mathbf{y}}) = \sum_i (\hat{y}_i - \overline{y})^2$

The residual sum of squares is: $SS_{RES} = (\mathbf{y} - \hat{\mathbf{y}})'(\mathbf{y} - \hat{\mathbf{y}}) = \sum_i (y_i - \hat{y}_i)^2$

The total sum of squares is: $SS_{TOT} = (\mathbf{y} - \overline{\mathbf{y}})'(\mathbf{y} - \overline{\mathbf{y}}) = \sum_i (y_i - \overline{y})^2$

or shortly, using the computed vector $\mathbf{b}$:

$$SS_{REG} = \mathbf{b'X'y} - n\overline{y}^2$$

$$SS_{RES} = \mathbf{y'y} - \mathbf{b'X'y}$$

$$SS_{TOT} = \mathbf{y'y} - n\overline{y}^2$$

Degrees of freedom for the total, regression and residual sums of squares are:

$$n - 1 = (p - 1) + (n - p)$$

Here, n is the number of observations and p is the number of parameters.

Mean squares are obtained by dividing the sums of squares by their corresponding degrees of freedom:

Regression mean square: $MS_{REG} = \dfrac{SS_{REG}}{p - 1}$

Residual mean square: $MS_{RES} = \dfrac{SS_{RES}}{n - p}$

If H_0 holds, the statistic:

$$F = \frac{MS_{REG}}{MS_{RES}}$$

has an F distribution with $(p-1)$ and $(n-p)$ degrees of freedom. For the α level of significance, H_0 is rejected if the calculated F is greater than the critical value $F_{\alpha,p-1,n-p}$.

The *ANOVA* table is:

Source	SS	df	MS = SS / df	F
Regression	SS_{REG}	$p-1$	$MS_{REG} = SS_{REG} / (p-1)$	$F = MS_{REG} / MS_{RES}$
Residual	SS_{RES}	$n-p$	$MS_{RES} = SS_{RES} / (n-p)$	
Total	SS_{TOT}	$n-1$		

The coefficient of multiple determination is:

$$R^2 = \frac{SS_{REG}}{SS_{TOT}} = 1 - \frac{SS_{RES}}{SS_{TOT}} \qquad 0 \le R^2 \le 1$$

R^2 describes the proportion of the total variance accounted for by the regression model.

Note that extension of the model to more than two independent variables is straight forward and follows the same logic as the model with two independent variables. Further, it is possible to define interaction between independent variables.

Example: For the example of weights, heart girths and heights of young bulls, test the null hypothesis H_0: $\beta_1 = \beta_2 = 0$ using an F distribution. The following were previously defined and calculated:

$n = 7$

$$\mathbf{y} = \begin{bmatrix} 480 \\ 450 \\ 480 \\ 500 \\ 520 \\ 510 \\ 500 \end{bmatrix}, \ \mathbf{X} = \begin{bmatrix} 1 & 175 & 128 \\ 1 & 177 & 122 \\ 1 & 178 & 124 \\ 1 & 175 & 128 \\ 1 & 186 & 131 \\ 1 & 183 & 130 \\ 1 & 185 & 124 \end{bmatrix}, \ \bar{y} = \frac{\sum_i y_i}{n} = \frac{3440}{7} = 491.43 \ \text{and} \ \mathbf{b} = \begin{bmatrix} b_0 \\ b_1 \\ b_2 \end{bmatrix} = \begin{bmatrix} -495.014 \\ 2.257 \\ 4.581 \end{bmatrix}$$

The sums of squares are:

$$SS_{REG} = \mathbf{b'X'y} - n\bar{y}^2 =$$

$$\begin{bmatrix} -495.014 & 2.257 & 4.581 \end{bmatrix} \begin{bmatrix} 1 & 1 & 1 & 1 & 1 & 1 & 1 \\ 175 & 177 & 178 & 175 & 186 & 183 & 185 \\ 128 & 122 & 124 & 128 & 131 & 130 & 124 \end{bmatrix} \begin{bmatrix} 480 \\ 450 \\ 480 \\ 500 \\ 520 \\ 510 \\ 500 \end{bmatrix} - (7)(491.43) = 2727.655$$

$$SS_{TOT} = \mathbf{y'y} - n\bar{y}^2 = \begin{bmatrix} 480 & 450 & 480 & 500 & 520 & 510 & 500 \end{bmatrix} \begin{bmatrix} 480 \\ 450 \\ 480 \\ 500 \\ 520 \\ 510 \\ 500 \end{bmatrix} - (7)(491.43) = 3285.714$$

$$SS_{RES} = SS_{TOT} - SS_{REG} = 3285.714 - 2727.655 = 558.059$$

The *ANOVA* table:

Source	SS	df	MS	F
Regression	2727.655	2	1363.828	9.78
Residual	558.059	4	139.515	
Total	3285.714	6		

The critical value of the F distribution for $\alpha = 0.05$ and 2 and 4 degrees of freedom is $F_{0.05,2,4} = 6.94$ (See Appendix B: Critical values of the F distributions). Since the calculated $F = 9.78$ is greater than the critical value, H_0 is rejected at the 5% level of significance.

The coefficient of determination is:

$$R^2 = \frac{2727.655}{3285.714} = 0.83$$

This value indicates that 83% of the variation in weight of bulls in this population can be accounted for by variation in height and heart girth.

9.2 Partial and Sequential Sums of Squares

Recall that the total sum of squares can be partitioned into regression and residual sums of squares. The regression sum of squares can further be partitioned into sums of squares corresponding to parameters in the model. By partitioning the sums of squares, the importance of adding or dropping particular parameters from the model can be tested. For example, consider a model with three independent variables and four parameters:

$$y = \beta_0 + \beta_1 x_1 + \beta_2 x_2 + \beta_3 x_3 + \varepsilon$$

Now, assume that this model contains a maximum number of parameters. This model can be called a full model. Any model with fewer parameters than the full model is called a reduced model. All possible reduced models derived from the full model with four parameters are:

$$y = \beta_0 + \beta_1 x_1 + \beta_2 x_2 + \varepsilon$$
$$y = \beta_0 + \beta_1 x_1 + \beta_3 x_3 + \varepsilon$$
$$y = \beta_0 + \beta_2 x_2 + \beta_3 x_3 + \varepsilon$$
$$y = \beta_0 + \beta_1 x_1 + \varepsilon$$
$$y = \beta_0 + \beta_2 x_2 + \varepsilon$$
$$y = \beta_0 + \beta_3 x_3 + \varepsilon$$
$$y = \beta_0 + \varepsilon$$

Let $SS_{REG}(\beta_0,\beta_1,\beta_2,\beta_3)$ denote the regression sum of squares when all parameters are in the model. Analogously, $SS_{REG}(\beta_0,\beta_1,\beta_2)$, $SS_{REG}(\beta_0,\beta_2,\beta_3)$, $SS_{REG}(\beta_0,\beta_1,\beta_3)$, $SS_{REG}(\beta_0,\beta_1)$, $SS_{REG}(\beta_0,\beta_2)$, $SS_{REG}(\beta_0,\beta_3)$ and $SS_{REG}(\beta_0)$ are the regression sums of squares for reduced models with corresponding parameters. A decrease of the number of parameters in a model always yields a decrease in the regression sum of squares and a numerically equal increase in the residual sum of squares. Similarly, adding new parameters to a model gives an increase in the regression sum of squares and a numerically equal decrease in the residual sum of squares. This difference in sums of squares is often called extra sums of squares. Let $R(*|\#)$ denote the extra sum of squares when the parameters * are added to a model already having parameters #, or analogously, the parameters * are dropped from the model leaving parameters #. For example, $R(\beta_2|\beta_0,\beta_1)$ depicts the increase in SS_{REG} when β_2 is included in the model already having β_0 and β_1:

$$R(\beta_2|\beta_0,\beta_1) = SS_{REG}(\beta_0,\beta_1,\beta_2) - SS_{REG}(\beta_0,\beta_1)$$

Also, $R(\beta_2|\beta_0,\beta_1)$ equals the decrease of the residual sum of squares when adding β_2 to a model already having β_1 and β_0:

$$R(\beta_2|\beta_0,\beta_1) = SS_{RES}(\beta_0,\beta_1) - SS_{RES}(\beta_0,\beta_1,\beta_2)$$

Technically, the model with β_0, β_1 and β_2 can be considered a full model, and the model with β_0 and β_1 a reduced model.

According to the way they are calculated, there exist two extra sums of squares, the sequential and partial sums of squares, also called Type I and Type II sums of squares, respectively. Sequential extra sums of squares denote an increase of the regression sum of squares when parameters are added one by one in the model. Obviously, the sequence of parameters is important. For the example model with four parameters, and the sequence of parameters β_0, β_1, β_2, and β_3, the following sequential sums of squares can be written:

$$R(\beta_1|\beta_0) = SS_{REG}(\beta_0,\beta_1) - SS_{REG}(\beta_0) \qquad \text{(Note that } SS_{REG}(\beta_0) = 0)$$
$$R(\beta_2|\beta_0,\beta_1) = SS_{REG}(\beta_0,\beta_1,\beta_2) - SS_{REG}(\beta_0,\beta_1)$$
$$R(\beta_3|\beta_0,\beta_1,\beta_2) = SS_{REG}(\beta_0,\beta_1,\beta_2,\beta_3) - SS_{REG}(\beta_0,\beta_1,\beta_2)$$

The regression sum of squares for the full model with four parameters is the sum of all possible sequential sums of squares:

$$SS_{REG}(\beta_0,\beta_1,\beta_2,\beta_3) = R(\beta_1|\beta_0) + R(\beta_2|\beta_0,\beta_1) + R(\beta_3|\beta_0,\beta_1,\beta_2)$$

The partial sums of squares denote an increase of regression sums of squares when a particular parameter is added to the model, and all other possible parameters are already in the model. For the current example there are three partial sums of squares:

$$R(\beta_1|\beta_0,\beta_2,\beta_3) = SS_{REG}(\beta_0,\beta_1,\beta_2,\beta_3) - SS_{REG}(\beta_0,\beta_2,\beta_3)$$
$$R(\beta_2|\beta_0,\beta_1,\beta_3) = SS_{REG}(\beta_0,\beta_1,\beta_2,\beta_3) - SS_{REG}(\beta_0,\beta_1,\beta_3)$$
$$R(\beta_3|\beta_0,\beta_1,\beta_2) = SS_{REG}(\beta_0,\beta_1,\beta_2,\beta_3) - SS_{REG}(\beta_0,\beta_1,\beta_2)$$

Note that the partial sums of squares do not sum to anything meaningful.

Sequential sums of squares are applicable when variation of one independent variable should be removed before testing the effect of the independent variable of primary interest. In other words, the values of the dependent variable are adjusted for the first independent variable. The variable used in adjustment is usually preexisting in an experiment. Thus, the order in which variables enter the model is important. For example, consider an experiment in which weaning weight of lambs is the dependent variable and inbreeding coefficient is the independent variable of primary interest. Lambs are weaned on a fixed date so vary in age on the day of weaning. Age at weaning is unaffected by inbreeding coefficient and the effect of age at weaning should be removed before examining the effect of inbreeding. Age at weaning serves only as an adjustment of weaning weight in order to improve the precision of testing the effect of inbreeding.

Partial sums of squares are used when all variables are equally important in explaining the dependent variable and the objective is testing and estimating regression parameters for all independent variables in the model. For example, if weight of bulls is fitted to a model including the independent variables height and heart girth, both variables must be tested and their order is not important.

Partial and sequential sums of squares can be used to test the suitability of adding particular parameters to a model. If the extra sum of squares is large enough, the added parameters account for significant variation in the dependent variable. The test is conducted with an F test by dividing the mean extra sum of squares by the residual mean square for the full model. For example, to test whether β_3 and β_4 are needed in a full model including β_0, β_1, β_2, β_3 and β_4:

$$F = \frac{R(\beta_3,\beta_4 \mid \beta_0,\beta_1,\beta_2)/(5-2)}{SS_{RES}(\beta_0,\beta_1,\beta_2,\beta_3,\beta_4)/(n-5)} = \frac{\left[SS_{RES}(\beta_0,\beta_1,\beta_2) - SS_{RES}(\beta_0,\beta_1,\beta_2,\beta_3,\beta_4)\right]/(5-2)}{SS_{RES}(\beta_0,\beta_1,\beta_2,\beta_3,\beta_4)/(n-5)}$$

An analogous test can be used for any set of parameters in the model. The general form of the test for adding some set of parameters to the model is:

$$F = \frac{(SS_{RES\_REDUCED} - SS_{RES\_FULL})/(p_{FULL} - p_{REDUCED})}{SS_{RES\_FULL}/(n - p_{FULL})}$$

where:

$p_{REDUCED}$ = the number of parameters in the reduced model
p_{FULL} = the number of parameters in the full model
n = number of observations on the dependent variable
$SS_{RES\_FULL} / (n - p_{FULL}) = MS_{RES\_FULL}$ = the residual mean square for the full model

Example: For the example of weights, heart girths and heights of seven young bulls, the sequential and partial sums of squares will be calculated. Recall that β_1 and β_2 are parameters explaining the influence of the independent variables heart girth and height, respectively, on the dependent variable weight. The following sums of squares for the full model have already been calculated:

$$SS_{TOT} = 3285.714$$
$$SS_{REG\_FULL} = 2727.655$$
$$SS_{RES\_FULL} = 558.059$$
$$MS_{RES\_FULL} = 139.515$$

The sequential sums of squares are:

$$R(\beta_1|\beta_0) = SS_{REG}(\beta_0,\beta_1) = 1400.983$$
$$R(\beta_2|\beta_0,\beta_1) = SS_{REG}(\beta_0,\beta_1,\beta_2) - SS_{REG}(\beta_0,\beta_1) = 2727.655 - 1400.983 = 1326.672$$

The same values are obtained when the residual sums of squares are used:

$$R(\beta_1|\beta_0) = SS_{RES}(\beta_0) - SS_{RES}(\beta_0,\beta_1) = 3285.714 - 1884.731 = 1400.983$$
$$R(\beta_2|\beta_0,\beta_1) = SS_{RES}(\beta_0,\beta_1) - SS_{RES}(\beta_0,\beta_1,\beta_2) = 1884.731 - 558.059 = 1326.672$$

The partial sums of squares are:

$$R(\beta_1|\beta_0,\beta_2) = SS_{REG}(\beta_0,\beta_1,\beta_2) - SS_{REG}(\beta_0,\beta_2) = 2727.655 - 2111.228 = 616.427$$
$$R(\beta_2|\beta_0,\beta_1) = SS_{REG}(\beta_0,\beta_1,\beta_2) - SS_{REG}(\beta_0,\beta_1) = 2727.655 - 1400.983 = 1326.672$$

The same values are obtained when the residual sums of squares are used:

$$R(\beta_1|\beta_0,\beta_2) = SS_{RES}(\beta_0,\beta_2) - SS_{RES}(\beta_0,\beta_1,\beta_2) = 1174.486 - 558.059 = 616.427$$
$$R(\beta_2|\beta_0,\beta_1) = SS_{RES}(\beta_0,\beta_1) - SS_{RES}(\beta_0,\beta_1,\beta_2) = 1884.731 - 558.059 = 1326.672$$

To test the parameters the following F statistics are calculated. For example, for testing H_0: $\beta_1 = 0$, vs. H_1: $\beta_1 \neq 0$, using the partial sum of squares, the value of F statistic is:

$$F = \frac{[SS_{RES}(\beta_0,\beta_2) - SS_{RES}(\beta_0,\beta_1,\beta_2)]/(3-2)}{SS_{RES}(\beta_0,\beta_1,\beta_2)/(7-3)} = \frac{[1174.486 - 558.059]/(1)}{558.059/(4)} = 4.42$$

or

$$F = \frac{R(\beta_1|\beta_0,\beta_2)/(3-2)}{SS_{RES}(\beta_0,\beta_1,\beta_2)/(7-3)} = \frac{616.427/1}{558.059/4} = 4.42$$

The critical value of the F distribution for $\alpha = 0.05$ and 1 and 4 degrees of freedom is $F_{0.05,1,4} = 7.71$ (See Appendix B: Critical values of F distributions). Since the calculated $F = 4.42$ is not greater than the critical value, H_0 is not rejected.

The sequential and partial sums of squares with corresponding degrees of freedom and F values can be summarized in *ANOVA* tables.

The *ANOVA* table with sequential sums of squares:

Source	SS	df	MS	F
Heart girth	1400.983	1	1400.983	10.04
Height	1326.672	1	1326.672	9.51
Residual	558.059	4	139.515	
Total	3285.714	6		

The *ANOVA* table with partial sums of squares:

Source	SS	df	MS	F
Heart girth	616.427	1	616.427	4.42
Height	1326.672	1	1326.672	9.51
Residual	558.059	4	139.515	
Total	3285.714	6		

9.3   Testing Model Fit using a Likelihood Ratio Test

The adequacy of a reduced model relative to the full model can be determined by comparing their likelihood functions. The values of the likelihood functions for both models are computed using their estimated parameters. When analyzing the ratio of the reduced to the full model,

$$\frac{L(reduced)}{L(full)}$$

values close to 1 indicate adequacy of the reduced model. The distribution of the logarithm of this ratio multiplied by (–2) has an approximate *chi*-square distribution with degrees of freedom equal to the difference between the number of parameters of the full and reduced models:

$$\chi^2 = -2log\frac{L(reduced)}{L(full)} = 2\left[-logL(reduced) + logL(full)\right]$$

This expression is valid when variances are either known or are estimated from a sample.

Example: For the example of weights, heart girths and heights of young bulls, likelihood functions will be used to test the necessity for inclusion of the variable height (described with the parameter β_2) in the model.

The full model is:

$$y_i = \beta_0 + \beta_1 x_{1i} + \beta_2 x_{2i} + \varepsilon_i$$

The reduced model is:

$$y_i = \beta_0 + \beta_1 x_{1i} + \varepsilon_i$$

where:

y_i = the weight of bull i
x_{1i} = the heart girth of bull i
x_{2i} = the height of bull i
$\beta_0, \beta_1, \beta_2$ = regression parameters
ε_i = random error

The parameters were estimated by finding the maximum of the corresponding likelihood functions. Recall that the equations for estimating the parameters are equal for both maximum likelihood and least squares. However, the maximum likelihood estimators of the variances are as follows. For the full model:

$$s^2_{ML\_FULL} = \frac{1}{n}\sum_i (y_i - b_0 - b_1 x_{1i} - b_2 x_{2i})^2 = \frac{1}{n} SS_{RES\_FULL}$$

For the reduced model:

$$s^2_{ML\_REDUCED} = \frac{1}{n}\sum_i (y_i - b_0 - b_1 x_{1i})^2 = \frac{1}{n} SS_{RES\_REDUCED}$$

Estimates of the parameters are given in the following table:

	Estimates				
	b_0	b_1	b_2	s^2_{ML}	SS_{RES}
Full model	−495.014	2.257	4.581	79.723	558.059
Reduced model	−92.624	3.247		269.247	1884.731

The value of the log likelihood function for the full model is:

$$logL(b_0, b_1, b_2, s^2_{ML\_FULL} \mid y) = -\frac{n}{2}log(s^2_{ML}) - \frac{n}{2}log(2\pi) - \frac{\sum_i (y_i - b_0 - b_1 x_{1i} - b_2 x_{2i})^2}{2s^2_{ML\_FULL}}$$

$$= -\frac{n}{2}log(s^2_{ML\_FULL}) - \frac{n}{2}log(2\pi) - \frac{SS_{RES\_FULL}}{2s^2_{ML\_FULL}}$$

$$= -\frac{7}{2}log(79.723) - \frac{7}{2}log(2\pi) - \frac{558.059}{2(79.723)} = -25.256$$

The value of log likelihood function for the reduced model is:

$$logL(b_0, b_1, s^2_{ML\_REDUCED} \mid y) = -\frac{n}{2}log(s^2_{ML\_REDUCED}) - \frac{n}{2}log(2\pi) - \frac{\sum_i (y_i - b_0 - b_1 x_{1i})^2}{2s^2_{ML\_REDUCED}}$$

$$= -\frac{n}{2}log(s^2_{ML\_REDUCED}) - \frac{n}{2}log(2\pi) - \frac{SS_{RES\_REDUCED}}{2s^2_{ML\_REDUCED}}$$

$$= -\frac{7}{2}log(269.247) - \frac{7}{2}log(2\pi) - \frac{1884.731}{2(269.247)} = -29.516$$

The value of the χ^2 statistic is:

$$\chi^2 = -2log\frac{L(b_0, b_1, s^2_{ML\_REDUCED} \mid y)}{L(b_0, b_1, b_2, s^2_{ML\_FULL} \mid y)}$$

$$= 2\left[- logL(b_0, b_1, s^2_{ML\_REDUCED} \mid y) + logL(b_0, b_1, b_2, s^2_{ML\_FULL} \mid y)\right] = 2(29.516 - 25.256) = 8.52$$

The critical value of the *chi*-square distribution for 1 degree of freedom (the difference between the number of parameters of the full and reduced models) and a significance level of 0.05, is $\chi^2_{0.05,1}$ = 3.841. The calculated value is greater than the critical value, thus the variable height is needed in the model.

Assuming that the variances are equal regardless of the model, the likelihood functions of the full and reduced models differ only for the expression of the residual sums of squares. Then, with the known variance:

$$- 2log\frac{L(reduced)}{L(full)} = \frac{SS_{RES\_REDUCED} - SS_{RES\_FULL}}{\sigma^2}$$

For large *n* the distribution of this expression is approximately *chi*-square. Further, assuming normality of *y*, the distribution is exactly *chi*-square.

The variance σ^2 can be estimated by dividing the residual sum of squares from the full model by $(n - p_{FULL})$ degrees of freedom. Then:

$$- 2log\frac{L(reduced)}{L(full)} = \frac{SS_{RES\_REDUCED} - SS_{RES\_FULL}}{SS_{RES\_FULL}/(n - p_{FULL})}$$

has a *chi*-square distribution with $(n - p_{FULL})$ degrees of freedom. Assuming normality, if the expression is divided by $(p_{FULL} - p_{REDUCED})$, then:

$$F = \frac{SS_{RES\_REDUCED} - SS_{RES\_FULL}/(p_{FULL} - p_{REDUCED})}{SS_{RES\_FULL}/(n - p_{FULL})}$$

has an *F* distribution with $(p_{FULL} - p_{REDUCED})$ and $(n - p_{FULL})$ degrees of freedom. Note that this is exactly the same expression derived from the extra sums of squares if *y* is a normal variable.

9.4   SAS Example for Multiple Regression

The SAS program for the example of weights, heart girths and heights of young bulls is as follows. Recall the data:

Bull:	1	2	3	4	5	6	7
Weight, kg (y):	480	450	480	500	520	510	500
Heart girth, cm (x_1):	175	177	178	175	186	183	185
Height, cm (x_2):	128	122	124	128	131	130	124

SAS program:

```
DATA bulls;
INPUT wt h_girth height;
DATALINES;
480 175 128
450 177 122
480 178 124
500 175 128
520 186 131
510 183 130
500 185 124

;
PROC GLM;
MODEL wt = h_girth height ;
RUN;
QUIT;
```

Explanation: Either the GLM or REG procedure can be used. The statement, *MODEL wt = h_girth height* defines *wt* as the dependent, and *h_girth* and *height* as independent variables.

SAS output:

Dependent Variable: wt

Source	DF	Sum of Squares	Mean Square	F Value	Pr > F
Model	2	2727.655201	1363.827601	9.78	0.0288
Error	4	558.059085	139.514771		
Corrected Total	6	3285.714286			

R-Square	Coeff Var	Root MSE	wt Mean
0.830156	2.403531	11.81164	491.4286

Source	DF	Type I SS	Mean Square	F Value	Pr > F
h_girth	1	1400.983103	1400.983103	10.04	0.0339
height	1	1326.672098	1326.672098	9.51	0.0368

Source	DF	Type III SS	Mean Square	F Value	Pr > F
h_girth	1	616.426512	616.426512	4.42	0.1034
height	1	1326.672098	1326.672098	9.51	0.0368

Parameter	Estimate	Standard Error	t Value	Pr > |t|
Intercept	-495.0140313	225.8696150	-2.19	0.0935
h_girth	2.2572580	1.0738674	2.10	0.1034
height	4.5808460	1.4855045	3.08	0.0368

Explanation: First is an *ANOVA* table for the dependent variable *wt*. The sources of variation are *Model*, *Error* (residual) and *Corrected Total*. In the table are listed: degrees of freedom (*DF*), *Sum of Squares*, *Mean Square*, calculated *F* (*F Value*) and *P* value (*Pr > F*). For the model, $F = 9.78$ with a *P* value $= 0.0288$. Below the *ANOVA* table, the coefficient of determination (*R-square*) $= 0.830156$ and the standard deviation of the dependent variable (*Root MSE*) $= 11.81164$ are given. In the next two tables *F* tests for *h_girth* and *height* are given. Here the *F* values and corresponding *P* values describe the significance of *h_girth* and *height* in the model. The first table is based on the sequential (*Type I SS*), the second on the partial sums of squares (*Type III SS*). The sequential sums of squares are sums of squares corrected for the effects of the variables preceding the observed effect. The partial sums of squares are sums of squares corrected for all other effects in the model, and indicate the significance of a particular independent variable in explaining variation of the dependent variable. The same can be seen in the next table, in which parameter estimates (*Estimate*) with corresponding standard errors (*Standard Error*), *t* and *P* values (*Pr > |T|*) are shown. The *t value* tests whether the estimates are significantly different than zero. The *P* values for b_1 (*h_girth*) and b_2 (*height*) are 0.1034 and 0.0368. Since *P* values are relatively small, it seems that both independent variables should be included in the model.

9.5 Power of Multiple Regression

Power of test for a multiple linear regression can be calculated by using *t* or *F* central and noncentral distributions. Recall that the null and alternative hypotheses are $H_0: \beta_1 = \beta_2 = \ldots = \beta_{p-1} = 0$ and H_1: at least one $\beta_i \neq 0$, where $i = 1$ to $p - 1$, when *p* is the number of parameters. As the alternative hypotheses for particular parameters, the estimates from a sample can be used and then $H_1: \beta_i = b_i$. The *t* distribution is used analogously as previously shown for simple linear regression. Here the use of an *F* distribution for the whole model and for particular regression parameters using the sum of squares for regression and partial sums of squares will be shown. If H_0 holds, the test statistic *F* follows a central *F* distribution with corresponding numerator and denominator degrees of freedom. However, if H_1 holds, the *F* statistic has a noncentral *F* distribution with a noncentrality parameter $\lambda = \dfrac{SS}{MS_{RES}}$ and the corresponding degrees of freedom. Here, *SS* denotes the corresponding regression sum of squares or partial sum of squares. The power is a probability:

$$Power = P\ (F > F_{\alpha, df1, df2} = F_\beta)$$

that uses a noncentral *F* distribution for H_1, where, $F_{\alpha, df1, df2}$ is the critical value with α level of significance, and *df1* and *df2* are degrees of freedom, typically those for calculating the regression (or partial regression) and residual mean squares.

Example: Calculate the power of test for the example of weights, heart girths and heights of young bulls. Recall that in this example β_1 and β_2 are parameters explaining the influence of heart girth and height, respectively, on weight. The following were previously calculated:

$SS_{REG\_FULL} = 2727.655 =$ the regression sum of squares for the full model

$s^2 = MS_{RES\_FULL} = 139.515 =$ the residual mean square for the full model

The partial sums of squares are:

$$R(\beta_1|\beta_0,\beta_2) = 616.427$$

$$R(\beta_2|\beta_0,\beta_1) = 1326.672$$

The partial sums of squares with corresponding means squares, degrees of freedom and F values are shown in the following *ANOVA* table:

Source	SS	df	MS	F
Heart girth	616.427	1	616.427	4.42
Height	1326.672	1	1326.672	9.51
Residual	558.059	4	139.515	
Total	3285.714	6		

The estimated noncentrality parameter for the full model is:

$$\lambda = \frac{SS_{REG\_FULL}}{MS_{RES}} = \frac{2727.655}{139.515} = 19.551$$

Using a noncentral F distribution with 2 and 4 degrees of freedom and the noncentrality parameter $\lambda = 19.551$, the power is 0.745.

The estimate of the noncentrality parameter for heart girth is:

$$\lambda = \frac{R(\beta_1|\beta_2,\beta_0)}{MS_{RES}} = \frac{616.427}{139.515} = 4.418$$

Using a noncentral F distribution with 1 and 3 degrees of freedom and the noncentrality parameter $\lambda = 4.418$, the power is 0.364.

The estimate of the noncentrality parameter for height is:

$$\lambda = \frac{R(\beta_2|\beta_1,\beta_0)}{MS_{RES}} = \frac{1326.672}{139.515} = 9.509$$

Using a noncentral F distribution with 1 and 4 degrees of freedom and the noncentrality parameter $\lambda = 9.509$, the power is 0.642.

9.5.1 SAS Example for Calculating Power

To compute the power of test with SAS, the following statements can be used:

```
DATA a;
alpha=0.05;
n=7;
ssreg0=2727.655;
ssreg1=616.427;
ssreg2=1326.672;
msres =139.515;
```

```
df=n-3;
lambda0=ssreg0/msres;
lambda1=ssreg1/msres;
lambda2=ssreg2/msres;
Fcrit0=FINV(1-alpha,2,df);
Fcrit=FINV(1-alpha,1,df);
power0=1-PROBF(Fcrit0,2,df,lambda0);
power1=1-PROBF(Fcrit,1,df,lambda1);
power2=1-PROBF(Fcrit,1,df,lambda2);
PROC PRINT;
RUN;
```

Explanation: The terms used above are: *alpha* = significance level, *n* = sample size, *ssreg0* = regression sum of squares, *ssreg1* = sum of squares for heart girth, *ssreg2* = sum of squares for height, *msres* = residual mean square, *df* = residual degrees of freedom. Then presented are the corresponding noncentrality parameter estimates, *lambda0*, *lambda1* and *lambda2*, and the critical values, *Fcrit0* for the full model regression and *Fcrit* for the partial regressions. The critical value is computed by using the FINV function, which must have the cumulative values of percentiles $(1 - \alpha = 0.95)$ and degrees of freedom 1 or 2 and *df* defined. The PROBF function is the cumulative function of the *F* distribution which needs critical values, degrees of freedom and the noncentrality parameters *lambda*. Instead of *PROBF(Fcrit,1,df,lambda)* the alternative *CDF('F',Fcrit,1,df,lambda)* can be used. The power is calculated as *power0*, *power1*, and *power2*, for the full regression, heart girth, and height, respectively. The PRINT procedure results in the following SAS output:

alpha	n	df	ssreg0	ssreg1	ssreg2	msres	lambda0	lambda1
0.05	7	4	2727.66	616.427	1326.67	139.515	19.5510	4.41836

lambda2	Fcrit0	Fcrit	power0	power1	power2
9.50917	6.94427	7.70865	0.74517	0.36381	0.64182

9.6   Problems with Regression

Recall that a set of assumptions must be satisfied in order for a regression analysis to be valid. If these assumptions are not satisfied, inferences can be incorrect. Also, there can be other difficulties, which are summarized as follows: 1) some observations are unusually extreme; 2) model errors do not have constant variance; 3) model errors are not independent; 4) model errors are not normally distributed; 5) a nonlinear relationship exists between the independent and dependent variables; 6) one or more important independent variables are not included in the model; 7) the model is predefined, that is, it contains too many independent variables; 8) there is multicolinearity, that is, there is a strong correlation between independent variables.

These difficulties can lead to the use of the wrong model, poor regression estimates, failure to reject the null hypothesis when relationship exists, or imprecise parameter estimation due to large variance. These problems should be diagnosed and if possible eliminated.

9.6.1 Analysis of Residuals

The analysis of residuals can be informative of possible problems or unsatisfied assumptions. Recall that a residual is the difference between observed and estimated values of the dependent variable:

$$e_i = y_i - \hat{y}_i$$

The simplest method to inspect residuals is by using graphs. The necessary graphs include that of the residuals e_i plotted either against estimated values of the dependent variable $\hat{y}_i$, or observed values of the independent variable x_i. The following figures illustrate correctness or incorrectness of a regression model.

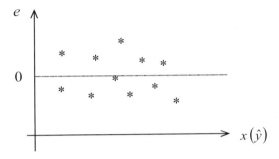

This model is correct. There is no systematic dispersion of residuals. The variance of e is constant across all values of $x(\hat{y})$. No unusually extreme observations are apparent.

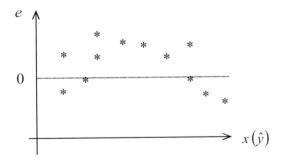

This figure shows a nonlinear influence of the independent on the dependent variable. Including x_i^2 or x_i^3 is probably required in the model. It is also possible that the relationship follows a log, exponential or some other nonlinear function.

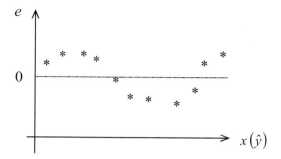

This figure illustrates errors that are not independent. This is called autocorrelation.

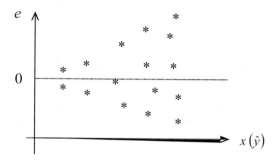

This figure illustrate variance that is not homogeneous (constant). Increasing values of the independent variable are associated with an increase in the variance. Transformation of either the *x* or *y* variable is needed. It may also be necessary to define a different variance structure. Normality of errors should be checked. Nonnormality may invalidate the *F* or *t* tests. One way to deal with such problems is to apply a generalized linear model, which can use distributions other than normal, to define a function of the mean of the dependent variable, and to correct the models for heterogeneous variance.

9.6.2   Extreme Observations

Some observations can be extreme compared either to the postulated model or to the mean of values of the independent variable(s). An extreme observation which opposes the postulated model is often called an *outlier*. An observation which is far from the mean of the *x* variable(s) is said to have *high leverage*. Extreme observations can, but do not always have, high influence on regression estimation. Figure 9.1 shows typical cases of extreme values.

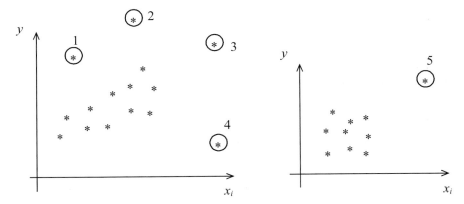

Figure 9.1 Extreme observations in regression analysis. Extremes are encircled and enumerated: a) high leverage extremes are: 3, 4 and 5; b) outliers are: 1, 2 and 4; c) extremes that influence regression estimation are: 2, 4 and 5

These extreme values should be checked to determine their validity. If an error in recording or a biological cause can be determined, there may be justification for deleting them from the dataset.

The simplest way to detect outliers is by inspection of graphs or tables of residuals. This approach can be very subjective. A better approach is to express residuals as standardized or Studentized residuals. Recall that a residual is $e_i = y_i - \hat{y}_i$. A standardized residual is:

$$r_i = \frac{e_i}{s}$$

in which $s = \sqrt{MS_{RES}}$ = estimated residual standard deviation.

A Studentized residual is:

$$r_i = \frac{e_i}{s_e}$$

where $s_e = \sqrt{MS_{RES}(1 - h_{ii})}$, and h_{ii} = diagonal element of the matrix $\mathbf{H} = [\mathbf{X(X'X)}^{-1}\mathbf{X'}]$.

When residual values are standardized or Studentized, a value $r_i > 2$ (greater than two standard deviations) implies considerable deviation.

Observations with high leverage can be detected by examining the h_{ii} value, that is, the corresponding diagonal element of $[\mathbf{X(X'X)}^{-1}\mathbf{X'}]$. This value itself is often referred to as leverage in statistical literature. Properties of h_{ii} are:

a) $^1/_n \leq h_{ii} \leq 1$

b) $\sum_i h_{ii} = p$

where p is the number of parameters in the model.

An observation i has a high leverage if $h_{ii} \geq \dfrac{2p}{n}$, where n is the number of observations.

Statistics used in detecting possible undue influence of a particular observation i on the estimated regression (influence statistics) are Difference in Fit (*DFITTS*), Difference in Betas (*DFBETAS*) and Cook's distance.

The *DFITTS* statistic determines the influence of an observation i on the estimated or fitted value $\hat{y}_i$, and is defined as:

$$DFITTS_i = \frac{\hat{y}_i - \hat{y}_{i,-i}}{s_{-i}\sqrt{h_{ii}}}$$

where:

$\hat{y}_i$ = the estimated value of the dependent variable for a given value of the independent variable x_i, with the regression estimated using all observations

$\hat{y}_{i,-i}$ = the predicted value of the dependent variable for a given value of the independent variable x_i, with the regression estimated without including observation i

$s_{-i} = \sqrt{MS_{RES}}$ not including observation i

h_{ii} = the value of a diagonal element i of the matrix $[\mathbf{X}(\mathbf{X'X})^{-1}\mathbf{X'}]$

The observation i excessively influences the estimation of the regression parameters if $|DFITTS_i| \geq 2\sqrt{\dfrac{p}{n}}$, where p is the number of parameters and n is the number of observations.

The statistic that determines the influence of an observation i on the estimated parameter b_k is *DFBETAS*, defined as:

$$DFBETAS_i = \frac{b_k - b_{k,-i}}{s_{-i}\sqrt{c_{kk}}}$$

where:

b_k = the estimate of parameter β_k including all observations

$b_{k,-i}$ = the estimate of parameter β_k wothout including observation i

$s_{-i} = \sqrt{MS_{RES}}$ without including observation i

c_{kk} = the value of the diagonal element k of the matrix $(\mathbf{X'X})^{-1}$

Observation i influences the estimation of parameter β_k if $|DFBETAS_i| \geq \dfrac{2}{\sqrt{n}}$, where n is the number of observations.

Cook's distance (D_i) determines the influence of an observation i on estimation of the vector of parameters $\mathbf{b}$, and consequently, on estimation of the regression:

$$D_i = \frac{(\mathbf{b} - \mathbf{b}_{-i})'(\mathbf{X'X})(\mathbf{b} - \mathbf{b}_{-i})}{ps^2}$$

where:

$\mathbf{b}$ = the vector of estimated parameters including all observations
$\mathbf{b}_{-i}$ = the vector of estimated parameters without including observation i
$s^2 = MS_{RES}$ = the residual mean square
p = the number of parameters in the model

Observation i excessively influences the estimation of the $\mathbf{b}$ vector if $D_i > 1$.

A statistic that also can be used to determine the influence of observations on estimation of parameters is *COVRATIO*. This is a ratio of generalized variances. A generalized variance is the determinant of the covariance matrix:

$$GV = |var(\mathbf{b})| = |\sigma^2(\mathbf{X'X})^{-1}|$$

COVRATIO is defined as:

$$COVRATIO_i = \frac{\left|s_{-i}^2\left(\mathbf{X}_{-i}{}'\mathbf{X}_{-i}\right)^{-1}\right|}{\left|s^2\left(\mathbf{X'X}\right)^{-1}\right|}$$

where:

$s^2(\mathbf{X'X})^{-1}$ = the covariance matrix for estimated parameters including all observations
$s^2_{-i}(\mathbf{X}_{-i}{}'\mathbf{X}_{-i})^{-1}$ = the covariance matrix for estimated parameters without including the observation i

An observation i may have excessive influence on estimation of the parameter vector $\mathbf{b}$ if

$$COVRATIO_i < 1 - \frac{3p}{n} \text{ or } COVRATIO_i > 1 + \frac{3p}{n}.$$

How should observations classified as outliers, with high leverage, and especially influential observations be treated? If it is known that specific observations are extreme due to a mistake in measurement or recording a malfunction of a measurement device, or some unusual environmental effect, there is justification for deleting them from the analysis. On the other hand, extreme observations may be the consequence of an incorrectly postulated model, for example, in a model where an important independent variable has been omitted, deletion of data would result in misinterpretation of the results. Thus, caution should be exercised before deleting extreme observations from an analysis.

9.6.3 Multicollinearity

Multicollinearity exists when there is high correlation between independent variables. In that case parameter estimates are unreliable, because the variance of parameter estimates is large. Recall that the estimated variance of the $\mathbf{b}$ vector is equal to:

$$Var(\mathbf{b}) = s^2(\mathbf{X'X})^{-1}$$

Multicollinearity means that columns of $(\mathbf{X'X})$ are nearly linearly dependent, which indicates that $(\mathbf{X'X})^{-1}$ and consequently the $Var(\mathbf{b})$ is not stable. The result is that slight changes of observations in a sample can lead to quite different parameter estimates. It is obvious that inferences based on such a model are not very reliable.

Multicollinearity can be determined using a Variance Inflation Factor (*VIF*) statistic, defined as:

$$VIF = \frac{1}{1 - R_k^2}$$

where R_k^2 is the coefficient of determination of the regression of independent variable k on all other independent variables in the postulated model.

If all independent variables are orthogonal, which means totally independent of each other, then $R_k^2 = 0$ and *VIF* = 1. If one independent variable can be expressed as a linear combination of the other independent variables (the independent variables are linearly dependent), then $R_k^2 = 1$ and *VIF* approaches infinity. Thus, a large *VIF* indicates low precision of estimation of the parameter β_k. A practical rule is that a *VIF* > 10 suggests multicollinearity.

Multicollinearity can also be determined by inspection of sequential and partial sums of squares. If the sequential sum of squares is much larger than the partial sum of squares for a particular independent variable, multicollinearity may be the cause. Further, if the partial parameter estimates are significant and the regression in whole is not, multicollinearity is very likely.

Possible remedies for multicollinearity are: a) drop unnecessary independent variables from the model; b) define several correlated independent variables as one new variable; c) drop problematic observations; or d) use an advanced statistical methods such as ridge regression (see section 9.7) or principal components analysis.

9.6.4   SAS Example for Detecting Problems with Regression

The SAS program for detecting extreme observations and multicollinearity will be shown using an example with measurements of weights, heart girths, withers heights and rump heights of 10 young bulls:

Weight (kg)	Heart girth (cm)	Height at withers (cm)	Height at rump (cm)
480	175	128	126
450	177	122	120
480	178	124	121
500	175	128	125
520	186	131	128
510	183	130	127
500	185	124	123
480	181	129	127
490	180	127	125
500	179	130	127

SAS program:

```
DATA bull;
INPUT wt  h_girth  ht_w  ht_r ;
DATALINES;
480 175 128 126
450 177 122 120
480 178 124 121
500 175 128 125
520 186 131 128
510 183 130 127
500 185 124 123
480 181 129 127
490 180 127 125
500 179 130 127
;
PROC REG DATA = bull;
MODEL wt = h_girth  ht_w  ht_r / SS1 SS2 INFLUENCE R VIF ;
PLOT (STUDENT.)*P.
     (STUDENT.)*h_girth
     (STUDENT.)*ht_w
     (STUDENT.)*ht_r
     (H.)*(OBS.);
RUN;
QUIT;
```

Explanation: The REG procedure was used. The statement, MODEL *wt = h_girth ht_w ht_r* denotes *wt* as the dependent variable and *h_girth* (heart girth), *ht_w* (height at withers) and *ht_r* (height at rump) as independent variables. Options used in the MODEL statement are *SS*1 (computes sequential sums of squares), *SS*2 (computes partial sums of squares), INFLUENCE (analyzes extreme observations), R (analyzes residuals) and VIF (variance inflation factor, analyzes multicollinearity). The PLOT statement requests plots of Studentized residuals (STUDENT.) on predicted values (P.) and independent variables, and leverage values (H.) on observation number (OBS.) as listed in the data. Other useful computed values that can be used in constructing plots include (COOKD.), (DFFITS.) and (COVRATIO.). For other plot options readers can consult SAS literature.

SAS output:

```
                    Dependent Variable: wt

                    Analysis of Variance

                        Sum of        Mean
Source            DF    Squares       Square    F Value   Pr > F
Model              3   2522.23150    840.74383    5.21    0.0415
Error              6    967.76850    161.29475
Corrected Total    9   3490.00000

        Root MSE              12.70019   R-Square     0.7227
        Dependent Mean       491.00000   Adj R-Sq     0.5841
        Coeff Var              2.58660
```

Parameter Estimates

Variable	DF	Parameter Estimate	Stand Error	t Value	Pr > |t|	Type I SS
Intercept	1	-382.75201	239.24982	-1.60	0.1608	2410810
h_girth	1	2.51820	1.21053	2.08	0.0827	1252.19422
ht_w	1	8.58321	6.65163	1.29	0.2444	1187.81454
ht_r	1	-5.37962	7.53470	-0.71	0.5021	82.22274

Parameter Estimates

Variable	DF	Type II SS	Variance Inflation
Intercept	1	412.81164	0
h_girth	1	697.99319	1.22558
ht_w	1	268.57379	22.52057
ht_r	1	82.22274	23.54714

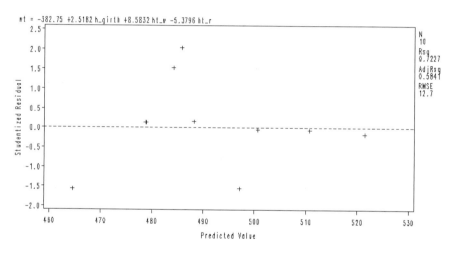

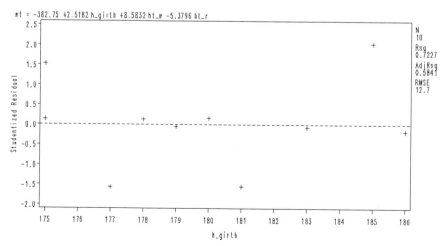

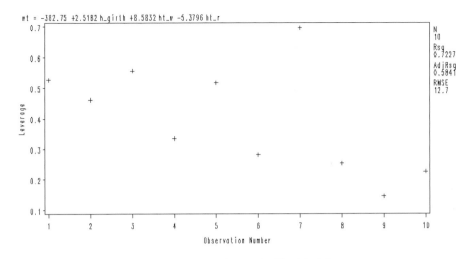

wt = -382.75 +2.5182 h_girth +8.5832 ht_w -5.3796 ht_r

Output Statistics

Obs	Dep Var wt	Predicted Value	Std Error Mean Predict	Residual	Std Error Residual	Student Residual
1	480.0000	478.7515	9.2109	1.2485	8.744	0.143
2	450.0000	464.5664	8.6310	-14.5664	9.317	-1.563
3	480.0000	478.8714	9.4689	1.1286	8.464	0.133
4	500.0000	484.1311	7.3592	15.8689	10.351	1.533
5	520.0000	521.4421	9.1321	-1.4421	8.826	-0.163
6	510.0000	510.6839	6.7483	-0.6839	10.759	-0.0636
7	500.0000	485.7395	10.5958	14.2605	7.002	**2.037**
8	480.0000	497.0643	6.3929	-17.0643	10.974	-1.555
9	490.0000	488.1389	4.8402	1.8611	11.742	0.159
10	500.0000	500.6111	6.0434	-0.6111	11.170	-0.0547

Obs	-2-1 0 1 2	Cook's D	RStudent	Hat Diag H	Cov Ratio	DFFITS
1	| | |	0.006	0.1306	0.5260	**4.3155**	0.1375
2	| ***| |	0.524	-1.8541	0.4619	0.4752	**-1.7176**
3	| | |	0.006	0.1219	0.5559	**4.6139**	0.1364
4	| |*** |	0.297	1.7945	0.3358	0.4273	**1.2759**
5	| | |	0.007	-0.1495	0.5170	**4.2175**	-0.1547
6	| | |	0.000	-0.0580	0.2823	**2.8816**	-0.0364
7	| |**** |	**2.375**	**3.3467**	0.6961	0.0619	**5.0647**
8	| ***| |	0.205	-1.8372	0.2534	0.3528	-1.0703
9	| | |	0.001	0.1450	0.1452	**2.3856**	0.0598
10	| | |	0.000	-0.0500	0.2264	**2.6752**	-0.0270

```
-----------------DFBETAS-----------------
Obs Intercept     h_girth      ht_w        ht_r
 1    0.0321      -0.1118     -0.0763      0.0867
 2   -1.3150       0.1374      0.0444      0.2591
 3    0.0757       0.0241      0.0874     -0.1033
 4    0.5872      -0.8449      0.3882     -0.3001
 5    0.1153      -0.1059     -0.0663      0.0545
 6    0.0208      -0.0181     -0.0193      0.0162
 7   -0.8927       2.3516     -2.9784      2.3708
 8    0.4682       0.1598      0.6285     -0.7244
 9   -0.0025      -0.0087     -0.0333      0.0328
10    0.0054       0.0067     -0.0099      0.0059
```

Explanation: The first table shows the analysis of variance. The next table is *Parameter Estimates*, in which the *Parameter Estimate*, degrees of freedom (*DF*), *Standard Error*, *t Value*, *P* value (*Pr > |t|*), sequential sums of squares (*Type I SS*), partial sums of squares (*Type II SS*), and *VIF* statistics (*Variance Inflation*) are given. For *ht_w* and *ht_r* VIF values are greater than 10. The *VIF* for these variables indicates that both are not necessary in the model. There is collinearity between them. Next, the plots of Studentized residuals vs. predicted values and the independent variable *h_girth* are shown (plots for other independent variables are not shown here). The last plot shows leverages (h values) vs. observation number as is listed in the data. Graphs which include observation number can help us to more easily identify a problematic observation. In the next table, the *Output statistics* for detection of extreme observations are shown. Listed are: the dependent variable (*Dep Var*), *Predicted Value*, standard error of prediction (*Std Error Mean Predict*), *Residual*, standard error of residual (*Std Error Residual*), Studentized residuals (*Student Residual*), simple graphical presentations of deviations of observations from the estimated values (*–2 –1 0 1 2*), Cook's distance (*Cook's D*), Studentized residuals estimated using $s_{-i} = \sqrt{MS_{RES}}$ not including observation *i* (*RStudent*), *h* value (*Hat Diag H*), *CovRatio*, *DFFITS* and *DFBETAS*.

SAS leaves it to the researcher to decide which observations are extreme and/or influential. For this example $p = 4$ and $n = 10$, and the calculated critical values are:

$$h_{ii} \geq \frac{2p}{n} = 0.8$$

$$|DFITTS_i| \geq 2\sqrt{\frac{p}{n}} = 1.26$$

$$|DFBETAS_i| \geq \frac{2}{\sqrt{n}} = 0.63$$

$$Cook's\ D_i > 1$$

$$COVRATIO_i < 1 - \frac{3p}{n} = -0.2 \text{ or}$$

$$COVRATIO_i > 1 + \frac{3p}{n} = 2.2$$

The values in the SAS output can be compared with the critical values shown above. In this output observations exceeding the critical values are emphasized with bold letters. The

Studentized residual was greater than 2 for observation 7. No h_{ii} was greater than 0.8, that is, no high leverage was detected. The Cook's D exceeded 1 for observation 7. The covariance ratios of observations, 1, 3, 5, 6, 9 and 10 exceeded the critical values which also can raise questions about the validity of the chosen model. The *DFFITS* for observations 2, 4 and 7 exceed the critical values. The *DFBETAS* exceeded the critical values for observations 2, 4, 7 and 8. Obviously, observation 7 is an influential outlier and it should be considered for removal. However, a more serious problem with this analysis is collinearity. To check for outliers and values with high leverages without the collinearity, the model was fit with the variable rump height (*ht_r*) removed. The following output was obtained.

Parameter Estimates

Variable	DF	Parameter Estimate	Standard Error	t Value	Pr > |t|	Type I SS
Intercept	1	-410.78486	227.59203	-1.80	0.1141	2410810
h_girth	1	2.21823	1.09481	2.03	0.0824	1252.19422
ht_w	1	3.94914	1.40337	2.81	0.0260	1187.81454

Obs	Dependent Variable	Predicted Value	Std Error Mean Predict	Residual	Std Error Residual	Student Residual
1	480.0000	482.8951	6.8976	-2.8951	10.120	-0.286
2	450.0000	463.6367	8.2281	-13.6367	9.072	-1.503
3	480.0000	473.7532	5.9660	6.2468	10.696	0.584
4	500.0000	482.8951	6.8976	17.1049	10.120	1.690
5	520.0000	519.1430	8.2408	0.8570	9.060	0.094
6	510.0000	508.5392	5.8275	1.4608	10.772	0.136
7	500.0000	489.2808	9.0294	10.7192	8.275	1.295
8	480.0000	500.1536	4.5384	-20.1536	11.376	-1.772
9	490.0000	490.0371	3.9005	-0.0371	11.610	-0.003
10	500.0000	499.6663	5.6865	0.3337	10.847	0.030

Obs	-2-1 0 1 2	Cook's D	RStudent	Hat Diag H	Cov Ratio	DFFITS
1	| | |	0.013	-0.2664	0.3172	**2.2450**	-0.1816
2	| ***| |	0.620	-1.6912	0.4513	0.8989	**-1.5339**
3	| |* |	0.035	0.5544	0.2373	1.7922	0.3092
4	| |*** |	0.442	2.0339	0.3172	0.4823	**1.3862**
5	| | |	0.002	0.0876	0.4527	**2.8906**	0.0797
6	| | |	0.002	0.1257	0.2264	2.0366	0.0680
7	| |** |	0.666	1.3755	0.5435	1.5287	**1.5010**
8	| ***| |	0.167	-2.2085	0.1373	0.3089	-0.8811
9	| | |	0.000	-0.00296	0.1014	1.7672	-0.0010
10	| | |	0.000	0.0285	0.2156	**2.0235**	0.0149

```
             ------------DFBETAS------------
Obs   Intercept      h_girth            ht_w

  1     -0.0799        0.1482          -0.0638
  2     -1.1653        0.2190           1.2274
  3      0.2065       -0.0433          -0.2110
  4      0.6096       -1.1312           0.4873
  5     -0.0694        0.0511           0.0329
  6     -0.0496        0.0277           0.0336
  7     -0.1712        1.1352          -1.0194
  8      0.3960       -0.1093          -0.4003
  9     -0.0001       -0.0001           0.0001
 10     -0.0036       -0.0053           0.0106
```

Due to reduced number of parameters some critical values are slightly changed:

$$h_{ii} \geq \frac{2p}{n} = 0.6$$

$$|DFITTS_i| \geq 2\sqrt{\frac{p}{n}} = 1.095$$

$$|DFBETAS_i| \geq \frac{2}{\sqrt{n}} = 0.63$$

$$Cook's\ D_i > 1$$

$$COVRATIO_i < 1 - \frac{3p}{n} = 0.1 \ \text{or}\ COVRATIO_i > 1 + \frac{3p}{n} = 1.9$$

This output confirms that a serious problem was collinearity and not leverages or outliers. Some observations slightly exceeded criteria values for influence of observations on estimation of parameters, but generally we can conclude that this data set is free from serious and influential outliers or values with high leverage.

9.7   Ridge Regression

Ridge regression is a method that can account for multicollinearity in multiple regression. Multicollinearity exists when there is high correlation between independent variables, and the columns of **X** (the matrix of observations of independent variables) are nearly linearly dependent. When variables are nearly linearly dependent one variable can be approximated with a linear function of other variables. If in **X** there is near dependency then it also exists in (**X'X**). Consequently, the inverse (**X'X**)$^{-1}$ contains very small numbers. The estimated variance of the solution vector **b** is equal to:

$$Var\ (\mathbf{b}) = s^2\ (\mathbf{X'X})^{-1}$$

In case of multicollinearity, the *Var*(**b**) is large and inferences based on such a model are not reliable.

Ridge regression introduces a number, designated λ, which is added to the diagonal of $(\mathbf{X'X})$. Then, the vector of estimates of the multiple regression parameters is:

$$\mathbf{b} = (\mathbf{X'X} + \lambda\mathbf{I})^{-1} \, \mathbf{X'y}$$

The variance of $\mathbf{b}$ is:

$$Var\,(\mathbf{b}) = s^2 \, (\mathbf{X'X} + \lambda\mathbf{I})^{-1}$$

The variances of the elements of $\mathbf{b}$ are smaller when calculated with the addition of λ.

Note that now the estimates of the regression coefficients are biased, as the expectation of the estimates, $E(\mathbf{b})$, is not equal to true parameter vector $\boldsymbol{\beta}$. However, the resulting bias is small and the noteworthy reduction of variance can result in parameters that have better utility. An important question is what value to use for λ? A simple way to determine an appropriate value is to define a grid of values, compute regression parameter estimates for each λ, and graph the resulting estimates to determine which one gives a reduction of variance and smaller bias. Values of λ can range from 0 to 1, and one should choose a value with a small effect on estimates (see SAS output below).

9.7.1 SAS Example for Ridge Regression

An example with measurements of weights, heart girths, and heights at withers and rump of 10 young bulls is shown using SAS. Previously we have shown collinearity between withers and rump heights (section 9.6.4). The regression estimates and their standard errors were 8.58321 ± 6.65163 and -5.37962 ± 7.53470 for withers and rump heights, respectively. The large standard errors indicate large variances and unreliability of the estimates.

SAS program:

```
DATA bull;
INPUT wt h_girth ht_w ht_r ;
DATALINES;
480  175  128  126
450  177  122  120
480  178  124  121
500  175  128  125
520  186  131  128
510  183  130  127
500  185  124  123
480  181  129  127
490  180  127  125
500  179  130  127
;
PROC REG DATA = bull OUTEST = bullout RIDGE=0 TO 0.1 BY 0.005 OUTSEB OUTVIF;
MODEL wt = h_girth  ht_w  ht_r;
PLOT / RIDGEPLOT;
RUN;
QUIT;
```

Explanation: The REG procedure was used. The option *OUTEST* = *bullout* specifies the name of the file in which statistics will be saved. RIDGE requests ridge regression analysis with λ values from 0 to 0.1 by increments of 0.005. The options OUTSEB and OUTVIF compute standard errors and variance inflation factors, respectively, for each λ. These results are saved to the *bullout* file. The statement, *MODEL wt* = *h_girth ht_w ht_r* denotes *wt* as the dependent variable, and *h_girth* (heart girth), *ht_w* (height at withers) and *ht_r* (height at rump) as independent variables. The PLOT statement specifies a plot of the parameter estimates (coefficients) over the λ grid.

SAS output:

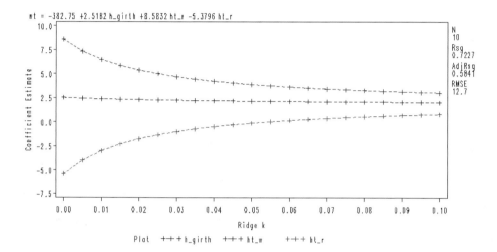

Explanation: First, the usual analysis of variance table and Parameter Estimates are given for the model without λ (not shown here). Next is the plot of regression parameter estimates (coefficients) for each λ. The top line correspond to *ht_w*, middle to *h_girth* and bottom line to *ht_r*. A value of λ should be identified at which the change in estimates becomes small. For this example the value 0.04 seems to be a reasonable choice. Regression coefficients for $\lambda = 0$ and $\lambda = 0.04$ can be obtained by using the PRINT procedure:

```
PROC PRINT DATA=bullout;
WHERE _RIDGE_ = 0 OR _RIDGE_ = 0.04;
RUN;
```

The following values are printed from the datafile *bullout*:

TYPE	_DEPVAR_	_RIDGE_	_RMSE_	Intercept	h_girth	ht_w	ht_r
RIDGEVIF	wt	0.00	.	.	1.22558	22.5206	23.5471
RIDGE	wt	0.00	12.7002	-382.752	2.51820	8.5832	-5.3796
RIDGESEB	wt	0.00	12.7002	239.250	1.21053	6.6516	7.5347
RIDGEVIF	wt	0.04	.	.	1.01767	3.0670	3.1658
RIDGE	wt	0.04	13.1561	-379.112	2.18918	4.2196	-0.4874
RIDGESEB	wt	0.04	13.1561	241.078	1.14268	2.5428	2.8619

Explanation: _RMSE_ is the residual mean square, *h_girth*, *ht_w* and *ht_r* are the variables in the model. For each value of λ are given the variance inflation factors (rows with _TYPE_ = RIDGEVIF), the parameter estimates (rows with _TYPE_ = RIDGE), and the standard errors of the estimates (rows with _TYPE_ = RIDGESEB). In this table are values for ordinary regression (_RIDGE_ = 0) and for $\lambda = 0.04$ (_RIDGE_ = 0.04).

By using $\lambda = 0.04$ the variance inflation factors decrease from 22.5206 and 23.5471 to 3.0670 and 3.1658, respectively, for *ht_w* and *ht_r*, showing that with ridge regression the multicollinearity problem is reduced. The standard errors are also reduced. Further, note that for *h_girth,* which was not affected by the collinearity, the estimates and standard errors change very little with the addition of λ.

9.8　Robust Regression

Robust regression is used to estimate parameters and test hypotheses while minimizing the effects of extreme observations. Observations that are extreme compared to the proposed model (or dependent variable) are referred to as outliers. Observations that are extreme compared to the means of the independent variables are said to have *high leverage*. The ROBUSTREG procedure of SAS has four methods which deal with outliers or observations with high leverage or both. It fits the regression by using an optimization function which minimizes the effects of extreme observations. More information about the methods can be found in Chen (2002). Here we will outline the use of the method called MM estimation. This method is robust in respect to both outliers and high leverage, and uses a maximum likelihood approach combined with a breakdown value. It gives weight or level of importance to observations depending on their level of extremity, and then minimizes the robust function. The cutoff point for outliers is defined as:

$$k\, s_e$$

in which s_e is the standard error of residuals and k is some positive integer. An observation is considered an outlier if:

$$|e_i| > k\, s_e$$

and e_i is the residual computed from the robust regression. For example, if we choose $k = 3$, the cutoff point is set at three standard errors from the estimated regression. Any observation more distant than three standard errors will be deleted.

The cutoff point to determine high leverage is:

$$\sqrt{\chi^2_{p,1-\alpha}}$$

where $\chi^2_{p,1-\alpha}$ is the value from the *chi*-square distribution with p degrees of freedom, with p being the number of parameters in the model, and α usually defined as 0.025.

An observation i is considered to have a high leverage if its robust distance $RD(x_i)$, calculated from the robust procedure, is greater than the cutoff point:

$$RD(x_i) > \sqrt{\chi^2_{p,1-\alpha}}$$

Using the cutoffs, the optimization function is iteratively run until the optimum importance of coefficients or weights for observations are obtained, depending on how distant they are from the estimated regression. The values of those weights range from 0 to 1, with zero meaning the observation is excluded from further analysis, and one meaning the observation is equal to the estimated value of the dependent variable. The robust regression parameters are then estimated by using those coefficients of importance as weights in a weighted regression.

Weighted regression is a regression in which a value of specific importance (the weight) is assigned to each observation. The parameters are estimated by minimizing the weighted residual sum of squares:

$$\sum_i w_i (y_i - \hat{y}_i)^2$$

where $\hat{y}_i$ is the estimated value of the dependent variable, and w_i is the weight for observation i.

9.8.1   SAS Example for Robust Regression

The SAS program for robust regression will be shown using an example with measurements of weights, heart girths, and withers heights of 10 young bulls:

SAS program:

```
DATA bull;
INPUT wt  h_girth  ht_w;
DATALINES;
480 175 128
450 177 122
480 178 124
500 175 128
520 186 131
510 183 130
500 185 124
480 181 129
490 180 127
500 179 130
;
PROC ROBUSTREG DATA = bull METHOD = MM;
MODEL wt = h_girth  ht_w
            / DIAGNOSTICS CUTOFF=3 LEVERAGE (CUTOFFALPHA = 0.025);
OUTPUT OUT = bullout LEVERAGE=leverage OUTLIER=outlier PREDICTED=predicted
STDP=stdp RESIDUAL=residual SRESIDUAL=sresidual WEIGHT=obswght;
RUN;
QUIT;

PROC PRINT DATA = bullout;
RUN;
```

Explanation: The ROBUSTREG procedure was used with *METHOD = MM*. The statement, *MODEL wt = h_girth ht_w* denotes *wt* as the dependent variable and *h_girth* (heart girth) and *ht_w* (height at withers) as independent variables. Options are: DIAGNOSTICS which yields information specifying which observations are possible outliers or have high leverage. The cutoff point for outliers was defined as 3, specifying those observations which exceed –3 or 3 in Studentized distance will be considered serious outliers. For leverage the cutoff was determined by specifying $\alpha = 0.025$. Useful statistics were saved to datafile *bullout*, with columns denoted *leverage, outlier, predicted, stdp* (standard error of prediction), *residual, sresidual* (standardized residual), and *obswght,* which is the coefficient of importance for each observation in the analysis. Note that WEIGHT is an option and not the dependent variable *wt*. The PROC PRINT prints the observations in the datafile *bullout*.

SAS output:

Parameter Estimates

Parameter	DF	Estimate	Standard Error	95% Confidence Limits		Chi-Square	Pr > ChiSq
Intercept	1	-432.353	242.8511	-908.333	43.6264	3.17	0.0750
h_girth	1	2.3836	1.1737	0.0831	4.6840	4.12	0.0423
ht_w	1	3.8882	1.5117	0.9253	6.8511	6.62	0.0101
Scale	0	13.5027					

Diagnostics

Obs	Mahalanobis Distance	Robust MCD Distance	Leverage	Standardized Robust Residual	Outlier
7	1.9980	2.9389	*	0.6855	

Diagnostics Summary

Observation Type	Proportion	Cutoff
Outlier	0.0000	3.0000
Leverage	0.1000	2.7162

Goodness-of-Fit

Statistic	Value
R-Square	0.5623
AICR	7.6257
BICR	11.9205
Deviance	913.9265

Obs	wt	h_girth	ht_w	predicted	stdp	residual	sresidual	out	lev	obswght
1	480	175	128	482.461	7.50066	-2.4613	-0.18228	0	0	0.99439
2	450	177	122	463.899	8.90762	-13.8992	-1.02937	0	0	0.82893
3	480	178	124	474.059	6.47598	5.9408	0.43997	0	0	0.96755
4	500	175	128	482.461	7.50066	17.5387	1.29890	0	0	0.73518
5	520	186	131	520.345	8.68232	-0.3452	-0.02556	0	0	0.99989
6	510	183	130	509.306	6.14673	0.6937	0.05138	0	0	0.99955
7	500	185	124	490.744	9.68313	9.2558	0.68548	0	1	0.92216
8	480	181	129	500.651	4.82520	-20.6509	-1.52939	0	0	0.64375
9	490	180	127	490.491	4.19920	-0.4909	-0.03636	0	0	0.99978
10	500	179	130	499.772	6.12314	0.2280	0.01689	0	0	0.99995

Explanation: The first table is *Parameter Estimates*, showing *Parameter*s, degrees of freedom (*DF*), *Estimate, Standard Error*, 95% *Confidence Limits, Chi-Square* and *P* value (*Pr > ChiSq*). Next is information about *Diagnostics*. Observations which are potential outliers or with high leverage are listed. *Mahalanobis* and *Robust* (*MCD*) *Distances*, and *Standardized Robust Residual* are given. In columns titled *Leverage* and *Outlier*, a star denotes if the observation has high leverage or is an outlier. In this example observation 7 has high leverage. The next table includes a *Diagnostics Summary* about proportions of outliers and observations with high leverage in the data and lists *Cutoff* points. The subsequent table describes the *Goodness-of-Fit* of the robust model with the coefficient of determination (*R-Square*), Akaike Criteria (*AICR*), Bayesian Criteria (*BICR*), and *Deviance* which is the robust residual sum of squares.

The results from the PRINT include values of all variables (*wt, h_girth* and *ht_w*), *predicted* value, standard error of prediction (*stdp*), *residual*, standardized residual (*sresidual*), outlier (1 if outlier), leverage (1 if high leverage) and weights of observations (*obswght*). The values in the column *obswght* are used to calculate robust regression parameter estimates using weighted regression. Again, observation 7 is marked with high leverage; however, it is not seriously influential, since its weight or importance factor (*obswght*) is 0.92216, which is close to 1. Observations 2, 4, and 8 are given less weight in estimating parameters due to their larger residuals, but overall there were no serious problems with these data.

In the table below parameter estimates and their standard errors from ordinary (PROC REG) and robust regression (PROC ROBUSTREG) analyses are compared:

	Ordinary Regression		Robust regression	
Variable	Estimate	Standard Error	Estimate	Standard Error
Intercept	-410.78486	227.59203	-432.353	242.8511
h_girth	2.21823	1.09481	2.3836	1.1737
ht_w	3.94914	1.40337	3.8882	1.5117

Note that differences are small, which confirms there are no serious problems with extreme observations in the original data set.

9.8.2 SAS Example for Comparing Methods for Detecting Extreme Observations

The tools to detect extreme observations and provide remedies were introduced in previous sections. The following example with measurements of weights and heart girths of cows will be used to check for extreme observations, compare various methods, and discuss possible actions and remedies:

Observation	1	2	3	4	5	6	7	8
Weight (*y*):	600	640	640	660	650	700	650	670
Heart girth (*x*):	205	214	215	215	216	216	217	217

Observation	9	10	11	12	13	14	15
Weight (*y*):	650	670	670	680	690	680	640
Heart girth (*x*):	218	218	219	219	220	221	225

SAS program:

```
DATA cows;
INPUT wt h_girth @@;
DATALINES;
600 205   640 214   640 215   660 215   650 216
700 216   650 217   670 217   650 218   670 218
670 219   680 219   690 220   680 221   640 225
;
PROC REG DATA = cows;
MODEL wt = h_girth / INFLUENCE R;
PLOT wt*h_girth;
RUN;
QUIT;
```

First the REG procedure is used with syntax explained in the previous example. The following excerpt from the SAS output shows information important for outlier and leverage analysis:

SAS output:

Parameter Estimates

Variable	DF	Parameter Estimate	Standard Error	t Value	Pr > |t|
Intercept	1	-52.95674	284.93972	-0.19	0.8554
h_girth	1	3.28244	1.31284	2.50	0.0266

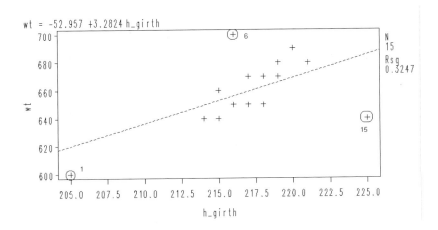

The graph from the PLOT statement shows three possible extreme observations. They are circled and marked with the observation number. Observation 1 is a possible high leverage observation, observation 6 an outlier, and observation 15 has high leverage and is an outlier. Calculated critical values of statistics for detecting extreme observations are shown below. For this example, the number of parameters is $p = 2$ and the number of observations is $n = 15$.

$$h_{ii} \geq \frac{2p}{n} = 0.267$$

$$|DFITTS_i| \geq 2\sqrt{\frac{p}{n}} = 0.73$$

$$|DFBETAS_i| \geq \frac{2}{\sqrt{n}} = 0.516$$

$$Cook's\ D_i > 1$$

$$COVRATIO_i < 1 - \frac{3p}{n} = 0.6\ \text{ or}$$

$$COVRATIO_i > 1 + \frac{3p}{n} = 1.4$$

Output for observations 1, 6 and 15 is shown in the subsequent tables.

Obs	Dependent Variable	Predicted Value	Std Error Mean Predict	Residual	Std Error Residual	Student Residual
1	600.0000	619.9440	16.6822	-19.9440	13.163	-1.515
6	700.0000	656.0509	5.6417	43.9491	20.488	2.145
15	640.0000	685.5929	11.8496	-45.5929	17.640	-2.585

Obs	-2-1 0 1 2	Cook's D	RStudent	Hat Diag H	Cov Ratio	DFFITS
1	| ***| |	**1.843**	-1.6042	**0.6163**	**2.0737**	**-2.0330**
6	| |**** |	0.174	**2.5642**	0.0705	**0.5269**	0.7061
15	| *****| |	**1.507**	**-3.5617**	0.3109	**0.4025**	**-2.3926**

Obs	Intercept	h_girth
	-------DFBETAS-------	
1	**-1.9324**	**1.9199**
6	0.1775	-0.1643
15	**2.0989**	**-2.1206**

Values that exceed critical values are shown in **bold**. Both observations 1 and 15 show influence on the estimation of regression parameters as indicated by *Cook's D*, *DFITTS*, *DFBETAS* and *COVRATIO* exceeding critical values. Those two observations are also marked as having high leverage. Observations 6 and 15 are further from the estimated regression line than are other observations. Observation 15 is an influential observation, as indicated by its *rstudent* (residual of the fitted regression without that observation) of –3.5617. Observations 1 and 15 should be considered for deletion.

The ROBUSTREG procedure was run on data from the previous example with the syntax as explained in section 9.8.1.

```
PROC ROBUSTREG DATA=cows METHOD = MM;
MODEL wt = h_girth  / DIAGNOSTICS CUTOFF=3 LEVERAGE
          (CUTOFFALPHA = 0.025);
OUTPUT OUT=cowsout LEVERAGE=leverage OUTLIER=outlier
       PREDICTED=predicted RESIDUAL=residual
       SRESIDUAL=sresidual STDP=stdp WEIGHT=obswght;
RUN;
QUIT;

PROC PRINT DATA=cowsout;
RUN;
```

The following excerpt from the SAS output shows important information for outlier and leverage analysis:

	Mahalanobis	Robust MCD		Standardized Robust	
Obs	Distance	Distance	Leverage	Residual	Outlier
1	2.7739	4.5222	*	0.2077	
15	1.8493	2.8473	*	-3.9622	*

Obs	wt	h_girth	predicted	stdp	residual	sresidual	outlier	leverage	obswght
1	600	205	596.631	9.99719	3.3689	0.20765	0	1	0.99273
2	640	214	645.074	3.79849	-5.0742	-0.31276	0	0	0.98354
3	640	215	650.457	3.43458	-10.4567	-0.64453	0	0	0.93102
4	660	215	650.457	3.43458	9.5433	0.58823	0	0	0.94238
5	650	216	655.839	3.24623	-5.8393	-0.35992	0	0	0.97823
6	700	216	655.839	3.24623	44.1607	2.72198	0	0	0.13979
7	650	217	661.222	3.26398	-11.2219	-0.69169	0	0	0.92077
8	670	217	661.222	3.26398	8.7781	0.54107	0	0	0.95113
9	650	218	666.604	3.48469	-16.6044	-1.02346	0	0	0.83080
10	670	218	666.604	3.48469	3.3956	0.20930	0	0	0.99261
11	670	219	671.987	3.87381	-1.9870	-0.12247	0	0	0.99747
12	680	219	671.987	3.87381	8.0130	0.49391	0	0	0.95920
13	690	220	677.370	4.38676	12.6305	0.77852	0	0	0.90019
14	680	221	682.752	4.98546	-2.7521	-0.16963	0	0	0.99514
15	640	225	704.282	7.81270	-64.2824	-3.96223	1	1	0.00000

Parameter Estimates

Parameter	DF	Estimate	Standard Error	95% Confidence Limits		Chi-Square	Pr > ChiSq
Intercept	1	-506.795	179.3994	-858.411	-155.178	7.98	0.0047
h_girth	1	5.3826	0.8288	3.7581	7.0070	42.18	<.0001

Output from the ROBUST REG procedure shows diagnostics of observations detected as outliers or with high leverage. Observation 1 was identified as having high leverage and observation 15 as both having high leverage and an outlier. It is given a weight of zero (*obswght*) which means that it is excluded from the robust regression estimation. Note the *obswght* of observation 6 (0.13979) is also a small value, although this observation is retained in the analysis. The standard error of slope is reduced, and the *P*-value is much smaller after robust regression was applied and observation 15 dropped (0.8288 vs. 1.31284, and *P* < 0.0001 vs. *P* = 0.0266). The next plot shows observations and the fitted robust regression.

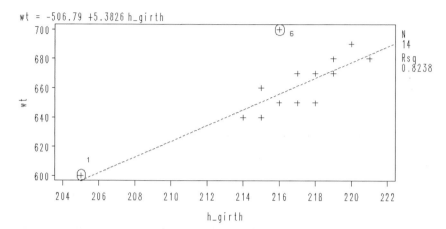

The plot is not satisfactory with respect to observation 1. The robust regression analysis did not exclude it, as it was not considered an outlier, because it is close to the estimated line. Recall that it had high leverage and was considered influential according to *Cook's D*, *DFITTS*, *DFBETAS* and *COVRATIO*. This single observation, because of its greater distance from the other observations, may change the slope to make the line closer to it (high leverage). Rerunning the REG and ROBUSTREG procedures without observations 1 and 15 allows us to determine the influence of these observations.

Since the number of observations used in the REG procedure is now 13, the critical values for influential statistics change slightly:

$$h_{ii} \geq \frac{2p}{n} = 0.308$$

$$|DFITTS_i| \geq 2\sqrt{\frac{p}{n}} = 0.784$$

$$|DFBETAS_i| \geq \frac{2}{\sqrt{n}} = 0.555$$

$$Cook's\ D_i > 1$$

$$COVRATIO_i < 1 - \frac{3p}{n} = 0.54 \ \text{or} \ COVRATIO_i > 1 + \frac{3p}{n} = 1.46$$

The next table shows output from the REG procedure:

Parameter Estimates

Variable	DF	Parameter Estimate	Standard Error	t Value	Pr > |t|
Intercept	1	-481.34111	481.84456	-1.00	0.3393
h_girth	1	5.27697	2.21724	2.38	0.0365

```
                          Output Statistics

         Dependent  Predicted     Std Error              Std Error   Student
   Obs   Variable     Value    Mean Predict   Residual   Residual   Residual
    6    700.0000    658.4840     5.3256       41.5160    15.201     2.731

                         Cook's              Hat Diag   Cov
   Obs   -2-1 0 1 2       D      RStudent       H       Ratio    DFFITS
    6  |   |*****  |     0.458    4.5901      0.1093    0.1407    1.6082

                 -------DFBETAS-------
         Obs  Intercept    h_girth
          6    0.8880      -0.8755
```

The standard errors of the estimates are now higher. This is due to discarding observation 1 which had high leverage. High leverage artificially reduces standard errors.

Observation 6 is now an influential outlier since observations 1 and 15 have been deleted. Note, one serious outlier or high leverage observation can hide the importance of others. Outliers on the opposite sides of the estimated line can nullify each others' influence.

By running the ROBUSTREG procedure without observations 1 and 15, the following output is obtained:

```
                            Diagnostics
                    Robust                  Standardized
         Mahalanobis   MCD                     Robust
   Obs    Distance   Distance    Leverage     Residual    Outlier
    6      0.6236     0.4437                    3.5855        *

   Obs  wt h_girth predicted   stdp   residual sresidual outlier leverage obswght
    6   700   216     653.345  3.71430  46.6549  3.58547     1       0      0.000

                        Parameter Estimates

                     Parameter    Standard
   Variable    DF    Estimate       Error     t Value    Pr > |t|
   Intercept   1     -753.19118   277.47321    -2.71      0.0218
   h_girth     1        6.51174     1.27619     5.10      0.0005
```

Observation 6 is identified as a serious outlier and its weight is *obswght* = 0.000, which means that it is excluded from the robust regression analysis.

The final estimated regression line without observations 1, 6 and 15, is:
$wt = -753.19 + 6.5117 \, h\_girth$. The plot of the estimated regression is:

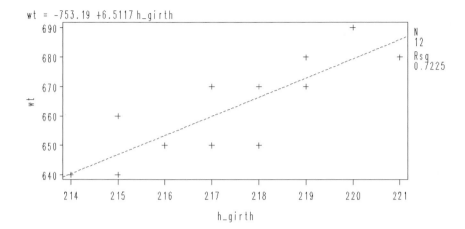

Final comments: The researcher must be aware of the origin of an extreme observation. If it is within normal biological variation, it could mean that data are missing between that observation and the rest. In any case, there should not be a notable gap between one observation and others. If so, the observation may have high leverage and unduly influence the estimated parameters, but not appear as an outlier.

9.9   Choosing the Best Model

In most cases where regression analyses are applied, there are several independent variables that could be included in the model. An ideal situation would be that the model is known in advance. However, it is often not easy to decide which independent variables are really needed in the model. Two errors can happen. First, the model has fewer variables than it should have. Here, the precision of the model would be less than possible. Second, the model has too many variables. This can lead to multicollinearity and its consequences which have already been discussed. For a regression model to be optimal it must have the best set of parameters. Several models with different sets of parameters might all be shown to be relatively good. In addition to statistical considerations for a model to be useful in explaining a problem, it should be easy to explain and use. There are several criteria widely used for selecting an optimal model.

a) Coefficient of determination (R^2)
The coefficient of determination always increases as new variables are added to the model. The question is when added to the model, which variables will notably increase the R^2 ?

b) Residual mean square (MS_{RES})
The residual mean square usually decreases when new variables are added to the model. There is a risk to choosing too large a model. The decrease in error degrees of freedom can offset the decrease in the error sum of squares and the addition of unnecessary effects to a model can increase the residual mean square.

c) Partial F tests

The significance of particular variables in the model are independently tested using partial F tests. However, those tests do not indicate anything about prediction and the optimal model. Due to multicollinearity, variables tested separately can look important; however, the total model may not be very accurate.

d) Cp criterion

Cp stands for Conceptual predictive criterion. It is used to determine a model maximizing explained variability with as few variables as possible. A model candidate is compared with the 'true' model. The formula for Cp is:

$$Cp = p + \frac{\left(MS_{RES} - \hat{\sigma}_0^2\right)(n-p)}{\hat{\sigma}_0^2}$$

where:

MS_{RES} = residual mean square for the candidate model

$\hat{\sigma}_0^2$ = variance estimate of the true model

n = the number of observations

p = the number of parameters of the candidate model

The problem is to determine the 'true' model. Usually, the estimate of the variance from the full model, that is, the model with the maximal number of parameters, is used. Then:

$$\hat{\sigma}_0^2 \cong MS_{RES\_FULL}$$

If the candidate model is too small, because some important independent variables are not in the model, then $Cp \gg p$. If the candidate model is large enough, that is, all important independent variables are included in the model, then Cp is less than p. Note that for the full model $Cp = p$.

e) Akaike information criterion (AIC)

The main characteristic of this criterion is that it is not necessary to define the largest model to compute the criterion. Each model has its own AIC regardless of other models. The model with the smallest AIC is considered optimal. For a regression model AIC is:

$$AIC = n \ log(SS_{RES}/n) + 2p$$

where:

SS_{RES} = residual sum of squares

n = the number of observations

p = the number of parameters of the model

9.9.1 SAS Example for Model Selection

The SAS program for defining an optimal model will be shown using the example of measurements of weight, heart girth, withers height and rump height of 10 young bulls:

Weight (kg)	Heart girth (cm)	Height at withers (cm)	Height at rump (cm)
480	175	128	126
450	177	122	120
480	178	124	121
500	175	128	125
520	186	131	128
510	183	130	127
500	185	124	123
480	181	129	127
490	180	127	125
500	179	130	127

SAS program:

```
DATA bull;
INPUT wt h_girth ht_w ht_r;
DATALINES;
480 175 128 126
450 177 122 120
480 178 124 121
500 175 128 125
520 186 131 128
510 183 130 127
500 185 124 123
480 181 129 127
490 180 127 125
500 179 130 127
;
PROC REG DATA = bull;
MODEL wt=h_girth ht_w ht_r / SSE CP AIC SELECTION=CP ;
PLOT (CP.)*(NP.) / CMALLOWS=red HAXIS= 1 2 3 4 VAXIS=1 TO 8 BY 1;
RUN;
QUIT;
```

Explanation: The REG procedure was used. The statement, *MODEL wt = h_girth ht_w ht_r* denotes *wt* as a dependent variable, and *h_girth* (heart girth), *ht_w* (height at withers) and *ht_r* (height at rump) as independent variables. Options used in the MODEL statement are SSE (computes SS_{RES} for each model), CP (*Cp* statistics), AIC (Akaike criterion), *SELECTION = CP* (model selection is done according to the CP criterion). The PLOT statement specifies a plot of CP. vs. NP., where NP is number of parameters in the models, including the intercept. The options *CMALLOWS = red* draws a reference line *cp=np*. The options HAXIS and VAXIS define reference points on the horizontal and vertical axes and are not essential but can be used to make the plot more easily interpreted.

SAS output:

No. in Model	C(p)	R-Square	AIC	SSE	Variables in Model
2	2.5098	0.6991	52.5395	1049.991	h_girth ht_w
2	3.6651	0.6457	54.1733	1236.342	h_girth ht_r
3	4.0000	0.7227	53.7241	967.768	h_girth ht_w ht_r
1	4.3275	0.5227	55.1546	1665.773	ht_w
1	4.8613	0.4980	55.6585	1751.868	ht_r
2	6.3274	0.5227	57.1545	1665.762	ht_w ht_r
1	7.8740	0.3588	58.1067	2237.806	h_girth

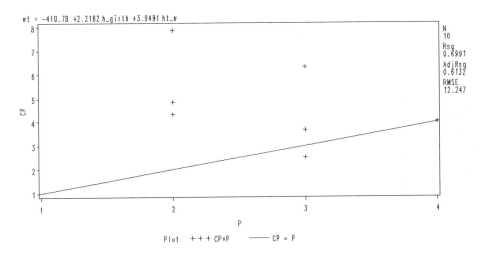

Explanation: The table presents the number of independent variables in the model (*No. in Model*), Cp statistic (*C(p)*), coefficient of determination (*R-Square*), Akaike criterion (*AIC*), residual sum of squares (*SSE*) and a list of variables included in the model (*Variables in Model*). Since the maximum number of independent variables is three, there are seven possible different models. The models are ranked according to *Cp*. The number of parameters for each model is the number of independent variables +1, $p = (No.\ in\ Model)$ +1. The value of *Cp* for the model with *h_girth* (heart girth) and *ht_w* (height at withers) is smaller than the number of parameters in that model, which implies that this is an optimal model. Also, there is a relatively small increase in R^2 for the model with *h_girth*, *ht_w* and *ht_r* compared to the model with *h_girth* and *ht_w*. The *AIC* criterion is the smallest for the model with *h_girth* and *ht_w*. It can be concluded that the model including *h_girth* and *ht_w* is optimal and sufficient to explain *wt*.

The optimal model based on the *Cp* criterion can be identified from a plot of the *Cp* values on the number parameters (*P*) in the model. Points below the line $CP = P$ denote good models. Note that *Cp* for the model with *h_girth* and *ht_w* lies below the line, that is $Cp = 2.5098 < 3 = P$. The *Cp* for the full model, *h_girth*, *ht_w*, *ht_r*, lies exactly on the line. On the graph the equation of the optimal model is shown:

$$wt = -410.78 + 2.2182\ h\_girth + 3.9491\ ht\_w$$

Chapter 10

Curvilinear Regression

In some situations the influence of an independent on a dependent variable is not linear. The simple linear regression model is not suitable for such problems, not only would the prediction be poor, but the assumptions of the model would likely not be satisfied. Three approaches will be described for evaluating curvilinear relationships: polynomial, nonlinear and segmented regression. Often these three approaches are used to fit individual data, for example lactation or growth curves of individual animals. It should be noted that in that case curvilinear regression is applied only to fit the data not for inference about the population parameters. For valid conclusions about the populations, a sample of animals each with measurements must be analyzed. This kind of analysis, called repeated measures analysis, will be shown in chapter 22.

10.1 Polynomial Regression

A curvilinear relationship between the dependent variable y and independent variable x can sometimes be described by using a polynomial regression of second or higher order. For example, a model for a polynomial regression of second degree, or quadratic regression, for n observations is:

$$y_i = \beta_0 + \beta_1 x_i + \beta_2 x^2_i + \varepsilon_i \qquad i = 1,..., n$$

where:

y_i = observation i of dependent variable y
x_i = observation i of independent variable x
$\beta_0, \beta_1, \beta_2$ = regression parameters
ε_i = random error

In matrix notation the model is:

$$\mathbf{y} = \mathbf{X}\boldsymbol{\beta} + \boldsymbol{\varepsilon}$$

The matrices and vectors are defined as:

$$\mathbf{y} = \begin{bmatrix} y_1 \\ y_2 \\ ... \\ y_n \end{bmatrix} \quad \mathbf{X} = \begin{bmatrix} 1 & x_1 & x_1^2 \\ 1 & x_2 & x_2^2 \\ ... & ... & ... \\ 1 & x_n & x_n^2 \end{bmatrix} \quad \boldsymbol{\beta} = \begin{bmatrix} \beta_0 \\ \beta_1 \\ \beta_2 \end{bmatrix} \quad \boldsymbol{\varepsilon} = \begin{bmatrix} \varepsilon_1 \\ \varepsilon_2 \\ ... \\ \varepsilon_n \end{bmatrix}$$

Note, that although the relationship between x and y is not linear, the polynomial model is still considered to be a linear model. A linear model is defined as a model that is linear in the parameters, regardless of relationships of the y and x variables. Consequently, a quadratic regression model can be considered as a multiple linear regression with two 'independent' variables x and x^2, and further estimation and tests are analogous as with multiple regression with two independent variables. For example, the estimated regression model is:

$$\hat{\mathbf{y}} = \mathbf{Xb}$$

and the vector of parameter estimators is:

$$\mathbf{b} = \begin{bmatrix} b_0 \\ b_1 \\ b_2 \end{bmatrix} = (\mathbf{X'X})^{-1}\mathbf{X'y}$$

The null and alternative hypotheses of the quadratic regression are:

$H_0: \beta_1 = \beta_2 = 0$
H_1: at least one $\beta_i \neq 0$, $i = 1$ and 2

If H_0 is true, the statistic

$$F = \frac{MS_{REG}}{MS_{RES}}$$

has an F distribution with 2 and $(n-3)$ degrees of freedom. Here, MS_{REG} and $MS_{RES} = s^2$ are the regression and residual means squares, respectively. The H_0 is rejected with α level of significance if the calculated F is greater than the critical value ($F > F_{\alpha,2,n-3}$).

The F test determines if b_1 or b_2 are significantly different from zero. Of primary interest is to determine if the parameter β_2 is needed in the model, that is, whether linear regression is adequate. A way to test the $H_0: \beta_2 = 0$ is by using a t statistic:

$$t = \frac{b_2}{s(b_2)}$$

where $s(b_2)$ is the standard deviation of b_2. Recall that the variance-covariance matrix for b_0, b_1 and b_2 is:

$$\mathbf{s^2(b)} = s^2(\mathbf{X'X})^{-1}$$

Example: Describe the growth of Zagorje turkeys with a quadratic function. Data are shown in the following table:

Weight, g (y):	44	66	100	150	265	370	455	605	770
Age, days (x):	1	7	14	21	28	35	42	49	56
Age^2 (x^2)	1	49	196	441	784	1225	1764	2401	3136

The **y** vector and **X** matrix are:

$$
\mathbf{y} = \begin{bmatrix} 44 \\ 66 \\ 100 \\ 150 \\ 265 \\ 370 \\ 455 \\ 605 \\ 770 \end{bmatrix}
\qquad
\mathbf{X} = \begin{bmatrix}
1 & 1 & 1 \\
1 & 7 & 49 \\
1 & 14 & 196 \\
1 & 21 & 441 \\
1 & 28 & 784 \\
1 & 35 & 1225 \\
1 & 42 & 1764 \\
1 & 49 & 2401 \\
1 & 56 & 3136
\end{bmatrix}
$$

The vector of parameter estimates is:

$$\mathbf{b} = (\mathbf{X'X})^{-1}\mathbf{X'y}$$

The **X'X** and **X'y** matrices are:

$$
\mathbf{X'X} = \begin{bmatrix}
1 & 1 & 1 & 1 & 1 & 1 & 1 & 1 & 1 \\
1 & 7 & 14 & 21 & 28 & 35 & 42 & 49 & 56 \\
1 & 49 & 196 & 441 & 784 & 1225 & 1764 & 2401 & 3136
\end{bmatrix}
\begin{bmatrix}
1 & 1 & 1 \\
1 & 7 & 49 \\
1 & 14 & 196 \\
1 & 21 & 441 \\
1 & 28 & 784 \\
1 & 35 & 1225 \\
1 & 42 & 1764 \\
1 & 49 & 2401 \\
1 & 56 & 3136
\end{bmatrix}
$$

$$
= \begin{bmatrix}
9 & 253 & 9997 \\
253 & 9997 & 444529 \\
9997 & 444529 & 21061573
\end{bmatrix}
= \begin{bmatrix}
n & \sum_i x_i & \sum_i x_i^2 \\
\sum_i x_i & \sum_i x_i^2 & \sum_i x_i^3 \\
\sum_i x_i^2 & \sum_i x_i^3 & \sum_i x_i^4
\end{bmatrix}
$$

$$
\mathbf{X'y} = \begin{bmatrix}
1 & 1 & 1 & 1 & 1 & 1 & 1 & 1 & 1 \\
1 & 7 & 14 & 21 & 28 & 35 & 42 & 49 & 56 \\
1 & 49 & 196 & 441 & 784 & 1225 & 1764 & 2401 & 3136
\end{bmatrix}
\begin{bmatrix} 44 \\ 66 \\ 100 \\ 150 \\ 265 \\ 370 \\ 455 \\ 605 \\ 770 \end{bmatrix}
= \begin{bmatrix} 2825 \\ 117301 \\ 5419983 \end{bmatrix}
= \begin{bmatrix}
\sum_i y_i \\
\sum_i x_i y_i \\
\sum_i x_i^2 y_i
\end{bmatrix}
$$

$$(\mathbf{X'X})^{-1} = \begin{bmatrix} 0.7220559 & -0.0493373 & 0.0006986 \\ -0.0493373 & 0.0049980 & -0.0000820 \\ 0.0006986 & -0.0000820 & 0.0000014 \end{bmatrix}$$

The **b** vector is:

$$\mathbf{b} = \begin{bmatrix} b_0 \\ b_1 \\ b_2 \end{bmatrix} = \begin{bmatrix} 0.7220559 & 0.0493373 & 0.0006986 \\ 0.0493373 & 0.0049980 & -0.0000820 \\ 0.0006986 & -0.0000820 & 0.0000014 \end{bmatrix} \begin{bmatrix} 2825 \\ 117301 \\ 5419983 \end{bmatrix} = \begin{bmatrix} 38.86 \\ 2.07 \\ 0.195 \end{bmatrix}$$

The estimated function is:

$$y = 38.86 + 2.07x + 0.195x^2$$

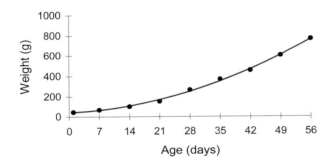

Figure 10.1 Growth of Zagorje turkeys described with a quadratic function. Observed values are shown as points relative to the fitted quadratic regression line (•)

The *ANOVA* table is:

Source	SS	df	MS	F
Regression	523870.4	2	261935.2	1246.8
Residual	1260.5	6	210.1	
Total	525130.9	8		

The estimated regression is significant. To test appropriateness of the quadratic term in the model, a *t* statistic can be used:

$$t = \frac{b_2}{s(b_2)}$$

The variance estimate is $s^2 = 210.1$. The inverse of $(\mathbf{X'X})$ is:

$$(\mathbf{X'X})^{-1} = \begin{bmatrix} 0.7220559 & 0.0493373 & 0.0006986 \\ 0.0493373 & 0.0049980 & -0.0000820 \\ 0.0006986 & -0.0000820 & 0.0000014 \end{bmatrix}$$

The variance-covariance matrix of the estimates is:

$$\mathbf{s^2(b)} = s^2(\mathbf{X'X})^{-1} = 210.1 \begin{bmatrix} 0.7220559 & 0.0493373 & 0.0006986 \\ 0.0493373 & 0.0049980 & -0.0000820 \\ 0.0006986 & -0.0000820 & 0.0000014 \end{bmatrix}$$

It follows that the estimated variance for b_2 is:

$$s^2(b_2) = (210.1)(0.0000014) = 0.000304$$

The standard deviation for b_2 is:

$$s(b_2) = \sqrt{0.000304} = 0.0174$$

The calculated t from the sample is:

$$t = \frac{0.195}{0.0174} = 11.207$$

The critical value is $t_{0.025,6} = 2.447$ (see Appendix B: Critical values of Student t distributions). Since the calculated t is more extreme than the critical value, H_0 is rejected and it can be concluded that a quadratic function is appropriate for describing the growth of Zagorje turkeys.

10.1.1 SAS Example for Quadratic Regression

The SAS program for the example of turkey growth data is as follows.

SAS program:

```
DATA turkey;
INPUT wt day @@;
DATALINES;
44 1   66 7   100 14   150 21   265 28   370 35   455 42   605 49   770 56
;
PROC GLM DATA = turkey ;
MODEL wt=day day*day/ ;
RUN;
QUIT;
```

Explanation: The GLM procedure is used. The statement *MODEL wt = day day*day* defines *wt* as the dependent variable, *day* as a linear component and *day*day* as a quadratic component of the independent variable.

SAS output:

Dependent Variable: WT

Source	DF	Sum of Squares	Mean Square	F Value	Pr > F
Model	2	523870.39532	261935.19766	1246.82	0.0001
Error	6	1260.49357	210.08226		
Corrected Total	8	525130.88889			

R-Square	C.V.	Root MSE	WT Mean
0.997600	4.617626	14.494215	313.88889

Source	DF	Type I SS	Mean Square	F Value	Pr > F
DAY	1	497569.66165	497569.66165	2368.45	0.0001
DAY*DAY	1	26300.73366	26300.73366	125.19	0.0001

Source	DF	Type III SS	Mean Square	F Value	Pr > F
DAY	1	859.390183	859.390183	4.09	0.0896
DAY*DAY	1	26300.733664	26300.733664	125.19	0.0001

Parameter	Estimate	T for H0: Parameter=0	Pr > |T|	Std Error of Estimate
INTERCEPT	38.85551791	3.15	0.0197	12.31629594
DAY	2.07249024	2.02	0.0896	1.02468881
DAY*DAY	0.19515458	11.19	0.0001	0.01744173

Explanation: In the *ANOVA* table there is a large *F* value (1246.82) and analogously small *P* value (*Pr > F*). This is not surprising for growth in time. The question is if the quadratic parameter is needed or if the linear component alone is enough to explain growth. The table with sequential (*Type I SS*) is used to determine if the quadratic component is needed after fitting the linear effect. The *P* value for *DAY*DAY* = 0.0001, indicating the quadratic component is needed. The same conclusion is reached by looking at the table of parameter estimates and *t* tests. The estimates are: b_0 (*INTERCEPT*) = 38.85551791, b_1 (*DAY*) = 2.07249024 and b_2 (*DAY*DAY*) = 0.19515458.

10.2   Nonlinear Regression

Explanation of a curvilinear relationship between a dependent variable *y* and an independent variable *x* sometimes requires a true nonlinear function. Recall that linear models are linear in the parameters. A nonlinear regression model is a model that is not linear in the parameters. Assuming additive errors a nonlinear model is:

$$y = f(x, \theta) + \varepsilon$$

where:

      y = the dependent variable
      $f(x, \theta)$ = a nonlinear function of the independent variable *x* with parameters θ
      ε = random error

Examples of nonlinear functions commonly used to fit biological phenomena include exponential, logarithmic and logistic functions and their families. The exponential regression model can be expressed as:

$$y_i = \beta_0 - \beta_1 e^{\beta_2 x_i} + \varepsilon_i \qquad i = 1,\dots, n$$

where β_0, β_1 and β_2 are parameters, and e is the base of the natural logarithm. This is not a linear model as the parameter β_2 is not linear in y. Figure 10.2 shows four exponential functions with different combinations of positive and negative β_1 and β_2 parameters.

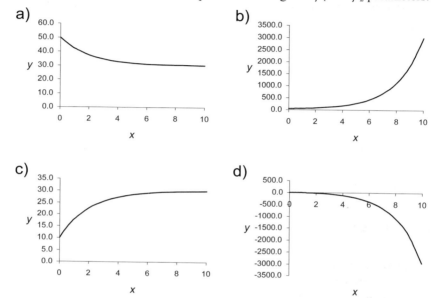

Figure 10.2 Exponential functions with parameter $\beta_0 = 30$ and: a) $\beta_1 = -20$, and $\beta_2 = -0.5$; b) $\beta_1 = -20$, and $\beta_2 = +0.5$; c) $\beta_1 = +20$, and $\beta_2 = -0.5$; and d) $\beta_1 = +20$, and $\beta_2 = +0.5$

Often the parameters are transformed in order to have biological meaning. For example, the following exponential function:

$$y_i = A - \left(A - y_0\right)e^{-k(x_i - x_0)} + \varepsilon_i$$

has parameters defined as follows:

A = asymptote, the maximum function value
y_0 = the function value at the initial value x_0 of the independent variable x
k = rate of increase of function values

When used to describe growth, this function is usually referred to as the Brody curve.

Another commonly applied model is the logistic regression model:

$$y_i = \frac{\beta_0}{1 + \beta_1 e^{\beta_2 x_i}} + \varepsilon_i$$

The logistic model has parameters defined as:

$\beta_0 = $ asymptote, the maximum function value

$\dfrac{\beta_0}{1 + \beta_1} = $ the initial value at $x_i = 0$

$\beta_2 = $ a parameter influencing the shape of the curve

This model is used as a growth model, but also is widely applied in analyses of binary dependent variables. A logistic model with the parameters: $\beta_0 = 30$, $\beta_1 = 20$, and $\beta_2 = -1$ is shown in Figure 10.3.

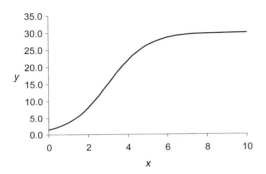

Figure 10.3 Logistic function with the parameters: $\beta_0 = 30$, $\beta_1 = 20$, and $\beta_2 = -1$

There are many functions that are used to describe growth, lactation or changes in concentration of a substance over time. Parameters of nonlinear functions can be estimated by using various numerical iterative methods. The NLIN procedure of SAS will be used to fit a Brody growth curve to weights of an Angus cow.

10.2.1 SAS Example for Nonlinear Regression

The SAS program for nonlinear regression is as follows. Data represent weights of an Angus cow at ages from 8 to 108 months:

Weight, kg:	280	340	430	480	550	580	590	600	590	600
Age, months:	8	12	24	36	48	60	72	84	96	108

The Brody curve was fitted to the data:

$$Wt_i = A - \left(A - Wt_0\right)e^{-k\left(Age_i - Age_0\right)}$$

where:

$A = $ the asymptotic (mature) weight

$Wt_0 = $ the estimated initial weight at $Age_0 = 8$ months

$k = $ the maturing rate index

SAS program:

```
DATA a;
INPUT age wt @@;
DATALINES;
   8   280      12   340      24   430      36   480      48   550
  60   580      72   590      84   600      96   590     108   600
;
PROC NLIN DATA = a ;
   PARMS A=600 wt0=280 k=0.05;
   MODEL wt=A-(A-wt0)*exp(-k*(age-8));
   RUN;
```

Explanation: The NLIN procedure is used. The PARMS statement defines parameters with their priors. Priors are guesses of the values of the parameters. Priors are needed to start the iterative numerical computation. The MODEL statement defines the model: *wt* is the dependent and *age* is an independent variable, and *A*, *wt0* and *k* are the parameters to be estimated.

SAS output:

Dependent Variable wt, Method: Gauss-Newton

Iterative Phase

Iter	A	wt0	k	Sum of Squares
0	600.0	280.0	0.0500	2540.5
1	610.2	285.8	0.0355	1388.7
2	612.2	283.7	0.0381	966.9
3	612.9	283.9	0.0379	965.9
4	612.9	283.9	0.0380	965.9
5	612.9	283.9	0.0380	965.9

NOTE: Convergence criterion met.

Source	DF	Sum of Squares	Mean Square	F Value	Approx Pr > F
Model	2	123274	61637.1	446.69	<.0001
Error	7	965.9	138.0		
Corrected Total	9	124240			

Parameter	Estimate	Approx Std Error	Approximate 95% Confidence Limits	
A	612.9	9.2683	590.9	634.8
wt0	283.9	9.4866	261.5	306.3
k	0.0380	0.00383	0.0289	0.0470

Approximate Correlation Matrix

	A	wt0	k
A	1.0000000	0.2607907	-0.8276063
wt0	0.2607907	1.0000000	-0.4940824
k	-0.8276063	-0.4940824	1.0000000

Explanation: The title of the output indicates that the numerical method of estimation is by default Gauss-Newton. The first table describes iterations with the current estimates together with residual sums of squares. The statement *Convergence criterion met* indicates that computation was successful. The next table presents an analysis of variance table including sources of variation (*Model, Error, Corrected Total*), degrees of freedom (*DF*), *Sum of Squares, Mean Square, F Value* and approximate *P* value (*Approx Pr > F*). The word '*approx*' warns that for a nonlinear model the *F* test is approximate, but asymptotically valid. Based on the large *F* and small *P*-value it can be concluded that the model effectively explains the growth of the cow. The next table shows the parameter *estimates* together with their approximate *Standard Errors* and *Confidence Limits*. The last table presents approximate correlations among the parameter estimates. The estimated curve is:

$$Wt_i = 612.9 - (612.9 - 283.9)e^{-0.038(Age_i - 8)}$$

Note that the curve and parameters are estimates for this particular cow. To estimate population parameters, a sample of cows with growth data must be used.

Figure 10.4 presents a graph of the function with observed weights and the fitted curve representing estimate values.

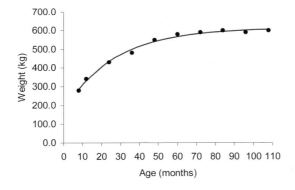

Figure 10.4 Weights over time of an Angus cow fitted to a Brody function; the line represents estimated values and the points (•) observed weights

10.3 Segmented Regression

Another way to describe a curvilinear relationship between a dependent and an independent variable is by defining two or more polynomials, each for a particular segment of values of the independent variable. The functions are joined at points separating the segments. The abscissa values of the joining points are usually called *knots*, and this approach is often called segmented, piecewise, or spline regression. The new curve can be defined to be continuous and smooth in such a way that in addition to the function values, the first $p - 1$ derivatives also agree at the knots (p being the order of the polynomial). Knots allow the new curve to bend and more closely follow the data. Such smooth segmented regressions are often known as spline regressions. Segmented regressions are useful under some circumstances such as when the relationship between variables is irregular and cannot be adequately represented with just one function. They tend to have more stable parameters and better predictions compared to, for example, higher order polynomials.

10.3.1 Segmented Linear Regression

In the simplest situation, assume an event which can be described with two simple linear functions that are joined at one point. The models of two simple regressions are:

$$y_i = \beta_{01} + \beta_{11}x_i + \varepsilon_i \qquad \text{for } x_i \leq x_c$$
$$y_i = \beta_{02} + \beta_{12}x_i + \varepsilon_i \qquad \text{for } x_i \geq x_c$$

Here, x_c denotes a knot such that the expected value $E(y_i|x_c)$ at that point is the same for both functions. These two models can be written as one multiple regression model if another independent variable u is defined:

$$u_i = 0 \qquad \text{for } x_i \leq x_c$$
$$u_i = (x_i - x_c) \qquad \text{for } x_i > x_c$$

The new model is:

$$y_i = \gamma_0 + \gamma_1 x_i + \gamma_2 u_i + \varepsilon_i$$

where γ_0, γ_1 and γ_2 are parameters. Expressing this function in terms of the original independent variable x and the knot x_c, the model is:

$$y_i = \gamma_0 + \gamma_1 x_i + \gamma_2(x_i - x_c)_+ + \varepsilon_i$$

where $(x_i - x_c)_+$ indicates that $(x_i - x_c)$ can take only nonnegative values, that is:

$$(x_i - x_c)_+ = 0 \text{ for } x_i \leq x_c$$
$$(x_i - x_c)_+ = (x_i - x_c) \text{ for } x_{1i} > x_c$$

The parameter γ_2 indicates the change in slope after the knot relative to the line before the knot. The test for a change in slope involves the testing of the following null and alternative hypotheses:

$H_0: \gamma_2 = 0$ only one regression line is needed, the slope does not change at the knot
$H_1: \gamma_2 \neq 0$ two regression lines are needed, the slope has changed at the knot

Using parameters of the new model (γ_0, γ_1, γ_2) and the value of the knot x_c, the previous simple regression models can be expressed as:

$$y_i = \gamma_0 + \gamma_1 x_i + \varepsilon_i \qquad\qquad \text{for } x_i \leq x_c$$
$$y_i = (\gamma_0 - \gamma_2 x_c) + (\gamma_1 + \gamma_2)\, x_i + \varepsilon_i \qquad \text{for } x_i \geq x_c$$

The parameters β are expressed as combinations of the new parameters γ and the knot x_c:

$$\beta_{01} = \gamma_0$$
$$\beta_{11} = \gamma_1$$
$$\beta_{02} = \gamma_0 - \gamma_2 x_c$$
$$\beta_{12} = \gamma_1 + \gamma_2$$

With this parameterization it is assured that the two regression lines intersect at the value x_c; however, it is not possible to obtain a smooth curve. The lines join at the knot at an angle, thus popular terms for segmented linear regression with just one knot are broken-stick or hockey-stick regression.

The model with two simple regression lines joined at one knot is applicable when we expect treatment to cause the slope to change at the value of the independent variable equal to the knot. For example, we may be interested if the slope of growth rate changes when a treatment is applied to animals at a particular point in time. If the slope changes, the treatment has had an effect on growth rate. If there is no change, only one regression line is sufficient, i.e. the treatment has no effect on growth rate. Note that we need to have repeated measures on animals, before and after treatment.

It is evident that the knot x_c can be defined a-priori. It can also be unknown and estimated from a sample. For example, we might expect some change in slope but not know exactly when the change occurs, and desire to estimate that from a sample. Several combinations of simple regressions with different knots can be estimated and the combination chosen such that the best fitting segmented line is obtained. Alternatively, a nonlinear approach and iterative numerical methods can be used since the segmented regression is nonlinear with respect to the parameters γ_2 and x_c.

Segmented regression can often be more easily interpretable if the intercept is defined at the knot. This can be done by defining the independent variable as a deviation from the knot. The model with one knot is:

$$y_i = \gamma_0 + \gamma_1(x_i - x_c) + \gamma_2(x_i - x_c)_+ + \varepsilon_i$$

10.3.1.1   SAS Example for Segmented Linear Regression

Example: Describe the growth of Zagorje turkeys by using two simple linear regression functions. Weight is the mean of several birds in a pen.

Weight, g (y):	44	66	100	150	265	370	455	605	770
Age, days (x):	1	7	14	21	28	35	42	49	56

By inspection of the data, the knot is defined at age 21 days, that is, $x_c = 21$. We define a new independent variable such that:

$$u_i = 0 \qquad \text{for } x_i \le 21$$
$$u_i = (x_i - 21) \qquad \text{for } x_i > 21$$

The new independent variable u has values:

$$0 \quad 0 \quad 0 \quad 0 \quad 7 \quad 14 \quad 21 \quad 28 \quad 35$$

paired with values of the variable x. Now a multiple regression with three parameters must be estimated:

$$y_i = \gamma_0 + \gamma_1 x_i + \gamma_2 u_i + \varepsilon_i$$

or by describing variable u as $(x - 21)_+$ the function is:

$$y_i = \gamma_0 + \gamma_1 x_i + \gamma_2(x_i - 21)_+ + \varepsilon_i$$

SAS program:

```
DATA turkey;
INPUT wt age @@;
u = (age>21)*(age-21);
DATALINES;
  44   1      66   7      100 14
 150 21      265 28      370 35
 455 42      605 49      770 56
 ;
PROC REG DATA = turkey;
MODEL wt = age u;
RUN;
```

Explanation: A new variable u is defined with $u = (age>21)*(age–21)$ in the data step. Here, $(age>21)$ is a logical term which takes the value 1 when the expression is true, and zero when not, thus the term $(age–21)$ is defined only if age > 21, and is zero otherwise.
The REG procedure was used to fit the model. The statement, *MODEL wt = age u* denotes *wt* as the dependent variable, and *age* and *u* as independent variables.

SAS output:

Analysis of Variance

Source	DF	Sum of Squares	Mean Square	F Value	Pr > F
Model	2	521837	260919	475.31	<.0001
Error	6	3293.67978	548.94663		
Corrected Total	8	525131			

Parameter Estimates

Variable	DF	Parameter Estimate	Standard Error	t Value	Pr > |t|
Intercept	1	36.52317	20.05141	1.82	0.1184
age	1	4.65995	1.34694	3.46	0.0135
u	1	12.54593	1.88693	6.65	0.0006

Explanation: The first table presents an *ANOVA* for the dependent variable *weight*. The sources of variation are *Model*, *Error* and *Corrected Total*. In the table are listed: degrees of freedom (*DF*), *Sum of Squares*, *Mean Square*, calculated *F* (*F Value*) and *P* value (*Pr > F*). The parameter estimates (*Estimate*) with corresponding standard errors (*Standard Error*), *t* and *P* values (*Pr > |t|*) are shown. The *t Value* tests whether the estimates are significantly different than zero. The *P* value for *u* is 0.0006 which indicates that segmented regression better explained growth of the turkeys than linear regression.

The parameter estimates are:

Parameter	Estimate	Standard error
γ_0	36.52	20.05
γ_1	4.66	1.35
γ_2	12.55	1.89

The estimated segmented linear regression is:

$$wt = 36.52317 + 4.65995\ age + 12.54593\ (age - 21)_+$$

and it is shown in figure 10.5.

Using parameters of the new model (γ_0, γ_1, γ_2) and the value of the knot x_c, the model can also be presented as two simple linear regression functions:

$$wt = 36.52 + 4.66\ age \qquad \text{for age} \leq 21$$
$$wt = -227.03 + 17.21\ age \qquad \text{for age} \geq 21$$

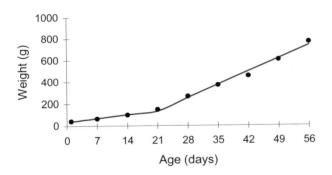

Figure 10.5 Growth of Zagorje turkeys shown with two linear regression functions and a fixed knot: observed (•) and estimated (__) values

The optimal position of a knot can be calculated from the data by using nonlinear regression. The SAS program is as follows:

SAS program:

```
DATA turkey;
INPUT wt age @@;
DATALINES;
 44   1    66   7    100 14
150 21    265 28    370 35
455 42    605 49    770 56
;
```

```
PROC NLIN DATA = turkey;
PARMS gama0 = 36 gama1 = 4 gama2 = 12 xc = 21;
MODEL wt = gama0 + gama1*age + gama2*(age>xc)*(age-xc);
RUN;
```

Explanation: The NLIN procedure is used for fitting nonlinear regression. Recall that equivalently two simple regressions are estimated:

$$wt_i = gama0 + gama1\ age_i \qquad \text{for } age_i \le x_c$$
$$wt_i = (gama0 - gama2\ x_c) + (gama1 + gama)\ age_i \qquad \text{for } age_i > x_c$$

which are joined at the knot x_c. Here, $gama0$, $gama1$, $gama2$ and x_c denote unknown parameters which will be estimated from the data. Note that this specifies a nonlinear regression with four unknowns. The PARMS statement defines the parameters with their priors, which are needed to start the iterative numerical computation. The model is defined by the MODEL statement:

$$MODEL\ wt = gama0 + gama1*age + gama2*(age>xc)*(age-xc);$$

here, *(age >xc)* is a logical term which takes the value 1 if the expression is true and zero if not, that is, the term *gama2*(age-xc)* is defined only if age > xc.

SAS output:

Source	DF	Sum of Squares	Mean Square	F Value	Pr > F
Model	3	522170	174057	293.91	<.0001
Error	5	2961.0	592.2		
Corrected Total	8	525131			

Parameter	Estimate	Approx Std Error	Approx 95% Confid Limits	
gama0	33.2725	21.2732	-21.4112	87.9563
gama1	5.2770	1.6232	1.1043	9.4496
gama2	12.5087	1.9605	7.4692	17.5483
xc	22.9657	2.6485	16.1577	29.7738

Explanation: The first table presents an analysis of variance including sources of variation (*Model, Error, Corrected Total*), degrees of freedom (*DF*), *Sum of Squares, Mean Square, F Value* and *P* value (*Pr > F*). The high *F* value suggests that the model effectively explains variation in these data. The next table shows the *Parameter* and their *Estimates* together with *Approximate Standard Errors* and *Confidence Limits*. Note that the optimal knot, *xc*, was estimated to be at 22.9657 days. Figure 10.6 presents a graph of the weight of Zagorje turkeys from 1 to 56 days of age with the segmented regression line fitted using parameters from the SAS program.

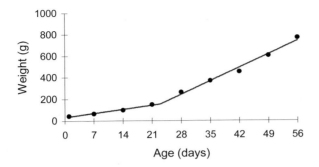

Figure 10.6 Weight of Zagorje turkeys over time with the fitted segmented regression and estimated knot: observed (•) and estimated (_) values

10.3.2 Segmented Cubic Regression – Cubic Splines

In the previous section segmented linear regression was defined as a multiple regression on an expanded independent variable x derived by defining 'new' variables as deviations from knots of the existing independent variable x. However, that function is not smooth; there is some angle at the knot where the linear regression functions meet. Smoothness can be obtained by defining polynomials of order higher than linear. In most cases cubic polynomials give satisfactory smoothness. As before, the expected values agree at the knot. The smoothness is assured by defining functions from which the first and second derivatives also agree at the knot. Segmented cubic regression is often called cubic spline regression or just cubic splines.

Similarly to the segmented linear regression model, the cubic spline model can be derived from several polynomials that are expressed as a multiple regression of variables derived from the original independent variable x. For example, for two knots the segmented cubic regression model is:

$$y_i = \gamma_0 + \gamma_1 x_i + \gamma_2 x^2_i + \gamma_3 x^3_i + \gamma_4(x_i - x_{c1})^3_+ + \gamma_5(x_i - x_{c2})^3_+ + \varepsilon_i$$

where $(x_i - x_{ck})_+$, {where $ck = c1$ or $c2$} indicates that $(x_i - x_{ck})$ can take only non-negative values, that is:

$$(x_i - x_{ck})_+ = 0 \text{ for } x_i \leq x_{ck}$$
$$(x_i - x_{ck})_+ = (x_i - x_{ck}) \text{ for } x_i > x_{ck}$$

Equivalently, this function can be presented as three cubic functions, one for each region:

$$y_i = \gamma_0 + \gamma_1 x_i + \gamma_2 x^2_i + \gamma_3 x^3_i + \varepsilon_i \qquad \text{for } x_i \leq x_{c1}$$

$$y_i = (\gamma_0 - \gamma_4 x_{c1}) + (\gamma_1 + 3\gamma_4 x_{c1}^2) x_i + (\gamma_2 - 3\gamma_4 x_{c1}) x^2_i + (\gamma_3 + \gamma_4) x^3_i + \varepsilon_i$$
$$\text{for } x_{c1} \leq x_{1i} \leq x_{c2}$$

$$y_i = (\gamma_0 - \gamma_4 x_{c1} - \gamma_5 x_{c2}) + (\gamma_1 + 3\gamma_4 x_{c1}^2 + 3\gamma_5 x_{c2}^2) x_i + (\gamma_2 - 3\gamma_4 x_{c1} - 3\gamma_5 x_{c2}) x^2_i +$$
$$(\gamma_3 + \gamma_4 + \gamma_5) x^3_i + \varepsilon_i \qquad \text{for } x_i \geq x_{c2}$$

These adjoined equations satisfy the condition that the expected values agree at their knot, and that their first and second derivatives agree at that knot.

A test for the need of three versus one polynomial is based on the H_0 hypothesis:

$$H_0: \gamma_4 = \gamma_5 = 0$$

Generally, the cubic spline function with m_k knots is:

$$y_i = \gamma_0 + \gamma_1 x_i + \gamma_2 x_i^2 + \gamma_3 x_i^3 + \sum_{k=1}^{m_k} \gamma_{3+k} \left(x_i - x_{ck}\right)_+^3 + \varepsilon_i$$

10.3.2.1   SAS Example for Cubic Spline Regression

Example: A cubic spline will be used to fit a growth curve to weights of an Angus cow. Data are 19 weights of an Angus cow at ages 0 to 108 months. The spline can be fitted by using the REG procedure. By inspection of data (figure 10.7), knots are set at 8, 30 and 60 months. The proposed model is:

$$wt_i = \gamma_0 + \gamma_1 x_i + \gamma_2 x^2{}_i + \gamma_3 x^3{}_i + \gamma_4 (x_i - 8)^3{}_+ + \gamma_5 (x_i - 30)^3{}_+ + \gamma_6 (x_i - 60)^3{}_+ + \varepsilon_i$$

where *wt* is weight, and *x* is age.

SAS program:

```
DATA cow;
INPUT age wt @@;
DATALINES;
  0    30     2    60     4   110     6   160     8   250
 10   300    12   330    14   370    16   390    18   410
 24   440    30   480    36   500    48   550    60   580
 72   590    84   600    96   590   108   600
;
DATA cowb;
SET cow;
age2=age**2;
age3=age**3;
u8=(age>8)*((age-8)**3);
u30=(age>30)*((age-30)**3);
u60=(age>60)*((age-60)**3);
RUN;

PROC REG DATA = cowb;
MODEL wt = age age2 age3 u8 u30 u60;
OUTPUT OUT = cowout PREDICTED = predwt;
RUN;
QUIT;
```

Explanation: A new data set named *cowb* is defined. New variables *age2* and *age3* are quadratic and cubic transformations, respectively, of the variable *age*. The new variable *u8* is defined with *u8=(age>8)*(age-8)**3* in the data step. Here, *(age>8)* is a logical term which takes the value 1 if the expression is true and zero if not, thus the term *(age-8)**3* is defined only for age > 8, and is zero otherwise. This also applies for variables *u30* and *u60*.

The REG procedure is used to fit the spline. The statement *MODEL wt = age age2 age3 u8 u30 u60* defines *wt* as the dependent variable, and *age*, *age2*, *age3*, *u8*, *u30* and *u60* as independent variables. The OUTPUT OUT statement defines a new dataset *cowout*, which includes the PREDICTED variable *predwt* from the analysis.

SAS output:

Source	DF	Sum of Squares	Mean Square	F Value	Pr > F
Model	6	653143	108857	1188.02	<.0001
Error	12	1099.54370	91.62864		
Corrected Total	18	654242			

Root MSE		9.57229	R-Square	0.9983
Dependent Mean		386.31579	Adj R-Sq	0.9975
Coeff Var		2.47784		

Parameter Estimates

Variable	DF	Parameter Estimate	Standard Error	t Value	Pr > |t|
Intercept	1	28.62722	9.07535	3.15	0.0083
age	1	10.28453	5.52573	1.86	0.0874
age2	1	3.52431	0.86593	4.07	0.0016
age3	1	-0.19863	0.03904	-5.09	0.0003
u8	1	0.21872	0.04044	5.41	0.0002
u30	1	-0.02222	0.00247	-8.99	<.0001
u60	1	0.00356	0.00119	3.00	0.0111

Explanation: The first table presents an *ANOVA* for the dependent variable *wt*. The sources of variability are *Model*, *Error* and *Corrected Total*. The table lists degrees of freedom (*DF*), *Sum of Squares*, *Mean Square*, calculated F (*F Value*) and P value (*Pr>F*). Note the high value of the coefficient of determination (*R-square*) = 0.9983 which indicates good fit of the model. In the table titled Parameter Estimates, the *Parameter Estimates* with their *Standard errors* and corresponding *t* tests and P values are presented.

The cubic spline function is:

$$wt_i = 28.62722 + 10.28453\ age_i + 3.52431\ age^2_i - 0.19863\ age^3_i + 0.21872\ (age_i - 8)^3_+ - 0.02222\ (age_i - 30)^3_+ + 0.00356\ (age_i - 60)^3_+$$

The graph of the function fitted to the data is shown in figure 10.7.

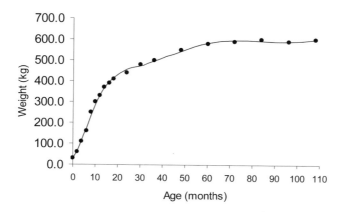

Figure 10.7 Weights of an Angus cow with a fitted cubic spline regression

Both nonlinear and polynomial functions can be used to explain patterns of change in data of this type. Which is preferable depends of the desired interpretation of the data. Nonlinear functions usually use parameters which have biological meaning. Polynomials and polynomial splines often yield better fit, but the parameters are not as easily interpreted; however, as they are linear models, the usual F and t tests are more easily applicable, especially when the effect of some treatment is tested. Note that this example fitted growth data of an individual and the coefficient estimates are not estimates of population parameters. To estimate population parameters or test effects, a sample of animals with growth data must be used.

10.3.3 Using Segmented Regression to Estimate a Plateau

Segmented regression is often used for estimating nutrient requirements. Commonly, values of the dependent variable y reach a plateau, beyond which further changes in the values of the independent variable x result in no changes to the value of y. For example, an increase in dietary methionine increases growth rate of broilers up to some point. After that limit, additional increases in methionine did not yield any further increase in daily gain. The objective of the analysis is to estimate the point at which the plateau begins, the knot. Two functions are used:

$$y_i = \beta_{01} + \beta_{11}x_i + \varepsilon_i \qquad \text{for } x_i \leq x_c$$
$$y_i = \beta_{02} + \varepsilon_i \qquad \text{for } x_i \geq x_c$$

where x_c is a knot and β_{02} is a plateau. The expectations of these two equations must agree at the knot, that is:

$$\beta_{01} + \beta_{11}x_c = \beta_{02}$$

Thus, the second equations can be presented as:

$$y_i = \beta_{01} + \beta_{11}x_c + \varepsilon_i \qquad\qquad \text{for } x_i \geq x_c$$

The knot can be presented in terms of regression parameters and plateau as:

$$x_c = \frac{\beta_{02} - \beta_{01}}{\beta_{11}}$$

A slightly more complicated example describes a quadratic increase to a plateau. Once again two functions are used:

$$y_i = \beta_{01} + \beta_{11}x_i + \beta_{21}x_i^2 + \varepsilon_i \qquad \text{for } x_i \le x_c$$
$$y_i = \beta_{02} + \varepsilon_i \qquad \text{for } x_i \ge x_c$$

The regression curve is continuous because the two segments are joined at x_c. That is, the expected value for x_c, $E(y_i | x_c)$, is the same for both functions:

$$E(y_i | x_c) = \beta_{01} + \beta_{11}x_c + \beta_{21}x^2{}_c = \beta_{02}$$

Thus, two functions are:

$$y_i = \beta_{01} + \beta_{11}x_i + \beta_{21}x_i^2 + \varepsilon_i \qquad \text{for } x_i \le x_c$$
$$y_i = \beta_{01} + \beta_{11}x_c + \beta_{21}x^2{}_c + \varepsilon_i \qquad \text{for } x_i \ge x_c$$

Also, to assure that the whole regression curve is smooth, the first derivatives with respect to x of two segments are by definition equal at x_c:

$$\beta_{11} + 2\beta_{21}x_c = 0$$

From this it follows:

$$x_c = \frac{-\beta_{11}}{2\beta_{21}}$$

and

$$\beta_{02} = \beta_{01} - \frac{\beta_{11}^2}{4\beta_{21}}$$

Thus, the segmented regression can be expressed with three parameters (β_{01}, β_{11} and β_{21}):

$$E(y_i | x_i) = \beta_{01} + \beta_{11}x_i + \beta_{21}x^2{}_i \qquad \text{for } x_i \le x_c$$
$$E(y_i | x_i) = \beta_{01} - \frac{\beta_{11}^2}{4\beta_{21}} \qquad \text{for } x_i \ge x_c$$

Note that this segmented regression is nonlinear with respect to those parameters and their estimation requires a nonlinear approach. The iterative numerical method for obtaining the parameters will be shown using SAS.

10.3.3.1   SAS Example for Segmented Regression with Plateau

In the following example, segmented regression will be used to estimate a nutrient requirement. The requirement is expected to be at the knot (x_c) or joint of the regression segments.

Example: The requirement for methionine will be estimated from measurements of 0–3 week gain of turkey poults.

Gain, g/d:	102	108	125	133	140	141	142	137	138
Methionine, % of NRC:	80	85	90	95	100	105	110	115	120

First, we will fit a model with a linear increase to a plateau. The proposed functions are:

$$y_i = \beta_{01} + \beta_{11}x_i + \varepsilon_i \qquad \text{for } x_i \leq x_c$$
$$y_i = \beta_{01} + \beta_{11}x_c + \varepsilon_i \qquad \text{for } x_i \geq x_c$$

which are joined at the knot x_c. The knot, x_c, is also unknown and can be presented in terms of regression parameters and plateau as:

$$x_c = \frac{plateau - \beta_{01}}{\beta_{11}}$$

In order to start the iterative computation, prior (guessed) values of the parameter estimates must be defined. This can be done by inspection of the data. The possible knot is observed at a methionine value around 100% of NRC and the corresponding gain (plateau) is about 140 g/d.

To estimate priors of the linear function, any two points below the knot can be used, say the values of methionine of 80 and 100% of NRC with corresponding gains of 102 and 140 g/d, respectively. Those values are entered into the proposed linear function resulting in two equations with b_0 and b_1 as the two unknowns:

$$102 = b_0 + b_1 (80)$$
$$140 = b_0 + b_1 (100)$$

The solutions of those equations are:

$$b_0 = -50; \text{ and } b_1 = 1.9. \text{ Those can be used as priors.}$$

SAS program:

```
DATA a;
INPUT met gain @@;
DATALINES;
 80   102     85   115     90   125     95   133    100   140
105   141    110   142    115   140    120   142
;
PROC NLIN DATA = a;
    PARMS b0 = -50  b1= .9 plateau=140;
    xc = (plateau-b0)/b1;
IF met < xc THEN
    MODEL gain = b0 + b1*met;
```

```
    ELSE
      MODEL gain = b0 + b1*xc;

    IF _obs_ =1 and _iter_ =. THEN DO;
      plateau = b0+ b1*xc;
      PUT / xc = plateau= ;
      END;
  RUN;
```

Explanation: The NLIN procedure is used. The PARMS statement defines parameters with their priors, which are needed to start the iterative numerical computation.
The block of statements:

```
IF met < xc THEN
     MODEL gain = b0 + b1*met;
   ELSE
     MODEL gain = b0 + b1*xc;
```

defines two models conditional on the estimated value x_c. Here, *gain* is the dependent variable and *met* is the independent variable, and *b0*, *b1* and *plateau* (or *xc*) are parameters to be estimated. The last block of statements outputs values of the estimated knot and plateau.

SAS output:

Source	DF	Sum of Squares	Mean Square	F Value	Approx Pr > F
Model	2	1630.6	815.3	474.93	<.0001
Error	6	10.3000	1.7167		
Corrected Total	8	1640.9			

Parameter	Estimate	Approx Std Error	Approximate 95% Confidence Limits	
b0	-61.5000	10.2750	-86.6420	-36.3580
b1	2.0600	0.1172	1.7732	2.3468
plateau	141.0	0.5859	139.6	142.4

```
xc=98.300970874 plateau=141
```

Explanation: First is an analysis of variance table including sources of variation (*Model, Error,* and *Corrected Total*), degrees of freedom (*DF*), *Sums of Squares, Mean Square, F Value* and approximated *P* value (*Approx Pr > F*). The high *F* value suggests that the model explains gain of turkeys well. The next table shows the *Parameter* and their *Estimates* together with *Approximate Standard Errors* and 95% *Confidence Limits*. Note that the optimal knot, x_c, was estimated to be at 98.30097% of NRC and the plateau was at 141.0 g. Figure 10.8 presents a graph of the segmented regression of gain of turkey poults using the parameters from the SAS program. The functions are:

$$gain_i = -61.5 + 2.06 \, (met_i) \qquad \text{for } met_i \le 98.30097$$
$$gain_i = 141.0 \qquad \text{for } met_i > 98.30097$$

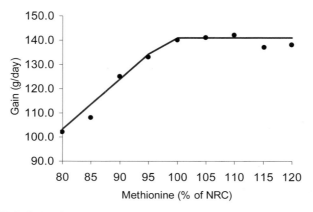

Figure 10.8 Gain of turkey poults on methionine level shown with linear and plateau functions and estimated knot: observed (•) and estimated (__) values

For a quadratic increase to a plateau the proposed functions are:

$$y_i = \beta_{01} + \beta_{11}x_i + \beta_{21}x_i^2 + \varepsilon_i \qquad \text{for } x_i \leq x_c$$
$$y_i = \beta_{01} + \beta_{11}x_c + \beta_{21}x_c^2 + \varepsilon_i \qquad \text{for } x_i \geq x_c$$

which are joined in the knot x_c. The knot, x_c, is unknown and will be estimated from the data, but must satisfy:

$$x_c = \frac{-\beta_{11}}{2\beta_{21}} .$$

Priors (guessed values of the parameter estimates) must be defined. This can be done by inspecting a plot of the data. The plateau begins at a methionine value around 100% of NRC, and the corresponding gain is about 140 g/d, thus giving $x_c = 100$ and *plateau* = 140 g/d.

To estimate priors of the quadratic function, any three points below the estimated value of x_c can be used, say the values of methionine of 80, 90 and 100% of NRC, with corresponding gains of 102, 125 and 140 g/d, respectively. Those values are entered into the proposed quadratic function resulting in three equations with b_0, b_1 and b_2 as the three unknowns:

$$102 = b_0 + b_1 (80) + b_2 (80)^2$$
$$125 = b_0 + b_1 (90) + b_2 (90)^2$$
$$140 = b_0 + b_1 (100) + b_2 (100)^2$$

The solutions of those equations are:

$$b_0 = -370; \; b_1 = 9.1; \text{ and } b_2 = -0.04.$$

Those can be used as priors.

SAS program:

```
DATA a;
INPUT met gain @@;
DATALINES;
 80   102    85   115    90   125    95   133   100   140
105   141   110   142   115   140   120   142
;
PROC NLIN;
    PARMS b0 = -370  b1 = 9.1  b2 = -0.04;
    xc = -.5* b1 / b2;
IF met < xc THEN
     MODEL gain = b0 + b1*met + b2*met*met;
     ELSE
      MODEL gain = b0 + b1*xc + b2*xc*xc;

    IF _obs_ =1 and _iter_ =. THEN DO;
     plateau = bc+ b1*xc+ b2*xc*xc;
     PUT /  xc = plateau=  ;
     END;
RUN;
```

Explanation: The NLIN procedure is used. The PARMS statement defines parameters, and their priors which are needed to start the iterative numerical computation.

The block of statements:

```
IF met<xc THEN
     MODEL gain = b0 + b1*met + b2*met*met;
     ELSE
      MODEL gain = b0 + b1*xc + b2*xc*xc;
```

defines two models conditional on the estimated value x_c. Here, *gain* is the dependent variable and *met* is the independent variable, and *b0*, *b1*, *b2* and *xc* are parameters to be estimated. Note that *xc* is expressed in terms of *b1* and *b2*. The last block of statements outputs the estimated knot and plateau values.

SAS output:

Source	DF	Sum of Squares	Mean Square	F Value	Approx Pr > F
Model	2	1635.8	817.9	957.58	<.0001
Error	6	5.1247	0.8541		
Corrected Total	8	1640.9			

Parameter	Estimate	Approx Std Error	Approximate 95% Confidence Limits	
b0	-474.4	39.3412	-570.7	-378.2
b1	11.4890	0.8365	9.4422	13.5358
b2	-0.0536	0.00440	-0.0644	-0.0428

xc=107.2147301 plateau=141.44982037

Explanation: First is an analysis of variance including sources of variation (*Model, Error,* and *Corrected Total*), degrees of freedom (*DF*), *Sum of Squares, Mean Square, F Value* and approximated *P* value (*Approx Pr > F*). The high *F* value suggests that the model effectively explains variation in gain of turkeys. The next table shows the *Parameters* and their *Estimates*, together with *Approximate Standard Errors* and *95% Confidence Limits*. Note that the optimal knot, x_c, was estimated to be at 107.214% of NRC, and the plateau at 141.4498 g. Figure 10.9 is a plot of gain of turkey poults with a segmented regression graphed using the parameters from the SAS program. The functions are:

$$gain_i = -474.4 + 11.4890 \ (met_i) - 0.0536 \ (met_i)^2 \qquad \text{for } met_i \le 107.214$$
$$gain_i = 141.44 \qquad \text{for } met_i > 107.214$$

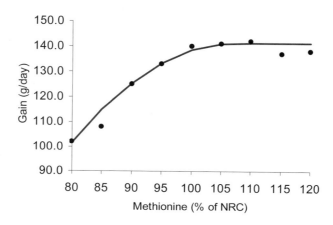

Figure 10.9 Gain of turkey chicks plotted against methionine level shown with quadratic and plateau functions and estimated knot: observed (•) and estimated (___) values

Chapter 11

Fixed Effects One-way Analysis of Variance

Perhaps the most common use of statistics in animal sciences is for testing hypotheses about differences between two or more categorical treatment groups. Each treatment group represents a population. Recall that in a statistical sense a population is a group of units with common characteristics. For example, by feeding three diets three populations are defined, each made up of those animals that will be fed those diets. Analysis of variance is used to determine whether those three populations differ in some characteristics like daily gain, variability, or severity of digestive problems.

In testing differences among populations, a model is used in which measurements or observations are described with a dependent variable, and the way of grouping by an independent variable. The independent variable is thus a qualitative, categorical or classification variable, and is often called a factor. For example, consider a study investigating the effect of several diets on the daily gain of steers and the steers can be fed and measured individually. Daily gain is the dependent, and diet the independent variable. In order to test the effect of diets, random samples must be drawn. The preplanned procedure by which samples are drawn is called an experimental design. A possible experimental design in the example with steers could be: choose a set of steers and assign diets randomly to them. That design is called a completely randomized design. Groups were determined corresponding to the different diets, but note that this does not necessarily mean physical separation into groups. Those groups are often called treatments, because in different groups the animals are treated differently.

Consider an experiment with 15 animals and three treatments. Three treatment groups must be defined, each with five animals. The treatment to which an animal is assigned is determined randomly. Often it is difficult to avoid bias in assigning animals to treatments. The researcher may subconsciously assign better animals to the treatment he thinks is superior. To avoid this, it is good to assign numbers to the animals, for example from 1 to 15, and then randomly choose numbers for each particular treatment. The following scheme describes a completely randomized design with three treatments and 15 animals. The treatments are denoted with T_1, T_2 and T_3:

Steer number	1	2	3	4	5	6	7	8
Treatment	T_2	T_1	T_3	T_2	T_3	T_1	T_3	T_2
Steer number	9	10	11	12	13	14	15	
Treatment	T_1	T_2	T_3	T_1	T_3	T_2	T_1	

For clarity, the data can be sorted by treatment:

Treatments					
T_1		T_2		T_3	
Steer	Measurement	Steer	Measurement	Steer	Measurement
2	y_{11}	1	y_{21}	3	y_{31}
6	y_{12}	4	y_{22}	5	y_{32}
9	y_{13}	8	y_{23}	7	y_{33}
12	y_{14}	10	y_{24}	11	y_{34}
15	y_{15}	14	y_{25}	13	y_{35}

Here, $y_{11}, y_{12},..., y_{35}$, or generally y_{ij} denotes experimental unit j in treatment i.

In the example of diets and steers, each sample group fed a different diet (treatment) represents a sample from an imaginary population fed with the same diet. Differences among arithmetic means of treatment groups will be calculated, and it will be projected whether the differences can be expected in a large number of similar experiments. If the differences between treatments on experimental animals are significant, it can be concluded that the differences will be expected between populations, that is, on future groups of animals fed those diets.

We may be also interested in testing differences among the means of several categories that do not come necessarily from the controlled experiment. For example, a simple study is conducted to determine if there are differences among three regions in milk yield of Holstein cows. A random sample of cows from each region is chosen to determine if differences among sample means are large enough to conclude that regions are different.

In applying a completely randomized design or when groups indicate a natural way of classification, the objectives can be:

1. Estimating the means
2. Testing the difference between groups

Analysis of variance is used for testing differences among group means by comparing explained variability, caused by differences among groups, with unexplained variability among the measured units within groups. If explained variability is much greater than unexplained, it can be concluded that the treatments or groups have significantly influenced the variability and that the arithmetic means of the groups are significantly different (Figure 11.1). The analysis of variance partitions the total variability to its sources, that among groups versus that remaining within groups, and analyzes the significance of the explained variability.

When data are classified into groups according to just one categorical variable, the analysis is called one-way analysis of variance. Data can also be classified according to two or more categorical variables. These analyses are called two-way, three-way, ..., multi-way analyses of variance.

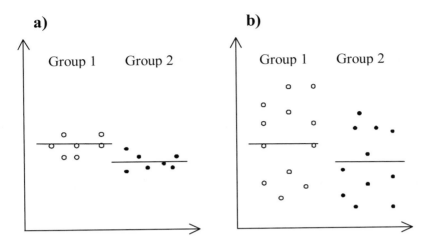

Figure 11.1 Differences between means of group 1 and group 2: a) variability within groups is relatively small; b) variability within groups is relatively large. The difference between groups is more obvious when the variability within groups is small compared to the variability between groups

11.1 The Fixed Effects One-way Model

The fixed effects one-way model is most often applied when the goal is to test differences among means of two or more populations. Populations are represented by groups or treatments each with its own population mean. The effects of groups are said to be fixed because they are specifically chosen or defined by some nonrandom process. The effect of the particular group is fixed for all observations in that group. Differences among observations within group are random. These inferences about the populations are made based on random samples drawn from those populations. The one-way model is:

$$y_{ij} = \mu + \tau_i + \varepsilon_{ij} \qquad i = 1,..., a; \ j = 1,..., n$$

where:

y_{ij} = observation j in group or treatment i
μ = the overall mean
τ_i = the fixed effect of group or treatment i (denotes an unknown parameter)
ε_{ij} = random error with mean 0 and variance σ^2

The independent variable τ, often called a factor, represents the effects of different treatments. The factor influences the values of the dependent variable y.

The model has the following assumptions:

$E(\varepsilon_{ij}) = 0$, the expectations of errors are zero
$Var(\varepsilon_{ij}) = \sigma^2$, the variance of errors is constant across groups (homogeneous)
Usually, it is also assumed that errors have a normal distribution

From the assumptions it follows:

$E(y_{ij}) = \mu + \tau_i = \mu_i$, the expectation of an observation y_{ij} is its group mean μ_i

$Var(y_{ij}) = \sigma^2$, the variance of y_i is constant across groups (homogeneous)

Let the number of groups be a. In each group there are n measurements. Thus, there is a total of $N = (n\ a)$ units divided into a groups of size n. A model that has an equal number of observations in each group is called balanced. For the unbalanced case, there is an unequal number of observations per group, n_i denotes the number of observations in group i, and the total number of observations is $N = \sum_i n_i$, $(i = 1,..., a)$.

For example, for three groups of five observations each, observations can be shown schematically:

	Group	
G1	G2	G3
y_{11}	y_{21}	y_{31}
y_{12}	y_{22}	y_{32}
y_{13}	y_{23}	y_{33}
y_{14}	y_{24}	y_{34}
y_{15}	y_{25}	y_{35}

It can be shown, by using either least squares or maximum likelihood estimation, that population means are estimated by arithmetic means of the sample groups ($\bar{y}_{i.}$). The estimated or fitted values of the dependent variable are:

$$\hat{y}_{ij} = \hat{\mu}_i = \hat{\mu} + \hat{\tau}_i = \bar{y}_{i.}. \qquad i = 1,..., a;\ j = 1,..., n$$

where:

$\hat{y}_{ij}$ = the estimated (fitted) value of the dependent variable

$\hat{\mu}_i$ = the estimated mean of group or treatment i

$\hat{\mu}$ = the estimated overall mean

$\hat{\tau}_i$ = the estimated effect of group or treatment i

$\bar{y}_{i.}$ = the arithmetic mean of group or treatment i

While $\hat{\mu}_i$ has a unique solution ($\bar{y}_{i.}$), there are no separate unique solutions for $\hat{\mu}$ and $\hat{\tau}_i$. A reasonable solution can be obtained by using a constraint $\sum_i n_i \hat{\tau}_i = 0$, in which n_i is the number of observations of group or treatment i. Then:

$$\hat{\mu} = \bar{y}..$$

$$\hat{\tau}_i = \bar{y}_{i.} - \bar{y}..$$

Also, $e_{ij} = y_{ij} - \hat{\mu}_i$ = residual

Thus, each measurement j in group i in the samples can be represented as:

$$y_{ij} = \hat{\mu}_i + e_{ij}$$

11.2 Partitioning Total Variability

Analysis of variance is used to partition total variability into that which is explained by group versus that which is unexplained, and the relative magnitude of the two measures of variability is used to test significance. For a one-way analysis, three sources of variability are defined and measured with corresponding sums of squares:

Source of variability	Sum of squares
Total variability - spread of observations about the overall mean	$SS_{TOT} = \sum_i \sum_j (y_{ij} - \bar{y}..)^2$ = total sum of squares equal to the sum of squared deviations of observations from the overall mean. Here, $\bar{y}.. = \dfrac{\sum_i \sum_j y_{ij}}{N}$ = mean of all observations ij, N = total number of observations
Variability between groups or treatments - explained variability - spread of group or treatment means about the overall mean	$SS_{TRT} = \sum_i \sum_j (\bar{y}_i. - \bar{y}..)^2 = \sum_i n_i (\bar{y}_i. - \bar{y}..)^2$ = sum of squares between groups or treatments known as the group or treatment sum of squares, equal to the sum of squared deviations of group or treatment means from the overall mean. Here, $\bar{y}_i. = \dfrac{\sum_j y_{ij}}{n_i}$ = mean of group i, n_i = number of observations of group i
Variability within groups or treatments - variability among observations - unexplained variability - spread of observations about the group or treatment means	$SS_{RES} = \sum_i \sum_j (y_{ij} - \bar{y}_i.)^2$ = sum of squares within groups or treatments, known as the residual sum of squares or error sum of squares, equal to the sum of squared deviations of observations from the group or treatment means

The deviations of individual observations from the overall mean can be partitioned into the deviation of the group mean from the overall mean plus the deviation of the individual observation from the group mean:

$$(y_{ij} - \bar{y}..) = (\bar{y}_i. - \bar{y}..) + (y_{ij} - \bar{y}_i.)$$

Analogously, it can be shown that the overall sum of squares can be partitioned into the sum of squares of group means around the overall mean plus the sum of squares of the individual observations around the group means:

$$\sum_i \sum_j (y_{ij} - \bar{y}..)^2 = \sum_i \sum_j (\bar{y}_i. - \bar{y}..)^2 + \sum_i \sum_j (y_{ij} - \bar{y}_i.)^2$$

By defining:

$$SS_{TOT} = \sum_i \sum_j (y_{ij} - \bar{y}..)^2$$

$$SS_{TRT} = \sum_i \sum_j (\bar{y}_i. - \bar{y}..)^2$$

$$SS_{RES} = \sum_i \sum_j (y_{ij} - \bar{y}_i.)^2$$

it can be written:

$$SS_{TOT} = SS_{TRT} + SS_{RES}$$

Similarly, the degrees of freedom can be partitioned:

Total		Group or treatment		Residual
$(N-1)$	$=$	$(a-1)$	$+$	$(N-a)$

where:

N = the total number of observations
a = the number of groups or treatments

Sums of squares can be calculated by using a shortcut calculation presented here in five steps:

1. Total sum = sum of all observations:

$$\sum_i \sum_j y_{ij}$$

2. Correction for the mean:

$$C = \frac{\left(\sum_i \sum_j y_{ij}\right)^2}{N} = \frac{(total\ sum)^2}{total\ number\ of\ observations}$$

3. Total (corrected) sum of squares:

$$SS_{TOT} = \sum_i \sum_j y_{ij}^2 - C = \text{Sum of all squared observations minus } C$$

4. Group or treatment sum of squares:

$$SS_{TRT} = \sum_i \frac{\left(\sum_j y_{ij}\right)^2}{n_i} - C = \text{Sum of } \frac{(group\ sum)^2}{group\ size} \text{ for each group minus } C$$

5. Residual sum of squares:

$$SS_{RES} = SS_{TOT} - SS_{TRT}$$

By dividing the sums of squares by their corresponding degrees of freedom, mean squares are obtained:

Group or treatment mean square: $MS_{TRT} = \dfrac{SS_{TRT}}{a-1}$

Residual mean square: $MS_{RES} = \dfrac{SS_{RES}}{N-a} = s^2$, which is the estimator of $Var(\varepsilon_{ij}) = \sigma^2$, the variance of errors in the population.

The variance estimator (s^2) is also equal to the mean of the estimated group variances $(s^2{}_i)$:

$$s^2 = \frac{\sum_i s_i^2}{a}$$

For unequal numbers of observations per group (n_i):

$$s^2 = \frac{\sum_i (n_i - 1)s_i^2}{\sum_i (n_i - 1)}$$

11.3 Hypothesis Test – *F* Test

Hypotheses of interest are about the differences between population means. A null hypothesis H_0 and an alternative hypothesis H_1 are stated:

H_0: $\mu_1 = \mu_2 = ... = \mu_a$, the population means are equal
H_1: $\mu_i \neq \mu_{i'}$, for at least one pair (i,i'), the means are not equal

The hypotheses can also be stated:

H_0: $\tau_1 = \tau_2 = ... = \tau_a$, there is no difference among treatments, i.e. there is no effect
of treatments
H_1: $\tau_i \neq \tau_{i'}$, for at least one pair (i,i') a difference between treatments exists

An F statistic is defined using sums of squares and their corresponding degrees of freedom. It is used to test whether the variability among observations is of magnitude to be expected from random variation or is influenced by a systematic effect of group or treatment. In other words, is the variability between treatments groups significantly greater than the variability within treatments? The test is conducted with an F statistic that compares the ratio of explained and unexplained variability:

F = (explained variability) / (unexplained variability) =
(variability between treatments) / (variability within treatments)

To justify using an F statistic, the variable y must have a normal distribution. Then the ratio:

$$\frac{SS_{RES}}{\sigma^2}$$

has a *chi*-square distribution with $(N - a)$ degrees of freedom. The ratio:

$$\frac{SS_{TRT}}{\sigma^2}$$

has *chi*-square distribution with $(a-1)$ degrees of freedom if there is no difference between treatments (H_0 holds). Also, it can be shown that SS_{TRT} and SS_{RES} are independent. A ratio of two *chi*-square variables divided by their respective degrees of freedom gives an F statistic:

$$F = \frac{(SS_{TRT}/\sigma^2)/(a-1)}{(SS_{RES}/\sigma^2)/(N-a)}$$

with an F distribution if H_0 holds.

Recall that:

$$\frac{SS_{TRT}}{a-1} = MS_{TRT} = \text{treatment mean square}$$

$$\frac{SS_{RES}}{N-a} = MS_{RES} = \text{residual mean square}$$

Thus, the F statistic is:

$$F = \frac{MS_{TRT}}{MS_{RES}}$$

with an F distribution with $(a-1)$ and $(N-a)$ degrees of freedom if H_0 holds.

It can be shown that the expectations of the mean squares are:

$$E(MS_{RES}) = \sigma^2$$

$$E(MS_{TRT}) \begin{cases} = \sigma^2 & \text{if } H_0 \\ > \sigma^2 & \text{if not } H_0 \end{cases}$$

With a constraint that $\sum_i \tau_i = 0$:

$$E(MS_{TRT}) = \sigma^2 + \frac{n\sum_i \tau_i^2}{a-1}$$

Thus, MS_{RES} is an unbiased estimator of σ^2 regardless of H_0, and MS_{TRT} is an unbiased estimator of σ^2 only if H_0 holds.

If H_0 is true, then $MS_{TRT} \approx \sigma^2$ and $F \approx 1$. If H_1 is true, then $MS_{TRT} > \sigma^2$ and $F > 1$. This also indicates that MS_{TRT} is much greater than MS_{RES}. The H_0 is rejected if the calculated F is 'large', that is if the calculated F is much greater than 1. For the α level of significance, H_0 is rejected if the calculated F from the sample is greater than the critical value, $F > F_{\alpha,(a-1),(N-a)}$ (Figure 11.2).

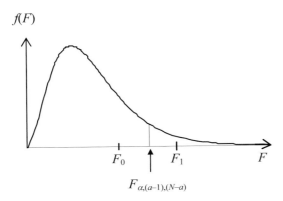

Figure 11.2 Test of hypotheses using an F distribution. If the calculated $F = F_0 < F_{\alpha,(a-1),(N-a)}$, H_0 is not rejected. If the calculated $F = F_1 > F_{\alpha,(a-1),(N-a)}$, H_0 is rejected with α level of significance

Usually, the sums of squares, degrees of freedom, mean squares and calculated F are written in a table called an analysis of variance or *ANOVA* table:

Source	SS	df	MS = SS / df	F
Group or Treatment	SS_{TRT}	$a - 1$	$MS_{TRT} = SS_{TRT} / (a - 1)$	$F = MS_{TRT} / MS_{RES}$
Residual	SS_{RES}	$N - a$	$MS_{RES} = SS_{RES} / (N - a)$	
Total	SS_{TOT}	$N - 1$		

Example: An experiment was conducted to investigate the effects of three different diets on daily gains (g) in pigs. The diets are denoted with TR_1, TR_2 and TR_3. Five pigs were fed each diet. Data, sums and means are presented in the following table:

	TR_1	TR_2	TR_3	
	270	290	290	
	300	250	340	
	280	280	330	
	280	290	300	
	270	280	300	Total
Σ	1400	1390	1560	4350
n	5	5	5	15
$\bar{y}$	280	278	312	290

For calculation of sums of squares the short method is shown:

1. Total sum:

$$\sum_i \sum_j y_{ij} = (270 + 300 + ...+ 300) = 4350$$

2. Correction for the mean:

$$C = \frac{\left(\sum_i \sum_j y_{ij}\right)^2}{N} = \frac{(4350)^2}{15} = 1261500$$

3. Total (corrected) sum of squares:

$$SS_{TOT} = \sum_i \sum_j y_{ij}^2 - C = (270^2 + 300^2 + \ldots + 300^2) - C$$
$$= 1268700 - 1261500 = 7200$$

4. Treatment sum of squares:

$$SS_{TRT} = \sum_i \frac{\left(\sum_j y_{ij}\right)^2}{n_i} - C =$$

$$= \frac{1400^2}{5} + \frac{1390^2}{5} + \frac{1560^2}{5} - C = 1265140 - 1261500 = 3640$$

5. Residual sum of squares:

$$SS_{RES} = SS_{TOT} - SS_{TRT} = 7200 - 3640 = 3560$$

ANOVA table:

Source	SS	df	MS	F
Treatment	3640	3 − 1 = 2	1820.00	6.13
Residual	3560	15 − 3 = 12	296.67	
Total	7200	15 − 1 = 14		

$$F = \frac{MS_{TRT}}{MS_{RES}} = \frac{1820.0}{296.67} = 6.13$$

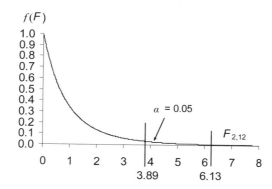

Figure 11.3 *F* test for the example of effect of pig diets

The critical value of *F* for 2 and 12 degrees of freedom and $\alpha = 0.05$ level of significance is $F_{0.05,2,12} = 3.89$ (see Appendix B: Critical values of *F* distributions). Since the calculated $F = 6.13$ is greater (more extreme) than the critical value, the H_0 is rejected, supporting the conclusion that a difference exists between at least two treatments means (Figure 11.3).

11.4   Estimation of Group Means

Estimators of the population means (μ_i) are arithmetic means of groups or treatments ($\bar{y}_{i\cdot}$). Estimators can be obtained by least squares or maximum likelihood methods, as previously shown for linear regression.

According to the central limit theorem, estimators of the means are normally distributed with mean μ_i and standard deviation $s_{\bar{y}_1} = \sqrt{\dfrac{MS_{RES}}{n_i}}$. Here, MS_{RES} is the residual mean square, which is an estimate of the population variance, and n_i is the number of observations in treatment i. The standard deviation of estimators of the mean is usually called the standard error of the mean. Confidence intervals for the means can be calculated by using a Student t distribution with $N - a$ degrees of freedom. A $100(1 - \alpha)\%$ confidence interval for the group or treatment i is:

$$\bar{y}_{i\cdot} \pm t_{\alpha/2, N-a} \sqrt{\frac{MS_{RES}}{n_i}}$$

Example: From the example with pig diets, calculate a confidence interval for diet TR_1. As previously shown: $MS_{RES} = 296.67$; $n_i = 5$; $\bar{y}_1. = 280$

The standard error is:

$$s_{\bar{y}_1} = \sqrt{\frac{MS_{RES}}{n_i}} = \sqrt{\frac{296.67}{5}} = 7.70$$

$t_{\alpha/2, N-a} = t_{0.025, 12} = 2.179$ (Appendix B, Critical values of Student t distributions)

The 95% confidence interval is:

$280 \pm (2.179)(7.70)$ which is equal to:
280 ± 16.78

11.5   Maximum Likelihood Estimation

The parameters μ_i and σ^2 can alternatively be estimated by using maximum likelihood (ML) . Under the assumption of normality, the likelihood function is a function of the parameters for a given set of N observations:

$$L(\mu_i, \sigma^2 \mid y) = \frac{1}{\left(\sqrt{2\pi\sigma^2}\right)^N} e^{-\sum_i \sum_j (y_{ij} - \mu_i)^2 / 2\sigma^2}$$

The log likelihood is:

$$logL(\mu_i, \sigma^2 \mid y) = -\frac{N}{2}log(\sigma^2) - \frac{N}{2}log(2\pi) - \frac{\sum_i\sum_j(y_{ij} - \mu_i)^2}{2\sigma^2}$$

A set of estimators chosen to maximize the log likelihood function is called the maximum likelihood estimators. The maximum of the function can be determined by taking the partial derivatives of the log likelihood function with respect to the parameters:

$$\frac{\partial\, logL(\mu_i, \sigma^2 \mid y)}{\partial\mu_i} = -\frac{1}{2\sigma^2}\sum_j(y_{ij} - \mu_i)(-2)$$

$$\frac{\partial\, logL(\mu_i, \sigma^2 \mid y)}{\partial\sigma^2} = -\frac{N}{2\sigma^2} + \frac{1}{2\sigma^4}\sum_i\sum_j(y_{ij} - \mu_i)^2$$

These derivatives are equated to zero in order to find the estimators $\hat{\mu}_i$ and $\hat{\sigma}^2_{ML}$. Note that the second derivative must be negative when parameters are replaced with solutions. The *ML* estimators are:

$$\hat{\mu}_i = \bar{y}_i.$$

$$\hat{\sigma}^2 = s^2_{ML} = \tfrac{1}{N}\sum_i\sum_j(y_{ij} - \bar{y}_i.)^2$$

The *ML* estimator for the variance is biased, i.e. $E(s^2_{ML}) \neq \sigma^2$. An unbiased estimator is obtained when the maximum likelihood estimator is multiplied by $N/(N-a)$, that is:

$$s^2 = \frac{N}{N-a}s^2_{ML}$$

11.6 Likelihood Ratio Test

The hypothesis H_0: $\mu_1 = \mu_2 = ... = \mu_a$ can be tested by using likelihood functions. The values of likelihood functions are compared using estimates for H_0 and H_1. Those estimates are maximums of the corresponding likelihood functions.

The likelihood function under H_0 is:

$$L(\mu, \sigma^2 \mid y) = \frac{1}{\left(\sqrt{2\pi\sigma^2}\right)^N}e^{-\sum_i\sum_j(y_{ij} - \mu)^2/2\sigma^2}$$

and the corresponding maximum likelihood estimators are:

$$\hat{\mu}_{ML\_0} = \frac{\sum_i\sum_j y_{ij}}{n} = \bar{y}..$$

$$\hat{\sigma}^2_{ML\_0} = s^2_{ML\_0} = \frac{\sum_i\sum_j(y_{ij} - \bar{y}..)^2}{N}$$

Using the estimators for H_0, the maximum of the likelihood function is:

$$L(\bar{y}..,s^2_{ML\_0} \mid y) = \frac{1}{\left(\sqrt{2\pi s^2_{ML\_0}}\right)^N} e^{-\sum_i \sum_j (y_{ij}-\bar{y}..)^2 / 2s^2_{ML\_0}}$$

The likelihood function when H_0 is not true is:

$$L(\mu_i, \sigma^2 \mid y) = \frac{1}{\left(\sqrt{2\pi\sigma^2}\right)^N} e^{-\sum_i \sum_j (y_{ij}-\mu_i)^2 / 2\sigma^2}$$

and the corresponding maximum likelihood estimators are:

$$\mu_i = \bar{y}_i.$$
$$\hat{\sigma}^2_{ML\_1} = s^2_{ML\_1} = \tfrac{1}{N}\sum_i \sum_j (y_{ij}-\bar{y}_i.)^2$$

Using the estimators for H_1, the maximum of the likelihood function is:

$$L(\bar{y}_i.,s^2_{ML\_1} \mid y) = \frac{1}{\left(\sqrt{2\pi s^2_{ML\_1}}\right)^N} e^{-\sum_i \sum_j (y_{ij}-\bar{y}_i.)^2 / 2s^2_{ML\_1}}$$

The likelihood ratio is:

$$\Lambda = \frac{L(\bar{y}..,s^2_{ML\_0} \mid y)}{L(\bar{y}_i.,s^2_{ML\_1} \mid y)}$$

Further, the logarithm of this ratio multiplied by (-2) has an approximate *chi*-square distribution with $N-a$ degrees of freedom, where N and a are the total number of observations and number of groups, respectively:

$$-2log\Lambda = -2log\frac{L(\bar{y}..,s^2_{ML\_0} \mid y)}{L(\bar{y}_i.,s^2_{ML\_1} \mid y)} = -2\big[logL(\bar{y}..,s^2_{ML\_0} \mid y) - logL(\bar{y}_i.,s^2_{ML\_1} \mid y)\big]$$

For the significance level α, H_0 is rejected if $-2log\Lambda > \chi^2_{N-a}$, where χ^2_{N-a} is a critical value. Assuming the variance σ^2 is known, then:

$$-2log\Lambda = -2\big[logL(\bar{y}.. \mid \sigma^2, y) - logL(\bar{y}_i. \mid \sigma^2, y)\big]$$

$$-2log\Lambda = -2\left[-\frac{n}{2}log(\sigma^2) - \frac{\sum_i \sum_j (y_{ij}-\bar{y}..)^2}{2\sigma^2} + \frac{n}{2}log(\sigma^2) + \frac{\sum_i \sum_j (y_{ij}-\bar{y}_i.)^2}{2\sigma^2}\right]$$

$$-2log\Lambda = \left[\frac{\sum_i \sum_j (y_{ij}-\bar{y}..)^2}{\sigma^2} - \frac{\sum_i \sum_j (y_{ij}-\bar{y}_i.)^2}{\sigma^2}\right]$$

And as shown previously:

$$\sum_i \sum_j (y_{ij}-\bar{y}..)^2 = SS_{TOT} = \text{the total sum of squares}$$

$$\sum_i \sum_j \left(y_{ij} - \bar{y}_{i.}\right)^2 = SS_{RES} = \text{the residual sum of squares}$$

$$SS_{TRT} = SS_{TOT} - SS_{RES} = \text{the treatment sum of squares}$$

Thus:

$$-2log\Lambda = \left[\frac{SS_{TRT}}{\sigma^2}\right]$$

Estimating σ^2 from the one-way model as $s^2 = MS_{RES} = \dfrac{SS_{RES}}{N-a}$, and having $MS_{TRT} = \dfrac{SS_{TRT}}{a-1}$, note that asymptotically $-2log\Lambda$ divided by the degrees of freedom $(a-1)$ is equivalent to the F statistic as shown before.

11.7 Multiple Comparisons among Group Means

An F test is used to conclude if there is a significant difference among groups or treatments. If H_0 is not rejected, it is not necessary or appropriate to further analyze the problem, although the researcher must be aware of the possibility of a type II error. If, as a result of the F test, H_0 is rejected, it is appropriate to further question which treatment(s) caused the effect, that is, between which groups the significant difference exists.

Let $\mu_i = \mu + \tau_i$ and $\mu_{i'} = \mu + \tau_{i'}$ be the means of populations represented by the group designations i and i'. The question is whether the means of the two populations i and i', represented by the sampled groups i and i', are different. For an experiment with a groups or treatments there is a total of $\binom{a}{2}$ pair-wise comparisons of means. For each comparison there is a possibility of making a type I or type II error. Recall that a type I error occurs when H_0 is rejected and actually $\mu_i = \mu_{i'}$. A type II error occurs when H_0 is not rejected but actually $\mu_i \neq \mu_{i'}$. Looking at the experiment as a whole, the probability of making a type I error in conclusion is defined as the experimental error rate (EER):

$$EER = P(\text{at least one conclusion } \mu_i \neq \mu_{i'}, \text{ but actually all } \mu_i \text{ are equal})$$

There are many procedures for pair-wise comparisons of means. These procedures differ in EER. Here, two procedures, the Least Significance Difference and Tukey tests, will be described. Others, not covered here include Bonferoni, Newman-Keuls, Duncan, Dunnet, etc. (see for example, Snedecor and Cochran, 1989 or Sokal and Rohlf, 1995).

11.7.1 Least Significance Difference (LSD)

The aim of this procedure is to determine the least difference between a pair of means that will be significant and to compare that value with the calculated differences between all pairs of group means. If the difference between two means is greater than the least significant difference (LSD), it can be concluded that the difference between this pair of means is significant. The LSD is computed:

$$LSD_{ii'} = t_{\alpha/2, N-a}\sqrt{MS_{RES}\left(\frac{1}{n_i} + \frac{1}{n_{i'}}\right)}$$

Note that $\sqrt{MS_{RES}\left(\dfrac{1}{n_i}+\dfrac{1}{n_{i'}}\right)}=s_{\bar{y}_i-\bar{y}_{i'}}=$ standard error of the estimator of the difference between the means of two groups or treatments i and i'.

An advantage of the *LSD* is that it has a low level of type II error and will most likely detect a difference if a difference really exists. A disadvantage of this procedure is that it has a high level of type I error. Because of the probability of type I error, a significant F test must precede the *LSD* in order to ensure a level of significance α for any number of comparisons. The whole procedure of testing differences is as follows:

1. F test (H_0: $\mu_1 = \ldots = \mu_a$, H_1: $\mu_i \neq \mu_{i'}$ for at least one pair i, i')
2. if H_0 is rejected then the $LSD_{ii'}$ is calculated for all pairs i, i'
3. conclude $\mu_i \neq \mu_{i'}$ if $\left|\bar{y}_i - \bar{y}_{i'}\right| \geq LSD_{ii'}$

11.7.2 Tukey Test

The Tukey test uses a q statistic which has a Q distribution (the Studentized range between the highest and lowest mean). The q statistic is defined as:

$$q = \frac{\bar{y}_{Max} - \bar{y}_{min}}{s/\sqrt{n}}$$

A critical value of this distribution, $q_{\alpha,a,N-a}$, is determined with a level of significance α, the number of groups a, and error degrees of freedom $N-a$ (see Appendix B: Critical values of the Studentized range). A Tukey critical difference, also known as the honestly significant difference (*HSD*), is computed:

$$HSD = q_{\alpha,a,N-a}\sqrt{\frac{MS_{RES}}{n_t}}$$

Here, MS_{RES} is the residual mean square and n_t is the group size. It can be concluded that the difference between means of any two groups $\bar{y}_i$ and $\bar{y}_{i'}$ is significant if the difference is equal to or greater than the *HSD* ($\mu_i \neq \mu_{i'}$ when $\left|\bar{y}_i - \bar{y}_{i'}\right| \geq HSD_{ii'}$). To ensure an experimental error rate less or equal to α, an F test must precede a Tukey test. Adjustment of a Tukey test for multiple comparisons will be shown using SAS in section 11.9.

If the number of observations per group (n_t) is not equal (this is usually denoted as unbalanced data), then a weighted number can be used for n_t:

$$n_t = \frac{1}{a-1}\left(N - \frac{\sum_i n_i^2}{N}\right)$$

where N is the total number of observations, a is the number of groups and n_i is the number of observations for group i. Alternatively, the harmonic mean of n_i can be used for n_t.

An advantage of the Tukey test is that it has fewer incorrect conclusions of $\mu_i \neq \mu_{i'}$ (type I errors) compared to the *LSD*; a disadvantage is that there are more incorrect $\mu_i = \mu_{i'}$ conclusions (type II errors).

Example: Continuing with the example using three diets for pigs, we concluded that a significant difference exists between group means, leading to the question of which of the diets is best. By the Tukey method:

$$HSD = q_{\alpha,a,N-a}\sqrt{\frac{MS_{RES}}{n_t}}$$

Taking:

$q_{3,12} = 3.77$ (see Appendix B: Critical values of the Studentized range)
$MS_{RES} = 296.67$
$n_t = 5$

The critical difference is:

$$HSD = 3.77\sqrt{\frac{296.67}{5}} = 29.0$$

For convenience all differences can be listed in a table.

Treatments	$\bar{y}_i$	TR_1 280	TR_2 278
TR_3	312	32	34
TR_1	280	-	2
TR_2	278	-	-

The differences between means of treatments TR_3 and TR_1, and TR_3 and TR_2, are 32.0 and 34.0, respectively, both of which are greater than the critical value $HSD = 29.0$. Therefore, diet TR_3 yields higher gains than either diets TR_1 or TR_2 with $\alpha = 0.05$ level of significance.

This result can be presented graphically in the following manner. The group means are ranked and all groups not found to be significantly different are connected with a line:

TR_3 TR_1 TR_2

Alternatively, superscripts can be used. Means with no superscript in common are significantly different with $\alpha = 0.05$.

	Treatment		
	TR_1	TR_2	TR_3
Mean daily gain (g)	280[a]	278[a]	312[b]

11.7.3 Contrasts

The analysis of contrasts is also a way to compare group or treatment means. Contrasts can also be used to test the difference of the mean of a group of several treatments against the mean of one or more other treatments. For example, suppose the objective of an experiment was to test the effects of two new rotational grazing systems on total pasture yield. Also, as a control, a standard grazing system was used. Thus, a total of three treatments were defined, a control and two new treatments. It may be of interest to determine if the rotational systems are better than the standard system. A contrast can be used to compare the mean of the standard against the combined mean of the rotational systems. In addition, the two rotational systems can be compared to each other. Consider a model:

$$y_{ij} = \mu + \tau_i + \varepsilon_{ij} \qquad i = 1,\dots, a \quad j = 1,\dots, n$$

a contrast is defined as:

$$\Gamma = \sum_i \lambda_i \, \tau_i$$

or

$$\Gamma = \sum_i \lambda_i \, \mu_i$$

where:

τ_i = the effect of group or treatment i
$\mu_i = \mu + \tau_i$ = the mean of group or treatment i
λ_i = contrast coefficients which define a comparison

The contrast coefficients must sum to zero:

$$\sum_i \lambda_i = 0$$

For example, for a model with three treatments in which the mean of the first treatment is compared with the mean of the other two, the contrast coefficients are:

$\lambda_1 = 2$
$\lambda_2 = -1$
$\lambda_3 = -1$

An estimate of the contrast is:

$$\hat{\Gamma} = \sum_i \lambda_i \hat{\mu}_i$$

Since in a one-way *ANOVA* model, the treatment means are estimated by arithmetic means, the estimator of the contrast is:

$$\hat{\Gamma} = \sum_i \lambda_i \bar{y}_i.$$

Hypotheses for contrast are:

$H_0: \Gamma = 0$
$H_1: \Gamma \neq 0$

The hypotheses can be tested using an F statistic:

$$F = \frac{SS_{\hat{\Gamma}}/1}{MS_{RES}}$$

which has an $F_{1,(N-a)}$ distribution. Here, $SS_{\hat{\Gamma}} = \dfrac{\hat{\Gamma}^2}{\sum_i \lambda_i^2 / n_i}$ is the contrast sum of squares, and

$SS_{\hat{\Gamma}}/1$ is the contrast mean square with 1 degree of freedom.

Example: In the example of three diets for pigs, the arithmetic means were calculated:
$$\bar{y}_1. = 280, \quad \bar{y}_2. = 278 \quad \text{and} \quad \bar{y}_3. = 312$$
A contrast can be used to compare the third diet against the first two.

The contrast coefficients are:

$$\lambda_1 = -1, \; \lambda_2 = -1 \text{ and } \lambda_3 = 2$$

The estimated contrast is:

$$\hat{\Gamma} = \sum_i \lambda_i \bar{y}_i. = (-1)(280) + (-1)(278) + (2)(312) = 66$$

The $MS_{RES} = 296.67$ and $n_i = 5$. The contrast sum of squares is:

$$SS_{\hat{\Gamma}} = \frac{\hat{\Gamma}^2}{\sum_i \lambda_i^2 / n_i} = \frac{(66)^2}{(-1)^2/5 + (-1)^2/5 + (2)^2/5} = 3630$$

The calculated F value is:

$$F = \frac{SS_{\hat{\Gamma}}/1}{MS_{RES}} = \frac{3630/1}{296.67} = 12.236$$

The critical value for 0.05 level of significance is $F_{0.05,1,12} = 4.75$. Since the calculated F is greater than the critical value, H_0 is rejected. This test provides evidence that the third diet yields greater gain than the first two.

11.7.4 Orthogonal contrasts

Let $\hat{\Gamma}_1$ and $\hat{\Gamma}_2$ be two contrasts with coefficients λ_{1i} and λ_{2i}. The contrasts $\hat{\Gamma}_1$ and $\hat{\Gamma}_2$ are orthogonal if:

$$\sum_i \lambda_{1i} \lambda_{2i} = 0$$

Generally, a model with a groups or treatments and $(a-1)$ degrees of freedom can be partitioned to $(a-1)$ orthogonal contrasts such that:

$$SS_{TRT} = \sum_i SS_{\hat{\Gamma}_i} \qquad i = 1,\ldots,(a-1)$$

that is, the sum of a complete set of orthogonal contrast sum of squares is equal to the treatment sum of squares. From this it follows that if a level of significance α is used in the F test for all treatments, then the level of significance for singular orthogonal contrasts will not exceed α, thus, type I error is controlled.

Example: In the example of three diets for pigs the following orthogonal contrasts can be defined: the third diet against the first two, and the first diet against the second. Previously it was computed: $MS_{RES} = 296.67$; $SS_{TRT} = 3640$; $n_i = 5$

The contrast coefficients are:

	TR_1	TR_2	TR_3
$\bar{y}_i.$	280	278	312
Contrast1	$\lambda_{11} = 1$	$\lambda_{12} = 1$	$\lambda_{13} = -2$
Contrast2	$\lambda_{21} = 1$	$\lambda_{22} = -1$	$\lambda_{23} = 0$

The contrasts are orthogonal because:

$$\sum_i \lambda_{1i}\,\lambda_{2i} = (1)(1) + (1)(-1) + (-2)(0) = 0$$

The contrasts are:

$$\hat{\Gamma}_1 = (1)(280) + (1)(278) + (-2)(312) = -66$$
$$\hat{\Gamma}_2 = (1)(280) + (-1)(278) + (0)(312) = 2$$

The contrast sums of squares:

$$SS_{\hat{\Gamma}_1} = \frac{\hat{\Gamma}_1^2}{\sum_i \lambda_i^2 / n_i} = \frac{(-66)^2}{(1)^2/5 + (1)^2/5 + (-2)^2/5} = 3630$$

$$SS_{\hat{\Gamma}_2} = \frac{\hat{\Gamma}_2^2}{\sum_i \lambda_i^2 / n_i} = \frac{(2)^2}{(1)^2/5 + (-1)^2/5} = 10$$

Thus:

$$SS_{\hat{\Gamma}_1} + SS_{\hat{\Gamma}_2} = SS_{TRT} = 3630 + 10 = 3640$$

The corresponding calculated F values are:

$$F_1 = \frac{SS_{\hat{\Gamma}_1}/1}{MS_{RES}} = \frac{3630/1}{296.67} = 12.236$$

$$F_2 = \frac{SS_{\hat{\Gamma}_2}/1}{MS_{RES}} = \frac{10/1}{296.67} = 0.034$$

ANOVA table:

Source	SS	df	MS	F
Diet	3640	3 − 1 = 2	1820.00	6.13
(TR_1, TR_2) vs. TR_3	3630	1	3630.00	12.23
TR_1 vs. TR_2	10	1	10.00	0.03
Residual	3560	15 − 3 = 12	296.67	
Total	7200	15 − 1 = 14		

The critical value for 1 and 12 degrees of freedom and $\alpha = 0.05$ is $F_{.05,1,12} = 4.75$. Since the calculated F for TR_1 and TR_2 vs. TR_3, $F = 12.23$, is greater than the critical value, the null hypothesis is rejected, the third diet results in higher gain than the first two. The second contrast, representing the hypothesis that the first and second diets are the same, is not rejected, as the calculated F for TR_1 vs. TR_2, $F = 0.03$ is less than $F_{.05,1,12} = 4.75$.

In order to retain the probability of type I error equal to the α used in the tests, contrasts should be constructed a-priori. Contrasts should be preplanned and not constructed based on examination of treatment means. Further, although multiple sets of orthogonal contrasts can be constructed in analysis with three or more treatment degrees of freedom, only one set of contrasts can be tested to retain the probability of type I error equal to α. In the example above one of the three sets of orthogonal contrast can be defined but not more than one:

$$\begin{bmatrix} TR_1 \text{ vs. } TR_2, TR_3 \\ TR_2 \text{ vs. } TR_3 \end{bmatrix} \quad \text{or} \quad \begin{bmatrix} TR_2 \text{ vs. } TR_1, TR_3 \\ TR_1 \text{ vs. } TR_3 \end{bmatrix} \quad \text{or} \quad \begin{bmatrix} TR_3 \text{ vs. } TR_1, TR_2 \\ TR_1 \text{ vs. } TR_2 \end{bmatrix}$$

11.7.5 Scheffe Test

By defining a set of orthogonal contrasts it is ensured that the probability of a type I error (an incorrect conclusion that a contrast is different than zero) is not greater than the level of significance α for the overall test of treatment effects. However, if a number of contrasts greater than the treatment degrees of freedom are tested at the same time using the test statistic $F = \dfrac{SS_{\hat{\Gamma}}/1}{MS_{RES}}$, the contrasts are not orthogonal, and the probability of type I error is greater than α. The Scheffe test ensures that the level of significance is still α by defining the following statistic:

$$F = \frac{SS_{\hat{\Gamma}}/(a-1)}{MS_{RES}}$$

which has an F distribution with $(a-1)$ and $(N-a)$ degrees of freedom. Here, a is the number of treatments, N is the total number of observations, $SS_{\hat{\Gamma}} = \dfrac{\hat{\Gamma}^2}{\sum_i \lambda_i^2 / n_i}$ is the contrast sum of squares, MS_{RES} is the residual mean square, λ_i are the contrast coefficients that define comparisons, and $\hat{\Gamma} = \sum_i \lambda_i \bar{y}_i.$ is the contrast. If the calculated F value is greater than the critical value $F_{\alpha,(a-1)(N-a)}$, the null hypothesis that the contrast is equal to zero is rejected. This test is valid for any number of contrasts.

Example: Using the previous example of pig diets, test the following contrasts: first diet vs. second, first diet vs. third, and second diet vs. third. The following were calculated and defined: $MS_{RES} = 296.67$; $SS_{TRT} = 3640$; $n_i = 5$, $a = 3$.

The following contrast coefficients were defined:

	TR_1	TR_2	TR_3
$\bar{y}_i.$	280	278	312
Contrast1	$\lambda_{11} = 1$	$\lambda_{12} = -1$	$\lambda_{13} = 0$
Contrast2	$\lambda_{11} = 1$	$\lambda_{12} = 0$	$\lambda_{13} = -1$
Contrast3	$\lambda_{21} = 0$	$\lambda_{22} = 1$	$\lambda_{23} = -1$

The contrasts are:

$$\hat{\Gamma}_1 = (1)(280) + (-1)(278) = 2$$
$$\hat{\Gamma}_2 = (1)(280) + (-1)(312) = -32$$
$$\hat{\Gamma}_3 = (1)(278) + (-1)(312) = -34$$

The contrast sums of squares are:

$$SS_{\hat{\Gamma}_1} = \frac{\hat{\Gamma}_1^2}{\sum_i \lambda_i^2 / n_i} = \frac{(2)^2}{(1)^2/5 + (-1)^2/5} = 10$$

$$SS_{\hat{\Gamma}_2} = \frac{\hat{\Gamma}_2^2}{\sum_i \lambda_i^2 / n_i} = \frac{(-32)^2}{(1)^2/5 + (-1)^2/5} = 2560$$

$$SS_{\hat{\Gamma}_3} = \frac{\hat{\Gamma}_3^2}{\sum_i \lambda_i^2 / n_i} = \frac{(-34)^2}{(1)^2/5 + (-1)^2/5} = 2890$$

Note that $\sum_i SS_{\hat{\Gamma}_1} \neq SS_{TRT}$ because of lack of orthogonality.

The F statistic is:

$$F = \frac{SS_{\hat{\Gamma}} / (a-1)}{MS_{RES}}$$

The values of F statistics for contrasts are:

$$F_1 = \frac{10/2}{296.67} = 0.017$$

$$F_2 = \frac{2560/2}{296.67} = 4.315$$

$$F_3 = \frac{2890/2}{296.67} = 4.871$$

The critical value for 2 and 12 degrees of freedom with $\alpha = 0.05$ level of significance is $F_{.05,2,12} = 3.89$. The calculated F statistics for TR_1 versus TR_3 and TR_2 versus TR_3 are greater than the critical value, supporting the conclusion that diet 3 yields higher gain than either diet 1 or 2.

11.8 Test of Homogeneity of Variance

Homogeneity of variance in two groups or treatments, assuming normal distributions of observations, can be tested by using an F statistic:

$$F = \frac{s_1^2}{s_2^2}$$

as shown in section 6.8. For more than two groups or treatments, also assuming a normal distribution of observations, the Bartlett test can be used. The Bartlett formula is as follows:

$$B = log(\overline{s^2})\sum_i (n_i - 1) - \sum_i (n_i - 1) \, log(s_i^2) \quad i = 1, ..., a$$

where:

$\overline{s^2}$ = average of estimated variances of all groups
s_i^2 = estimated variance for group i
n_i = the number of observations in group i
a = the number of groups or treatments

For unequal group sizes the average of estimated group variances is replaced by:

$$\overline{s^2} = \frac{\sum_i SS_i}{\sum_i (n_i - 1)}$$

where SS_i = sum of squares for group i

For small group sizes (less than 10), it is necessary to correct B by dividing it by a correction factor C_B:

$$C_B = 1 + \frac{1}{3(a-1)}\left[\sum_i \frac{1}{n_i - 1} - \frac{1}{\sum_i (n_i - 1)}\right]$$

Both B and B/C_B have approximate *chi*-square distributions with $(a - 1)$ degrees of freedom. To test the significance of difference of variances, the calculated value of B or B/C_B is compared to a critical value of the *chi*-square distribution (see Appendix B).

A *Levene* test can also be used to test homogeneity of variance as shown in section 6.8.

11.9   SAS Example for the Fixed Effects One-way Model

The SAS program for the example comparing three diets for pigs is as follows. Recall the data:

TR_1	TR_2	TR_3
270	290	290
300	250	340
280	280	330
280	290	300
270	280	300

SAS program:

```
DATA pigs;
INPUT  diet $ d_gain @@;
DATALINES;
TR1 270    TR2 290    TR3 290
TR1 300    TR2 250    TR3 340
TR1 280    TR2 280    TR3 330
TR1 280    TR2 290    TR3 300
TR1 270    TR2 280    TR3 300
;
```

```
PROC GLM DATA = pigs;
CLASS diet;
MODEL d_gain = diet  ;
LSMEANS diet / STDERR PDIFF ADJUST=TUKEY;
CONTRAST 'TR1,TR2 : TR3' diet 1 1 -2;
CONTRAST 'TR1 : TR2' diet 1 -1 0;
RUN;
QUIT;
```

Explanation: The GLM procedure is used. The CLASS statement defines the classification (categorical) independent variable. The MODEL statement defines dependent and independent variables: *d_gain = diet* indicates that *d_gain* is the dependent, and *diet* the independent variable. The LSMEANS statement calculates means of diets. Options after the slash (*STDERR PDIFF ADJUST=TUKEY*) specify calculation of standard errors and tests of differences between means using the Tukey test adjusted for multiple comparisons of means. Alternatively, for a preplanned comparison of groups, the CONTRAST statements can be used. The first contrast is *TR1* and *TR2* vs. *TR3*, and the second contrast is *TR1* vs. *TR2*. The text between apostrophes ' ' are labels for the contrasts, *diet* denotes the variable for which contrasts are computed, and at the end the contrast coefficients are listed.

SAS output:

Dependent Variable: d_gain

Source	DF	Sum of Squares	Mean Square	F Value	Pr > F
Model	2	3640.0000	1820.0000	6.13	0.0146
Error	12	3560.0000	296.6667		
Corrected Total	14	7200.0000			

R-Square	Coeff Var.	Root MSE	d_gain Mean
0.505556	5.939315	17.224014	290.00000

Least Squares Means
Adjustment for multiple comparisons: Tukey

diet	d_gain LSMEAN	Standard Error	Pr > |t|	LSMEAN Number
TR1	280.000000	7.702813	0.0001	1
TR2	278.000000	7.702813	0.0001	2
TR3	312.000000	7.702813	0.0001	3

Least Squares Means for effect diet
Pr > \|t\| for H0: LSMean(i)=LSMean(j)

i/j	1	2	3
1		0.9816	0.0310
2	0.9816		0.0223
3	0.0310	0.0223	

```
Dependent Variable: d_gain
```

Contrast	DF	Contrast SS	Mean Square	F Value	Pr > F
TR1,TR2 : TR3	1	3630.0000	3630.0000	12.24	0.0044
TR1 : TR2	1	10.0000	10.0000	0.03	0.8574

Explanation: First is an *ANOVA* table for the *Dependent Variable d_gain*. The *Sources* of variability are *Model*, *Error* and *Corrected Total*. In the table are listed degrees of freedom (*DF*), *Sum of Squares*, *Mean Square*, calculated *F* (*F value*) and *P* value (*Pr > F*). For this example $F = 6.13$ and the *P* value is 0.0146, thus it can be concluded that an effect of diets exists. Below the ANOVA table descriptive statistics are listed, including the *R square* (0.505556), coefficient of variation (*Coeff Var.* = 5.939315), standard deviation (*Root MSE* = 17.224014) and overall mean (*d-gain Mean* = 290.000). In the table titled *Least Squares Means* are estimates (*LSMEAN*) with *Standard Errors*. In the next table *P* values for differences among treatments are shown. For example, the number in the first row and third column (0.0310) is the *P* value for testing the difference between diets *TR*1 and *TR*3. The *P* value = 0.0310 indicates that the difference is significant. Finally, contrasts with contrast sum of squares (*Contrast SS*), *Mean Squares*, *F* and *P* values (*F Value*, *Pr > F*) are shown. The means compared in the first contrast are significantly different, as shown with a *P* value = 0.0044, but the contrast between *TR*1 and *TR*2 is not, *P* value = 0.8574. Note the difference in that *P* value comparing to the Tukey test (0.8574 vs. 0.9816) due to different tests.

11.10 Power of the Fixed Effects One-way Model

Recall that the power of test is the probability that a false null hypothesis is correctly rejected or a true difference is correctly declared different. The estimation of power for a particular sample can be achieved by setting as an alternative hypothesis the measured difference between samples. Using that difference and variability estimated from the samples, the theoretical distribution is set for H_1, and the deviation compared to the assumed critical value. The power of test is the probability that the deviation is greater than the critical value, but using the H_1 distribution. In the one-way analysis of variance, the null and alternative hypotheses are:

$H_0: \tau_1 = \tau_2 = ... = \tau_a$
$H_1: \tau_i \neq \tau_{i'}$ for at least one pair (i, i')

where τ_i are the treatment effects and a is the number of groups.

Under H_0, the F statistic has a central F distribution with $(a - 1)$ and $(N - a)$ degrees of freedom. For the α level of significance, H_0 is rejected if $F > F_{\alpha,(a-1),(N-a)}$, that is, if the calculated F from the sample is greater that the critical value $F_{\alpha,(a-1),(N-a)}$. When at least one treatment effect is nonzero, the F test statistic follows a non-central F distribution with a noncentrality parameter $\lambda = \dfrac{n \sum_i \tau_i^2}{\sigma^2}$, and degrees of freedom $(a - 1)$ and $(N - a)$. The power of the test is given by:

$$Power = P\left(F > F_{\alpha,(a-1),(N-a)} = F_\beta\right)$$

using a noncentral F distribution for H_1.

Using samples, $n \sum_i \tau_i^2$ can be estimated with SS_{TRT}, and σ^2 with $s^2 = MS_{RES}$. Then the noncentrality parameter is:

$$\lambda = \frac{SS_{TRT}}{MS_{RES}}$$

Regardless of the complexity of the model, the power for treatments can be computed in a similar way by calculating SS_{TRT}, estimating the variance, and defining appropriate degrees of freedom.

The level of significance and power of an F test are shown graphically in Figure 11.4. The areas under the central and noncentral curves to the right of the critical value are the significance level and power, respectively. Note the relationship between significance level (α), power, difference between treatments (explained with SS_{TRT}) and variability within treatments (explained with $MS_{RES} = s^2$). If a more stringent α is chosen, which means that critical value will be shifted to the right, the power will decrease. A larger SS_{TRT} and smaller MS_{RES} means a larger noncentrality parameter λ, and the noncentrality curve is shifted to the right. This results in a larger area under the noncentrality curve to the right of the critical value and consequently more power.

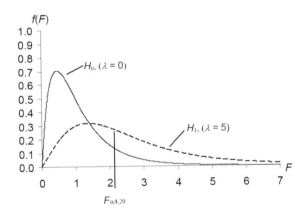

Figure 11.4 Significance and power of the F test. Under H_0 the F statistic has a central F distribution and under H_1 it has a noncentral F distribution. The distributions with 4 and 20 degrees of freedom and noncentrality parameters $\lambda = 0$ and 5 are shown. The critical value for an α level of significance is $F_{\alpha,4,20}$. The area under the H_0 curve to the right of the critical value is the level of significance (α). The area under the H_1 curve to the right of the critical value is the power ($1 - \beta$). The area under the H_1 curve on the left of the critical value is the type II error (β)

Example: Calculate the power of test using the example of effects of three diets on daily gains (g) in pigs. There were five pigs in each group. The *ANOVA* table was:

Source	SS	df	MS	F
Treatment	3640	3 – 1 = 2	1820.00	6.13
Residual	3560	15 – 3 = 12	296.67	
Total	7200	15 – 1 = 14		

The calculated *F* value was:

$$F = \frac{MS_{TRT}}{MS_{RES}} = \frac{1820.0}{296.67} = 6.13$$

The critical value for 2 and 12 degrees of freedom and $\alpha = 0.05$ level of significance is $F_{0.05,2,12} = 3.89$. The calculated $F = 6.13$ is greater (more extreme) than the critical value and H_0 is rejected.

The power of test is calculated using the critical value $F_{0.05,2,12} = 3.89$, and the noncentral F distribution for H_1 with the noncentrality parameter $\lambda = \dfrac{SS_{TRT}}{MS_{RES}} = \dfrac{3640}{296.67} = 12.27$ and 2 and 12 degrees of freedom. The power is:

$$Power = 1 - \beta = P[F > 3.89] = 0.79$$

Calculation of power using a noncentral F distribution with SAS will be shown in section 11.10.1. The level of significance and power for this example are shown graphically in Figure 11.5.

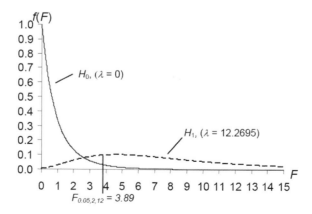

Figure 11.5 Power for the example with pigs. The critical value is 3.89. The area under the H_0 curve to the right of the critical value 3.89 is the level of significance $\alpha = 0.05$. The area under the H_1 curve to the right of 3.89 is the power $1 - \beta = 0.792$

11.10.1 SAS Example for Calculating Power

To compute power of test with SAS, the following statements are used:

```
DATA a;
alpha=0.05;
a=3;
n=5;
df1=a-1;
df2=a*n-a;
sstrt=3640;
msres=296.67;
lambda=sstrt/msres;
Fcrit=FINV(1-alpha,df1,df2);
power=1-CDF('F',Fcrit,df1,df2,lambda);
PROC PRINT;
RUN;
```

Explanation: First the following are defined: *alpha* = significance level, *a* = number of treatments, *n* = number of replications per treatment, *df1* = treatment degrees of freedom, *df2* = residual degrees of freedom, *sstrt* = treatment sum of squares, *msres* = residual (error) mean square, the estimated variance. Then, the noncentrality parameter (*lambda*), and the critical value (*Fcrit*) for the given degrees of freedom and level of significance are calculated. The critical value is computed with the FINV function, which must have as inputs the cumulative value of percentiles ($1 - \alpha = 0.95$) and degrees of freedom *df1* and *df2*. The *power* is calculated with the CDF function. This is a cumulative function of the *F* distribution that needs as inputs the critical value, degrees of freedom and the noncentrality parameter *lambda*. As an alternative to *CDF('F',Fcrit,df1,df2,lambda)*, the statement *PROBF(Fcrit,df1,df2,lambda)* can be used. The PRINT procedure gives the following SAS output:

```
alpha   a   n   df1   df2   sstrt    mse     lambda    Fcrit      power
 0.05   3   5    2     12   3640    296.67   12.2695   3.88529    0.79213
```

Thus, the power is 0.79213.

The same analysis can be obtained by using the POWER procedure (SAS version 9.1 and higher).

SAS program:

```
PROC POWER;
   ONEWAYANOVA
   TEST      = OVERALL
   ALPHA     = 0.05
   GROUPMEANS = 280 | 278 | 312
   STDDEV  =  17.224
   NPERGROUP = 5
   POWER     = .
   ;
RUN;
```

Explanation: The POWER procedure is used with ONEWAYANOVA and *TEST* = *OVERALL* statements indicating one-way ANOVA *F* test. Values are from the example of effects of three diets on daily gains (g) in pigs, with sample means of 280, 278 and 312 kg and model residual standard deviation of 17.224 kg. Alpha of the test = 0.05, and number of observations per group is *NPERGROUP* = 5. We put a dot to direct POWER to be calculated by the program.

SAS output:

```
Method                        Exact
Alpha                          0.05
Group Means             280 278 312
Standard Deviation           17.224
Sample Size Per Group             5


                  Power
                  0.792
```

Again, the power is calculated to be 0.792.

The calculation of sample size required to obtain a desired power will be shown in section 14.6, Required number of replications.

11.11 Matrix Approach to the Fixed Effects One-way Model

11.11.1 Linear Model

Recall that the scalar one-way model with equal numbers of observations per group is:

$$y_{ij} = \mu + \tau_i + \varepsilon_{ij} \qquad i = 1,..., a; \ j = 1,..., n$$

where:

y_{ij} = observation j in group or treatment i
μ = the overall mean
τ_i = the fixed effect of group or treatment i (denotes an unknown parameter)
ε_{ij} = random error with mean 0 and variance σ^2

Thus, each observation y_{ij} can be expressed as:

$$y_{11} = \mu + \tau_1 + \varepsilon_{11} = 1\mu + 1\tau_1 + 0\tau_2 + ... \, 0\tau_a + \varepsilon_{11}$$
$$y_{12} = \mu + \tau_1 + \varepsilon_{12} = 1\mu + 1\tau_1 + 0\tau_2 + ... \, 0\tau_a + \varepsilon_{12}$$
$$...$$
$$y_{1n} = \mu + \tau_1 + \varepsilon_{1n} = 1\mu + 1\tau_1 + 0\tau_2 + ... \, 0\tau_a + \varepsilon_{1n}$$
$$y_{21} = \mu + \tau_2 + \varepsilon_{21} = 1\mu + 0\tau_1 + 1\tau_2 + ... \, 0\tau_a + \varepsilon_{21}$$
$$...$$
$$y_{2n} = \mu + \tau_2 + \varepsilon_{2n} = 1\mu + 0\tau_1 + 1\tau_2 + ... \, 0\tau_a + \varepsilon_{2n}$$
$$...$$
$$y_{a1} = \mu + \tau_a + \varepsilon_{a1} = 1\mu + 0\tau_1 + 0\tau_2 + ... \, 1\tau_a + \varepsilon_{a1}$$
$$...$$
$$y_{an} = \mu + \tau_a + \varepsilon_{an} = 1\mu + 0\tau_1 + 0\tau_2 + ... \, 1\tau_a + \varepsilon_{an}$$

The set of equations can be shown using vectors and matrices:

$$\mathbf{y} = \mathbf{X}\boldsymbol{\beta} + \boldsymbol{\varepsilon}$$

where:

$$
\mathbf{y} = \begin{bmatrix} y_{11} \\ y_{12} \\ \cdots \\ y_{1n} \\ y_{21} \\ \cdots \\ y_{2n} \\ \cdots \\ \cdots \\ y_{a1} \\ \cdots \\ y_{an} \end{bmatrix}_{an \times 1}
\quad
\mathbf{X} = \begin{bmatrix} 1 & 1 & 0 & \cdots & 0 \\ 1 & 1 & 0 & \cdots & 0 \\ \cdots & \cdots & \cdots & \cdots & \cdots \\ 1 & 1 & 0 & \cdots & 0 \\ 1 & 0 & 1 & \cdots & 0 \\ \cdots & \cdots & \cdots & \cdots & \cdots \\ 1 & 0 & 1 & \cdots & 0 \\ \cdots & \cdots & \cdots & \cdots & \cdots \\ \cdots & \cdots & \cdots & \cdots & \cdots \\ 1 & 0 & 0 & \cdots & 1 \\ \cdots & \cdots & \cdots & \cdots & \cdots \\ 1 & 0 & 0 & \cdots & 1 \end{bmatrix}_{an \times (a+1)}
\quad
\boldsymbol{\beta} = \begin{bmatrix} \mu \\ \tau_1 \\ \tau_2 \\ \cdots \\ \tau_a \end{bmatrix}_{(a+1) \times 1}
\quad
\boldsymbol{\varepsilon} = \begin{bmatrix} \varepsilon_{11} \\ \varepsilon_{12} \\ \cdots \\ \varepsilon_{1n} \\ \varepsilon_{21} \\ \cdots \\ \varepsilon_{2n} \\ \cdots \\ \cdots \\ \varepsilon_{a1} \\ \cdots \\ \varepsilon_{an} \end{bmatrix}_{an \times 1}
$$

$\mathbf{y}$ = vector of observations
$\mathbf{X}$ = design matrix which relates $\mathbf{y}$ to $\boldsymbol{\beta}$
$\boldsymbol{\beta}$ = vector of parameters
$\boldsymbol{\varepsilon}$ = vector of random errors with mean $E(\boldsymbol{\varepsilon}) = \mathbf{0}$ and variance $Var(\boldsymbol{\varepsilon}) = \sigma^2 \mathbf{I}$

Also, vector $\mathbf{0}$ is a vector with all zero elements, and $\mathbf{I}$ is an identity matrix. The dimensions of each vector or matrix is shown at its lower right.

The expectation and variance of $\mathbf{y}$ are:

$$E(\mathbf{y}) = \mathbf{X}\boldsymbol{\beta} \qquad Var(\mathbf{y}) = \sigma^2 \mathbf{I}$$

11.11.2 Estimating Parameters

Assuming a normal model, $\mathbf{y}$ is a vector of independent normal random variables with a multivariate normal distribution with mean $\mathbf{X}\boldsymbol{\beta}$ and variance $\mathbf{I}\sigma^2$.

The parameters can be estimated by using either least squares or maximum likelihood estimation. To calculate the solutions for vector $\boldsymbol{\beta}$, the normal equations are obtained:

$$\mathbf{X'X}\tilde{\boldsymbol{\beta}} = \mathbf{X'y}$$

in which:

$$
\mathbf{X'X} = \begin{bmatrix} an & n & n & \cdots & n \\ n & n & 0 & \cdots & 0 \\ n & 0 & n & \cdots & 0 \\ \cdots & \cdots & \cdots & \cdots & \cdots \\ n & 0 & 0 & \cdots & n \end{bmatrix}_{(a+1) \times (a+1)}
\quad
\tilde{\boldsymbol{\beta}} = \begin{bmatrix} \tilde{\mu} \\ \tilde{\tau}_1 \\ \tilde{\tau}_2 \\ \cdots \\ \tilde{\tau}_a \end{bmatrix}_{(a+1) \times 1}
\quad
\mathbf{X'y} = \begin{bmatrix} \sum_i \sum_j y_{ij} \\ \sum_j y_{1j} \\ \sum_j y_{2j} \\ \cdots \\ \sum_j y_{aj} \end{bmatrix}_{(a+1) \times 1}
$$

These equations are often called ordinary least squares (*OLS*) equations. The $\mathbf{X'X}$ matrix does not have an inverse, since its columns are not linearly independent. The first column is equal to sum of all other columns. To solve for $\widetilde{\boldsymbol{\beta}}$, a generalized inverse $(\mathbf{X'X})^{-}$ is used. The solution vector is:

$$\widetilde{\boldsymbol{\beta}} = (\mathbf{X'X})^{-}\mathbf{X'y}$$

with the mean:

$$E\left(\widetilde{\boldsymbol{\beta}}\right) = (\mathbf{X'X})^{-}\mathbf{X'X}\boldsymbol{\beta}$$

and variance:

$$Var(\widetilde{\boldsymbol{\beta}}) = (\mathbf{X'X})^{-}\mathbf{X'X}(\mathbf{X'X})^{-}\sigma^2$$

In this case there are many solutions and the vector of solutions is denoted by $\widetilde{\boldsymbol{\beta}}$. However, this model gives unique solutions for the differences of groups or treatments and their means. Specific generalized inverse matrices are used to provide constraints to obtain meaningful solutions. A useful constraint is to set the sum of all group effects to be zero. Alternatively, one of the group effects may be set to zero and the others expressed as deviations from it. If $\widetilde{\mu} = 0$ then the effect of group or treatment i is:

$$\widetilde{\tau}_i + \widetilde{\mu}$$

which denotes an estimator of the group mean:

$$\widetilde{\mu} + \widetilde{\tau}_i = \hat{\mu} + \hat{\tau}_i = \hat{\mu}_i$$

Such solutions are obtained by setting the first row and the first column of $\mathbf{X'X}$ to zero. Then its generalized inverse is:

$$(\mathbf{X'X})^{-} = \begin{bmatrix} 0 & 0 & 0 & \dots & 0 \\ 0 & \frac{1}{n} & 0 & \dots & 0 \\ 0 & 0 & \frac{1}{n} & \dots & 0 \\ \dots & \dots & \dots & \dots & \dots \\ 0 & 0 & 0 & \dots & \frac{1}{n} \end{bmatrix}$$

The solution vector is:

$$\widetilde{\boldsymbol{\beta}} = \begin{bmatrix} \widetilde{\mu} \\ \widetilde{\tau}_1 \\ \widetilde{\tau}_2 \\ \dots \\ \widetilde{\tau}_a \end{bmatrix} = \begin{bmatrix} 0 \\ \hat{\mu} + \hat{\tau}_1 \\ \hat{\mu} + \hat{\tau}_2 \\ \dots \\ \hat{\mu} + \hat{\tau}_a \end{bmatrix} \quad \text{which are mean estimators.}$$

A vector of the fitted values is:

$$\hat{\mathbf{y}} = \mathbf{X}\widetilde{\boldsymbol{\beta}}$$

This is a linear combination of **X** and parameter estimates. The variance of the fitted values is:

$$Var(\mathbf{X}\widetilde{\boldsymbol{\beta}}) = \mathbf{X}(\mathbf{X'X})^{-}\mathbf{X'}\sigma^2$$

Estimates of interest can be calculated by defining a vector λ such that $\lambda'\boldsymbol{\beta}$ is defined and estimable.

The following vector λ is used to define the mean of the first group or treatment (population):

$$\lambda' = [1\ 1\ 0\ 0\ \ldots\ 0]$$

Then the mean is:

$$\lambda'\boldsymbol{\beta} = \begin{bmatrix} 1 & 1 & 0 & \ldots & 0 \end{bmatrix} \begin{bmatrix} \mu \\ \tau_1 \\ \tau_2 \\ \ldots \\ \tau_a \end{bmatrix} = \mu + \tau_1$$

An estimator of the mean is:

$$\lambda'\widetilde{\boldsymbol{\beta}} = \begin{bmatrix} 1 & 1 & 0 & \ldots & 0 \end{bmatrix} \begin{bmatrix} \widetilde{\mu} \\ \widetilde{\tau}_1 \\ \widetilde{\tau}_2 \\ \ldots \\ \widetilde{\tau}_a \end{bmatrix} = \widetilde{\mu} + \widetilde{\tau}_1$$

Similarly, the difference between two groups or treatments can be defined. For example, to define the difference between the first and second groups the vector λ is:

$$\lambda' = [1\ 1\ 0\ 0\ \ldots\ 0] - [1\ 0\ 1\ 0\ \ldots\ 0] = [0\ 1\ -1\ 0\ \ldots\ 0]$$

The difference is:

$$\lambda'\boldsymbol{\beta} = \begin{bmatrix} 0 & 1 & -1 & 0 & \ldots & 0 \end{bmatrix} \begin{bmatrix} \mu \\ \tau_1 \\ \tau_2 \\ \ldots \\ \tau_a \end{bmatrix} = \tau_1 - \tau_2$$

An estimator of the difference is:

$$\lambda'\widetilde{\boldsymbol{\beta}} = \begin{bmatrix} 0 & 1 & -1 & 0 & \ldots & 0 \end{bmatrix} \begin{bmatrix} \widetilde{\mu} \\ \widetilde{\tau}_1 \\ \widetilde{\tau}_2 \\ \ldots \\ \widetilde{\tau}_a \end{bmatrix} = \widetilde{\tau}_1 - \widetilde{\tau}_2$$

Generally, the variances of such estimators are:

$$Var(\boldsymbol{\lambda'}\widetilde{\boldsymbol{\beta}}) = \boldsymbol{\lambda'}(\mathbf{X'X})^{-}\boldsymbol{\lambda'}\sigma^2$$

As shown before, an unknown variance σ^2 can be replaced by the estimated variance $s^2 = MS_{RES}$ = residual mean square. The square root of the variance of the estimator is the standard error of the estimator.

The sums of squares needed for hypothesis testing using an F test can be calculated as:

$$SS_{TRT} = \widetilde{\boldsymbol{\beta}}'\mathbf{X'y} - an(\bar{y}..)^2$$
$$SS_{RES} = \mathbf{y'y} - \widetilde{\boldsymbol{\beta}}'\mathbf{X'y}$$
$$SS_{TOT} = \mathbf{y'y} - an(\bar{y}..)^2$$

Example: A matrix approach is used to calculate sums of squares for the example of pig diets. Recall the problem: an experiment was conducted in order to investigate the effects of three different diets on daily gains (g) in pigs. The diets are denoted with TR_1, TR_2 and TR_3. Daily gains of five different pigs fed each of three diets are shown in the following table:

TR_1	TR_2	TR_3
270	290	290
300	250	340
280	280	330
280	290	300
270	280	300

The model is:

$$\mathbf{y} = \mathbf{X}\boldsymbol{\beta} + \boldsymbol{\varepsilon}$$

where:

$$\mathbf{y} = \begin{bmatrix} 270 \\ \cdots \\ 270 \\ 290 \\ \cdots \\ 280 \\ 290 \\ \cdots \\ 300 \end{bmatrix} \quad \mathbf{X} = \begin{bmatrix} 1 & 1 & 0 & 0 \\ \cdots & \cdots & \cdots & \cdots \\ 1 & 1 & 0 & 0 \\ 1 & 0 & 1 & 0 \\ \cdots & \cdots & \cdots & \cdots \\ 1 & 0 & 1 & 0 \\ 1 & 0 & 0 & 1 \\ \cdots & \cdots & \cdots & \cdots \\ 1 & 0 & 0 & 1 \end{bmatrix} \quad \boldsymbol{\beta} = \begin{bmatrix} \mu \\ \tau_1 \\ \tau_2 \\ \tau_3 \end{bmatrix} \quad \boldsymbol{\varepsilon} = \begin{bmatrix} \varepsilon_{11} \\ \cdots \\ \varepsilon_{15} \\ \varepsilon_{21} \\ \cdots \\ \varepsilon_{25} \\ \varepsilon_{31} \\ \cdots \\ \varepsilon_{35} \end{bmatrix}$$

The normal equations are:

$$\mathbf{X'X}\widetilde{\boldsymbol{\beta}} = \mathbf{X'y}$$

where:

$$\mathbf{X'X} = \begin{bmatrix} 15 & 5 & 5 & 5 \\ 5 & 5 & 0 & 0 \\ 5 & 0 & 5 & 0 \\ 5 & 0 & 0 & 5 \end{bmatrix} \qquad \tilde{\boldsymbol{\beta}} = \begin{bmatrix} \tilde{\mu} \\ \tilde{\tau}_1 \\ \tilde{\tau}_2 \\ \tilde{\tau}_3 \end{bmatrix} \qquad \mathbf{X'y} = \begin{bmatrix} 4350 \\ 1400 \\ 1390 \\ 1560 \end{bmatrix}$$

The solution vector is:

$$\tilde{\boldsymbol{\beta}} = (\mathbf{X'X})^{-} \mathbf{X'y}$$

By defining the generalized inverse as:

$$(\mathbf{X'X})^{-} = \begin{bmatrix} 0 & 0 & 0 & 0 \\ 0 & \frac{1}{5} & 0 & 0 \\ 0 & 0 & \frac{1}{5} & 0 \\ 0 & 0 & 0 & \frac{1}{5} \end{bmatrix}$$

The solution vector is:

$$\tilde{\boldsymbol{\beta}} = \begin{bmatrix} 0 \\ \hat{\mu} + \hat{\tau}_1 \\ \hat{\mu} + \hat{\tau}_2 \\ \hat{\mu} + \hat{\tau}_3 \end{bmatrix} = \begin{bmatrix} 0 \\ 280 \\ 278 \\ 312 \end{bmatrix}$$

The sums of squares needed for testing hypotheses are:

$$SS_{TRT} = \tilde{\boldsymbol{\beta}}'\mathbf{X'y} - an(\bar{y}..)^2 = \begin{bmatrix} 0 & 280 & 278 & 312 \end{bmatrix} \begin{bmatrix} 4350 \\ 1400 \\ 1390 \\ 1560 \end{bmatrix} - (3)(5)(290)^2$$

$$= 1265140 - 1261500 = 3640$$

$$SS_{RES} = \mathbf{y'y} - \tilde{\boldsymbol{\beta}}'\mathbf{X'y} = \begin{bmatrix} 270 & \ldots & 270 & 290 & \ldots & 280 & 290 & \ldots & 300 \end{bmatrix} \begin{bmatrix} 270 \\ \ldots \\ 270 \\ 290 \\ \ldots \\ 280 \\ 290 \\ \ldots \\ 300 \end{bmatrix} - 1265140$$

$$= 1268700 - 1265140 = 3560$$

$$SS_{TOT} = \mathbf{y'y} - an(\bar{y}..)^2 = 1268700 - 1261500 = 7200$$

Construction of the *ANOVA* table and testing is the same as already shown with the scalar model in section 11.3.

11.11.3 Maximum Likelihood Estimation

Assuming a multivariate normal distribution, $y \sim N(\mathbf{X}\boldsymbol{\beta}, \sigma^2\mathbf{I})$, the likelihood function is:

$$L(\boldsymbol{\beta}, \sigma^2 \mid \mathbf{y}) = \frac{e^{-\frac{1}{2}(\mathbf{y}-\mathbf{X}\boldsymbol{\beta})'(\mathbf{I}\sigma^2)^{-1}(\mathbf{y}-\mathbf{X}\boldsymbol{\beta})}}{\sqrt{(2\pi\sigma^2)}^N}$$

The log likelihood is:

$$logL(\boldsymbol{\beta}, \sigma^2 \mid \mathbf{y}) = -\frac{1}{2}N\,log(2\pi) - \frac{1}{2}N\,log(\sigma^2) - \frac{1}{2\sigma^2}(\mathbf{y}-\mathbf{X}\boldsymbol{\beta})'(\mathbf{y}-\mathbf{X}\boldsymbol{\beta})$$

To find the estimator that will maximize the log likelihood function, partial derivatives are taken and equated to zero. The following normal equations are obtained:

$$\mathbf{X}'\mathbf{X}\widetilde{\boldsymbol{\beta}} = \mathbf{X}'\mathbf{y}$$

and the maximum likelihood estimator of the variance is:

$$\hat{\sigma}^2_{ML} = \frac{1}{N}(\mathbf{y}-\mathbf{X}\boldsymbol{\beta})'(\mathbf{y}-\mathbf{X}\boldsymbol{\beta})$$

11.11.4 Regression Model for the One-way Analysis of Variance

A one-way analysis of variance can be expressed as a multiple linear regression model in the following way. For a groups define $a - 1$ independent variables such that the value of a variable is one if the observation belongs to the group and zero if the observation does not belong to the group. For example, the one-way model with three groups and n observations per group is:

$$y_i = \beta_0 + \beta_1 x_{1i} + \beta_2 x_{2i} + \varepsilon_i \qquad i = 1,\dots, 3n$$

where:

y_i = observation i of dependent variable y

x_{1i} = an independent variable with the value 1 if observation i is in the first group, 0 if observation i is not in the first group

x_{2i} = an independent variable with the value 1 if observation i is in the second group, 0 if observation i is not in the second group

$\beta_0, \beta_1, \beta_2$ = regression parameters

ε_i = random error

Note that it is not necessary to define a regression parameter for the third group since if the values for both independent variables are zero that will denote observation is in the third group.

We can show the model to be equivalent to the one-way model with one categorical independent variable with groups defined as levels.

The regression model in matrix notation is:

$$\mathbf{y} = \mathbf{X}_r\boldsymbol{\beta}_r + \boldsymbol{\varepsilon}$$

where:

$\mathbf{y}$ = the vector of observations of a dependent variable

$$\boldsymbol{\beta}_r = \begin{bmatrix} \beta_0 \\ \beta_1 \\ \beta_2 \end{bmatrix} = \text{the vector of parameters}$$

$$\mathbf{X}_r = \begin{bmatrix} \mathbf{1}_n & \mathbf{1}_n & \mathbf{0}_n \\ \mathbf{1}_n & \mathbf{0}_n & \mathbf{1}_n \\ \mathbf{1}_n & \mathbf{0}_n & \mathbf{0}_n \end{bmatrix}_{3n \times 3} = \text{the matrix of observations of independent variables,}$$

$\mathbf{1}_n$ is a vector of ones, $\mathbf{0}_n$ is a vector of zeros

$\boldsymbol{\varepsilon}$ = the vector of random errors

Recall that the vector of parameter estimates is:

$$\hat{\boldsymbol{\beta}}_r = (\mathbf{X}_r{}'\mathbf{X}_r)^{-1}\mathbf{X}_r{}'\mathbf{y}$$

where:

$$\hat{\boldsymbol{\beta}}_r = \begin{bmatrix} \hat{\beta}_0 \\ \hat{\beta}_1 \\ \hat{\beta}_2 \end{bmatrix}$$

The $\mathbf{X}_r{}'\mathbf{X}_r$ matrix and its inverse are:

$$(\mathbf{X}_r{}'\mathbf{X}_r) = \begin{bmatrix} 3n & n & n \\ n & n & 0 \\ n & 0 & n \end{bmatrix} \text{ and } (\mathbf{X}_r{}'\mathbf{X}_r)^{-1} = \begin{bmatrix} \frac{1}{n} & -\frac{1}{n} & -\frac{1}{n} \\ -\frac{1}{n} & \frac{2}{n} & \frac{1}{n} \\ -\frac{1}{n} & \frac{1}{n} & \frac{2}{n} \end{bmatrix}$$

The one-way model with a categorical independent variable is:

$$y_{ij} = \mu + \tau_i + \varepsilon_{ij} \qquad i = 1,..., a; \ j = 1,..., n$$

where:

y_{ij} = observation j in group or treatment i

μ = the overall mean

τ_i = the fixed effect of group or treatment i

ε_{ij} = random error

In matrix notation the model is:

$$\mathbf{y} = \mathbf{X}_{ow}\boldsymbol{\beta}_{ow} + \boldsymbol{\varepsilon}$$

where:

$$\boldsymbol{\beta}_{ow} = \begin{bmatrix} \mu \\ \tau_1 \\ \tau_2 \\ \tau_3 \end{bmatrix} = \text{the vector of parameters}$$

$$\mathbf{X}_{ow} = \begin{bmatrix} \mathbf{1}_n & \mathbf{1}_n & \mathbf{0}_n & \mathbf{0}_n \\ \mathbf{1}_n & \mathbf{0}_n & \mathbf{1}_n & \mathbf{0}_n \\ \mathbf{1}_n & \mathbf{0}_n & \mathbf{0}_n & \mathbf{1}_n \end{bmatrix}_{3n \times 4} = \text{the matrix of observations of independent variable,}$$

$\mathbf{1}_n$ is a vector of ones, $\mathbf{0}_n$ is a vector of zeros

The solution vector for the one-way model is:

$$\tilde{\boldsymbol{\beta}}_{ow} = (\mathbf{X}_{ow}'\mathbf{X}_{ow})^{-}\mathbf{X}_{ow}'\mathbf{y}$$

where:

$$\tilde{\boldsymbol{\beta}}_{ow} = \begin{bmatrix} \tilde{\mu} \\ \tilde{\tau}_1 \\ \tilde{\tau}_2 \\ \tilde{\tau}_3 \end{bmatrix} \quad \text{and} \quad \mathbf{X}_{ow}'\mathbf{X}_{ow} = \begin{bmatrix} 3n & n & n & n \\ n & n & 0 & 0 \\ n & 0 & n & 0 \\ n & 0 & 0 & n \end{bmatrix}$$

The columns of $\mathbf{X}_{ow}'\mathbf{X}_{ow}$ are linearly dependent since the first column is equal to the sum of the second, third and fourth columns. Also, $\mathbf{X}_{ow}'\mathbf{X}_{ow}$ being symmetric the rows are linearly dependent as well. Consequently, for finding a solution only three rows and three columns are needed. A solution for $\tilde{\boldsymbol{\beta}}_{ow}$ can be obtained by setting $\tilde{\tau}_3$ to zero, that is by setting the last row and the last column of $\mathbf{X}_{ow}'\mathbf{X}_{ow}$ to zero. This will give:

$$\mathbf{X}_{ow}'\mathbf{X}_{ow} = \begin{bmatrix} 3n & n & n & 0 \\ n & n & 0 & 0 \\ n & 0 & n & 0 \\ 0 & 0 & 0 & 0 \end{bmatrix}$$

Its generalized inverse is:

$$(\mathbf{X}_{ow}'\mathbf{X}_{ow})^{-} = \begin{bmatrix} \frac{1}{n} & -\frac{1}{n} & -\frac{1}{n} & 0 \\ -\frac{1}{n} & \frac{2}{n} & \frac{1}{n} & 0 \\ -\frac{1}{n} & \frac{1}{n} & \frac{2}{n} & 0 \\ 0 & 0 & 0 & 0 \end{bmatrix}$$

The solution vector is:

$$\tilde{\boldsymbol{\beta}}_{ow} = \begin{bmatrix} \tilde{\mu} \\ \tilde{\tau}_1 \\ \tilde{\tau}_2 \\ 0 \end{bmatrix}$$

Since $\mathbf{X}_{ow}'\mathbf{X}_{ow}$ and $\mathbf{X}_r'\mathbf{X}_r$ matrices are equivalent, giving equivalent inverses, it follows that:

$$\tilde{\mu} = \hat{\beta}_0$$
$$\tilde{\tau}_1 = \hat{\beta}_1$$
$$\tilde{\tau}_2 = \hat{\beta}_2$$

and the effect for the third group is zero.

As stated before, the difference between the group means will be the same regardless of using regression model or any generalized inverse in a one-way model.

The equivalence of parameter estimates of the models defined above can be shown in the following table:

	Models		Equivalence of solution
	$y_i = \beta_0 + \beta_1 x_{1i} + \beta_2 x_{2i} + \varepsilon_i$	$y_{ij} = \mu + \tau_i + \varepsilon_{ij}$	
Group 1	$x_1 = 1; x_2 = 0$ then $\hat{y}_i = \hat{\beta}_0 + \hat{\beta}_1$	$\hat{y}_i = \tilde{\mu} + \tilde{\tau}_1$	$\hat{\beta}_1 = \tilde{\tau}_1$
Group 2	$x_1 = 0; x_2 = 1$ then $\hat{y}_i = \hat{\beta}_0 + \hat{\beta}_2$	$\hat{y}_i = \tilde{\mu} + \tilde{\tau}_2$	$\hat{\beta}_2 = \tilde{\tau}_2$
Group 3	$x_1 = 0; x_2 = 0$ then $\hat{y}_i = \hat{\beta}_0$	$\hat{y}_i = \tilde{\mu} + \tilde{\tau}_3$	$0 = \tilde{\tau}_3$ and $\hat{\beta}_0 = \tilde{\mu}$

11.12 Nonparametric Tests

It is appropriate to use an F distribution for testing hypotheses of differences among several treatment groups only if it is assumed that the populations from which the observations are drawn have a normal distribution. In cases that samples are from another known distribution, a generalized model can often be used as will be shown in chapter 24. In some situations, either the distribution is not known or the samples are from a known distribution, but not chosen at random. In chapter 6, a nonparametric approach using ranks was shown as an alternative to a t test for testing the difference between two treatment groups. An equivalent approach will be shown here for testing differences among two or more treatment groups. The use of ranks diminishes the importance of the distribution and the influence of extreme values in samples. The rank of observations is usually determined in the following manner: the observations of the combined groups are sorted in ascending order and ranks are assigned to them. If some observations have the same value, then the mean of their ranks is assigned to them. For example, if the 10th and 11th observations have the same value, say 20, their ranks are $(10 + 11) / 2 = 10.5$. Such ranks are known as Wilcoxon rank scores.

The following test statistic can be used for testing the difference between ranks for several treatment groups:

$$KW = \sum_i \frac{(T_i - E(T_i))^2}{s_R^2 n_i}$$

where T_i is the sum of ranks in treatment group i, $E(T_i) = n_i \bar{R}$ is the expected sum of ranks in treatment group i assuming no difference among groups, n_i is the number of observations in treatment group i, $\bar{R}$ is the mean rank of all treatment groups, and s_R^2 is the sample variance of the ranks for all treatment groups. The distribution of this statistic is approximated by a *chi*-square with $(a-1)$ degrees of freedom, with a being the number of treatment groups. This test for differences among treatment groups using a *chi*-square distribution is known as a Kruskal-Wallis test. If the calculated statistic, KW, is more extreme than the critical value at the chosen level of significance α, then the null hypothesis that the treatment groups are equal is rejected, and we can conclude that a difference exists among treatment groups.

Example: The effect of two gonadotropin preparations on ovulation rate of sows was investigated. Randomly chosen sows were injected with gonadotropin 1, gonadotropin 2, or saline (control group). The aim of the experiment was to determine if gonadotropin would result in higher ovulation rate. The following ovulation rates were measured:

Gonadotropin 1	14	14	7	45	18	36	15
Gonadotropin 2	12	11	12	12	14	13	9
Saline	10	7	9	7	12	19	8

The observations were sorted regardless of the treatment and ranked:

Treatment	Ovulation rate	Rank
Gonadotropin 1	7	2.0
Saline	7	2.0
Saline	7	2.0
Saline	8	4.0
Gonadotropin 2	9	5.5
Saline	9	5.5
Saline	10	7.0
Gonadotropin 2	11	8.0
Gonadotropin 2	12	10.5
Gonadotropin 2	12	10.5
Gonadotropin 2	12	10.5
Saline	12	10.5
Gonadotropin 2	13	13.0
Gonadotropin 1	14	15.0
Gonadotropin1	14	15.0
Gonadotropin 2	14	15.0
Gonadotropin1	15	17.0
Gonadotropin1	18	18.0
Saline	19	19.0
Gonadotropin 1	36	20.0
Gonadotropin 1	45	21.0
	Mean rank $\overline{R} =$	11.0

It is convenient to write a table with ranks by treatment group:

	Gonadotropin 1	Gonadotropin 2	Saline
	2.0	5.5	2.0
	15.0	8.0	2.0
	15.0	10.5	4.0
	17.0	10.5	5.5
	18.0	10.5	7.0
	20.0	13.0	10.5
	21.0	15.0	19.0
Sum	108.0	73.0	50.0

$n_1 = n_2 = n_3 = 7$
$T_1 = 108.0, \ T_2 = 73.0, \ T_3 = 50.0$

$$E(T_1) = n_1 \, \overline{R} = (7)(11.0) = 77.0$$
$$E(T_2) = n_2 \, \overline{R} = (7)(11.0) = 77.0$$
$$E(T_3) = n_3 \, \overline{R} = (7)(11.0) = 77.0$$

$$s^2_R = 38.025$$

$$KW = \sum_i \frac{(T_i - E(T_i))^2}{s^2_R n_i} = \frac{(108 - 77)^2}{(38.025)(7)} + \frac{(73 - 77)^2}{(38.025)(7)} + \frac{(50 - 77)^2}{(38.025)(7)} = 6.4093$$

The critical value of the *chi*-square distribution with $3 - 1 = 2$ degrees of freedom (number of groups $- 1$) and level of significance $\alpha = 0.05$ is 5.991. Since the calculated value of $KW = 6.4093$ is more extreme than the critical value, 5.991, the null hypothesis is rejected at the $\alpha = 0.05$ level of significance. We can conclude there is a difference among treatment groups.

Another way to conduct a nonparametric test is to use an approximate F test on the ranks. An F test using Wilcoxon ranks is equivalent to a Kruskal-Wallis test. The rationale is that if any population variable is transformed to ranks, the distribution of the ranks is approximately normal. Further, there are methods to further transform ordinary ranks to make them more closely approximate a normal distribution.

$$y_i = \Phi^{-1} \frac{(r_i - 3/8)}{(n + 1/4)}$$

in which Φ^{-1} is the inverse of the cumulative normal function, r_i is the rank of observation i, and n is number of observations.

A series of SAS examples of a Kruskal-Wallis test, an F test using Wilcoxon ranks, and an F test using Blom ranks is shown below.

11.12.1 SAS Example for Nonparametric Tests

The SAS program for the example of the effect of gonadotropin treatments on ovulation rate applying a Kruskal-Wallis test is as follows:

SAS program:

```
DATA superov;
INPUT trmt $ or @@;
DATALINES;
G1  14   G1  14   G1   7   G1  45   G1  18   G1  36   G1  15
G2  12   G2  11   G2  12   G2  12   G2  14   G2  13   G2   9
 S  10    S   7    S    9 S   7    S  12    S  19    S   8
;
PROC NPAR1WAY DATA = superov WILCOXON;
CLASS trmt;
VAR or;
EXACT;
RUN;
```

Explanation: The program uses the NPAR1WAY procedure with the WILCOXON option for Wilcoxon ranks. The CLASS statement defines the variable that classifies observations to a particular treatment. The VAR statement defines the dependent variable *or* (ovulation rate). The EXACT statement specifies use of the EXACT test.

SAS output:

```
           Wilcoxon Scores (Rank Sums) for Variable or
                   Classified by Variable trmt

                   Sum of      Expected      Std Dev        Mean
    trmt     N     Scores      Under H0      Under H0       Score
    G1       7     108.0         77.0        13.321036    15.428571
    G2       7      73.0         77.0        13.321036    10.428571
    S        7      50.0         77.0        13.321036     7.142857

                       Kruskal-Wallis Test

          Chi-Square                         6.4093
          DF                                      2
          Asymptotic Pr >  Chi-Square        0.0406
          Exact        Pr >= Chi-Square      0.0342
```

Explanation: The groups (*trmt*), number of observations (*N*), sum of ranks (*Sum of Scores*), expected rank scores under the null hypothesis, standard deviations, and mean ranks for each treatment group are presented. The calculated value of the *Kruskal-Wallis* statistic is 6.4093. Using the asymptotic (approximate) *chi*-square distribution gives a probability of 0.0406 that the null hypothesis is falsely rejected. The *Exact* probability is calculated on the basis of the actual data and is suitable when sample size is small.

The same example will be used to illustrate use of an *F* test on ranks. First, the Wilcoxon rank scores will be calculated.

SAS program:

```
PROC RANK DATA = superov OUT = rankings ;
VAR or;
RANKS rankor;
RUN;

PROC GLM DATA = rankings;
CLASS trmt;
MODEL rankor = trmt;
RUN;
QUIT;
```

Explanation: The RANK procedure computes Wilcoxon ranks on the variable *or* and saves them to the file as a new variable named *rankor*. The GLM procedure applies the usual *F* test on the ranks. The MODEL statement defines the dependent and independent variables: *rankor* = *trmt*. An LSMEANS statement (not shown here) could be used to test differences between particular treatments. The ANOVA table from the output is shown below.

SAS output:

Source	DF	Sum of Squares	Mean Square	F Value	Pr > F
Model	2	243.7142857	121.8571429	4.24	0.0309
Error	18	516.7857143	28.7103175		
Corrected Total	20	760.5000000			

Explanation: The *Sources* of variability are *Model* (corresponding to *trmt* here), *Error* and *Corrected Total*. The table lists degrees of freedom (*DF*), *Sum of Squares*, *Mean Square*, calculated *F* (*F Value*) and *P* value (*Pr > F*). For this example $F = 4.24$ and the *P* value is 0.0309, thus it can be concluded that treatments differ for ovulation rate. Note the very similar *P*-value to that from the NPAR1WAY procedure.

For the ranks to more closely approximate a normal distribution, an option *NORMAL = BLOM* can be added to the first line of the RANK procedure, giving:

PROC RANK DATA = superov OUT = rankings NORMAL= BLOM;

The GLM procedure gives the following output:

SAS output:

Source	DF	Sum of Squares	Mean Square	F Value	Pr > F
Model	2	5.31423011	2.65711506	3.71	0.0449
Error	18	12.90496486	0.71694249		
Corrected Total	20	18.21919497			

Note, the *P*-value is similar to those of previous outputs. For this example, all three approaches were satisfactory.

Exercises

11.1. Four lines of chickens (*A*, *B*, *C* and *D*) were crossed to obtain four crosses *AB*, *AC*, *BC* and *BD*. Egg weights of those crosses were compared. The weights (g) are as follows:

AB	58	51	56	52	54	57	58	60		
AC	59	62	64	60	62					
BC	56	57	56	55						
BD	59	55	50	64	57	53	57	53	56	55

Test the significance of the difference of arithmetic means.

11.2. Hay was stored using three different methods and its nutrition value was measured. Are there significant differences among different storage methods?

TRT1	TRT2	TRT3
17.3	22.0	19.0
14.0	16.9	20.2
14.8	18.9	18.8
12.2	17.8	19.6

Chapter 12

Random Effects One-way Analysis of Variance

In a random effects model, groups or treatments are defined as levels of a random variable with some theoretical distribution. In estimation of variability and effects of groups a random sample of groups from a population of groups is used. For example, data from few farms can be thought of as a sample from the population of 'all' farms, or, if an experiment is conducted on several locations, locations are a random sample of 'all' locations.

The main characteristics and differences between fixed and random effects are the following. An effect is defined as fixed if: there is a small (finite) number of groups or treatments; groups represent distinct populations, each with its own mean; and the variability between groups is not explained by some distribution. The effect can be defined as random if there exists a large (even infinite) number of groups or treatments; the groups investigated are a random sample drawn from a single population of groups; and the effect of a particular group is a random variable with some probability or density distribution. The sources of variability for fixed and random models of the one-way analysis of variance are shown in Figures 12.1 and 12.2.

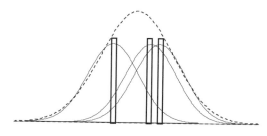

Figure 12.1 Sources of variability for the fixed effects one-way model: total variability, variability within groups, variability between groups

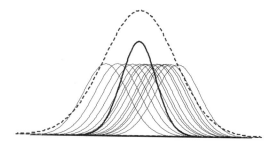

Figure 12.2 Sources of variability for the random effects one-way model: total variability, variability within groups, variability between groups

12.1   The Random Effects One-way Model

The random effects one-way model is:

$$y_{ij} = \mu + \tau_i + \varepsilon_{ij} \qquad\qquad i = 1,..., a; \; j = 1,..., n$$

where:

y_{ij} = an observation of unit j in group or treatment i
μ = the overall mean
τ_i = the random effect of group or treatment i with mean 0 and variance σ^2_τ
ε_{ij} = random error with mean 0 and variance σ^2

For the unbalanced case, that is unequal numbers of observations per group, n_i denotes the number of observations in group i, and the total number of observations is $N = \sum_i n_i$, $(i = 1,..., a)$.

The assumptions of the random model are:

$E(\tau_i) = 0$
$E(\varepsilon_{ij}) = 0$
$Var(\tau_i) = \sigma^2_\tau$
$Var(\varepsilon_{ij}) = \sigma^2$
τ_i and ε_{ij} are independent, that is $Cov(\tau_i, \varepsilon_{ij}) = 0$

Usually it is also assumed that τ_i and ε_{ij} are normal:

$\tau_i \sim N(0, \sigma^2_\tau)$
$\varepsilon_{ij} \sim N(0, \sigma^2)$

The variances σ^2_τ and σ^2 are between and within group variance components, respectively. From the assumptions it follows:

$$E(y_{ij}) = \mu \quad \text{and} \quad Var(y_{ij}) = \sigma^2_\tau + \sigma^2$$

That is:

$$y_{ij} \sim N(\mu, \sigma^2_\tau + \sigma^2)$$

Also:

$Cov(y_{ij}, y_{ij'}) = \sigma^2_\tau$
$Cov(\tau_i, y_{ij}) = \sigma^2_\tau$

The covariance between observations within a group is equal to the variance between groups (for proof, see section 12.5).

The expectation and variance of y for a given τ_i (conditional on τ_i) are:

$$E(y_{ij} | \tau_i) = \mu + \tau_i \quad \text{and} \quad Var(y_{ij} | \tau_i) = \sigma^2$$

The conditional distribution of y is:

$$y_{ij} \sim N(\mu + \tau_i, \sigma^2)$$

Possible aims of an analysis of a random model are:

1. A test of group or treatment effects, the test of

$$H_0: \sigma^2_\tau = 0$$
$$H_1: \sigma^2_\tau \neq 0$$

2. Prediction of effects $\tau_1,..., \tau_a$
3. Estimation of the variance components

12.2 Hypothesis Test

Hypotheses for the random effects model are used to determine whether there is variability between groups:

$$H_0: \sigma^2_\tau = 0$$
$$H_1: \sigma^2_\tau \neq 0$$

If H_0 is correct, the group variance is zero, all groups are equal since there is no variability among their means.

The expectations of the sums of squares are:

$$E(SS_{RES}) = \sigma^2(N - a)$$
$$E(SS_{TRT}) = (\sigma^2 + n\,\sigma^2_\tau)(a - 1)$$

The expectations of the mean squares are:

$$E(MS_{RES}) = \sigma^2$$

$$E(MS_{TRT}) = \begin{cases} = \sigma^2 & \text{if } H_0 \\ = \sigma^2 + n\sigma^2_\tau & \text{if not } H_0 \end{cases}$$

This indicates that the F test is analogous to that of the fixed model. The F statistic is:

$$F = \frac{MS_{TRT}}{MS_{RES}}$$

If H_0 is correct then $\sigma^2_\tau = 0$ and $F = 1$.

An *ANOVA* table is used to summarize the analysis of variance for a random model. It is helpful to add the expected mean squares $E(MS)$ to the table:

Source	SS	df	MS = SS / df	E(MS)
Between groups or treatment	SS_{TRT}	$a - 1$	MS_{TRT}	$\sigma^2 + n\,\sigma^2_\tau$
Residual (within groups or treatments)	SS_{RES}	$N - a$	MS_{RES}	σ^2

For unbalanced cases n is replaced with $\dfrac{1}{a-1}\left(N - \dfrac{\sum_i n_i^2}{N}\right)$.

12.3 Prediction of Group Means

Since the effects τ_i are random variables, they are not estimated, but their expectations are predicted, given the means estimated from the data $E(\tau_i \mid \bar{y}_i.)$. This expectation can be predicted by using the following function of the random variable y:

$$\hat{\tau}_i = b_{\tau \mid \bar{y}_i.}(\bar{y}_i. - \hat{\mu})$$

where:

$\hat{\mu} = \bar{y}.. $ = estimator of the overall mean

$$b_{\tau \mid \bar{y}_i.} = \frac{Cov(\tau_i, \bar{y}_i.)}{Var(\bar{y}_i.)} = \frac{\sigma_\tau^2}{\sigma_\tau^2 + \sigma^2 / n_i} = \text{regression coefficient of } \tau_i \text{ on the arithmetic}$$

mean $\bar{y}_i.$ of group i

If variance components are unknown and must also be estimated, the expression for $b_{\tau \mid \bar{y}_i.}$ is:

$$b_{\tau \mid \bar{y}_i.} = \frac{\hat{\sigma}_\tau^2}{\hat{\sigma}_\tau^2 + \hat{\sigma}^2 / n_i}$$

12.4 Variance Component Estimation

Recall the *ANOVA* table for the random effects model:

Source	SS	df	MS = SS / df	E(MS)
Between groups or treatment	SS_{TRT}	$a - 1$	MS_{TRT}	$\sigma^2 + n\,\sigma_\tau^2$
Residual (within groups or treatments	SS_{RES}	$N - a$	MS_{RES}	σ^2

Since from the *ANOVA* table:

$$E(MS_{TRT}) = \sigma^2 + n\,\sigma_\tau^2$$
$$E(MS_{RES}) = \sigma^2$$

the mean squares can be equated to the estimators of the variance components:

$$MS_{TRT} = \hat{\sigma}^2 + n\hat{\sigma}_\tau^2$$
$$MS_{RES} = \hat{\sigma}^2$$

Rearranging:

$$\hat{\sigma}^2 = MS_{RES}$$
$$\hat{\sigma}_\tau^2 = \frac{(MS_{TRT} - MS_{RES})}{n}$$

where:

$\hat{\sigma}^2$ and $\hat{\sigma}_\tau^2$ = estimators of variance components

n = the number of observations per treatment.

For unbalanced data:

$$\hat{\sigma}_\tau^2 = \frac{\left(MS_{TRT} - MS_{RES}\right)}{\dfrac{1}{a-1}\left(N - \dfrac{\sum_i n_i^2}{N}\right)}$$

where n_i denotes the number of observations in group i, and the total number of observations is $N = \sum_i n_i$, ($i = 1$ to a).

These estimators are called *ANOVA* estimators. If assumptions of the model are not satisfied, and above all, if variances across groups are not homogeneous, estimates of the variance components and inferences about them may be incorrect.

Example: Progesterone concentration (ng/ml) was measured for eight sows to estimate variability within and between sows, and to determine if variability between sows is significant. Samples were taken three times on each sow. Data are presented in the following table.

Measure	Sow							
	1	2	3	4	5	6	7	8
1	5.3	6.6	4.3	4.2	8.1	7.9	5.5	7.8
2	6.3	5.6	7.0	5.6	7.9	4.7	4.6	7.0
3	4.2	6.3	7.9	6.6	5.8	6.8	3.4	7.9
Sum	15.8	18.5	19.2	16.4	21.8	19.4	13.5	22.7

Total sum = 147.3

By computing the sums of squares and defining the degrees of freedom as for a fixed model, the following *ANOVA* table can be constructed:

Source	SS	df	MS	E(MS)
Between sows	22.156	7	3.165	$\sigma^2 + 3\,\sigma_\tau^2$
Within sows	23.900	16	1.494	σ^2

The estimated variance components are:

$$\hat{\sigma}^2 = 1.494$$

$$\hat{\sigma}_\tau^2 = \frac{(3.165 - 1.494)}{3} = 0.557$$

F test:

$$F = \frac{MS_{TRT}}{MS_{RES}} = \frac{3.165}{1.494} = 2.118$$

Predicted values for sows are:

$$\hat{\tau}_i = b_{\tau,\bar{y}_{i\cdot}}(\bar{y}_{i\cdot} - \hat{\mu})$$

The estimated overall mean is:

$$\hat{\mu} = \bar{y}.. = 6.138$$

The regression coefficient is:

$$b_{\tau y} = \frac{\hat{\sigma}_\tau^2}{\hat{\sigma}_\tau^2 + \hat{\sigma}^2 / n_i} = \frac{0.557}{0.557 + 1.494/3} = 0.528$$

The mean of sow 1 is:

$$\bar{y}_1. = 5.267$$

The effect of sow 1 is:

$$\hat{\tau}_1 = 0.528 \,(5.267 - 6.138) = -0.460$$

The mean of sow 2 is:

$$\bar{y}_2. = 6.167$$

The effect of sow 2 is:

$$\hat{\tau}_2 = 0.528 \,(6.167 - 6.138) = 0.015$$

Using the same formula, the effect of each sow can be predicted.

12.5 Intraclass Correlation

An intraclass correlation is a correlation between observations within a group or treatment. Recall that a correlation is a ratio of the covariance to the square root of the product of the variances:

$$\rho_t = \frac{Cov(y_{i,j}, y_{i,j'})}{\sqrt{Var(y_{i,j})Var(y_{i,j'})}}$$

Also recall that the covariance between observations within a group is equal to the variance component between groups:

$$Cov(y_{ij}, y_{ij'}) = Var(\tau_i) = \sigma_\tau^2$$

The variance of any observation y_{ij} is:

$$Var(y_{ij}) = Var(y_{ij'}) = Var(y) = \sigma_\tau^2 + \sigma^2$$

These can be easily verified. Assume two observations in a group i:

$$y_{ij} = \mu + \tau_i + \varepsilon_{ij}$$
$$y_{ij'} = \mu + \tau_i + \varepsilon_{ij'}$$

The covariance between two observations within the same group is:

$$Cov(y_{ij}, y_{ij'}) = Cov(\mu + \tau_i + \varepsilon_{ij}, \mu + \tau_i + \varepsilon_{ij'}) = Var(\tau_i) + Cov(\varepsilon_{ij}, \varepsilon_{ij'}) = \sigma_\tau^2 + 0 = \sigma_\tau^2$$

The variance of y_{ij} is:

$$Var(y_{ij}) = Var(\mu + \tau_i + \varepsilon_{ij}) = Var(\tau_i) + Var(\varepsilon_{ij}) = \sigma_\tau^2 + \sigma^2$$

Note that τ_i and ε_{ij} are independent, the covariance between them is zero. The intraclass correlation is:

$$\rho_t = \frac{Cov(y_{i,j}, y_{i,j'})}{\sqrt{Var(y_{i,j})Var(y_{i,j'})}} = \frac{\sigma_\tau^2}{\sqrt{(\sigma_\tau^2 + \sigma^2)(\sigma_\tau^2 + \sigma^2)}} = \frac{\sigma_\tau^2}{\sigma_\tau^2 + \sigma^2}$$

If the variance components are estimated from a sample, the intraclass correlation is:

$$r_t = \frac{\hat{\sigma}_\tau^2}{\hat{\sigma}_\tau^2 + \hat{\sigma}^2}$$

Example: For the example of progesterone concentration of sows, estimate the intraclass correlation.

The estimated variance components are:

$$\hat{\sigma}^2 = 1.464 \text{ and } \hat{\sigma}_\tau^2 = 0.557$$

The intraclass correlation or correlation between repeated measurements on a sow is:

$$r_t = \frac{\hat{\sigma}_\tau^2}{\hat{\sigma}_\tau^2 + \hat{\sigma}^2} = \frac{1.464}{1.464 + 0.557} = 0.724$$

12.6 Maximum Likelihood Estimation

Alternatively, parameters can be obtained by using maximum likelihood (*ML*) estimation. Under the assumption of normality, the likelihood function is a function of the parameters for a given set of *N* observations:

$$L(\mu, \sigma^2, \sigma_\tau^2 \mid y)$$

It can be shown that under the assumption of normality, the log likelihood of a random effects one-way model is:

$$logL(\mu, \sigma^2, \sigma_\tau^2 \mid y) = -\frac{N}{2}log(2\pi) - \frac{1}{2}\sum_i log(\sigma^2 + n_i\sigma_\tau^2) - \frac{(N-a)}{2}log(\sigma^2)$$
$$- \frac{\sum_i \sum_j (y_{ij} - \mu)^2}{2\sigma^2} - \frac{\sigma_\tau^2}{2\sigma^2}\sum_i \frac{(\sum_i y_{ij} - n_i\mu)^2}{\sigma^2 + n_i\sigma_\tau^2}$$

Writing $y_{ij} - \mu$ as $y_{ij} - \bar{y}_{i\cdot} + \bar{y}_{i\cdot} - \mu$ and simplifying, the log likelihood is:

$$logL(\mu,\sigma^2,\sigma_\tau^2 \mid y) = -\frac{N}{2}log(2\pi)-\frac{1}{2}\sum_i log(\sigma^2+n_i\sigma_\tau^2)-\frac{(N-a)}{2}log(\sigma^2)$$
$$-\frac{\sum_i\sum_j(y_{ij}-\bar{y}_{i\cdot})^2}{2\sigma^2}-\sum_i\frac{n_i(\bar{y}_{i\cdot}-\mu)^2}{2(\sigma^2+n_i\sigma_\tau^2)}$$

The maximum likelihood estimators are chosen to maximize the log likelihood function. The maximum of the function can be determined by taking partial derivatives of the log likelihood function with respect to the parameters:

$$\frac{\partial\,logL(\mu,\sigma^2,\sigma_\tau^2 \mid y)}{\partial\mu}=\frac{1}{\sigma^2+n_i\sigma_\tau^2}\sum_j(\bar{y}_{i\cdot}-\mu_i)$$

$$\frac{\partial\,logL(\mu,\sigma^2,\sigma_\tau^2 \mid y)}{\partial\sigma^2}=-\frac{N-a}{2\sigma^2}-\frac{1}{2}\sum_i\frac{1}{\sigma^2+n_i\sigma_\tau^2}+\frac{\sum_i\sum_j(y_{ij}-\mu)^2}{2\sigma^4}+\sum_i\frac{n_i(\bar{y}_{i\cdot}-\mu)^2}{2(\sigma^2+n_i\sigma_\tau^2)^2}$$

$$\frac{\partial\,logL(\mu,\sigma^2,\sigma_\tau^2 \mid y)}{\partial\sigma_\tau^2}=-\frac{1}{2}\sum_i\frac{n_i}{\sigma^2+n_i\sigma_\tau^2}+\frac{1}{2}\sum_i\frac{n_i^2(\bar{y}_{i\cdot}-\mu)^2}{(\sigma^2+n_i\sigma_\tau^2)^2}$$

These derivatives are equated to zero to find the estimators $\hat{\mu}$, $\hat{\sigma}^2_{\tau\_ML}$ and $\hat{\sigma}^2_{ML}$. Note that the second derivative must be negative when parameters are replaced with solutions. Also, maximum likelihood estimators must satisfy $\hat{\sigma}^2_{ML}>0$ and $\hat{\sigma}^2_{\tau\_ML}\geq 0$.

For $\hat{\mu}$:

$$\hat{\mu}=\frac{\sum_i\dfrac{n_i\bar{y}_{i\cdot}}{\hat{\sigma}^2_{ML}+n_i\hat{\sigma}^2_{\tau\_ML}}}{\sum_i\dfrac{n_i}{\hat{\sigma}^2_{ML}+n_i\hat{\sigma}^2_{\tau\_ML}}}=\frac{\sum_i\dfrac{\bar{y}_{i\cdot}}{\hat{Var}(\bar{y}_{i\cdot})}}{\sum_i\dfrac{1}{\hat{Var}(\bar{y}_{i\cdot})}}$$

For $\hat{\sigma}^2_{\tau\_ML}$ and $\hat{\sigma}^2_{ML}$ the equations are:

$$-\frac{N-a}{\hat{\sigma}^2_{ML}}-\sum_i\frac{1}{\hat{\sigma}^2_{ML}+n_i\hat{\sigma}^2_{\tau\_ML}}+\frac{\sum_i\sum_j(y_{ij}-\mu)^2}{\hat{\sigma}^4_{ML}}+\sum_i\frac{n_i(\bar{y}_{i\cdot}-\mu)^2}{(\hat{\sigma}^2_{ML}+n_i\hat{\sigma}^2_{\tau\_ML})^2}=0$$

$$-\sum_i\frac{n_i}{\hat{\sigma}^2_{ML}+n_i\hat{\sigma}^2_{\tau\_ML}}+\sum_i\frac{n_i^2(\bar{y}_{i\cdot}-\hat{\mu})^2}{(\hat{\sigma}^2_{ML}+n_i\hat{\sigma}^2_{\tau\_ML})^2}=0$$

Note that for unbalanced data there is not an analytical solution for these two equations, they must be solved iteratively. For balanced data, that is, when $n_i=n$, there is an analytical solution, the *log likelihood* simplifies to:

$$logL(\mu,\sigma^2,\sigma_\tau^2 \mid y)=-\frac{N}{2}log(2\pi)-\frac{1}{2}a\,log(\sigma^2+n\sigma_\tau^2)-\frac{a(n-1)}{2}log(\sigma^2)$$
$$-\frac{\sum_i\sum_j(y_{ij}-\bar{y}_{i\cdot})^2}{2\sigma^2}-\frac{\sum_i n(y_{i\cdot}-\bar{y}_{\cdot\cdot})^2}{2\sigma^2}-\frac{an(\bar{y}_{\cdot\cdot}-\mu)^2}{2(\sigma^2+n\sigma_\tau^2)}$$

After taking partial derivatives and equating them to zero the solutions are:

$$\hat{\mu} = \bar{y}..$$

$$\hat{\sigma}^2_{ML} = \frac{\sum_i \sum_j (y_{ij} - \bar{y}..)^2}{an - a}$$

$$\hat{\sigma}^2_{\tau\_ML} = \frac{\dfrac{\sum_i n(\bar{y}_{i.} - \bar{y}..)^2}{a} - \hat{\sigma}^2_{ML}}{n}$$

These solutions are the *ML* estimators if $\dfrac{\sum_i n(\bar{y}_{i.} - \bar{y}..)^2}{a} \geq \dfrac{\sum_i \sum_j (y_{ij} - \bar{y}..)^2}{an - a}$.

If $\dfrac{\sum_i n(\bar{y}_{i.} - \bar{y}..)^2}{a} < \dfrac{\sum_i \sum_j (y_{ij} - \bar{y}..)^2}{N - a}$ then $\hat{\sigma}^2_{\tau\_ML} = 0$ and $\hat{\sigma}^2_{ML} = \dfrac{\sum_i \sum_j (y_{ij} - \bar{y}..)^2}{an}$

Example: For the example of progesterone concentration in sows, estimate between and within sows variance components by maximum likelihood.

The following was computed previously:

$$SS_{WITHIN\ SOW} = \sum_i n(\bar{y}_{i.} - \bar{y}..)^2 = 22.156$$
$$SS_{SOW} = \sum_i \sum_j (y_{ij} - \bar{y}..)^2 = 23.900$$

Also, $a = 8$, and $n = 3$.

$$\hat{\sigma}^2_{ML} = \frac{\sum_i \sum_j (y_{ij} - \bar{y}..)^2}{an - a} = \frac{23.900}{(8)(3) - 8} = 1.494$$

$$\hat{\sigma}^2_{\tau\_ML} = \frac{\dfrac{\sum_i n(\bar{y}_{i.} - \bar{y}..)^2}{a} - \hat{\sigma}^2_{ML}}{n} = \frac{\dfrac{22.156}{8} - 1.494}{3} = 0.425$$

12.7 Restricted Maximum Likelihood Estimation

Restricted maximum likelihood (*REML*) estimation is a maximum likelihood estimation that does not involve μ, but takes into account the degrees of freedom associated with estimating the mean. The simplest example is estimation of the variance based on the n observations for which the *REML* estimator is:

$$\hat{\sigma}^2_{REML} = \frac{\sum_i (y_i - \bar{y})^2}{n - 1}$$

comparing to maximum likelihood estimator:

$$\hat{\sigma}^2_{ML} = \frac{\sum_i (y_i - \bar{y})^2}{n}$$

The *REML* estimator takes into account the degree of freedom needed for estimating μ.

For the one-way random model and balanced data *REML* maximizes the part of the likelihood which does not involve μ. This is the likelihood function of σ^2 and σ^2_τ given the means or deviations of means expressed as $\sum_i n(\bar{y}_{i.} - \bar{y}..)^2$ and $\sum_i \sum_j (y_{ij} - \bar{y}_{i.})^2$. The likelihood is:

$$L\left[\sigma^2, \sigma^2_\tau \mid \sum_i n(\bar{y}_{i.} - \bar{y}..)^2, \sum_i \sum_j (y_{ij} - \bar{y}_{i.})^2 > 0\right] = \frac{e^{-\frac{1}{2}\left(\frac{\sum_i \sum_j (y_{ij} - \bar{y}_{i.})^2}{\sigma^2} + \frac{\sum_i n(\bar{y}_{i.} - \bar{y}..)^2}{(\sigma^2 + n\sigma^2_\tau)}\right)}}{\sqrt{(2\pi)^{(an-1)} \sigma^{2(an-a)} (\sigma^2 + n\sigma^2_\tau)^{a-1} an}}$$

The log likelihood that is to be maximized is:

$$logL\left[\sigma^2, \sigma^2_\tau \mid \sum_i n(\bar{y}_{i.} - \bar{y}..)^2, \sum_i \sum_j (y_{ij} - \bar{y}_{i.})^2 > 0\right]$$

$$= \frac{1}{2}(an-1)log(2\pi) - \frac{1}{2}log(an) - \frac{1}{2}a(n-1)log(\sigma^2)$$

$$- \frac{1}{2}(a-1)log(\sigma^2 + n\sigma^2_\tau) - \frac{\sum_i \sum_j (y_{ij} - \bar{y}_{i.})}{2\sigma^2} - \frac{\sum_i n(\bar{y}_{i.} - \bar{y}..)^2}{2(\sigma^2 + n\sigma^2_\tau)}$$

By taking the first derivatives and equating them to zero the following estimators are obtained:

$$\hat{\sigma}^2_{REML} = \frac{\sum_i \sum_j (y_{ij} - \bar{y}_{i.})^2}{an - a}$$

$$\hat{\sigma}^2_{\tau\_REML} = \frac{\left(\frac{\sum_i n(\bar{y}_{i.} - \bar{y}..)^2}{a-1} - \hat{\sigma}^2\right)}{n}$$

It must also hold that $\hat{\sigma}^2_{REML} > 0$ and $\hat{\sigma}^2_{\tau\_REML} \geq 0$, that is, these are the *REML* estimators if

$$\frac{\sum_i n(\bar{y}_{i.} - \bar{y}..)^2}{a-1} \geq \frac{\sum_i \sum_j (y_{ij} - \bar{y}..)^2}{an - a}.$$

If $\dfrac{\sum_i n(\bar{y}_{i.} - \bar{y}..)^2}{a-1} < \dfrac{\sum_i \sum_j (y_{ij} - \bar{y}..)^2}{N - a}$ then $\hat{\sigma}^2_{\tau\_REML} = 0$ and $\hat{\sigma}^2_{REML} = \dfrac{\sum_i \sum_j (y_{ij} - \bar{y}..)^2}{an - 1}.$

Note that for balanced data (the same number of observations per group) these estimators are equal to *ANOVA* estimators, since:

$$\hat{\sigma}^2_{REML} = \frac{\sum_i \sum_j (y_{ij} - \bar{y}_{i.})^2}{an - a} = MS_{RES} = \text{residual mean square, and}$$

$$\frac{\sum_i n(\bar{y}_{i\cdot} - \bar{y}..)^2}{a-1} = MS_{TRT} = \text{treatment mean square. Thus:}$$

$$\hat{\sigma}^2_{\tau\_REML} = \frac{(MS_{TRT} - MS_{RES})}{n}$$

12.8   SAS Example for the Random Effects One-way Model

The SAS program for the example of progesterone concentration of sows is as follows. Recall the data:

Measure	Sow							
	1	2	3	4	5	6	7	8
1	5.3	6.6	4.3	4.2	8.1	7.9	5.5	7.8
2	6.3	5.6	7.0	5.6	7.9	4.7	4.6	7.0
3	4.2	6.3	7.9	6.6	5.8	6.8	3.4	7.9

SAS program:

```
DATA sow;
INPUT sow prog @@;
DATALINES;
1   5.3   1   6.3   1   4.2   2   6.6   2   5.6   2   6.3
3   4.3   3   7.0   3   7.9   4   4.2   4   5.6   4   6.6
5   8.1   5   7.9   5   5.8   6   7.9   6   4.7   6   6.8
7   5.5   7   4.6   7   3.4   8   7.8   8   7.0   8   7.9
;
PROC MIXED DATA = sow METHOD = REML;
   CLASS sow ;
   MODEL prog =  / SOLUTION DDFM = SATTERTH;
   RANDOM sow / SOLUTION;
RUN;
```

Explanation: The MIXED procedure is used, which is appropriate for analysis when using a random effects model, because it gives correct predictions of the random effects and correct estimates of standard errors. The default method for variance component estimation is restricted maximum likelihood (REML). It can be changed to maximum likelihood by defining METHOD = ML. The CLASS statement defines the independent categorical variable (*sow*). The MODEL statement defines the dependent variable (*prog*); MODEL *prog* = , that is, without any independent variable, indicates that in the model there is no fixed independent variable, only the overall mean is considered to be fixed. The RANDOM statement defines *sow* as a random variable. As standard errors of predictions consist of linear combination of error and sow variances, it is recommended that the degrees of freedom used for testing whether significance of solutions are corrected by using the option DDFM = SATTERTH or DDFM = KENWARDROGER with the MODEL statement. This option would also correct degrees of freedom accounting for possible unequal number of observations per sow. The SOLUTION options after the slash specify output of solutions (predictions of sow effects).

SAS output:

```
          Covariance Parameter Estimates

               Cov Parm     Estimate
               sow           0.5571
               Residual      1.4937
```

```
             Solution for Fixed Effects

                         Standard
   Effect      Estimate    Error     DF    t Value    Pr > |t|
   Intercept    6.1375     0.3632     7     16.90      <.0001
```

```
             Solution for Random Effects

                              Std Err
   Effect   sow   Estimate     Pred      DF    t Value    Pr > |t|
   sow       1    -0.4599     0.5475    5.49    -0.84     0.4360
   sow       2     0.0154     0.5475    5.49     0.03     0.9785
   sow       3     0.1386     0.5475    5.49     0.25     0.8093
   sow       4    -0.3542     0.5475    5.49    -0.65     0.5437
   sow       5     0.5963     0.5475    5.49     1.09     0.3216
   sow       6     0.1738     0.5475    5.49     0.32     0.7626
   sow       7    -0.8647     0.5475    5.49    -1.58     0.1698
   sow       8     0.7547     0.5475    5.49     1.38     0.2216
```

Explanation: Not shown are *Model Information, Class Level Information, Dimensions, Number of observations,* and *Iteration History.* The first table shows the variance components (*Covariance Parameter Estimates*). Variance components for *sow* and *Residual* are 0.5571 and 1.4937, respectively. The next table shows *Solution for Fixed Effects.* In this example, only the overall mean (*INTERCEPT*) is defined as a fixed effect. The *Estimate* is 6.1375 with the *Standard Error* 0.3632. In the next table are predictions for individual *sows.* For example *sow* 1 has the *Estimate* of –0.4599 with the prediction standard error (*SE Pred*) of 0.5475. The *t* tests (*Pr > |t|*) show that no sow effect is different than zero. This implies that sow variance is also not significantly different than zero. Note the corrected degrees of freedom (5.49), without correction they would be 16.

12.9 Matrix Approach to the One-way Analysis of Variance Model

12.9.1 Linear Model

The random effects model with equal numbers of observations per group can be presented using vectors and matrices as follows:

$$\mathbf{y} = \mathbf{1}\mu + \mathbf{Z}\mathbf{u} + \varepsilon$$

where:

$$
\mathbf{y} = \begin{bmatrix} y_{11} \\ y_{12} \\ \dots \\ y_{1n} \\ y_{21} \\ \dots \\ y_{2n} \\ \dots \\ \dots \\ y_{a1} \\ \dots \\ y_{an} \end{bmatrix}_{an \times 1}
\quad
\mathbf{1} = \begin{bmatrix} 1 \\ 1 \\ \dots \\ 1 \end{bmatrix}_{an \times 1}
\quad
\mathbf{Z} = \begin{bmatrix} 1 & 0 & \dots & 0 \\ 1 & 0 & \dots & 0 \\ \dots & \dots & \dots & \dots \\ 1 & 0 & \dots & 0 \\ 0 & 1 & \dots & 0 \\ \dots & \dots & \dots & \dots \\ 0 & 1 & \dots & 0 \\ \dots & \dots & \dots & \dots \\ \dots & \dots & \dots & \dots \\ 0 & 0 & \dots & 1 \\ \dots & \dots & \dots & \dots \\ 0 & 0 & \dots & 1 \end{bmatrix}_{an \times a}
\quad
\mathbf{u} = \begin{bmatrix} \tau_1 \\ \tau_2 \\ \dots \\ \tau_a \end{bmatrix}_{a \times 1}
\quad
\varepsilon = \begin{bmatrix} \varepsilon_{11} \\ \varepsilon_{12} \\ \dots \\ \varepsilon_{1n} \\ \varepsilon_{21} \\ \dots \\ \varepsilon_{2n} \\ \dots \\ \dots \\ \varepsilon_{a1} \\ \dots \\ \varepsilon_{an} \end{bmatrix}_{an \times 1}
$$

$\mathbf{y}$ = vector of observations
μ = the mean
$\mathbf{Z}$ = design matrix which relates $\mathbf{y}$ to $\mathbf{u}$
$\mathbf{u}$ = vector of random effects τ_i with mean $\mathbf{0}$ and variance $\mathbf{G} = \sigma_\tau^2 \mathbf{I}_a$
ε = vector of random errors with mean $\mathbf{0}$ and variance $\mathbf{R} = \sigma^2 \mathbf{I}_{an}$
a = the number of groups, n = the number of observations in each group

The expectations and (co)variances of the random variables are:

$$E(\mathbf{u}) = \mathbf{0} \text{ and } Var(\mathbf{u}) = \mathbf{G} = \sigma_\tau^2 \mathbf{I}_a$$
$$E(\varepsilon) = \mathbf{0} \text{ and } Var(\varepsilon) = \mathbf{R} = \sigma^2 \mathbf{I}_{an}$$
$$E(\mathbf{y}) = \mu \text{ and } Var(\mathbf{y}) = \mathbf{V} = \mathbf{ZGZ'} + \mathbf{R} = \sigma_\tau^2 \mathbf{ZZ'} + \sigma^2 \mathbf{I}_{an}$$

$$
\mathbf{V} = \begin{bmatrix} \sigma_\tau^2 \mathbf{J}_n & 0 & \dots & 0 \\ 0 & \sigma_\tau^2 \mathbf{J}_n & \dots & 0 \\ \dots & \dots & \dots & \dots \\ 0 & 0 & \dots & \sigma_\tau^2 \mathbf{J}_n \end{bmatrix}_{an \times an}
+
\begin{bmatrix} \sigma^2 \mathbf{I}_n & 0 & \dots & 0 \\ 0 & \sigma^2 \mathbf{I}_n & \dots & 0 \\ \dots & \dots & \dots & \dots \\ 0 & 0 & \dots & \sigma^2 \mathbf{I}_n \end{bmatrix}_{an \times an}
$$

$$
= \begin{bmatrix} \sigma_\tau^2 \mathbf{J}_n + \sigma^2 \mathbf{I}_n & 0 & \dots & 0 \\ 0 & \sigma_\tau^2 \mathbf{J}_n + \sigma^2 \mathbf{I}_n & \dots & 0 \\ \dots & \dots & \dots & \dots \\ 0 & 0 & \dots & \sigma_\tau^2 \mathbf{J}_n + \sigma^2 \mathbf{I}_n \end{bmatrix}_{an \times an}
$$

where $\mathbf{J}_n$ is a matrix of ones, and $\mathbf{I}_n$ is an identity matrix, both with dimension $n \times n$.

Here:

$$\sigma_\tau^2 \mathbf{J}_n + \sigma^2 \mathbf{I}_n = \begin{bmatrix} \sigma_\tau^2 + \sigma^2 & \sigma_\tau^2 & \cdots & \sigma_\tau^2 \\ \sigma_\tau^2 & \sigma_\tau^2 + \sigma^2 & \cdots & \sigma_\tau^2 \\ \cdots & \cdots & \cdots & \cdots \\ \sigma_\tau^2 & \sigma_\tau^2 & \cdots & \sigma_\tau^2 + \sigma^2 \end{bmatrix}_{n \times n}$$

12.9.2 Prediction of Random Effects

In order to predict the random vector **u**, it is often more convenient to use the following equations:

$$[\mathbf{1} \quad \mathbf{Z}]' \mathbf{V}^{-1} [\mathbf{1} \quad \mathbf{Z}] \begin{bmatrix} \hat{\mu} \\ \hat{\mathbf{u}} \end{bmatrix} = [\mathbf{1} \quad \mathbf{Z}]' \mathbf{y}$$

These equations are derived by exactly the same procedure as for the fixed effects model (i.e. by least squares), only they contain variance **V**. Because of that, these equations are called generalized least squares (*GLS*) equations.

Using $\mathbf{V} = \mathbf{ZGZ'} + \mathbf{R}$, the *GLS* equations are:

$$\begin{bmatrix} \mathbf{1}'\mathbf{R}^{-1}\mathbf{1} & \mathbf{1}'\mathbf{R}^{-1}\mathbf{1} \\ \mathbf{1}'\mathbf{R}^{-1}\mathbf{1} & \mathbf{1}'\mathbf{R}^{-1}\mathbf{1} + \mathbf{G}^{-1} \end{bmatrix} \begin{bmatrix} \hat{\beta} \\ \hat{\mathbf{u}} \end{bmatrix} = \begin{bmatrix} \mathbf{1}'\mathbf{R}^{-1}\mathbf{y} \\ \mathbf{1}'\mathbf{R}^{-1}\mathbf{y} \end{bmatrix}$$

Substituting the expressions of variances, $\mathbf{G} = \sigma_\tau^2 \mathbf{I}_a$ and $\mathbf{R} = \sigma^2 \mathbf{I}_{an}$, in the equations:

$$\begin{bmatrix} \frac{an}{\sigma^2} & \frac{1}{\sigma^2}\mathbf{1}'\mathbf{Z} \\ \frac{1}{\sigma^2}\mathbf{Z}'\mathbf{1} & \frac{1}{\sigma^2}\mathbf{Z}'\mathbf{Z} + \frac{1}{\sigma_\tau^2}\mathbf{I}_a \end{bmatrix} \begin{bmatrix} \hat{\mu} \\ \hat{\mathbf{u}} \end{bmatrix} = \begin{bmatrix} \frac{1}{\sigma^2}\mathbf{1}'\mathbf{y} \\ \frac{1}{\sigma^2}\mathbf{Z}'\mathbf{y} \end{bmatrix}$$

or simplified:

$$\begin{bmatrix} an & \mathbf{1}'\mathbf{Z} \\ \mathbf{Z}'\mathbf{1} & \mathbf{Z}'\mathbf{Z} + \frac{\sigma^2}{\sigma_\tau^2}\mathbf{I}_a \end{bmatrix} \begin{bmatrix} \hat{\mu} \\ \hat{\mathbf{u}} \end{bmatrix} = \begin{bmatrix} \mathbf{1}'\mathbf{y} \\ \mathbf{Z}'\mathbf{y} \end{bmatrix}$$

The solutions are:

$$\begin{bmatrix} \hat{\mu} \\ \hat{\mathbf{u}} \end{bmatrix} = \begin{bmatrix} an & \mathbf{1}'\mathbf{Z} \\ \mathbf{Z}'\mathbf{1} & \mathbf{Z}'\mathbf{Z} + \frac{\sigma^2}{\sigma_\tau^2}\mathbf{I}_a \end{bmatrix}^{-1} \begin{bmatrix} \mathbf{1}'\mathbf{y} \\ \mathbf{Z}'\mathbf{y} \end{bmatrix}$$

or written differently:

$$\hat{\mu} = (an)^{-1}\mathbf{1}'\mathbf{y} = \frac{\sum_i \sum_j y_{ij}}{an} = \bar{y}_{..}$$

$$\hat{\mathbf{u}} = \left(\mathbf{Z}'\mathbf{Z} + \frac{\sigma^2}{\sigma_\tau^2}\mathbf{I} \right)^{-1} \mathbf{Z}'(\mathbf{y} - \mathbf{1}\hat{\mu})$$

If the variances are known the solutions are obtained by simple matrix operations. If the variances are not known, they must be estimated, by using for example, maximum likelihood estimation.

Example: Calculate the solutions for the example of progesterone concentrations in sows by using matrices. Recall the data:

Measure	Sow							
	1	2	3	4	5	6	7	8
1	5.3	6.6	4.3	4.2	8.1	7.9	5.5	7.8
2	6.3	5.6	7.0	5.6	7.9	4.7	4.6	7.0
3	4.2	6.3	7.9	6.6	5.8	6.8	3.4	7.9

Assume the variance components are known, between sows $\sigma_\tau^2 = 1$ and within sows $\sigma^2 = 2$. The number of sows is $a = 8$ and the number of measurements per sow is $n = 3$.

$$\mathbf{Z'Z} = \begin{bmatrix} 3 & 0 & 0 & \dots & 0 \\ 0 & 3 & 0 & \dots & 0 \\ 0 & 0 & 3 & \dots & 0 \\ \dots & \dots & \dots & \dots & \dots \\ 0 & 0 & 0 & \dots & 3 \end{bmatrix}_{8\times8} \quad \hat{\mathbf{u}} = \begin{bmatrix} \hat{\tau}_1 \\ \hat{\tau}_2 \\ \hat{\tau}_3 \\ \dots \\ \hat{\tau}_8 \end{bmatrix}_{8\times1} \quad \mathbf{Z'y} = \begin{bmatrix} \sum_j y_{1j} \\ \sum_j y_{2j} \\ \sum_j y_{3j} \\ \dots \\ \sum_j y_{8j} \end{bmatrix}_{8\times1} = \begin{bmatrix} 15.8 \\ 18.5 \\ 19.2 \\ \dots \\ 22.7 \end{bmatrix}_{8\times1}$$

$$\mathbf{X'y} = \sum_i \sum_j y_{ij} = 147.3$$

$$\mathbf{1'1} = an = 24$$

$$\begin{bmatrix} an & \mathbf{1'Z} \\ \mathbf{Z'1} & \mathbf{Z'Z} + \frac{\sigma^2}{\sigma_\tau^2}\mathbf{I} \end{bmatrix} \begin{bmatrix} \hat{\mu} \\ \hat{\mathbf{u}} \end{bmatrix} = \begin{bmatrix} \mathbf{1'y} \\ \mathbf{Z'y} \end{bmatrix} \implies \begin{bmatrix} \hat{\mu} \\ \hat{\mathbf{u}} \end{bmatrix} = \begin{bmatrix} an & \mathbf{1'Z} \\ \mathbf{Z'1} & \mathbf{Z'Z} + \frac{\sigma^2}{\sigma_\tau^2}\mathbf{I} \end{bmatrix}^{-1} \begin{bmatrix} \mathbf{1'y} \\ \mathbf{Z'y} \end{bmatrix}$$

$$\begin{bmatrix} \hat{\mu} \\ \hat{\mathbf{u}} \end{bmatrix} = \begin{bmatrix} 24 & 3 & 3 & 3 & 3 & 3 & 3 & 3 & 3 \\ 3 & 3+2 & 0 & 0 & 0 & 0 & 0 & 0 & 0 \\ 3 & 0 & 3+2 & 0 & 0 & 0 & 0 & 0 & 0 \\ 3 & 0 & 0 & 3+2 & 0 & 0 & 0 & 0 & 0 \\ 3 & 0 & 0 & 0 & 3+2 & 0 & 0 & 0 & 0 \\ 3 & 0 & 0 & 0 & 0 & 3+2 & 0 & 0 & 0 \\ 3 & 0 & 0 & 0 & 0 & 0 & 3+2 & 0 & 0 \\ 3 & 0 & 0 & 0 & 0 & 0 & 0 & 3+2 & 0 \\ 3 & 0 & 0 & 0 & 0 & 0 & 0 & 0 & 3+2 \end{bmatrix}^{-1} \begin{bmatrix} 147.3 \\ 15.8 \\ 18.5 \\ 19.2 \\ 16.4 \\ 21.8 \\ 19.4 \\ 13.5 \\ 22.7 \end{bmatrix} = \begin{bmatrix} 6.1375 \\ -0.5225 \\ 0.0175 \\ 0.1575 \\ -0.4025 \\ 0.6775 \\ 0.1975 \\ -0.9825 \\ 0.8575 \end{bmatrix}$$

The vector $\begin{bmatrix} \hat{\mu} \\ \hat{\mathbf{u}} \end{bmatrix}$ contains estimates of the mean and individual effects of the sows. These estimates do not exactly match those from the SAS program in section 12.8 because simplified given values of the between and within sow variance components ($\sigma_\tau^2 = 1$ and $\sigma^2 = 2$) were used here, and in section 12.8 variance components were estimated from the data.

12.9.3 Maximum Likelihood Estimation

Assuming a multivariate normal distribution, $\mathbf{y} \sim N(\mathbf{1}\mu, \mathbf{V} = \sigma_\tau^2\,\mathbf{ZZ'} + \sigma^2\mathbf{I}_N)$, the density function of the $\mathbf{y}$ vector is:

$$f(\mathbf{y}\,|\,\mathbf{1}\mu,\mathbf{V}) = \frac{e^{-\frac{1}{2}(\mathbf{y}-\mathbf{1}\mu)'\mathbf{V}^{-1}(\mathbf{y}-\mathbf{1}\mu)}}{\sqrt{(2\pi)^N\,|\mathbf{V}|}}$$

where N is the total number of observations, $|\mathbf{V}|$ is the determinant of the $\mathbf{V}$ matrix, and $\mathbf{1}$ is a vector of ones. The likelihood function is:

$$L(\mathbf{1}\mu,\mathbf{V}\,|\,\mathbf{y}) = \frac{e^{-\frac{1}{2}(\mathbf{y}-\mathbf{1}\mu)'\mathbf{V}^{-1}(\mathbf{y}-\mathbf{1}\mu)}}{\sqrt{(2\pi)^N\,|\mathbf{V}|}}$$

The log likelihood is:

$$logL = -\frac{1}{2}N\,log(2\pi) - \frac{1}{2}log|\mathbf{V}| - \frac{1}{2}(\mathbf{y}-\mathbf{1}\mu)'\mathbf{V}^{-1}(\mathbf{y}-\mathbf{1}\mu)$$

To find the estimator which will maximize the log likelihood, partial derivatives are taken and equated to zero, to obtain the following:

$$\left(\mathbf{1}'\hat{\mathbf{V}}^{-1}\mathbf{1}\right)\hat{\mu} = \mathbf{1}'\hat{\mathbf{V}}^{-1}\mathbf{y}$$

$$tr\left(\hat{\mathbf{V}}^{-1}\right) = (\mathbf{y}-\mathbf{1}\hat{\mu})'\hat{\mathbf{V}}^{-1}\hat{\mathbf{V}}^{-1}(\mathbf{y}-\mathbf{1}\hat{\mu})$$

$$tr\left(\hat{\mathbf{V}}^{-1}\mathbf{ZZ'}\right) = (\mathbf{y}-\mathbf{1}\hat{\mu})'\hat{\mathbf{V}}^{-1}\mathbf{ZZ'}\hat{\mathbf{V}}^{-1}(\mathbf{y}-\mathbf{1}\hat{\mu})$$

where tr is the trace or the sum of the diagonal elements of the corresponding matrix.

Often those equations are expressed in a simplified form by defining:

$$\hat{\mathbf{V}}^{-1}(\mathbf{y}-\mathbf{1}\hat{\mu}) = \hat{\mathbf{P}}_1\mathbf{y}$$

Note that from the first likelihood equation:

$$\mathbf{1}\hat{\mu} = \mathbf{1}\left(\mathbf{1}'\hat{\mathbf{V}}^{-1}\mathbf{1}\right)^{-1}\mathbf{1}'\hat{\mathbf{V}}^{-1}\mathbf{y} \quad \text{and}$$

$$\hat{\mathbf{P}}_1\mathbf{y} = \hat{\mathbf{V}}^{-1}\mathbf{y} - \hat{\mathbf{V}}^{-1}\mathbf{1}\left(\mathbf{1}'\hat{\mathbf{V}}^{-1}\mathbf{1}\right)^{-1}\mathbf{1}'\hat{\mathbf{V}}^{-1}\mathbf{y}$$

Then the two variance equations are:

$$tr\left(\hat{\mathbf{V}}^{-1}\right) = \mathbf{y}'\hat{\mathbf{P}}_1\hat{\mathbf{P}}_1\mathbf{y}$$

$$tr\left(\hat{\mathbf{V}}^{-1}\mathbf{ZZ'}\right) = \mathbf{y}'\hat{\mathbf{P}}_1\mathbf{ZZ'}\hat{\mathbf{P}}_1\mathbf{y}$$

As shown in section 12.6, for balanced data there is an analytical solution of those equations. For unbalanced data the maximum likelihood equations must be solved iteratively using extensive computing methods such as Fisher scoring, Newton-Raphson, or an expectation maximization (*EM*) algorithm (see for example McCulloch and Searle, 2001).

12.9.4 Restricted Maximum Likelihood Estimation

Variance components are estimated with restricted maximum likelihood (*REML*) by using residuals after fitting the fixed effects in the model. In the one-way random model the fixed effect corresponds to the mean. Thus, instead of using a **y** data vector, *REML* uses linear combinations of **y**, say **K'y**, with **K** chosen such that **K'1** = **0**. The **K** matrix has $N - 1$ independent **k** vectors such that **k'1** = 0.

The transformed model is:

$$\mathbf{K'y} = \mathbf{K'Zu} + \mathbf{K'\varepsilon}$$

If **y** has a normal distribution, $\mathbf{y} \sim N(\mathbf{1}\mu, \mathbf{V} = \sigma_\tau^2 \mathbf{ZZ'} + \sigma^2\mathbf{I})$, and because **K'1** = **0**, the distribution of **K'y** is $N(0, \mathbf{K'VK})$, that is:

$$E(\mathbf{K'y}) = 0 \text{ and}$$
$$Var(\mathbf{K'y}) = \mathbf{K'VK} = \sigma_\tau^2 \mathbf{K'ZZ'K} + \mathbf{K'I}\sigma^2 \mathbf{I K}$$

The **K'y** are linear contrasts and they represent residual deviations from the estimated mean $\left(y_{ij} - \overline{y}..\right)$.

Following the same logic as for maximum likelihood, the *REML* equations are:

$$tr\left[\left(\mathbf{K'\hat{V}K}\right)^{-1}\mathbf{K'K}\right] = \mathbf{y'K}\left(\mathbf{K'\hat{V}K}\right)^{-1}\mathbf{K'K}\left(\mathbf{K'\hat{V}K}\right)^{-1}\mathbf{K'y}$$

$$tr\left[\left(\mathbf{K'\hat{V}K}\right)^{-1}\mathbf{K'ZZ'K}\right] = \mathbf{y'K}\left(\mathbf{K'\hat{V}K}\right)^{-1}\mathbf{K'ZZ'K}\left(\mathbf{K'\hat{V}K}\right)^{-1}\mathbf{K'y}$$

It can be shown that for any **K**:

$$\mathbf{K}\left(\mathbf{K'\hat{V}K}\right)^{-1}\mathbf{K'} = \hat{\mathbf{P}}_1$$

Recall that:

$$\hat{\mathbf{P}}_1 = \hat{\mathbf{V}}^{-1} - \hat{\mathbf{V}}^{-1}\mathbf{1}\left(\mathbf{1'}\hat{\mathbf{V}}^{-1}\mathbf{1}\right)^{-1}\mathbf{1'}\hat{\mathbf{V}}^{-1}$$

Then:

$$\mathbf{K}\left(\mathbf{K'\hat{V}K}\right)^{-1}\mathbf{K'y} = \hat{\mathbf{P}}_1\mathbf{y} = \hat{\mathbf{V}}^{-1}\left(\mathbf{y} - \mathbf{1}\hat{\mu}\right) = \hat{\mathbf{V}}^{-1}\mathbf{y} - \hat{\mathbf{V}}^{-1}\mathbf{1'}\left(\mathbf{1'}\hat{\mathbf{V}}^{-1}\mathbf{1}\right)^{-1}\mathbf{1'}\hat{\mathbf{V}}^{-1}\mathbf{y}$$

With rearrangement, the *REML* equations can be simplified to:

$$tr\left(\hat{\mathbf{P}}\right) = \mathbf{y'}\hat{\mathbf{P}}_1\hat{\mathbf{P}}_1\mathbf{y}$$
$$tr\left(\hat{\mathbf{P}}\mathbf{ZZ'}\right) = \mathbf{y'}\hat{\mathbf{P}}_1\mathbf{ZZ'}\hat{\mathbf{P}}_1\mathbf{y}$$

Exercise

12.1. Daily gains of heifers kept on two pastures were measured. There were 20 heifers on each pasture. The pastures are considered a random sample of the population of pastures. Estimate the intraclass correlation, that is, correlation between heifers within pastures. The mean squares, degrees of freedom and expected mean squares are shown in the following *ANOVA* table:

Source	df	MS	E(MS)
Between pasture	1	21220	$\sigma^2 + 20\,\sigma^2_\tau$
Within pasture	38	210	σ^2

Chapter 13

Mixed Models

In previous chapters one-way classification models have been introduced for estimation and tests. Logical development is to models with two or more classifications, with random, fixed, or combinations of fixed and random effects.

There are three categories of models with regard to types of effects:
1. Fixed effects model (all effects in the model are fixed)
2. Random effects model (all effects in the model are random)
3. Mixed effects model (some effects are fixed and some are random)

More complex models for particular applications will be introduced in later chapters. Here, some general aspects of mixed linear models will be briefly explained. Mixed models are models with both fixed and random effects. Fixed effects explain the mean, and random effects explain variance-covariance structure of the dependent variable. Consider a linear model with fixed effects $\boldsymbol{\beta}$, a random effect $\mathbf{u}$, and random error effects ε. Using matrix notation the model is:

$$\mathbf{y} = \mathbf{X}\boldsymbol{\beta} + \mathbf{Z}\mathbf{u} + \varepsilon$$

where:

$\mathbf{y}$ = vector of observations
$\mathbf{X}$ = design matrix which relates $\mathbf{y}$ to $\boldsymbol{\beta}$
$\boldsymbol{\beta}$ = vector of fixed effects
$\mathbf{Z}$ = design matrix which relates $\mathbf{y}$ to $\mathbf{u}$
$\mathbf{u}$ = vector of random effects with mean $\mathbf{0}$ and variance-covariance matrix $\mathbf{G}$
ε = vector of random errors with mean $\mathbf{0}$ and variance-covariance matrix $\mathbf{R}$

The expectations and (co)variances of the random variables are:

$$E \begin{bmatrix} \mathbf{y} \\ \mathbf{u} \\ \varepsilon \end{bmatrix} = \begin{bmatrix} \mathbf{X}\boldsymbol{\beta} \\ \mathbf{0} \\ \mathbf{0} \end{bmatrix} \qquad Var \begin{bmatrix} \mathbf{y} \\ \mathbf{u} \\ \varepsilon \end{bmatrix} = \begin{bmatrix} \mathbf{ZGZ'}+\mathbf{R} & \mathbf{ZG} & \mathbf{R} \\ \mathbf{GZ'} & \mathbf{G} & \mathbf{0} \\ \mathbf{R} & \mathbf{0} & \mathbf{R} \end{bmatrix}$$

Thus,

$E(\mathbf{y}) = \mathbf{X}\boldsymbol{\beta}$ and $Var(\mathbf{y}) = \mathbf{V} = \mathbf{ZGZ'} + \mathbf{R}$
$E(\mathbf{u}) = \mathbf{0}$ and $Var(\mathbf{u}) = \mathbf{G}$
$E(\varepsilon) = \mathbf{0}$ and $Var(\varepsilon) = \mathbf{R}$

Although the structure of the $\mathbf{G}$ and $\mathbf{R}$ matrices can be very complex, the usual structure is diagonal, for example:

$$\mathbf{G} = \sigma_\tau^2 \mathbf{I}_a$$
$$\mathbf{R} = \sigma^2 \mathbf{I}_N$$

with dimensions corresponding to the identity matrices $\mathbf{I}$, N being the total number of observations, and a the number of levels of $\mathbf{u}$.

Then $\mathbf{V}$ is:

$$\mathbf{V} = \mathbf{ZGZ'} + \mathbf{R} = \mathbf{ZZ'}\sigma_\tau^2 + \sigma^2 \mathbf{I}_N$$

Obviously, the model can contain more than one random effect. Nevertheless, the assumptions and properties of more complex models are a straightforward extension of the model with one random effect.

13.1 Prediction of Random Effects

In order to find solutions for $\boldsymbol{\beta}$ and $\mathbf{u}$ the following equations, called mixed model equations (*MME*), can be used:

$$\begin{bmatrix} \mathbf{X'R^{-1}X} & \mathbf{X'R^{-1}Z} \\ \mathbf{Z'R^{-1}X} & \mathbf{Z'R^{-1}Z} + \mathbf{G^{-1}} \end{bmatrix} \begin{bmatrix} \widetilde{\boldsymbol{\beta}} \\ \hat{\mathbf{u}} \end{bmatrix} = \begin{bmatrix} \mathbf{X'R^{-1}y} \\ \mathbf{Z'R^{-1}y} \end{bmatrix}$$

These equations are developed by maximizing the joint density of $\mathbf{y}$ and $\mathbf{u}$, which can be expressed as:

$$f(\mathbf{y}, \mathbf{u}) = f(\mathbf{y} \mid \mathbf{u}) f(\mathbf{u})$$

Generally, the solutions derived from the mixed model equations are:

$$\widetilde{\boldsymbol{\beta}} = \left(\mathbf{X'V^{-1}X}\right)^{-1} \mathbf{X'V^{-1}y}$$
$$\hat{\mathbf{u}} = \mathbf{GZ'V^{-1}}\left(\mathbf{y} - \mathbf{X}\widetilde{\boldsymbol{\beta}}\right)$$

The estimators $\widetilde{\boldsymbol{\beta}}$ are known as best linear unbiased estimators (*BLUE*), and the predictors $\hat{\mathbf{u}}$ are known as best linear unbiased predictors (*BLUP*). If the variances are known the solutions are obtained by simple matrix operations. If the variances are not known, they must be estimated from the data, using for example, maximum likelihood estimation.

Example: The one-way random model can be considered a mixed model with the overall mean μ a fixed effect and the vector $\mathbf{u}$ a random effect. Taking $\mathbf{G} = \sigma_\tau^2 \mathbf{I}$ and $\mathbf{R} = \sigma^2 \mathbf{I}$ we have:

$$\begin{bmatrix} \frac{1}{\sigma^2}\mathbf{X'X} & \frac{1}{\sigma^2}\mathbf{X'Z} \\ \frac{1}{\sigma^2}\mathbf{Z'X} & \frac{1}{\sigma^2}\mathbf{Z'Z} + \frac{1}{\sigma_\tau^2}\mathbf{I} \end{bmatrix} \begin{bmatrix} \hat{\boldsymbol{\beta}} \\ \hat{\mathbf{u}} \end{bmatrix} = \begin{bmatrix} \frac{1}{\sigma^2}\mathbf{X'y} \\ \frac{1}{\sigma^2}\mathbf{Z'y} \end{bmatrix}$$

Here $\mathbf{X} = \mathbf{1}$ yielding:

$$\begin{bmatrix} \frac{N}{\hat{\sigma}^2} & \frac{1}{\sigma^2}\mathbf{1'Z} \\ \frac{1}{\sigma^2}\mathbf{Z'1} & \frac{1}{\sigma^2}\mathbf{Z'Z} + \frac{1}{\sigma_\tau^2}\mathbf{I} \end{bmatrix} \begin{bmatrix} \hat{\mu} \\ \hat{\mathbf{u}} \end{bmatrix} = \begin{bmatrix} \frac{1}{\sigma^2}\mathbf{1'y} \\ \frac{1}{\sigma^2}\mathbf{Z'y} \end{bmatrix}$$

13.2 Maximum Likelihood Estimation

Assuming a multivariate normal distribution, $\mathbf{y} \sim N(\mathbf{X}\boldsymbol{\beta}, \mathbf{V})$, the density function of the $\mathbf{y}$ vector is:

$$f(\mathbf{y} \mid \boldsymbol{\beta}, \mathbf{V}) = \frac{e^{-\frac{1}{2}(\mathbf{y}-\mathbf{X}\boldsymbol{\beta})'\mathbf{V}^{-1}(\mathbf{y}-\mathbf{X}\boldsymbol{\beta})}}{\sqrt{(2\pi)^N |\mathbf{V}|}}$$

where N is the total number of observations, $|\mathbf{V}|$ is a determinant of the variance matrix $\mathbf{V}$. The $\mathbf{V}$ matrix in its simplest form is often defined as $\mathbf{V} = \sum_{j=0}^{m} \mathbf{Z}_j \mathbf{Z}_j' \sigma_j^2$, with $(m + 1)$ components of variance, and $\mathbf{Z}_0 = \mathbf{I}_N$.

The likelihood function is:

$$L(\boldsymbol{\beta}, \mathbf{V} \mid \mathbf{y}) = \frac{e^{-\frac{1}{2}(\mathbf{y}-\mathbf{X}\boldsymbol{\beta})'\mathbf{V}^{-1}(\mathbf{y}-\mathbf{X}\boldsymbol{\beta})}}{\sqrt{(2\pi)^N |\mathbf{V}|}}$$

The log likelihood is:

$$logL = -\frac{1}{2} N \, log(2\pi) - \frac{1}{2} log|\mathbf{V}| - \frac{1}{2}(\mathbf{y} - \mathbf{X}\boldsymbol{\beta})' \mathbf{V}^{-1} (\mathbf{y} - \mathbf{X}\boldsymbol{\beta})$$

By taking partial derivatives of *logL* with respect to the parameters and equating them to zero, the following equations are obtained:

$$\left(\mathbf{X}'\hat{\mathbf{V}}^{-1}\mathbf{X}\right)\tilde{\boldsymbol{\beta}} = \mathbf{X}'\hat{\mathbf{V}}^{-1}\mathbf{y}$$
$$tr\left(\hat{\mathbf{V}}^{-1}\mathbf{Z}_j\mathbf{Z}_j'\right) = \left(\mathbf{y} - \mathbf{X}\tilde{\boldsymbol{\beta}}\right)'\hat{\mathbf{V}}^{-1}\mathbf{Z}_j\mathbf{Z}_j'\hat{\mathbf{V}}^{-1}\left(\mathbf{y} - \mathbf{X}\tilde{\boldsymbol{\beta}}\right)$$

where *tr* is a trace or the sum of the diagonal elements of the corresponding matrix, and second expressions denote $(m + 1)$ different equations for each corresponding variance component.

Alternatively, those equations can be expressed in a simplified form defining:

$$\hat{\mathbf{V}}^{-1}\left(\mathbf{y} - \mathbf{X}\tilde{\boldsymbol{\beta}}\right) = \hat{\mathbf{P}}\mathbf{y}$$

Note that from the first likelihood equation:

$$\mathbf{X}\tilde{\boldsymbol{\beta}} = \mathbf{X}\left(\mathbf{X}'\hat{\mathbf{V}}^{-1}\mathbf{X}\right)^{-1}\mathbf{X}'\hat{\mathbf{V}}^{-1}\mathbf{y} \quad \text{and}$$
$$\hat{\mathbf{P}}\mathbf{y} = \hat{\mathbf{V}}^{-1}\mathbf{y} - \hat{\mathbf{V}}^{-1}\mathbf{X}\left(\mathbf{X}'\hat{\mathbf{V}}^{-1}\mathbf{X}\right)^{-1}\mathbf{X}'\hat{\mathbf{V}}^{-1}\mathbf{y}$$

Then the variance equations are:

$$tr\left(\hat{\mathbf{V}}^{-1}\mathbf{Z}_j\mathbf{Z}_j'\right) = \mathbf{y}'\hat{\mathbf{P}}\mathbf{Z}_j\mathbf{Z}_j'\hat{\mathbf{P}}\mathbf{y}$$

Generally, these equations must be solved with use of iterative numerical methods.

Example: For a normal distribution, $\mathbf{y} \sim N(\mathbf{X\beta},\mathbf{V} = \sigma_\tau^2 \mathbf{ZZ'} + \sigma^2 \mathbf{I})$, with two variance components, partial derivatives $\dfrac{\delta \log L}{\delta \boldsymbol{\beta}}$, $\dfrac{\delta \log L}{\delta \sigma_\tau^2}$ and $\dfrac{\delta \log L}{\delta \sigma^2}$ are taken and equated to zero, giving the following equations:

$$\left(\mathbf{X'\hat{V}^{-1}X}\right)\tilde{\boldsymbol{\beta}} = \mathbf{X'\hat{V}^{-1}y}$$
$$tr\left(\mathbf{\hat{V}^{-1}}\right) = \left(\mathbf{y} - \mathbf{X}\tilde{\boldsymbol{\beta}}\right)' \mathbf{\hat{V}^{-1}\hat{V}^{-1}}\left(\mathbf{y} - \mathbf{X}\tilde{\boldsymbol{\beta}}\right)$$
$$tr\left(\mathbf{\hat{V}^{-1}ZZ'}\right) = \left(\mathbf{y} - \mathbf{X}\tilde{\boldsymbol{\beta}}\right)' \mathbf{\hat{V}^{-1}ZZ'\hat{V}^{-1}}\left(\mathbf{y} - \mathbf{X}\tilde{\boldsymbol{\beta}}\right)$$

By using $\mathbf{\hat{V}^{-1}}(\mathbf{y} - \mathbf{X\beta}) = \mathbf{\hat{P}y}$ the equations for the variance components are:

$$tr\left(\mathbf{\hat{V}^{-1}}\right) = \mathbf{y'\hat{P}\hat{P}y}$$
$$tr\left(\mathbf{\hat{V}^{-1}ZZ'}\right) = \mathbf{y'\hat{P}ZZ'\hat{P}y}$$

13.3 Restricted Maximum Likelihood Estimation

Variance components are estimated with restricted maximum likelihood (*REML*) by using the residuals after having fitted the fixed effects part of the model. Thus, instead of using a $\mathbf{y}$ data vector, *REML* uses linear combinations of $\mathbf{y}$, say $\mathbf{K'y}$, with $\mathbf{K}$ chosen such that $\mathbf{K'X} = \mathbf{0}$. The $\mathbf{K}$ matrix has $N-1$ independent $\mathbf{k}$ vectors such that $\mathbf{k'X} = \mathbf{0}$.

The transformed model is:

$$\mathbf{K'y} = \mathbf{K'Zu} + \mathbf{K'\varepsilon}$$

If $\mathbf{y}$ has a normal distribution, $\mathbf{y} \sim N(\mathbf{X\beta}, \mathbf{V})$, then because $\mathbf{K'X} = \mathbf{0}$, the distribution of $\mathbf{K'y}$ is $N(0, \mathbf{K'VK})$, that is:

$$E(\mathbf{K'y}) = 0 \text{ and}$$
$$Var(\mathbf{K'y}) = \mathbf{K'VK}$$

The $\mathbf{K'y}$ are linear contrasts and they represent residual deviations from the estimated mean and fixed effects. Following the same logic as for *ML*, the *REML* equations are:

$$tr\left[\left(\mathbf{K'\hat{V}K}\right)^{-1}\mathbf{K'Z}_j\mathbf{Z}_j\mathbf{'K}\right] = \mathbf{y'K}\left(\mathbf{K'\hat{V}K}\right)^{-1}\mathbf{K'Z}_j\mathbf{Z}_j\mathbf{'K}\left(\mathbf{K'\hat{V}K}\right)^{-1}\mathbf{K'y}$$

It can be shown that for any $\mathbf{K}$:

$$\mathbf{K}\left(\mathbf{K'\hat{V}K}\right)^{-1}\mathbf{K'} = \mathbf{\hat{P}}$$

Recall that $\mathbf{\hat{P}} = \mathbf{\hat{V}^{-1}} - \mathbf{\hat{V}^{-1}X}\left(\mathbf{X'\hat{V}^{-1}X}\right)^{-1}\mathbf{X'\hat{V}^{-1}}$ and then:

$$\mathbf{K}\left(\mathbf{K'\hat{V}K}\right)^{-1}\mathbf{K'y} = \mathbf{\hat{P}y} = \mathbf{\hat{V}^{-1}}(\mathbf{y} - \mathbf{X\beta}) = \mathbf{\hat{V}^{-1}y} - \mathbf{\hat{V}^{-1}X'}\left(\mathbf{X'\hat{V}^{-1}X}\right)^{-1}\mathbf{X'\hat{V}^{-1}y}$$

After some rearrangement, the *REML* equations simplify to:

$$tr\left(\mathbf{\hat{P}Z}_j\mathbf{Z}_j\mathbf{'}\right) = \mathbf{y'\hat{P}Z}_j\mathbf{Z}_j\mathbf{'\hat{P}y}$$

Again, these equations must be solved by using iterative numerical methods.

Example: If $\mathbf{y}$ has a normal distribution $\mathbf{y} \sim N(\mathbf{1}\mu, \mathbf{V} = \sigma_\tau^2 \, \mathbf{ZZ'} + \sigma^2\mathbf{I}_N)$, then because $\mathbf{K'X} = \mathbf{0}$, the distribution for $\mathbf{K'y}$ is $N(0, \sigma_\tau^2 \, \mathbf{K'ZZ'K} + \sigma^2 \, \mathbf{K'K})$, that is:

$$E(\mathbf{K'y}) = 0 \text{ and}$$
$$Var(\mathbf{K'y}) = \mathbf{K'VK} = \sigma_\tau^2 \, \mathbf{K'ZZ'K} + \sigma^2 \, \mathbf{K'K}$$

The following equations are obtained:

$$tr\left[\left(\hat{\sigma}_\tau^2\mathbf{K'ZZ'K} + \hat{\sigma}^2\mathbf{K'K}\right)^{-1}\mathbf{K'K}\right] =$$
$$\mathbf{y'K}\left(\hat{\sigma}_\tau^2\mathbf{K'ZZ'K} + \hat{\sigma}^2\mathbf{K'K}\right)^{-1}\mathbf{K'K}\left(\hat{\sigma}_\tau^2\mathbf{K'ZZ'K} + \hat{\sigma}^2\mathbf{K'K}\right)^{-1}\mathbf{K'y}$$
$$tr\left[\left(\hat{\sigma}_\tau^2\mathbf{K'ZZ'K} + \hat{\sigma}^2\mathbf{K'K}\right)^{-1}\mathbf{K'ZZ'K}\right] =$$
$$\mathbf{y'K}\left(\hat{\sigma}_\tau^2\mathbf{K'ZZ'K} + \hat{\sigma}^2\mathbf{K'K}\right)^{-1}\mathbf{K'ZZ'K}\left(\hat{\sigma}_\tau^2\mathbf{K'ZZ'K} + \hat{\sigma}^2\mathbf{K'K}\right)^{-1}\mathbf{K'y}$$

Using:

$$\mathbf{K}\left(\hat{\sigma}_\tau^2\mathbf{K'ZZ'K} + \hat{\sigma}^2\mathbf{K'K}\right)^{-1}\mathbf{K'} = \hat{\mathbf{P}}$$

The two variance equations are:

$$tr\left(\hat{\mathbf{P}}\right) = \mathbf{y'}\hat{\mathbf{P}}\hat{\mathbf{P}}\mathbf{y}$$
$$tr\left(\hat{\mathbf{P}}\mathbf{ZZ'}\right) = \mathbf{y'}\hat{\mathbf{P}}\mathbf{ZZ'}\hat{\mathbf{P}}\mathbf{y}$$

Chapter 14

Concepts of Experimental Design

An experiment can be defined as planned research conducted to obtain new facts, or to confirm or refute the results of previous experiments. An experiment helps a researcher to get an answer to some question or to make an inference about some phenomenon. Most generally, observing, collecting or measuring data can be considered an experiment. In a narrow sense, an experiment is conducted in a controlled environment in order to study the effects of one or more categorical or continuous variables on observations. An experiment is usually planned and can be described in several steps: 1) introduction to the problem; 2) statement of the hypotheses; 3) description of the experimental design; 4) collection of data (running the experiment); 5) analysis of the data resulting from the experiment; and 6) interpretation of the results relative to the hypotheses.

The planning of an experiment begins with an introduction in which a problem is generally stated, and a review of the relevant literature including previous results and a statement of the importance of solving the problem. After that, the objective of the research is stated. The objective should be precise and can be a question to be answered, a hypothesis to be verified, or an effect to be estimated. All further work in the experiment should depend on the stated objective.

The next step is defining the materials and methods. One part of that is choosing and developing an experimental design. An experimental design defines how to obtain data. Data can come from observations of natural processes, or from controlled experiments. It is more efficient and easier to draw conclusions if it is clear what information (data) is sought, and the procedure that will be used to obtain it. This is true for both controlled experiments and observation of natural processes. It is also important to be open to unexpected information which may lead to new conclusions. This is especially true when observing natural processes.

For a statistician, an experimental design is a set of rules used to choose samples from populations. The rules are defined by the researcher himself, and should be determined in advance. In controlled experiments, the experimental design describes how to assign treatments to experimental units, but within the frame of the design must be an element of randomness of treatment assignment. In the experimental design it is necessary to define treatments (populations), size of samples, experimental units, sample units (observations), replications and experimental error. The definition of a population (usually some treatment) should be such that the results of the experiment will be applicable and repeatable. From the defined populations, random and representative samples must be drawn.

The statistical hypotheses usually follow the research hypothesis. Accepting or rejecting statistical hypotheses helps in finding answers to satisfy the objective of the research. In testing statistical hypotheses a statistician uses a statistical model. The statistical model follows the experimental design, often is explained with a mathematical formula, and includes three components: 1) definition of means (expectations); 2) definition of dispersion (variances and covariances); and 3) definition of distribution. Within these three components assumptions and restrictions must be defined in order to be able to design appropriate statistical tests.

Having defined the experimental design, the experiment or data collection is performed. Data collection must be carried out according to the experimental design. Once the data are collected, data analysis follows, which includes performing the statistical analysis, and describing and interpreting the results. The models used in the analysis are determined by the goals of the experiment and its design. Normally, the data analysis should be defined prior to data collection; however, sometimes it can be refined after data collection if the researcher has recognized an improved way of making inferences or identified new facts about the problem. Finally, the researcher should be able to make conclusions to fulfill the objective of the experiment. Conclusions and interpretations should be clear and precise. It is also useful to discuss practical implications of the research and possible future questions relating to similar problems.

14.1 Experimental Units and Replications

An experimental unit is a unit of material to which a treatment is applied. The experimental unit can be an animal, but could also be a group of animals, for example, 10 steers in a pen. The main characteristic of experimental units is that they must be independent of each other. If a treatment is applied to all steers in a pen, obviously the steers are not independent, and that is why the whole pen is considered the experimental unit. The effect of treatment is measured on a sample unit. The sample unit can be identical to the experimental unit or it can be a part of the experimental unit. For example, if we measure weights of independent calves at the age of 6 months, then a calf is a sample and an experimental unit. On the other hand, if some treatment is applied to 10 chicks in a cage, and each chick is weighed, then the cage is the experimental unit, and each chick is a sample unit.

When a treatment is applied to more than one experimental unit, the treatment is replicated. There is often neglected difference between replications, subsamples and repetitions. A characteristic of experimental units is that they are independent of one another. Replications are several experimental units, all treated alike. In some experiments it is impossible to measure the entire experimental unit. It is necessary to select subsamples from the unit. For example, consider an experiment designed to measure the effect of pasture treatments on the protein content in plants. Plots are defined as experimental units, and treatments assigned randomly to those plots. The protein content will not be measured on the whole plant mass from each plot, but rather subsamples will be drawn from each plot. Note that those subsamples are not experimental units and they are not replications, because dependency exists among them. Repetitions are repeated measurements on the same experimental unit. For example, in an experiment for testing the effects of two treatments on milk production of dairy cows, cows are chosen as experimental units. Milk yield can be measured daily for, say, 2 weeks. These single measurements are not replications, but repetitions, repeated measurements of the same experimental units. Obviously, repeated measurements are not independent of each other since they are measured on the same experimental unit.

Often in field research, the experiment is replicated across several years. Also, to test treatments in different environments, an experiment can be replicated at several locations. Those repeats of an experiment in time and space can be regarded as replications. The purpose of such experiments is to extend conclusions over several populations and different environments. Similarly in labs, the whole experiments can be repeated several times, of course with different experimental units, but often even with different technicians, in order to account for environmental or human factors in the experiment.

14.2 Experimental Error

A characteristic of biological material is variability. Recall that in randomly selected experimental samples, total variability can be partitioned to explained and unexplained causes. In terms of a single experimental unit (y_{ij}), each can be expressed simply as:

$$y_{ij} = \hat{\mu}_i + e_{ij}$$

where:

$\hat{\mu}_i$ = the estimated value describing a set of the explained effects i, treatments, farms, years, etc.

e_{ij} = unexplained effect

Therefore, observations y_{ij} differ because they belong to different explained groups i, and because of different unexplained effects e_{ij}. The term $\hat{\mu}_i$ estimates and explains the effects of the group i; however, there is no explanation, in experimental terms, for the differences between experimental units (replicates) within a group. Hence, this variation is often called experimental error. Common measures of experimental error are the mean square or square root of mean square, which are estimates of variance or standard deviation. For the simplest example, if some trait is measured on n experimental units and there is unexplained variability between units, the best estimate of the true value of that trait is the mean of the n measured values. A measure of experimental error can be the mean of the squared deviations of observations from the estimated mean.

In regression or one-way analysis of variance a measure of experimental error is the residual mean square (MS_{RES}), which is a measure of unexplained variability between experimental units after accounting for explained variability (the regression or treatment effect). Recall, that $MS_{RES} = s^2$ is an estimator of the population variance. In more complex designs the mean square for experimental error can be denoted by MS_E. Inferences about treatments or regression require a measure of the experimental error. Replication allows estimation of experimental error, without which there is no way of differentiating random variation from real treatment or regression effects.

Experimental error can consist of two types of errors: systematic and random. Systematic errors are effects which change the measurements under study in a consistent way and can be assigned to some source. They produce bias in estimation. This variability can come from lack of uniformity in conducting the experiment, from uncalibrated instruments, unaccounted temperature effects, biases in using equipment, etc. If they are recognized, correction should be made for their effect. They are particularly problematic if they are not recognized, because they affect measurements in systematic but unknown ways.

Random errors occur due to random, unpredictable, phenomena. They produce variability that cannot be explained. They have an expectation of zero, so over a series of replicates they will cancel out. In biological material there are always random errors in measurements. Their contribution to variance can be characterized by using replications in the experiment. For example, in an experiment with livestock, the individual animals will have different genetic constitution. This is random variability of experimental material.

Measurement error, the degree to which measurements are rounded, is also a source of random error.

Recall the difference between experimental units and sample units, and between replications, subsamples and repeated measurements. Their relationship is important to defining appropriate experimental error to test treatment effects. Recall again that experimental error is characterized by unexplained variability between *experimental units* treated alike. Here are some examples:

Example: The aim of an experiment is to test several dosages of injectable growth hormone for dairy cows. Cows are defined as *experimental units*. The variability among all cows consists of variability due to different growth hormone injections, but also due to unexplained differences between cows even when they are treated alike, which is the *experimental error*. To have a measure of the error it is necessary to have replicates, more than one cow per treatment, in the experiment. The trait milk yield can be measured repeatedly on the same cow. These are *repeated measures*. Although it is possible to take multiple milk samples, cows are still *the experimental units* because treatments are applied to the cows, not individual milk samples. The *experimental error* for testing treatments is still the unexplained variability between cows, not between repeated measurements on the cows.

Second example: The aim of an experiment is to test three rations for fattening steers. The treatments are applied randomly to nine pens each with ten steers. Here, each pen is an *experimental unit*, and pens within treatment are *replications*. Because a single treatment is applied to all animals in a pen, pen is the experimental unit even if animals are measured individually.

14.3 Precision of Experimental Design

When developing a useful experimental design and an appropriate statistical model, all potential sources of variability should be accounted for in the design and analysis. The design must provide sufficient experimental units for adequate power. To determine the appropriate size of the experiment preliminary estimates of variability should be obtained, either from similar experiments or from the literature. The level of significance, α, and power of test should be defined. These will be used to determine the minimum number of replicates sufficient to detect the smallest difference (effect size) of practical importance. Too many replicates incur unnecessary work and cost.

There are differences in meanings between accuracy and precision. Generally, accuracy indicates how close trials are to an aim, and precision how close together trials are to each other. In terms of an experiment, accuracy is often represented by how close the estimated mean of replicated measurements is to the true mean. The closer to the true mean, the more accurate the result. Precision is how close the measurements are to one another regardless of how close they are to the true mean, that is, it explains the repeatability of the results. Figure 14.1 shows the meaning of accuracy and precision of observations when estimating the true mean. Random errors affect precision of an experiment and to a lesser extent its accuracy. Smaller random errors mean greater precision. Systematic errors (bias) affect the accuracy of an experiment, but not the precision. Repeated trials and statistical analysis are of no use in eliminating the effects of systematic errors. In order to have a successful experiment systematic errors must be eliminated and random errors should be as small as possible. In other words, the experimental error must be reduced as much as possible and must be an unbiased estimate of random variability in the populations.

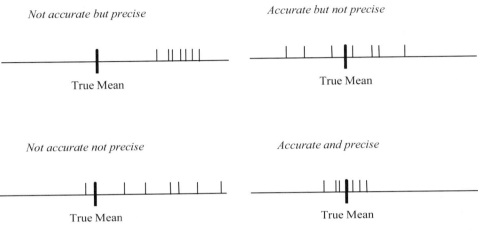

Figure 14.1 Accuracy and precision

In experiments precision is expressed as the amount of information (I):

$$I = \frac{n}{\sigma^2}$$

in which n is the number of observations in a group or treatment, and σ^2 is the variance between units in the population. Just as the estimator of the variance σ^2 is the mean square error $s^2 = MS_E$, the estimator of the amount of information is:

$$I = \frac{n}{MS_E}$$

Note that the reciprocal of I is the square of the estimator of the standard error of the mean ($s_{\bar{y}}$):

$$\frac{1}{I} = \frac{MS_E}{n} = s_{\bar{y}}^2$$

Clearly, more information results in a smaller standard error and estimation of the mean is more precise. More information and greater precision also results in easier detection of possible differences between means. Recall that the probability that an experiment will result in appropriate rejection of the null hypothesis is called power of test. Power is increased by reducing experimental error and/or increasing sample size. As long as experimental units are representative of the population, it is always more beneficial to decrease experimental error by controlling unexplained variability in the experiment than to increase the size of the experiment.

14.4 Controlling Experimental Error

An effective experimental design should: 1) yield unbiased results; 2) have high power and low likelihood of type I error; and 3) be representative of the population to which results will apply. The more the researcher knows about the treatments and experimental material, the easier it is to use appropriate statistical methods. For the experiment to be unbiased treatments must be randomly applied to experimental units. If the treatments are applied in a selective way bias can result. Samples should be drawn randomly, that is, experimental units should be representative of the population from which they originate. If an experiment is conducted with selected samples (for example, superior animals), the experimental error may be smaller than population variance. The consequence is that differences between treatments may be significant in the experiment, but the conclusion may not be applicable to the population. The power of the experiment depends on the number of replicates or degrees of freedom. Experimental error can be reduced by adding more replicates, or by grouping the experimental units into blocks according to sources of variability other than treatments. The treatments must be applied randomly within blocks. For example, treatments can be applied to experimental units on several farms. Farm can be used as a block to explain variation and consequently reduce experimental error.

In controlling the experimental error it is important to choose an optimal experimental design. Recall that the amount of information (I) is defined as:

$$I = \frac{n}{MS_E}$$

The efficiency of two experimental designs can be compared by calculating the relative efficiency (RE) of design 2 to design 1 (with design 2 expected to have an improvement in efficiency):

$$RE = \left(\frac{df_2 + 1}{(df_2 + 3)s_2^2} \right) \bigg/ \left(\frac{df_1 + 1}{(df_1 + 3)s_1^2} \right)$$

where s_1^2 and s_2^2 are experimental error mean squares, and df_1 and df_2 are the error degrees of freedom for designs 1 and 2, respectively. A value of RE close to one indicates no improvement in efficiency, values greater than one indicate that the design 2 is preferred.

The importance of properly conducting experiments is obvious. There is no statistical method that can account for mishandling animals or instruments, record mistakes, cheating, using improper or uncalibrated instruments, etc. That variability is not random and inappropriate conclusions will result from the statistical tests. These 'mistakes' may affect the whole experiment, a particular treatment or group, or even particular experimental units. The least damage will result if they affect the whole experiment (systematic errors). That will influence the estimation of means, but will not influence the estimation of experimental errors and conclusions about the differences between treatments. Mistakes that affect particular treatments lead to confounding. The effect of treatment may be under or over estimated, but again this will not affect experimental error. If mistakes are made in an unsystematic way, only on particular units, experimental error will increase and reduce precision of the experiment.

14.5 Unbalanced and Missing Data

The term 'unbalanced data' usually refers to data from an experiment in which there is an unequal number of observations per level or combination of levels of categorical independent variables. For example, in a study of the effects of treatment and farm on daily gain of pigs, the number of observations (pigs) per combination of farm × treatment can be unequal. Whole levels or combinations of levels may even contain no observations.

Unbalanced data can exist for several reasons: sample sizes may follow the proportions of groups in populations, an unbalanced design may be intended by a researcher, or observations may be lost in the process of collecting, storing or saving data.

Consider the following example. In a study of evaluating the effect of geographical region on milk production of cows, to improve the precision of results cows can be sampled from subpopulations such as year, season of birth, and age of cow. Those effects serve as adjustments allowing more precise evaluation of region means free from these nuisance effects. Naturally, the number of cows in different groups defined by the levels of the subpopulations is unlikely to be equal. If data are collected randomly in a population of cows, then the different group size according to year, season, etc. should follow the proportions in the population, and those proportions in subpopulations or particular groups are likely to be different. Unequal group sizes should not be considered a problem. This type of unbalance does not necessarily mean that observations have been lost.

The following is an example of an intentionally unbalanced design. The objective of the experiment is to test differences among six treatment levels on several farms. Typically, we would allocate all treatments to all farms. We first need to know how many animals are needed to obtain satisfactory power and precision to support our conclusions. However, we can have the same power if only four treatments, randomly chosen, are applied to animals on each farm, with restriction that all combinations of four out of the six treatments are equally applied. Some combinations of treatment × farm are deliberately left without observations, which allows us to use more animals for the remaining treatment × farm combinations. Note that here whole combinations are missing, but the number of observations for available combinations tends to be equal. Again, the unbalance does not imply lost observations or missing data.

In a planned and controlled experiment it is recommended that groups and combinations of groups contain equal numbers of experimental units and observations. Data from such designs have desirable statistical properties and conclusions from the analyses have maximum power and precision. However, in even the most carefully planned experiments something can go 'wrong', animals can get sick or even die, measuring instruments can break or malfunction, for some reason data cannot be collected on some animals, or on some particular day it is impossible to go to the field to collect data, etc. In short, a researcher may need to deal with unexpectedly missing or lost data which were expected to be present according to our rules of sampling or our experimental design. Such missing values may or may not influence estimation and inferences in statistical models. They may be important, and our analysis or even our statistical tools may not be satisfactory because of that lack of information.

There are three general types of missing observations: a) missing completely at random; b) conditionally missing at random; and c) missing not at random.

Missing completely at random denotes missing observations which do not depend on their own values, or on the values of any other observations or variables. For example, a test tube broke and a blood sample was lost, a paper record with cow measurements became dirty and unreadable, measuring equipment malfunctioned, or it was impossible to go on the range to take measurements due to bad weather. An observation missing completely at random is missing independent of its category and value. If data are missing completely at random the analysis and estimated parameters and inferences are still unbiased and valid, because the observed (nonmissing) data are considered a random sample from the complete data.

Conditionally missing at random are those observations missing as a result of some controlling process (usually defined by a classification variable). For example, measuring weight can be more difficult for bulls than for cows, causing missing observations to be related to sex. Within sex, the missing data are lost randomly, so the estimated means for both bulls and cows will be unbiased. The statistical analysis should account for the controlling variable, that is, the categorical variable separating bulls and cows must be defined in the statistical model. Another example of conditionally missing at random is when data are lost because an animal was removed from an experiment due to its performance being outside the expected range (it was an outlier). The data are actually classified to two groups: within the expected limit and outside the limit. Removal of the animal outside the limits will not affect randomness for observations falling within the expected limits. We may consider that animal as belonging to a different population than the rest of the animals and its removal is justified.

Values missing not at random are observations lost due to the value of the missing observation. For example, in estimating mean milk production of ewes from data collected directly by farmers, values for low producing ewes are less likely to be reported than those for high producing ewes. Obviously, the estimated mean will be biased upward toward the more productive ewes.

It can sometimes be difficult to distinguish whether or not observations are missing completely at random. The rule is not simple and depends on the goal and hypotheses of the experiment. For example, in an experiment with a goal of evaluating stress, if a cow is nervous and makes collecting blood samples impossible, that missing observation has not been lost completely at random. It is actually not random at all with respect to stress; however, if the goal is to evaluate milk production only, that particular cow may not be important, and loss of its data can be considered completely at random.

Ideally, we would not have missing data. Good planning and careful conduct of an experiment will largely avoid missing not at random data. Unbalanced data because of observations missing completely at random or conditionally missing at random are best dealt with through likelihood based methods of analysis. If some observations of some variables are missing, that does not mean that the complete record or all repeated measurements of an animal should be dropped. Likelihood based methods can compensate for that. We will further address missing data when explaining particular designs and statistical models, especially those for which it is a common problem.

14.6 Required Number of Replications

A very important factor in planning an experiment is determining the number of replications needed for rejecting the null hypothesis given a true difference between treatments of a specific size. Increasing the number of replications increases the precision of estimates; however, as the number of replicates increases, the experiment may require excessive space, cost and time. When a sufficiently large number of replications are used, any difference can be found statistically significant. The difference, although significant, may be too small to have practical meaning. For example, in an experiment comparing two diets for pigs, a difference in daily gain of several grams may be neither practically nor economically meaningful, although with a sufficiently large experiment, even that difference can be statistically significant.

Recall that in determining the number of replications the following must be considered:

1. The estimate of the variance
2. The effect size of practical importance and which should be found significant
3. The power of test $(1 - \beta)$ or probability of finding significant the effect of practical importance
4. The significance level (α), the probability of a type I error
5. The type of statistical test

For a test of difference between means, the number of replications (n) can be calculated from the following expression:

$$n \geq \frac{\left(z_{\alpha/2} + z_\beta\right)}{\delta^2} 2\sigma^2$$

where:

$z_{\alpha/2}$ = the value of a standard normal variable determined with α probability of type I error

z_β = the value of a standard normal variable determined with β probability of type II error

δ = the difference desired to be found significant

σ^2 = the variance

For more than two treatments, the level of significance must be adjusted for multiple comparisons. An alternative approach is to use a noncentral F distribution. The expression for the power of test for a given level of significance (α) and power $(1 - \beta)$ is used. The power of the test is given by:

$$Power = P\,(F > F_{\alpha,(a-1),(N-a)} = F_\beta\,)$$

using a noncentral F distribution for H_1 with a noncentrality parameter $\lambda = \dfrac{n\sum_i \tau_i^2}{\sigma^2}$, and degrees of freedom $(a - 1)$ and $(na - a)$. Here, τ_i are the treatment effects, σ^2 is the variance of the experimental units, n is the number of replications per treatment, a is the number of treatments and $F_{\alpha,(a-1),(N-a)}$ is the critical value. The necessary parameters can be estimated from samples as follows: $n\sum_i \tau_i^2$ with SS_{TRT} and σ^2 with $s^2 = MS_{RES}$. The noncentrality parameter is:

$$\lambda = \frac{SS_{TRT}}{MS_{RES}}$$

A simple way to determine the number of replications needed is to calculate the power for a series of numbers of replications n. The fewest n for which the calculated power is greater than the desired power is the appropriate number of replications.

14.6.1 SAS Example for the Number of Replications

The following SAS program can be used to calculate the number of replications needed to obtain the desired power. The example data are from the experiment examining the effects of pig diets on daily gains shown in section 11.3.

SAS program:

```
DATA a;
DO n=2 TO 50;
alpha=0.05;
a=3;
sst=728;
df1=a-1;
df2=a*n-a;
sstrt=n*sst;
mse=296.67;
lambda=sstrt/mse;
Fcrit=FINV(1-alpha,df1,df2);
power=1-PROBF(Fcrit,df1,df2,lambda);
OUTPUT;
END;
RUN;
PROC PRINT DATA=a (OBS=1 );
WHERE power > .80;
VAR alpha df1 df2 n power;
RUN;
```

Explanation: The statement, *DO n = 2 TO 50;* indicates computation of the power for 2 to 50 replications. The following are defined: *alpha* = significance level, *a* = number of treatments, $sst = n \sum_i \tau_i^2$ = sum of squared treatment effects (if SS_{TRT} = the treatment sum of squares from the samples, then $sst = SS_{TRT}/n_0$, where n_0 is the number of replications per treatment from the sample), *mse* = residual (error) mean square, *df1* = treatment degrees of freedom, *df2* = residual degrees of freedom, *sstrt* = treatment sum of squares, the estimated variance. Then, the noncentrality parameter (*lambda*), and the critical value (*Fcrit*) for the given degrees of freedom and level of significance are calculated. The critical value is computed with the FINV function, which must have as inputs the cumulative value of percentiles ($1 - \alpha = 0.95$) and degrees of freedom *df1* and *df2*. The *power* is calculated with the PROBF function. This is a cumulative function of the F distribution which needs the critical value, degrees of freedom and the noncentrality parameter *lambda*. As an alternative to *PROBF(Fcrit,df1,df2,lambda)*, the function *CDF('F',Fcrit,df1,df2,lambda)* can be used. This PRINT procedure reports only the least number of replications that results in a power greater than 0.80.

SAS output:

alpha	df1	df2	n	power
0.05	2	15	6	0.88181

To obtain power of test greater than 0.80 with a significance level of 0.05, at least six observations per treatment are required.

The same analysis can be done by using the POWER procedure (SAS version 9.1 and higher).

SAS program:

```
PROC POWER;
   ONEWAYANOVA
      TEST     = OVERALL
      ALPHA    = 0.05
      GROUPMEANS = 280 | 278 | 312
      STDDEV =  17.224
      NPERGROUP = .
      POWER    = 0.8
      ;
RUN;
```

Explanation: The POWER procedure is used with ONEWAYANOVA and TEST = OVERALL statements indicating one-way *ANOVA F* test. Values are from the example of effects of three diets on daily gains (g) in pigs, with sample means of 280, 278 and 312 kg and model residual standard deviation 17.224 kg. Alpha of the test = 0.05, The desired power is 0.8. We put a dot next to the number of observations per group NPERGROUP indicating it is to be calculated by the program.

SAS output:

```
        Overall F Test for One-Way ANOVA

            Fixed Scenario Elements

    Method                      Exact
    Alpha                        0.05
    Group Means             280 278 312
    Standard Deviation          17.224
    Nominal Power                0.8

        Computed N Per Group

        Actual     N Per
        Power      Group
        0.882        6
```

Chapter 15

Blocking

In many experiments it is recognized in advance that some experimental units will respond similarly, regardless of treatments. For example, neighboring plots will be more similar than those further apart, heavier animals will have different gain than lighter ones, measurement on the same day will be more similar than measurements taken on different days, etc. In these cases, experimental designs should be able to account for those known sources of variability by grouping homogeneous units in blocks, resulting in reduced experimental error, and improved possibility of finding a difference between treatments. Consider an experiment with the objective of comparing efficiency of utilization of several feeds for pigs in some region. It is known that several breeds are produced in that area. If it is known that breed does not influence efficiency of feed utilization, then the experiment can be designed in a simple way: randomly choose pigs and feed them with different feeds. However, if an effect of breed exists, variability between pigs will be greater than expected, because of variability between breeds. For a more precise and correct conclusion, it is necessary to determine the breed of each pig. Breeds can then be defined as blocks and pigs within each breed fed different feeds.

15.1 Randomized Complete Block Design

A randomized complete block design is used when experimental units can be grouped in blocks according to some defined source of variability before assignment of treatments. Blocks are groups that are used to explain another part of variability, but the test of their difference is usually not of primary interest. The number of experimental units in each block is equal to the number of treatments, and each treatment is randomly assigned to one experimental unit in each block. The precision of the experiment is increased because variation between blocks is removed in the analysis and the possibility of detecting treatment effects is increased. The characteristics of randomized complete block design are:

1. Experimental units are divided to a treatments and b blocks. Each treatment appears in each block only once.
2. The treatments are assigned to units in each block randomly.

This design is balanced, each experimental unit is grouped according to blocks and treatments, and there is the same number of blocks for each treatment. Data obtained from this design are analyzed with a two-way *ANOVA*, because two ways of grouping, blocks and treatments are defined.

Animals are most often grouped into blocks according to initial weight, body condition, breed, sex, stage of lactation, litter size, etc. Note, block does not necessarily indicate physical grouping. It is important that during the experiment all animals within a block receive the same conditions in everything except treatments. Every change of environment should be changed in all blocks, but must be changed for all animals within a block.

Example: The aim of this experiment was to determine the effect of three treatments (T_1, T_2 and T_3) on average daily gain of steers. Before the start of the experiment 12 steers were weighed, ranked according to weight, and assigned to four blocks. The three heaviest animals were assigned to block I, the three next heaviest to block II, etc. In each block there were three animals to which the treatments were randomly assigned. The identification numbers were assigned to steers in the following manner:

Block	Animal number
I	1,2,3
II	4,5,6,
III	7,8,9
IV	10,11,12

In each block the treatments were randomly assigned to steers.

	Block			
	I	II	III	IV
Steer No. (Treatment)	No. 1 (T_3)	No. 4 (T_1)	No. 7 (T_3)	No. 10 (T_3)
	No. 2 (T_1)	No. 5 (T_2)	No. 8 (T_1)	No. 11 (T_2)
	No. 3 (T_2)	No. 6 (T_3)	No. 9 (T_2)	No. 12 (T_1)

Upon completion of an experiment, for ease of computing, the data can be arranged as in the following table:

	Blocks			
Treatment	I	II	III	IV
T_1	y_{11}	y_{12}	y_{13}	y_{14}
T_2	y_{21}	y_{22}	y_{23}	y_{24}
T_3	y_{31}	y_{32}	y_{33}	y_{34}

Generally for a treatments and b blocks:

	Blocks			
Treatment	I	II	...	b
T_1	y_{11}	y_{12}	...	y_{1b}
T_2	y_{21}	y_{22}	...	y_{2b}
...	...	...	...	...
T_a	y_{a1}	y_{a2}	...	y_{ab}

Here, $y_{11}, y_{12}, ..., y_{34}$, or generally y_{ij}, denote experimental units in treatment i and block j.

The model for a randomized complete block design is:

$$y_{ij} = \mu + \tau_i + \beta_j + \varepsilon_{ij} \qquad i = 1,...,a; \ j = 1,...,b$$

where:

y_{ij} = an observation in treatment i and block j
μ = the overall mean
τ_i = the effect of treatment i
β_j = the fixed effect of block j
ε_{ij} = random error with mean 0 and variance σ^2
a = the number of treatments; b = the number of blocks

15.1.1 Partitioning Total Variability

In a randomized complete block design the total sum of squares can be partitioned to block, treatment and residual sums of squares:

$$SS_{TOT} = SS_{TRT} + SS_{BLK} + SS_{RES}$$

The corresponding degrees of freedom are:

$$(ab - 1) = (a - 1) + (b - 1) + (a - 1)(b - 1)$$

Note that $(a - 1)(b - 1) = ab - a - b + 1$.

Compared to a one-way *ANOVA*, the residual sum of squares in a two-way *ANOVA* is decreased by the block sum of squares. Namely:

$$SS'_{RES} = SS_{BLK} + SS_{RES}$$

where:

SS_{RES} = the two-way residual sum of squares (the experimental error for the randomized complete block design)
SS'_{RES} = the one-way residual sum of squares

The consequence of the decreased residual sum of squares is increased precision in determining possible differences among treatments.

The sums of squares are:

$$SS_{TOT} = \sum_i \sum_j (y_{ij} - \bar{y}..)^2$$

$$SS_{TRT} = \sum_i \sum_j (\bar{y}_{i.} - \bar{y}..)^2 = b \sum_i (\bar{y}_{i.} - \bar{y}..)^2$$

$$SS_{BLK} = \sum_i \sum_j (\bar{y}_{.j} - \bar{y}..)^2 = a \sum_i (\bar{y}_{.j} - \bar{y}..)^2$$

$$SS_{RES} = \sum_i \sum_j (y_{ij} - \bar{y}_{i.} - \bar{y}_{.j} + \bar{y}..)^2$$

The sums of squares can be computed using a short-cut computation:

1. Total sum:

$$\sum_i \sum_j y_{ij}$$

2. Correction for the mean:

$$C = \frac{\left(\sum_i \sum_j y_{ij}\right)^2}{(a)(b)} = \frac{(total\ sum)^2}{total\ number\ of\ observations}$$

3. Total (corrected) sum of squares:

$$SS_{TOT} = \sum_i \sum_j y_{ij}^2 - C = Sum\ of\ all\ squared\ observations\ minus\ C$$

4. Treatment sum of squares:

$$SS_{TRT} = \sum_i \frac{\left(\sum_j y_{ij}\right)^2}{b} - C = Sum\ of\ \frac{(treatment\ sum)^2}{no.of\ observations\ in\ treatment}\ for\ each$$

treatment minus C

Note that the number of observations in a treatment is equal to the number of blocks.

5. Block sum of squares:

$$SS_{BLK} = \sum_j \frac{\left(\sum_i y_{ij}\right)^2}{a} - C = Sum\ of\ \frac{(block\ sum)^2}{no.of\ observations\ in\ block}\ for\ each$$

block minus C.

Note that the number of observations in a block is equal to the number of treatments.

6. Residual sum of squares:

$$SS_{RES} = SS_{TOT} - SS_{TRT} - SS_{BLK}$$

By dividing the sums of squares with corresponding degrees of freedom mean squares are obtained:

Mean square for blocks: $MS_{BLK} = \dfrac{SS_{BLK}}{b-1}$

Mean square for treatments: $MS_{TRT} = \dfrac{SS_{TRT}}{a-1}$

Mean square for residual: $MS_{RES} = \dfrac{SS_{RES}}{(a-1)(b-1)}$

15.1.2 Hypotheses Test – *F* test

The hypotheses of interest are to determine if there are treatment differences. The null hypothesis H_0 and alternative hypothesis H_1 are stated as follows:

H_0: $\tau_1 = \tau_2 = ... = \tau_a$, there are no differences among treatments
H_1: $\tau_i \neq \tau_{i'}$, for at least one pair (i,i') a difference between treatments exists

To test these hypotheses an F statistic can be used which, if H_0 holds, has an F distribution with $(a - 1)$ and $(a - 1)(b - 1)$ degrees of freedom:

$$F = \frac{MS_{TRT}}{MS_{RES}}$$

The residual mean square, MS_{RES}, is also the mean square for experimental error, which is an estimator of the population variance. For an α level of significance H_0 is rejected if $F > F_{\alpha,(a-1),(a-1)(b-1)}$, that is, if the calculated F from the sample is greater than the critical value. The test for blocks is usually not of primary interest, but can be conducted analogously as for the treatments. The calculations can be summarized in an *ANOVA* table:

Source	SS	df	MS = SS/df	F
Block	SS_{BLK}	$b - 1$	MS_{BLK}	MS_{BLK}/MS_{RES}
Treatment	SS_{TRT}	$a - 1$	MS_{TRT}	MS_{TRT}/MS_{RES}
Residual	SS_{RES}	$(a - 1)(b - 1)$	MS_{RES}	
Total	SS_{TOT}	$ab - 1$		

When only one set of treatments is present in each block, the SS_{RES} is the same as the interaction of Blocks × Treatments. The block × treatment interaction is the appropriate experimental error term for treatment. A significant F test for treatments can be thought of as indicating that treatments rank consistently across blocks.

The estimators of treatments means are the sample arithmetic means. Estimation of standard errors depends on whether blocks are random or fixed. For fixed blocks the standard errors of treatment mean estimates are:

$$s_{\bar{y}_{i\cdot}} = \sqrt{\frac{MS_{RES}}{b}}$$

For random blocks the standard errors of estimates of treatment means are:

$$s_{\bar{y}_{i\cdot}} = \sqrt{\frac{\left(MS_{RES} + \hat{\sigma}^2_{BLK}\right)}{b}}$$

where $\hat{\sigma}^2_{BLK} = \dfrac{MS_{BLK} - MS_{RES}}{a}$ = variance component for blocks.

For both fixed and random blocks, the standard errors of estimates of the differences between treatment means are:

$$s_{\bar{y}_{i\cdot} - \bar{y}_{i'\cdot}} = \sqrt{MS_{RES}\left(\frac{2}{b}\right)}$$

Example: The objective of this experiment was to determine the effect of three treatments (T_1, T_2 and T_3) on average daily gain (g/d) of steers. Steers were weighed and assigned to four blocks according to initial weight. Treatments were randomly assigned to three animals in each block. Therefore, a total of 12 animals were used. Data with means and sums are shown in the following table:

| | \multicolumn{4}{c}{Blocks} | | |
	I	II	III	IV	Σ treatments	Treatment means
T_1	826	865	795	850	3336	834
T_2	827	872	721	860	3280	820
T_3	753	804	737	822	3116	779
Σ blocks	2406	2541	2253	2532	9732	
Block means	802	847	751	844		811

Short-cut computations of sums of squares:

1. Total sum:

$$\Sigma_i \Sigma_j y_{ij} = (826 + \ldots\ldots + 822) = 9732$$

2. Correction for the mean:

$$C = \frac{\left(\Sigma_i \Sigma_j y_{ij}\right)^2}{(a)(b)} = \frac{(9732)^2}{12} = 7892652$$

3. Total (corrected) sum of squares:

$$SS_{TOT} = \Sigma_i \Sigma_j y_{ij}^2 - C = (826^2 + \ldots + 822^2) - 7892652 = 7921058 - 7892652$$
$$= 28406$$

4. Treatment sum of squares:

$$SS_{TRT} = \Sigma_i \frac{\left(\Sigma_j y_{ij}\right)^2}{b} - C = \frac{3336^2}{4} + \frac{3280^2}{4} + \frac{3116^2}{4} - 7892652 = 6536$$

5. Block sum of squares:

$$SS_{BLK} = \Sigma_j \frac{\left(\Sigma_i y_{ij}\right)^2}{a} - C = \frac{2406^2}{3} + \frac{2541^2}{3} + \frac{2253^2}{3} + \frac{2532^2}{3} - 7892652 = 18198$$

6. Residual sum of squares:

$$SS_{RES} = SS_{TOT} - SS_{TRT} - SS_{BLK} = 28406 - 6536 - 18198 = 3672$$

The hypotheses are:

H_0: $\tau_1 = \tau_2 = \ldots = \tau_a$, there are no differences among treatments
H_1: $\tau_i \neq \tau_{i'}$, for at least one pair (i,i') a difference between treatments exists

The *ANOVA* table is:

Source	SS	df	MS	F
Block	18198	3	6066	9.91
Treatment	6536	2	3268	5.34
Residual	3672	6	612	
Total	28406	11		

The calculated F is:

$$F = \frac{MS_{TRT}}{MS_{RES}} = \frac{3268}{612} = 5.34$$

The critical value of F for testing treatments for 2 and 6 degrees of freedom and level of significance $\alpha = 0.05$ is $F_{0.05,2,6} = 5.14$ (See Appendix B: Critical values of F distributions). Since the calculated $F = 5.34$ is greater than the critical value, H_0 is rejected indicating that significant differences exist between sample treatment means.

Example: Compute the efficiency of using a randomized block design instead of completely randomized design.

Recall from chapter 14 the efficiency of two experimental designs can be compared by calculating the relative efficiency (*RE*) of design 2 to design 1 (with design 2 expected to have improved in efficiency):

$$RE = \left(\frac{df_2 + 1}{(df_2 + 3)s_2^2} \right) \bigg/ \left(\frac{df_1 + 1}{(df_1 + 3)s_1^2} \right)$$

defining the completely randomized design as design 1, and the randomized block design as design 2; s_1^2 and s_2^2 are experimental error mean squares, and df_1 and df_2 are the error degrees of freedom for the completely randomized design and the randomized block design, respectively.

For the block design:

$$SS_{RES} = 3672; \ s_2^2 = MS_{RES} = 612 \text{ and } df_2 = 6, \ SS_{BLK} = 18198$$

For the completely randomized design:

$$SS'_{RES} = SS_{BLK} + SS_{RES} = 18198 + 3672 = 21870$$
$$df_1 = 9$$
$$s_1^2 = SS'_{RES} / df_2 = 21870 / 9 = 2430$$

The relative efficiency is:

$$RE = \left(\frac{6+1}{(6+3)612} \right) \bigg/ \left(\frac{9+1}{(9+3)2430} \right) = 3.71$$

Since *RE* is much greater than one, the randomized block design is more efficient than the completely randomized design for this experiment.

15.1.3 SAS Example for Block Design

The SAS program for the example of average daily gain of steers is as follows. Recall the data:

		Blocks		
Treatments	I	II	III	IV
T_1	826	865	795	850
T_2	827	872	721	860
T_3	753	804	737	822

SAS program:

```
DATA steer;
  INPUT trt block $ d_gain @@;
  DATALINES;
  1 I 826    1 II 865    1 III 795    1 IV 850
  2 I 827    2 II 872    2 III 721    2 IV 860
  3 I 753    3 II 804    3 III 737    3 IV 822
;
PROC GLM DATA = steer;
CLASS block trt;
MODEL d_gain = block trt/ ;
LSMEANS trt / STDERR PDIFFADJUST=TUKEY;
RUN;
QUIT;
```

Explanation: The GLM procedure was used. The CLASS statement defines categorical (class) variables. The statement: *MODEL d_gain = block trt* defines *d_gain* as the dependent, and *block* and *trt* as the independent variables. The LSMEANS statement calculates least squares means for *trt* which are means corrected on all other effects in the model and possible missing observations. The options after the slash are for computing standard errors and tests of difference between treatment means by using a Tukey test with the adjustment for multiple comparisons.

SAS output:

```
Dependent Variable: d_gain
                        Sum of         Mean
Source            DF    Squares       Square    F Value   Pr > F
Model              5   24734.0000   4946.8000    8.08    0.0122
Error              6    3672.0000    612.0000
Corrected Total   11   28406.0000

R-Square      Coeff Var     Root MSE      d_gain Mean
0.870732      3.050386      24.7386         811.000

Source    DF    Type III SS    Mean Square   F Value   Pr > F
block      3     18198.0000     6066.0000      9.91    0.0097
trt        2      6536.0000     3268.0000      5.34    0.0465
```

```
                    Least Squares Means
              Adjustment for Multiple Comparisons: Tukey


                 d_gain           Standard                      LSMEAN
    trt          LSMEAN            Error          Pr > |t|      Number
     1         834.000000       12.369317          0.0001         1
     2         820.000000       12.369317          0.0001         2
     3         779.000000       12.369317          0.0001         3


          Least Squares Means for effect trt
        Pr > |t| for H0: LSMean(i)=LSMean(j)
            Dependent Variable: d_gain

        i/j           1              2              3
         1                         0.7165         0.0456
         2          0.7165                        0.1246
         3          0.0456         0.1246
```

Explanation: First is the *ANOVA* table for the *Dependent Variable d_gain*. The *Sources* of variability are *Model*, *Error* (residual) and *Corrected Total*. In the table are listed degrees of freedom (*DF*), *Sum of Squares*, *Mean Square*, calculated *F* (*F Value*) and *P* value (*Pr > F*). In the next table the explained sources of variability are partitioned to *block* and *trt*. For *trt*, the calculated *F* and *P* values are 5.34 and 0.0465, respectively. The effect of treatments is significant. At the end of the output are least squares means (*LSMEAN*) with their standard errors (*Standard Error*), and a Tukey test of difference between treatment groups. Least squares means are the means adjusted for other effects in the model. When there are equal numbers of observations per treatment and block × treatment combination, they are identical to arithmetic means. However, when the number of observations differs, the least squares means account for it. *P* values for the differences between least squares means are given. For example, in column 3 and row 1 the *P* value of 0.0456 denotes that the difference between treatments 1 and 3 is significant as it is smaller than 0.05.

15.2 Randomized Block Design – Two or More Units per Treatment and Block

In some situations there will be more experimental units in a block than there are treatments in the experiment. Treatments are repeated in each block. In the previous section there was just one experimental unit per treatment × block combination, and the experimental error was equal to the interaction between treatment × block. Consequently, it was impossible to test any effect of interaction between treatment and block. A way to test the interaction effect is to increase the number of experimental units to at least two per treatment × block combination. Consider again *a* treatments and *b* blocks, but with *n* experimental units per treatment × block combination. Thus, the number of experimental units within each block is (*n a*). Treatments are randomly assigned to those (*n a*) experimental units in each block. Each treatment is assigned to *n* experimental units within each block.

For example, consider an experiment with four blocks, three treatments, and six animals per block, that is, two animals per block × treatment combination. A design could be:

	Blocks			
	I	II	III	IV
	No. 1 (T_3)	No. 7 (T_3)	No. 13 ($T3$)	No. 19 ($T1$)
	No. 2 (T_1)	No. 8 (T_2)	No. 14 (T_1)	No. 20 (T_2)
Animal *No.*	No. 3 (T_3)	No. 9 (T_1)	No. 15 (T_2)	No. 21 (T_3)
(Treatment)	No. 4 (T_1)	No. 10 (T_1)	No. 16 (T_1)	No. 22 (T_3)
	No. 5 (T_2)	No. 11 (T_2)	No. 17 (T_3)	No. 23 (T_2)
	No. 6 (T_2)	No. 12 (T_3)	No. 18 (T_2)	No. 24 (T_1)

Observations can be shown sorted by treatments and blocks:

	Blocks			
Treatment	I	II	III	IV
T_1	y_{111}	y_{121}	y_{131}	y_{141}
	y_{112}	y_{122}	y_{132}	y_{142}
T_2	y_{211}	y_{221}	y_{231}	y_{241}
	y_{212}	y_{222}	y_{232}	y_{242}
T_3	y_{311}	y_{321}	y_{331}	y_{341}
	y_{312}	y_{322}	y_{332}	y_{342}

Here, $y_{111}, y_{121},..., y_{342}$, or generally y_{ijk} denotes experimental unit k in treatment i and block j.

The statistical model is:

$$y_{ijk} = \mu + \tau_i + \beta_j + \tau\beta_{ij} + \varepsilon_{ijk} \qquad i = 1,..., a;\ j = 1,..., b;\ k = 1,..., n$$

where:

y_{ijk} = observation k in treatment i and block j
μ = the overall mean
τ_i = the effect of treatment i
β_j = the effect of block j
$\tau\beta_{ij}$ = the interaction effect of treatment i and block j
ε_{ijk} = random error with mean 0 and variance σ^2
a = the number of treatments; b = the number of blocks; n = the number of observations in each treatment × block combination

15.2.1 Partitioning Total Variability and Test of Hypotheses

Again, the total variability is partitioned to sources of variability. In the samples, the total sum of squares can be partitioned to block sum of squares, treatment sum of squares, interaction sum of squares and residual sum of squares:

$$SS_{TOT} = SS_{TRT} + SS_{BLK} + SS_{TRT \times BLK} + SS_{RES}$$

The corresponding degrees of freedom are:

$$(abn - 1) = (a - 1) + (b - 1) + (a - 1)(b - 1) + ab(n - 1)$$

The sums of squares are:

$$SS_{TOT} = \sum_i \sum_j \sum_k (y_{ijk} - \bar{y}...)^2$$

$$SS_{TRT} = \sum_i \sum_j \sum_k (\bar{y}_i.. - \bar{y}...)^2 = bn \sum_i (\bar{y}_i.. - \bar{y}...)^2$$

$$SS_{BLK} = \sum_i \sum_j \sum_k (\bar{y}._j. - \bar{y}...)^2 = an \sum_i (\bar{y}._j. - \bar{y}...)^2$$

$$SS_{TRT \times BLK} = n \sum_i \sum_j (\bar{y}_{ij}. - \bar{y}...)^2 - SS_{BLK} - SS_{TRT}$$

$$SS_{RES} = \sum_i \sum_j \sum_k (y_{ijk} - \bar{y}_{ij}.)^2$$

The sums of squares can be computed by using the short-cut computations:

1. Total sum:

$$\sum_i \sum_j \sum_k y_{ijk}$$

2. Correction for the mean:

$$C = \frac{\left(\sum_i \sum_j \sum_k y_{ijk} \right)^2}{abn}$$

3. Total (corrected) sum of squares:

$$SS_{TOT} = \sum_i \sum_j \sum_k (y_{ijk})^2 - C$$

4. Treatment sum of squares:

$$SS_{TRT} = \sum_i \frac{\left(\sum_j \sum_k y_{ijk} \right)^2}{nb} - C$$

5. Block sum of squares:

$$SS_{BLK} = \sum_j \frac{\left(\sum_i \sum_k y_{ijk} \right)^2}{na} - C$$

6. Interaction sum of squares:

$$SS_{TRT \times BLK} = \sum_i \sum_j \frac{\left(\sum_k y_{ijk} \right)^2}{n} - SS_{TRT} - SS_{BLK} - C$$

7. Residual sum of squares:

$$SS_{RES} = SS_{TOT} - SS_{TRT} - SS_{BLK} - SS_{TRT \times BLK}$$

Dividing sums of squares by appropriate degrees of freedom, gives the mean squares:

Mean square for blocks: $MS_{BLK} = \dfrac{SS_{BLK}}{b-1}$

Mean square for treatments: $MS_{TRT} = \dfrac{SS_{TRT}}{a-1}$

Mean square for interaction: $MS_{TRT \times BLK} = \dfrac{SS_{TRT \times BLK}}{(a-1)(b-1)}$

Mean square for residuals: $MS_{RES} = \dfrac{SS_{RES}}{ab(n-1)}$

The sums of squares, degrees of freedom, and mean squares can be presented in an *ANOVA* table:

Source	SS	df	MS = SS/df
Block	SS_{BLK}	$b-1$	MS_{BLK}
Treatment	SS_{TRT}	$a-1$	MS_{TRT}
Treatment × Block	$SS_{TRT \times BLK}$	$(a-1)(b-1)$	$MS_{TRT \times BLK}$
Residual	SS_{RES}	$ab(n-1)$	MS_{RES}
Total	SS_{TOT}	$abn-1$	

The hypotheses about block × treatment interactions in the population are:

$H_0: \tau\beta_{11} = \tau\beta_{12} = \ldots = \tau\beta_{ab}$
$H_1: \tau\beta_{ij} \neq \tau\beta_{i'j'}$ for at least one pair $(ij, i'j')$

The hypotheses about treatment effects are:

$H_0: \tau_1 = \tau_2 = \ldots = \tau_a$, no treatment effect
$H_1: \tau_i \neq \tau_{i'}$, for at least one pair (i, i'), a difference between treatments exists

Recall that within blocks the treatments are assigned randomly on experimental units, and each animal is an experimental unit. Testing of hypotheses depends on whether blocks are defined as random or fixed. A block is defined as fixed if there is a small (finite) number of blocks and they represent distinct populations. A block is defined as random if blocks are considered a random sample from a population of blocks.

When blocks are fixed and a block × treatment interaction is fitted, it is necessary to test the interaction first. If the effect of the interaction is significant the test for the main treatment effects is meaningless; however, if the treatment mean square is large compared to the interaction mean square, it indicates that there is little reranking among treatments across blocks. The effect of block × treatment interaction is also fixed and it is possible to test the difference of estimates of particular combinations. On the contrary, when blocks are random the interaction is also assumed to be random, and it is hard to quantify different effects of treatments among blocks. If there is significant interaction, then it serves as an experimental error term in testing the difference among treatments.

The following table presents expected mean squares and appropriate tests of the effect of interaction and treatments when blocks are defined as fixed or random.

	Fixed blocks		Random blocks	
Source	E(MS)	F	E(MS)	F
Block	$\sigma^2 + \dfrac{a\sum_j \beta_j^2}{b-1}$	$\dfrac{MS_{BLK}}{MS_{RES}}$	$\sigma^2 + n\sigma_{TRT \times BLK}^2 + an\sigma_{BLK}^2$	-
Treatment	$\sigma^2 + \dfrac{b\sum_j \tau_j^2}{a-1}$	$\dfrac{MS_{TRT}}{MS_{RES}}$	$\sigma^2 + n\sigma_{TRT \times BLK}^2 + \dfrac{b\sum_j \tau_j^2}{a-1}$	$\dfrac{MS_{TRT}}{MS_{TRT \times BLK}}$
Trt × Block	$\sigma^2 + \dfrac{b\sum_i \sum_j (\tau\beta)_{ij}^2}{(a-1)(b-1)}$	$\dfrac{MS_{TRT \times BLK}}{MS_{RES}}$	$\sigma^2 + n\sigma_{TRT \times BLK}^2$	$\dfrac{MS_{TRT \times BLK}}{MS_{RES}}$
Residual	σ^2		σ^2	

σ_{BLK}^2 and $\sigma_{TRT \times BLK}^2$ are variance components for block and interaction.

If there is no evidence that a treatment × block interaction exists, the model can be reduced to include only the effects of blocks and treatments. The appropriate experimental error term for testing the effect of treatments will consist of the combined interaction and residual from the model shown above.

Estimators of the populations treatment means are the arithmetic means of treatment groups $\bar{y}_{i..}$, and estimators of the interaction means are the samples arithmetic means $\bar{y}_{ij.}$.

Estimation of the standard errors depend on whether blocks are random or fixed. For fixed blocks the standard errors of estimators of the treatment means are:

$$s_{\bar{y}_{i..}} = \sqrt{\frac{MS_{RES}}{nb}}$$

Standard errors of estimators of interaction means are:

$$s_{\bar{y}_{ij.}} = \sqrt{\frac{MS_{RES}}{n}}$$

Standard errors of estimators of the differences between treatment means are:

$$s_{\bar{y}_{i..}-\bar{y}_{i'..}} = \sqrt{2\left(\frac{MS_{RES}}{nb}\right)}$$

Standard errors of estimators of the differences between interaction means are:

$$s_{\bar{y}_{ij.}-\bar{y}_{i'j'.}} = \sqrt{2\left(\frac{MS_{RES}}{n}\right)}$$

For random blocks standard errors of estimators of the treatment means are:

$$s_{\bar{y}_{i..}} = \sqrt{\frac{MS_{RES} + n\hat{\sigma}_{BLK}^2 + n\hat{\sigma}_{TRT \times BLK}^2}{nb}}$$

where $\hat{\sigma}_{BLK}^2$ and $\hat{\sigma}_{TRT \times BLK}^2$ are the estimates of the variance components for block and interaction.

Standard errors of estimators of the differences between treatment means are:

$$s_{\bar{y}_{i}..-\bar{y}_{i'}..} = \sqrt{2\left(\frac{MS_{RES} + n\hat{\sigma}^2_{TRT \times BLK}}{nb}\right)}$$

Example: Recall that the objective of the experiment previously described was to determine the effect of three treatments (T_1, T_2 and T_3) on average daily gain of steers, and four blocks were defined. However, in this example six animals are assigned to each block. Therefore, a total of 4×3×2 = 24 steers were used. Treatments were randomly assigned to steers within block. The data are as follows:

| Treatments | Blocks | | | |
	I	II	III	IV
T_1	826	864	795	850
	806	834	810	845
T_2	827	871	729	860
	800	881	709	840
T_3	753	801	736	820
	773	821	740	835

Short-cut computations of sums of squares:

1. Total sum:

$$\sum_i \sum_j \sum_k y_{ijk} = (826 + 806 + \ldots\ldots + 835) = 19426$$

2. Correction for the mean:

$$C = \frac{\left(\sum_i \sum_j \sum_k y_{ijk}\right)^2}{abn} = \frac{19426^2}{24} = 15723728.17$$

3. Total (corrected) sum of squares:

$$SS_{TOT} = \sum_i \sum_j \sum_k (y_{ijk})^2 - C = 826^2 + 806^2 + \ldots + 835^2 = 15775768 - 15723728.17$$
$$= 52039.83$$

4. Treatment sum of squares:

$$SS_{TRT} = \sum_i \frac{\left(\sum_j \sum_k y_{ijk}\right)^2}{nb} - C = \frac{6630^2}{8} + \frac{6517^2}{8} + \frac{6279^2}{8} - 15723728.17 = 8025.58$$

5. Block sum of squares:

$$SS_{BLK} = \sum_j \frac{\left(\sum_i \sum_k y_{ijk}\right)^2}{na} - C = \frac{4785^2}{6} + \frac{5072^2}{6} + \frac{4519^2}{6} + \frac{5050^2}{6} - 15723728.17 = 33816.83$$

6. Interaction sum of squares:

$$SS_{TRT \times BLK} = \sum_i \sum_j \frac{\left(\sum_k y_{ijk}\right)^2}{n} - SS_{TRT} - SS_{BLK} - C =$$

$$\frac{(826+806)^2}{2} + \frac{(864+871)^2}{2} + \ldots + \frac{(820+835)^2}{2} - 8025.58 - 33816.83 - 15723728.17 = 8087.42$$

7) Residual sum of squares:

$$SS_{RES} = SS_{TOT} - SS_{TRT} - SS_{BLK} - SS_{TRT \times BLK}$$
$$= 52039.83 - 8025.58 - 33816.83 - 8087.42 = 2110.00$$

ANOVA table:

Source	SS	df	MS
Block	33816.83	3	11272.28
Treatment	8025.58	2	4012.79
Treatment x Block	8087.42	(2)(3) = 6	1347.90
Residual	2110.00	(3)(4)(2 − 1) = 12	175.83
Total	52039.83	23	

Tests for interaction and treatment effects when blocks are fixed include:

F test for interaction:

$$F = \frac{1347.90}{175.83} = 7.67$$

F test for treatments:

$$F = \frac{4012.79}{175.83} = 22.82$$

The critical value for testing the interaction is $F_{0.05,6,12} = 3.00$, and for testing treatments is $F_{0.05,2,12} = 3.89$ (see Appendix B: Critical values of F distributions). Thus, at the 0.05 level of significance, H_0 is rejected for both treatments and interaction. This indicates that there is an effect of treatments and that the treatment effects are different in different blocks. It is useful to further compare the magnitude of treatment effects against the block × treatment interaction. If the ratio is large it indicates that there is little reranking of treatments among blocks. For this example the ratio is $\frac{4012.79}{1347.90} = 2.98$. This ratio is not large, thus although there is an effect of treatment compared to residual, this effect is not large compared to the interaction. Thus, the effect of this treatment is likely to differ depending on the initial weights of steers to which it is applied.

Tests for interaction and treatment effects when blocks are random include:

F test for interaction:

$$F = \frac{1347.90}{175.83} = 7.67$$

F test for treatments:

$$F = \frac{4012.79}{1347.90} = 2.98$$

Note when blocks are defined as random there is no significant effect of treatments, since the critical value for this test is $F_{0.05,2,6} = 5.14$.

15.2.2 SAS Example for Two or More Experimental Units per Block × Treatment

The SAS program for the example of daily gain of steers with two experimental units per treatment by block combination is as follows. Two approaches will be presented: blocks defined as fixed using the GLM procedure and blocks defined as random using the MIXED procedure. Recall the data:

| Treatments | Blocks | | | |
	I	II	III	IV
T_1	826	864	795	850
	806	834	810	845
T_2	827	871	729	860
	800	881	709	840
T_3	753	801	736	820
	773	821	740	835

SAS program:

```
DATA d_gain;
INPUT trt  block $ d_gain @@;
DATALINES;
1     I 826   1     I 806   1 II 864   1 II 834
1 III 795   1 III 810   1 IV 850   1 IV 845
2     I 827   2     I 800   2 II 871   2 II 881
2 III 729   2 III 709   2 IV 860   2 IV 840
3     I 753   3     I 773   3 II 801   3 II 821
3 III 736   3 III 740   3 IV 820   3 IV 835
;
PROC GLM  DATA = d_gain;
CLASS block trt;
MODEL d_gain = block trt block*trt/;
LSMEANS trt / STDERR  PDIFF ADJUST=TUKEY;
LSMEANS block*trt / STDERR PDIFF ADJUST=TUKEY;
RUN;
QUIT;
```

```
PROC MIXED  DATA = d_gain;
CLASS block trt;
MODEL d_gain = trt/;
RANDOM block block*trt;
LSMEANS trt / DIFF ADJUST=TUKEY;
RUN;
```

Explanation: The GLM procedure was used to analyze the example with fixed blocks. The CLASS statement defines the categorical (class) variables. The MODEL statement defines *d_gain* as the dependent, and *block* and *trt* as the independent variables. Also, the *block*trt* interaction is defined. The LSMEANS statement calculates least squares means corrected on all other effects in the model. The options after the slash are for computing standard errors and tests of difference between treatment means by using a Tukey test. The second LSMEANS statement is for the interaction of *block*trt*. The test of difference can be done in the same way as for *trt*. It is not shown here because of the length of the output.

The MIXED procedure was used to analyze the example with blocks considered random. Most of the statements are similar to those in the GLM procedure, except that the model contains only *trt* and the RANDOM statement defines *block* and *block*trt* interaction as random effects. No least squares means of the random block *trt* effect can be requested.

SAS output *of the GLM procedure for fixed blocks:*

Dependent Variable: d_gain

Source	DF	Sum of Squares	Mean Square	F Value	Pr > F
Model	11	49929.83333	4539.07576	25.81	<.0001
Error	12	2110.00000	175.83333		
Corrected Total	23	52039.83333			

R-Square	Coeff Var	Root MSE	d_gain Mean
0.959454	1.638244	13.26022	809.4167

Source	DF	Type III SS	Mean Square	F Value	Pr > F
block	3	33816.83333	11272.27778	64.11	<.0001
trt	2	8025.58333	4012.79167	22.82	<.0001
block*trt	6	8087.41667	1347.90278	7.67	0.0015

Least Squares Means
Adjustment for Multiple Comparisons: Tukey

trt	d_gain LSMEAN	Standard Error	Pr > |t|	LSMEAN Number
1	828.750000	4.688194	<.0001	1
2	814.625000	4.688194	<.0001	2
3	784.875000	4.688194	<.0001	3

```
       Least Squares Means for Effect trt
    t for H0: LSMean(i)=LSMean(j) / Pr > |t|
           Dependent Variable: d_gain


 i/j          1            2            3
  1                     0.1251       <.0001
  2        0.1251                    0.0020
  3        <.0001       0.0020
```

```
                 Least Squares Means

                   d_gain      Standard
   block   trt     LSMEAN        Error     Pr > |t|
    I       1    816.000000    9.376389    <.0001
    I       2    813.500000    9.376389    <.0001
    I       3    763.000000    9.376389    <.0001
    II      1    849.000000    9.376389    <.0001
    II      2    876.000000    9.376389    <.0001
    II      3    811.000000    9.376389    <.0001
    III     1    802.500000    9.376389    <.0001
    III     2    719.000000    9.376389    <.0001
    III     3    738.000000    9.376389    <.0001
    IV      1    847.500000    9.376389    <.0001
    IV      2    850.000000    9.376389    <.0001
    IV      3    827.500000    9.376389    <.0001
```

Explanation: Following a summary of class level information (not shown) is an *ANOVA* table for the *Dependent Variable d_gain*. The *Sources* of variability are *Model, Error* (residual) and *Corrected Total*. The descriptive statistics listed next include the coefficient of determination (*R-Square = 0.959454*), the coefficient of variation (*Coeff Var = 1.638244*), standard deviation (*Root MSE = 13.26022*) and mean of the dependent variable (*d-gain Mean = 809.4167*). In the next table the explained sources of variability are partitioned to *block, trt* and *block*trt*. In the table are listed the degrees of freedom (*DF*), Sums of Squares (*Type III SS*), *Mean Square*, calculated *F* (*F Value*) and *P* value (*Pr > F*). For *trt*, the calculated *F* and *P* values are 22.82 and <0.0001, respectively. The effect of the interaction of *block*trt* is significant (*F* and *P* values are 7.67 and 0.0015, respectively). Next in the output is a table of least squares means (*LSMEAN*) with their standard errors (*Standard Error*), and then an array of Tukey tests between treatment groups. These indicate that *treatment* 1 is different than *treatment* 3 ($P<0.0001$) and *treatment* 2 is different than *treatment* 3 ($P=0.002$), but *treatment* 1 and *treatment* 2 are not significantly different ($P=0.1251$). The final table of the output shows the *block * trt LSMEAN* and their *Standard Error(s)*. The array of Tukey tests for differences between pairs of least squares means of treatment * block combination is not shown.

SAS output *of the MIXED procedure for random blocks:*

```
           Covariance Parameter Estimates

              Cov Parm        Estimate
              block           1654.06
              block*trt        586.03
              Residual         175.83

           Type 3 Tests of Fixed Effects

      Effect          Num DF   Den DF   F Value   Pr > F
      trt               2        6       2.98     0.1264

                 Least Squares Means

                        Stand
   Effect   trt   Est    Error   DF   t Val   Pr>|t|
   trt       1   828.75  24.1247  6   34.35   <.0001
   trt       2   814.62  24.1247  6   33.77   <.0001
   trt       3   784.87  24.1247  6   32.53   <.0001

          Differences of Least Squares Means

                     Stand
Effect  tr_tr   Est   Error   DF  t Val  Pr > |t|  Adjustment    Adj P
trt      1  2  14.125  18.3569  6  0.77   0.4708   Tukey-Kramer  0.7339
trt      1  3  43.875  18.3569  6  2.39   0.0540   Tukey-Kramer  0.1174
trt      2  3  29.750  18.3569  6  1.62   0.1562   Tukey-Kramer  0.3082
```

Explanation: The MIXED procedure gives (co)variance components (*Covariance Parameter Estimates*) and *F* tests for fixed effects (*Type 3 Tests of Fixed Effects*). In the table titled *Least Squares Means* are *Estimates* with *Standard Errors*. In the table *Differences of Least Squares Means* are listed the differences between all possible pairs of treatment levels (*Estimate*). Further, those differences are tested using the Tukey-Kramer procedure, which adjusts tests for multiple comparison and unbalanced designs. Thus, the correct *P* value is the adjusted *P* value (*Adj P*). For example, the *P* value for the difference between treatments 1 and 3 is 0.1174. The MIXED procedure calculates appropriate standard errors for the least squares means and differences between them.

15.3 Power of Test

Power of test for the randomized block design can be calculated in a manner similar to that shown for the one-way analysis of variance by using central and noncentral *F* distributions. Recall that if H_0 holds, then the *F* test statistic follows a central *F* distribution with corresponding degrees of freedom. However, if H_1 holds, then the *F* statistic has a noncentral *F* distribution with a noncentrality parameter $\lambda = \dfrac{SS_{TRT}}{MS_{RES}}$ and corresponding degrees of freedom. Here, SS_{TRT} denotes the treatment sum of squares, and MS_{RES} denotes the residual mean square. The power is a probability:

$$Power = P\left(F > F_{\alpha, df1, df2} = F_\beta\right)$$

using a noncentral F distribution for H_1. Here, α is the level of significance, $df1$ and $df2$ are degrees of freedom for treatment and residual, respectively, and $F_{\alpha,df1,df2}$ is the critical value.

15.3.1 SAS Example for Calculating Power

Example: Calculate the power of test of the example examining the effects of three treatments on average daily gain of steers from section 15.1.2.

The *ANOVA* table was:

Source	SS	df	MS	F
Block	18198	3	6066	9.91
Treatment	6536	2	3268	5.34
Residual	3672	6	612	
Total	28406	11		

The calculated F value was:

$$F = \frac{MS_{TRT}}{MS_{RES}} = \frac{3268}{612} = 5.34$$

The power of test is:

$$Power = P\,(F > F_{0.05,\,2,12} = F_\beta\,)$$

using a noncentral F distribution for H_1. The estimated noncentrality parameter is:

$$\lambda = \frac{SS_{TRT}}{MS_{RES}} = \frac{6536}{612} = 10.68$$

Using the noncentral F distribution with 2 and 12 degrees of freedom and the noncentrality parameter $\lambda = 10.68$, the power is 0.608. The power for blocks can be calculated in a similar manner, but is usually not of primary interest.

To compute the power of test with SAS, the following statements are used:

```
DATA a;
alpha=0.05;
df1=2;
df2=6;
sstrt=6536;
mse=612;
lambda=sstrt/mse;
Fcrit=FINV(1-alpha,df1,df2);
power=1-CDF('F',Fcrit,df1,df2,lambda);
PROC PRINT;
RUN;
```

Explanation: First, the following are defined: *alpha* = significance level, *df1* = treatment degrees of freedom, *df2* = residual degrees of freedom, *sstrt* = treatment sum of squares, *mse* = residual (error) mean square, the estimated variance. Then, the noncentrality parameter (*lambda*) and the critical value (*Fcrit*) for the given degrees of freedom and level of significance are calculated. The critical value is computed by using the FINV function, which must have as input the cumulative value of percentiles $(1 - \alpha = 0.95)$ and degrees of freedom, *df1* and *df2*. The *power* is calculated by using the CDF function. This is a cumulative function of the F distribution which needs as input the critical value, degrees of freedom and the noncentrality parameter *lambda*. As an alternative to using *CDF('F',Fcrit,df1,df2,lambda)*, the function *PROBF(Fcrit,df1,df2,lambda)* can be used. The PRINT procedure gives the following SAS output:

```
alpha    df1    df2    sstrt    mse    lambda     Fcrit      power
0.05      2      6     6536     612   10.6797    5.14325    0.60837
```

Thus, the power is 0.608.

15.4 Missing data in Randomized Block Designs

In chapter 14 definitions and examples of missing data were introduced. Here we will assume that missing observations are random, and some issues of computation will be shown. To deal with missing data, likelihood-based methods are preferred. Two cases concerning missing observations in randomized block designs will be discussed: a) when some observations are missing, but all treatment × block combinations have some observations, and b) when all observations in a treatment × block combination are missing (missing cell). Analysis of data with SAS will be shown on the example of daily gain of steers with two experimental units per treatment by block combination. The same data were used in section 15.2.1 except with some observations discarded to show analyses with missing data.

Some missing observations, but no missing cells

The data are as follows:

Treatments	Blocks			
	I	II	III	IV
T_1	806	864 834	795 810	850 845
T_2	827 800	871 881	729 709	860 840
T_3	753 773	801 821	736 740	820

SAS program for fixed blocks when some observations are missing, but with no missing treatment × block combinations

SAS program:

```
DATA d_gain;
INPUT trt  block $ d_gain @@;
DATALINES;
1    I   .    1    I 806    1 II 864    1 II 834
1 III 795    1 III 810    1 IV 850    1 IV 845
2    I 827    2    I 800    2 II 871    2 II 881
2 III 729    2 III 709    2 IV 860    2 IV 840
3    I 753    3    I 773    3 II 801    3 II 821
3 III 736    3 III 740    3 IV 820    3 IV  .
;

PROC MIXED DATA = d_gain;
CLASS block trt;
MODEL d_gain = block trt block*trt /;
LSMEANS trt / DIFF ADJUST=TUKEY;
LSMEANS block*trt / DIFF ADJUST=TUKEY;
RUN;
```

Explanation: The symbol '.' in the input data denotes a missing observation. The MIXED procedure was used because by default it uses restricted maximum likelihood in estimating parameters, and is thus preferred for analyzing data with missing observations. Note that for fixed blocks, the GLM procedure can also be used with the same syntax as when no data are missing; however, the MIXED procedure is generally the procedure of choice.

SAS output:

Covariance Parameter Estimates

Cov Parm	Estimate
Residual	179.75

Type 3 Tests of Fixed Effects

Effect	Num DF	Den DF	F Value	Pr > F
block	3	10	58.60	<.0001
trt	2	10	18.02	0.0005
block*trt	6	10	7.31	0.0033

```
                    Least Squares Means
                             Standard
Effect      block   trt   Estimate    Error    DF    t Value    Pr > |t|
trt                  1     826.25     5.2996    10    155.91     <.0001
trt                  2     814.63     4.7401    10    171.86     <.0001
trt                  3     783.00     5.2996    10    147.75     <.0001
block*trt    I       1     806.00    13.4071    10     60.12     <.0001
block*trt    I       2     813.50     9.4802    10     85.81     <.0001
block*trt    I       3     763.00     9.4802    10     80.48     <.0001
block*trt    II      1     849.00     9.4802    10     89.55     <.0001
block*trt    II      2     876.00     9.4802    10     92.40     <.0001
block*trt    II      3     811.00     9.4802    10     85.55     <.0001
block*trt    III     1     802.50     9.4802    10     84.65     <.0001
block*trt    III     2     719.00     9.4802    10     75.84     <.0001
block*trt    III     3     738.00     9.4802    10     77.85     <.0001
block*trt    IV      1     847.50     9.4802    10     89.40     <.0001
block*trt    IV      2     850.00     9.4802    10     89.66     <.0001
block*trt    IV      3     820.00    13.4071    10     61.16     <.0001
```

Explanation: An excerpt of the output is presented. The form of the output is the same as when no data are missing; however, a couple points should be emphasized. *Type 3* tests denote that the test for each effect in the model is corrected for all other effects in the model. Degrees of freedom for *F* tests (*Num DF* and *Den DF*, denote numerator and denominator degrees of freedom, respectively) are the same as when no data are missing since there is at least one observation in each *block*trt* combination. Least squares means are means adjusted for other effects in the model. Due to the missing observations, their values are different than arithmetic means. Standard errors of least squares means differ between treatments because of different number of observations per treatment. The same is true for the treatment × block combinations. The test of difference including Tukey adjustment for multiple comparison (following / *DIFF ADJUST = TUKEY* options) is not shown here, but standard errors again differ due to different number of observations.

SAS program for random blocks with missing observations, but with no missing treatment × block combinations

SAS program:

```
PROC MIXED DATA = d_gain;
CLASS block trt;
MODEL d_gain = trt / DDFM = KENWARDROGER;
RANDOM block block*trt;
LSMEANS trt / DIFF ADJUST=TUKEY;
RUN;
```

Explanation: The program is the same as that without missing data except for the *DDFM = KENWARDROGER* option in the MODEL statement. This is Kenward-Rogers estimation of degrees of freedom which is useful when accounting for possible missing values, and when the mean square of the error term or standard errors are linear combinations of variances.

SAS output:

```
Covariance Parameter Estimates

    Cov Parm        Estimate
    block           1622.28
    block*trt        586.48
    Residual         177.93

        Type 3 Tests of Fixed Effects

            Num      Den
    Effect   DF       DF    F Value    Pr > F
    trt       2      6.2      2.95     0.1261

            Least Squares Means

                        Standard
Effect   trt   Estimate   Error     DF   t Value   Pr > |t|
trt       1     826.50    24.0826   4.57   34.32    <.0001
trt       2     814.62    23.9672   4.49   33.99    <.0001
trt       3     782.70    24.0826   4.57   32.50    <.0001

            Differences of Least Squares Means

                        Standard
Effect  trt  _trt  Estimate  Error    DF   t Value  Pr>|t|  Adjust   Adj P
trt      1    2    11.8782   18.5274  6.14   0.64    0.5446  Tuk-Kr   0.8037
trt      1    3    43.8008   18.6772  6.32   2.35    0.0553  Tuk-Kr   0.1221
trt      2    3    31.9226   18.5274  6.14   1.72    0.1345  Tuk-Kr   0.2702
```

Explanation: An excerpt of the output is presented. Degrees of freedom for *F* tests (*Den DF*, denominator degrees of freedom), and the adjusted Tukey test for the differences between least squares means are corrected using Kenward-Rogers method. The degrees of freedom without correction would be 6. Here the correction was not considerable, but it is recommended that the correction is used, especially for more complicated models.

All observations in a treatment × block combination are missing (*missing cells*)

The data are as follows:

Treatments	Blocks			
	I	II	III	IV
T_1		864	795	850
		834	810	845
T_2	827	871	729	860
	800	881	709	840
T_3	753	801	736	
	773	821	740	

SAS program for fixed blocks, main effects and interaction with all observations in a treatment × block combination missing

SAS program:

```
DATA d_gain;
INPUT trt  block $ d_gain @@;
DATALINES;
1    I   .    1    I   .    1 II 864    1 II 834
1 III 795    1 III 810    1 IV 850    1 IV 845
2    I 827    2    I 800    2 II 871    2 II 881
2 III 729    2 III 709    2 IV 860    2 IV 840
3    I 753    3    I 773    3 II 801    3 II 821
3 III 736    3 III 740    3 IV  .     3 IV  .
;


PROC MIXED DATA = d_gain;
CLASS block trt;
MODEL d_gain = block trt block*trt /;
LSMEANS trt / DIFF ADJUST=TUKEY;
LSMEANS block*trt / DIFF ADJUST=TUKEY;
RUN;
```

Explanation: The SAS program is as before.

SAS output:

Type 3 Tests of Fixed Effects

Effect	Num DF	Den DF	F Value	Pr > F
block	3	10	53.77	<.0001
trt	2	10	19.07	0.0004
block*trt	4	10	10.61	0.0013

Least Squares Means

Effect	block	trt	Estimate	Standard Error	DF	t Value	Pr > |t|
trt		1	Non-est	.	.	.	.
trt		2	814.63	4.7401	10	171.86	<.0001
trt		3	Non-est	.	.	.	.
block*trt	I	2	813.50	9.4802	10	85.81	<.0001
block*trt	I	3	763.00	9.4802	10	80.48	<.0001
block*trt	II	1	849.00	9.4802	10	89.55	<.0001
block*trt	II	2	876.00	9.4802	10	92.40	<.0001
block*trt	II	3	811.00	9.4802	10	85.55	<.0001
block*trt	III	1	802.50	9.4802	10	84.65	<.0001
block*trt	III	2	719.00	9.4802	10	75.84	<.0001
block*trt	III	3	738.00	9.4802	10	77.85	<.0001

```
                    Least Squares Means

                             Standard
   Effect       block   trt   Estimate   Error    DF   t Value   Pr > |t|
   block*trt     IV      1     847.50    9.4802   10    89.40    <.0001
   block*trt     IV      2     850.00    9.4802   10    89.66    <.0001
```

Explanation: When *blocks* are defined as fixed and one or more *block* × *trt* cells is missing, the *trt* least squares means are not estimable if the *block* × *trt* interaction is present. The least squares means of *block* × *trt* combinations are estimable for those that have at least one observation. Note that least squares means for *block* I × *trt* 1 and *block* IV × *trt* 3 are absent from the output. The *F* tests for all effects in the model are still available.

SAS program with a model including only main effects for a dataset in which all observations in a treatment × block combination are missing (missing cells)

SAS program:

```
PROC MIXED DATA = d_gain;
CLASS block trt;
MODEL d_gain = block trt /;
LSMEANS trt / DIFF ADJUST=TUKEY;
RUN;
```

Explanation: The SAS program is as before except that the *block * trt* interaction has been deleted from the MODEL statement and LSMEANS for *block * trt* are not requested.

SAS output:

```
           Type 3 Tests of Fixed Effects

                  Num      Den
   Effect         DF       DF     F Value    Pr > F
   block           3       14      14.35     0.0001
   trt             2       14       5.09     0.0218


                    Least Squares Means

                           Standard
   Effect   trt   Estimate   Error    DF   t Value   Pr > |t|
   trt       1     830.07   11.2899   14    73.52    <.0001
   trt       2     814.63    9.1746   14    88.79    <.0001
   trt       3     779.47   11.2899   14    69.04    <.0001
```

Explanation: When *blocks* are defined as fixed and no interaction is fitted in the model, the *trt* least squares means are estimable. The *F* tests for both *trt* and *block* are valid.

SAS program for a model including random blocks for a dataset in which all observations in a treatment × block combination are missing (missing cells)

SAS program:

```
PROC MIXED DATA = d_gain;
CLASS block trt;
MODEL d_gain = trt / DDFM = KENWARDROGER;
RANDOM block block*trt;
LSMEANS trt / DIFF ADJUST=TUKEY;
RUN;
```

Explanation: The program is the same as for a dataset with no missing data, except for inclusion of the *DDFM = KENWARDROGER* option to estimate degrees of freedom.

SAS output:

Covariance Parameter Estimates

Cov Parm	Estimate
block	1460.51
block*trt	834.93
Residual	179.76

Type 3 Tests of Fixed Effects

Effect	Num DF	Den DF	F Value	Pr > F
trt	2	4.44	2.03	0.2365

Least Squares Means

Effect	trt	Estimate	Standard Error	DF	t Value	Pr > |t|
trt	1	831.06	26.9096	5.72	30.88	<.0001
trt	2	814.62	24.4198	4.82	33.36	<.0001
trt	3	777.23	26.9096	5.72	28.88	<.0001

Differences of Least Squares Means

Effect	trt	_trt	Estimate	Standard Error	DF	t Value	Pr>|t|	Adjust	Adj P
trt	1	2	16.4317	24.2941	4.38	0.68	0.5328	Tuk–Kr	0.7877
trt	1	3	53.8305	27.6661	4.6	1.95	0.1143	Tuk–Kr	0.2310
trt	2	3	37.3988	24.2941	4.38	1.54	0.1924	Tuk–Kr	0.3592

Explanation: The estimated variances and tests are valid for a dataset with missing cells. The least squares means and tests of difference between treatment means use degrees of freedom estimated by the Kenward-Roger method.

In summary, when data are missing:

Some observations missing, but no missing *block × treatment* cells

Block	Block × treatment	Use	Comments
fixed	fixed	GLM or MIXED	Both *treatment* and *block × treatment* estimable
random	random	MIXED with DDFM=KENWARDROGERS or SATTERTH (correction on missing values and heterogeneity of variance)	*Treatment* estimable Solution of *block × treatment* available with S options with the RANDOM statement

Some complete *block × treatment* cells missing

Block	Block × treatment	Use	Comments
fixed	not present	GLM or MIXED	*Treatment* estimable when only main effect (no interaction) in the model
fixed	fixed		*Treatment* not estimable when missing cell and fixed interaction present Combinations *block × treatment* estimable only for those that have at least one observation
random	random	MIXED with DDFM= KENWARDROGERS or SATTERTH	*Treatment* estimable Solution of *block × treatment* available only for those that have at least one observation with S options with the RANDOM statement

15.5 A Nonparametric Test

A nonparametric test within a randomized block design can be done by rank transformation. As blocks are a priori considered different populations, values are ranked separately within each block. Ranking nullifies the block effect. The usual analysis of variance with an F test is applicable, since we can assume that the ranks are drawn from a normal distribution, but only for the main effect of treatment. Since ranking within blocks nullifies the block effect, the interaction of block × treatment cannot be evaluated. The ranking can be done by using several methods. An application of normal ranking in SAS will be shown.

15.5.1 SAS Examples for Nonparametric Test

To illustrate a nonparametric test, data from the example of daily gain of steers with two experimental units per treatment by block combination from section 15.2.2 will be used.

SAS program:

```
DATA d_gain;
INPUT trt  block $ d_gain @@;
DATALINES;
1    I 826   1    I 806   1 II 864   1 II 834
1 III 795   1 III 810   1 IV 850   1 IV 845
2    I 827   2    I 800   2 II 871   2 II 881
2 III 729   2 III 709   2 IV 860   2 IV 840
3    I 753   3    I 773   3 II 801   3 II 821
3 III 736   3 III 740   3 IV 820   3 IV 835
;

PROC SORT DATA = d_gain;
BY block;
RUN;

PROC RANK DATA = d_gain OUT = ranking NORMAL= BLOM;
VAR d_gain;
RANKS rankd_gain;
BY block;
RUN;

PROC GLM DATA = ranking;
CLASS trt;
MODEL rankd_gain =  trt /;
RUN;
QUIT;
```

Explanation: Data are ranked within blocks, and the data must first be sorted by block using the SORT procedure. The RANK procedure computes normal ranks using the BLOM algorithm and saves it to the file *ranking*. Observations are ranked within block on the dependent variable *d_gain*, and the new ranked variable is called *rankd_gain*. The GLM procedure applies the usual F test to the data. The MODEL statement defines the dependent and independent variables: *rankd_gain = trt*.

SAS output:

Source	DF	Sum of Squares	Mean Square	F Value	Pr > F
Model	2	6.38507950	3.19253975	6.45	0.0065
Error	21	10.39115012	0.49481667		
Corrected Total	23	16.77622962			

Explanation: The *Sources* of variability are *Model* (corresponding to *trt* here), *Error* and *Corrected Total*. In the table are listed degrees of freedom (*DF*), *Sum of Squares*, *Mean Square*, calculated F (*F value*) and P value (*Pr > F*). For this example $F = 6.45$ and the P value is 0.0065, thus it can be concluded that an effect of treatments exists.

Exercise

15.1. The objective of an experiment was to analyze the effects of four treatments on ovulation rate in sows. The treatments are PG600, PMSG, FSH and saline. A sample of 20 sows was randomly chosen and they were assigned to five pens. The treatments were randomly assigned to the four sows in each pen. Are there significant differences between treatments? The data are as follows:

Treatment	Pens				
	I	II	III	IV	V
FSH	13	16	16	14	14
PG600	14	14	17	17	15
PMSG	17	18	19	19	16
Saline	13	11	14	10	13

Chapter 16

Change-over Designs

Change-over experimental designs have two or more treatments assigned to the same animal, but in different periods. Each animal is measured more than once, and each measurement corresponds to a different treatment. The order of treatment assignments is random. In effect, each animal is used as a block, and generally called a subject. Since treatments are exchanged on the same animal, this design is called a change-over or cross-over design. With two treatments the design is simple; animals are randomly assigned to two groups, to the first group the first treatment is applied, and to the second group the second treatment is applied. At the end of the treatment period, the treatments are exchanged. To the first group the second treatment is applied, and to the second group the first treatment is applied. In order to avoid after-effects of treatments, it may be good to rest the animals for some time between changing the treatments, and not to use measurements taken during that rest period. The number of treatments can be greater than the number of periods, thus, different animals receive different sets of treatments. The animal is then an incomplete block. However, such designs lose on precision. Here, only designs with equal numbers of treatments and periods will be described.

16.1   Simple Change-over Design

Consider an experiment for testing differences between treatments with all treatments applied on each subject or animal. The number of treatments, a, is equal to the number of measurements per subject, and the number of subjects is n. The order of treatment is random, but equal numbers of subjects should receive each treatment in every period. For example, for three treatments (T_1, T_2 and T_3) and n subjects a schema of an experiment can be:

Period	Subject 1	Subject 2	Subject 3	...	Subject n
1	T_2	T_1	T_2	...	T_3
2	T_1	T_3	T_3	...	T_2
3	T_3	T_2	T_1	...	T_1

Note that an experimental unit is not the subject or animal, but one measurement on the subject. In effect, subjects can be considered as blocks, and the model is similar to a randomized block design model, with the subject effect defined as random:

$$y_{ij} = \mu + \tau_i + SUB_j + \varepsilon_{ij} \qquad\qquad i = 1,...,a; \ \ j = 1,...,n;$$

where:

y_{ij} = observation on subject (animal) j in treatment i

336

μ = the overall mean

τ_i = the fixed effect of treatment i

SUB_j = the random effect of subject (animal) j with mean 0 and variance σ^2_S

ε_{ij} = random error with mean 0 and variance σ^2

a = number of treatments; n = number of subjects

Total sum of squares is partitioned to sums of squares between subjects and within subjects:

$$SS_{TOT} = SS_{SUB} + SS_{WITHIN\ SUBJECT}$$

Further, the sum of squares within subjects is partitioned to the treatment sum of squares and residual sum of squares:

$$SS_{WITHIN\ SUBJECT} = SS_{TRT} + SS_{RES}$$

Then, the total sum of squares is:

$$SS_{TOT} = SS_{SUB} + SS_{TRT} + SS_{RES}$$

with corresponding degrees of freedom:

$$(an - 1) = (n - 1) + (a - 1) + (n - 1)(a - 1)$$

The sums of squares are:

$$SS_{TOT} = \sum_i \sum_j (y_{ij} - \bar{y}..)^2$$

$$SS_{SUB} = \sum_i \sum_j (\bar{y}._j - \bar{y}..)^2 = a \sum_i (\bar{y}._j - \bar{y}..)^2$$

$$SS_{TRT} = \sum_i \sum_j (\bar{y}_i. - \bar{y}..)^2 = \sum_i n_i (\bar{y}_i. - \bar{y}..)^2$$

$$SS_{WITHIN\ SUBJECTS} = \sum_i \sum_j (y_{ij} - \bar{y}._j)^2$$

$$SS_{RES} = \sum_i \sum_j (y_{ij} - \bar{y}_i. - \bar{y}._{ji} + \bar{y}..)^2$$

By dividing the sums of squares by their corresponding degrees of freedom the mean squares are obtained:

Mean square for subjects: $MS_{SUB} = \dfrac{SS_{SUB}}{n - 1}$

Mean square within subjects: $MS_{WITHIN\ SUBJECT} = \dfrac{SS_{WITHIN\ SUBJECT}}{n(a - 1)}$

Mean square for treatments: $MS_{TRT} = \dfrac{SS_{TRT}}{a - 1}$

Mean square for experimental error (residual): $MS_{RES} = \dfrac{SS_{RES}}{(a - 1)(n - 1)}$

The null and alternative hypotheses are:

H_0: $\tau_1 = \tau_2 = ... = \tau_a$, no treatment effects

H_1: $\tau_i \neq \tau_{i'}$ for at least one pair (i, i'), a difference exists between treatments

The test statistic is:

$$F = \frac{MS_{TRT}}{MS_{RES}}$$

with an F distribution with $(a - 1)$ and $(a - 1)(n - 1)$ degrees of freedom, if H_0 holds. For α level of significance H_0 is rejected if $F > F_{\alpha,(a-1),(a-1)(n-1)}$, that is, if the calculated F from the sample is greater than the critical value.

The results can be summarized in an *ANOVA* table:

Source	SS	df	MS = SS/df	F
Between subjects	SS_{SUB}	$b - 1$	MS_{SUB}	
Within subjects	$SS_{WITHIN\,SUB}$	$n(a - 1)$	$MS_{WITHIN\,SUB}$	
Treatment	SS_{TRT}	$a - 1$	MS_{TRT}	MS_{TRT}/MS_{RES}
Residual	SS_{RES}	$(n - 1)(a - 1)$	MS_{RES}	

The estimators of treatments means are the sample arithmetic means. The standard errors of the treatment mean estimators are:

$$s_{\bar{y}_{i.}} = \sqrt{\frac{\left(MS_{RES} + \hat{\sigma}_S^2\right)}{n}}$$

where $\hat{\sigma}_S^2 = \dfrac{MS_{SUB} - MS_{RES}}{a}$ = variance component for subjects

The standard errors of estimators of the differences between treatment means are:

$$s_{\bar{y}_i - \bar{y}_{i'}} = \sqrt{MS_{RES}\left(\frac{2}{n}\right)}$$

The change-over design will have more power than a completely random design if the variability between subjects is large. The MS_{RES} will be smaller and consequently, it is more likely that a treatment effect will be detected.

Example: The effect of two treatments on milk yield of dairy cows was investigated. The experiment was conducted as a 'change-over' design, that is, each cow received both treatments in different periods. Ten cows in the third and fourth month of lactation were used. The order of treatments was randomly assigned. The following average milk yields were measured in kg:

BLOCK I						
Period	Treatment	Cow 1	Cow 4	Cow 5	Cow 9	Cow 10
1	1	31	34	43	28	25
2	2	27	25	38	20	19

BLOCK II		Cow 2	Cow 3	Cow 6	Cow 7	Cow 8
Period	Treatment	Cow 2	Cow 3	Cow 6	Cow 7	Cow 8
1	2	22	40	40	33	18
2	1	21	39	41	34	20

The hypotheses are:

H_0: $\tau_1 = \tau_2$, there is no difference between treatments
H_1: $\tau_1 \neq \tau_2$, there is a difference between treatments

The *ANOVA* table is:

Source	SS	df	MS	F
Between subjects	1234.800	9	137.200	
Within subjects	115.000	10	11.500	
Treatment	57.800	1	57.800	9.09
Residual	57.200	9	6.356	
Total	1349.800	19		

If H_0 holds, the F statistic has an F distribution with 1 and 9 degrees of freedom. The calculated F value from the samples is:

$$F = \frac{MS_{TRT}}{MS_{RES}} = \frac{57.200}{6.356} = 9.09$$

Since the calculated $F = 9.09$ is greater than the critical value $F_{0.05,1,9} = 5.12$, H_0 is rejected at the 0.05 level of significance.

This was a very simplified approach. Because of possible effects of period of lactation and/or order of treatment application, those effects should be tested as well.

16.2 Change-over Designs with the Effects of Periods

Period also can account for variability among measurements. For example, in an experiment with dairy cows, milk yield depends on stage of lactation. A way to improve the precision of the experiment is by including the effect of period in the change-over model. Further, the effect of order of treatment application can be included. A possible model is:

$$y_{ijkl} = \mu + \tau_i + \beta_k + SUB(\beta)_{jk} + t_l + \varepsilon_{ijkl}$$

$$i = 1,\ldots, a;\ j = 1,\ldots, n_k;\ k = 1,\ldots, b;\ l = 1,\ldots, a$$

where:

y_{ijkl} = observation on subject j with treatment i, order of treatment k and period l
μ = the overall mean
τ_i = the fixed effect of treatment i
β_k = the effect of order k of applying treatments
$SUB(\beta)_{jk}$ = the random effect of subject j within order k with mean 0 and variance σ^2_g
t_l = the effect of period l
ε_{ijkl} = random error with mean 0 and variance σ^2

a = number of treatments and periods; b = number of orders; n_k = number of subjects within order k; $n = \sum_k n_k$ = total number of subjects

The statistic for testing the effect of treatments is:

$$F = \frac{MS_{TRT}}{MS_{RES}}$$

which has an F distribution with $(a-1)$ and $(a-1)(n-2)$ degrees of freedom, if H_0 holds. For α level of significance H_0 is rejected if $F > F_{\alpha,(a-1),(a-1)(n-2)}$, that is, if the calculated F from the sample is greater than the critical value.

The test statistic for testing the effects of order is:

$$F = \frac{MS_{ORDER}}{MS_{SUB(ORDER)}}$$

The results can be summarized in the following *ANOVA* table:

Source	SS	df	MS = SS/df	F
Order	SS_{ORD}	$b-1$	MS_{ORD}	MS_{ORD}/MS_{SUB}
Subject within order	SS_{SUB}	$\sum_k(n_k-1) = n-b$	MS_{SUB}	
Period	SS_t	$a-1$	MS_t	MS_t/MS_{RES}
Treatment	SS_{TRT}	$a-1$	MS_{TRT}	MS_{TRT}/MS_{RES}
Residual	SS_{RES}	$(a-1)(n-2)$	MS_{RES}	
Total	SS_{TOT}	$an-1$		

Example: Using the previous example examining the effects of two treatments on milk yield, now with the effects of periods and order of treatment included in the model. Order is defined as order I if treatment 1 is applied first, and order II if treatment 1 is applied second. Recall the data:

ORDER I

Period	Treatment	Cow 1	Cow 4	Cow 5	Cow 9	Cow 10
1	1	31	34	43	28	25
2	2	27	25	38	20	19

ORDER II

Period	Treatment	Cow 2	Cow 3	Cow 6	Cow 7	Cow 8
1	2	22	40	40	33	18
2	1	21	39	41	34	20

The formulas for hand calculation of these sums of squares are lengthy and thus have not been shown. The SAS program for their calculation is presented in section 16.2.1.

The results are shown in the *ANOVA* table:

Source	SS	df	MS	F
Order	16.20	1	16.200	0.11
Subject within order	1218.60	8	152.325	
Period	45.00	1	45.000	29.51
Treatment	57.80	1	57.800	37.90
Residual	12.20	8	1.525	

The effects of treatment and period are significant, while the effect of treatment order has not affected the precision of the experiment. The residual mean square (experimental error) is smaller in the model with periods comparing to the model without periods. Inclusion of periods has increased the precision of the model and the possibility that the same conclusion could be obtained with fewer cows.

16.2.1 SAS Example for Change-over Designs with the Effects of Periods

The SAS program for the example with the effect of two treatments on milk yield of dairy cows is as follows. Recall the data:

ORDER I

Period	Treatment	Cow 1	Cow 4	Cow 5	Cow 9	Cow 10
1	1	31	34	43	28	25
2	2	27	25	38	20	19

ORDER II

Period	Treatment	Cow 2	Cow 3	Cow 6	Cow 7	Cow 8
1	2	22	40	40	33	18
2	1	21	39	41	34	20

SAS program:

```
DATA cows;
INPUT period trt order cow milk @@;
DATALINES;
1 1 1   1 31        1 2 2   2 22
2 2 1   1 27        2 1 2   2 21
1 1 1   4 34        1 2 2   3 40
2 2 1   4 25        2 1 2   3 39
1 1 1   5 43        1 2 2   6 40
2 2 1   5 38        2 1 2   6 41
1 1 1   9 28        1 2 2   7 33
2 2 1   9 20        2 1 2   7 34
1 1 1  10 25        1 2 2   8 18
2 2 1  10 19        2 1 2   8 20
;
```

```
PROC MIXED DATA = cows;
CLASS trt cow period order;
MODEL milk = order trt period;
RANDOM cow(order) ;
LSMEANS trt/ DIFF ADJUST=TUKEY ;
RUN;
```

Explanation: The MIXED procedure is used because of the defined random categorical variable included in the model. The CLASS statement defines the categorical (class) variables. The MODEL statement defines the dependent variable *milk*, and independent variables *trt*, *period* and *order*. The RANDOM statement indicates that *cow*(*order*) is defined as a random variable. The LSMEANS statement calculates treatment means. The DIFF option tests significance between all pairs of means.

Note: To account for missing values it is recommended that degrees of freedom and standard errors are corrected by using options *DDFM = SATTERTH* or *DDFM = KENWARDROGER* with the MODEL statement (*MODEL milk = order trt period / DDFM = SATTERTH;*)

SAS output:

Covariance Parameter Estimates

Cov Parm	Estimate
cow(order)	75.4000
Residual	1.5250

Type 3 Tests of Fixed Effects

Effect	Num DF	Den DF	F Value	Pr > F
order	1	8	0.11	0.7527
trt	1	8	37.90	0.0003
period	1	8	29.51	0.0006

Least Squares Means

Effect	trt	Estimate	Standard Error	DF	t Value	Pr > |t|
trt	1	31.6000	2.7735	8	11.39	<.0001
trt	2	28.2000	2.7735	8	10.17	<.0001

Differences of Least Squares Means

Effect	trt	_trt	Estimate	Standard Error	DF	t Value	Pr>|t|	Adjust.	Adj P
trt	1	2	3.4000	0.5523	8	6.16	0.0003	Tuk-Kr	0.0003

Explanation: The MIXED procedure gives estimates of variance components for random effects (*Covariance Parameter Estimates*). Here the random effects are *cow*(*order*) and *Residual*. Next, the *F* test for the fixed effects (*Type 3 Tests of Fixed Effects*) are given. In the table are listed *Effect*, degrees of freedom for the numerator (*Num DF*), degrees of freedom for the denominator (*Den DF*), *F Value* and *P* value (*Pr > F*). The *P* value for treatments is 0.0003. In the *Least Squares Means* table the least squares means (*Estimates*) together with their *Standard Error* are shown. In the *Differences of Least Squares Means* table the *Estimates* of mean differences with their *Standard Error* and P values (*Pr* > |*t*|) are shown.

Note: For balanced designs and no missing values, defining cows as random or fixed, yields the same *F* value for the treatment effect; however, the standard errors of least squares means are different due to presence or absence of the cow component of variance.

16.3 Latin Square

In the Latin square design treatments are assigned to blocks in two different ways, usually represented as columns and rows. Each column and each row is a complete block of all treatments. Hence, in a Latin square three explained sources of variability are defined: columns, rows and treatments. A particular treatment is assigned just once in each row and column. Often one of the blocks corresponds to animal and the other to period. Each animal will receive all treatment in different periods. In that sense, the Latin square is a change-over design. The number of treatments (*r*) is equal to the number of columns and rows. The total number of measurements (observations) is equal to r^2. With treatments denoted by capital letters (A, B, C, D, etc.), then examples of 3 × 3 and 4 × 4 Latin squares are:

A C B	C A B	A B D C	C D B A
B A C	A B C	C A B D	D B A C
C B A	B C A	B D C A	B A C D
		D C A B	A C D B

Example: Assume the number of treatments *r* = 4. Treatments are denoted T_1, T_2, T_3 and T_4. Columns and rows denote periods and animals, respectively. A possible design could be:

Rows (Periods)	Columns (Animals)			
	1	2	3	4
1	T_1	T_3	T_2	T_4
2	T_3	T_4	T_1	T_2
3	T_2	T_1	T_4	T_3
4	T_4	T_2	T_3	T_1

Let $y_{ij(k)}$ denote a measurement in row *i* and column *j*, and with the treatment *k*, then a possible design of Latin square is:

Rows (Periods)	Columns (Animals)			
	1	2	3	4
1	$y_{11(1)}$	$y_{12(3)}$	$y_{13(2)}$	$y_{14(4)}$
2	$y_{21(3)}$	$y_{22(4)}$	$y_{23(1)}$	$y_{24(2)}$
3	$y_{31(2)}$	$y_{32(1)}$	$y_{33(4)}$	$y_{34(3)}$
4	$y_{41(4)}$	$y_{42(2)}$	$y_{43(3)}$	$y_{44(1)}$

The model for a Latin square is:

$$y_{ij(k)} = \mu + ROW_i + COL_j + \tau_{(k)} + \varepsilon_{ij(k)} \qquad\qquad i, j, k = 1,...., r$$

where:

$y_{ij(k)}$ = observation $ij(k)$
μ = the overall mean
ROW_i = the effect of row i
COL_j = the effect of column j
$\tau_{(k)}$ = the fixed effect of treatment k
$\varepsilon_{ij(k)}$ = random error with mean 0 and variance σ^2
r = the number of treatments, rows and columns

The total sum of squares is partitioned to the sum of squares for columns, rows, treatments and residual:

$$SS_{TOT} = SS_{ROW} + SS_{COL} + SS_{TRT} + SS_{RES}$$

The corresponding degrees of freedom are:

$$r^2 - 1 = (r - 1) + (r - 1) + (r - 1) + (r - 1)(r - 2)$$

The sums of squares are:

$$SS_{TOT} = \sum_i \sum_j (y_{ij(k)} - \bar{y}..)^2$$

$$SS_{ROW} = r\sum_i (\bar{y}_i. - \bar{y}..)^2$$

$$SS_{COL} = r\sum_j (\bar{y}._j - \bar{y}..)^2$$

$$SS_{TRT} = r\sum_k (\bar{y}_k - \bar{y}..)^2$$

$$SS_{RES} = \sum_i \sum_j (\bar{y}_{ij} - \bar{y}_i. - \bar{y}._j - \bar{y}_k + 2\bar{y}..)^2$$

The sum of squares can be calculated with a short-cut computation:

1. Total sum:

$$\sum_i \sum_j y_{ij(k)}$$

2. Correction factor for the mean:

$$C = \frac{\left(\sum_i \sum_j y_{ij(k)}\right)^2}{r^2}$$

3. Total (corrected) sum of squares:

$$SS_{TOT} = \sum_i \sum_j (y_{ij(k)})^2 - C$$

4. Row sum of squares:

$$SS_{ROW} = \sum_i \frac{\left(\sum_j y_{ij(k)}\right)^2}{r} - C$$

5. Column sum of squares:

$$SS_{COL} = \sum_j \frac{\left(\sum_i y_{ij(k)}\right)^2}{r} - C$$

6. Treatment sum of squares:

$$SS_{TRT} = \sum_k \frac{\left(\sum_i \sum_j y_{ij(k)}\right)^2}{r} - C$$

7. Residual sum of squares:

$$SS_{RES} = SS_{TOT} - SS_{ROW} - SS_{COL} - SS_{TRT}$$

Dividing the sums of squares by their corresponding degrees of freedom yields the following mean squares:

Mean square for rows: $MS_{ROW} = \dfrac{SS_{ROW}}{r-1}$

Mean square for columns: $MS_{COL} = \dfrac{SS_{COL}}{r-1}$

Mean square for treatments: $MS_{TRT} = \dfrac{SS_{TRT}}{r-1}$

Mean square for experimental error: $MS_{RES} = \dfrac{SS_{RES}}{(r-1)(r-2)}$

The null and alternative hypotheses are:

H_0: $\tau_1 = \tau_2 = ... = \tau_a$, no treatment effects
H_1: $\tau_i \neq \tau_{i'}$, for at least one pair (i,i'), a difference exists between treatments

An F statistic is used for testing the hypotheses:

$$F = \frac{MS_{TRT}}{MS_{RES}}$$

which, if H_0 holds, has an F distribution with $(r-1)$ and $(r-1)(r-2)$ degrees of freedom. For the α level of significance H_0 is rejected if $F > F_{\alpha,(r-1),(r-1)(r-2)}$, that is, if the calculated F from the sample is greater than the critical value. Tests for columns and rows are usually not of primary interest, but can be done analogously as for the treatments.

The results can be summarized in an *ANOVA* table:

Source	SS	df	MS	F
Row	SS_{ROW}	$r-1$	MS_{ROW}	MS_{ROW}/MS_{RES}
Column	SS_{COL}	$r-1$	MS_{COL}	MS_{COL}/MS_{RES}
Treatment	SS_{TRT}	$r-1$	MS_{TRT}	MS_{TRT}/MS_{RES}
Residual	SS_{RES}	$(r-1)(r-2)$	MS_{RES}	
Total	SS_{TOT}	r^2-1		

It is possible to reduce the experimental error by accounting for column and row variability. Note that columns and rows can be defined as additional factors, but their interaction is impossible to separate from residual. If an interaction exists, the Latin square cannot be used because there is not an appropriate experimental error term for treatment. Similarly as with classical 'change-over' designs, one must be careful because carryover effects of treatments can be confounded with the effect of the treatment applied in the next period.

Example: The aim of this experiment was to test the effect of four different supplements (*A*, *B*, *C* and *D*) on hay intake of fattening steers. The experiment was designed as a Latin square with four animals in four periods of 20 days. The steers were housed individually. Each period consists of 10 days of adaptation and 10 days of measurements. The data in the following table are the means of 10 days:

Periods	Steers 1	2	3	4	Σ
1	10.0(*B*)	9.0(*D*)	11.1(*C*)	10.8(*A*)	40.9
2	10.2(*C*)	11.3(*A*)	9.5(*D*)	11.4(*B*)	42.4
3	8.5(*D*)	11.2(*B*)	12.8(*A*)	11.0(*C*)	43.5
4	11.1(*A*)	11.4(*C*)	11.7(*B*)	9.9(*D*)	44.1
Σ	39.8	42.9	45.1	43.1	170.9

The sums for treatments:

	A	*B*	*C*	*D*	Total
Σ	46.0	44.3	43.7	36.9	170.9

1. Total sum:

$$\sum_i \sum_j y_{ij(k)} = (10.0 + 9.0 + \ldots\ldots + 9.9) = 170.9$$

2. Correction factor for the mean:

$$C = \frac{\left(\sum_i \sum_j y_{ij(k)}\right)^2}{r^2} = \frac{(170.9)^2}{16} = 1825.4256$$

3. Total (corrected) sum of squares:

$$SS_{TOT} = \sum_i \sum_j (y_{ij(k)})^2 - C = (10.0)^2 + (9.0)^2 + \ldots\ldots + (9.9)^2 - 1825.4256 = 17.964$$

4. Row sum of squares:

$$SS_{ROW} = \sum_i \frac{\left(\sum_j y_{ij(k)}\right)^2}{r} - C = \frac{1}{4}\left[(40.9)^2 + \ldots + (44.1)^2\right] - C = 1.482$$

5. Column sum of squares:

$$SS_{COL} = \sum_j \frac{\left(\sum_i y_{ij(k)}\right)^2}{r} - C = \frac{1}{4}\left[(39.8)^2 + ... + (43.1)^2\right] - C = 3.592$$

6. Treatment sum of squares:

$$SS_{TRT} = \sum_k \frac{\left(\sum_i \sum_j y_{ij(k)}\right)^2}{r} - C = \frac{1}{4}\left[(46.0)^2 + ... + (36.9)^2\right] - C = 12.022$$

7. Residual sum of squares:

$$SS_{RES} = SS_{TOT} - SS_{ROW} - SS_{COL} - SS_{TRT} = 17.964375 - 1.481875 - 3.591875 - 12.021875 =$$
$$= 0.868$$

The *ANOVA* table:

Source	SS	df	MS	F
Rows (periods)	1.482	3	0.494	3.41
Columns (steers)	3.592	3	1.197	8.26
Treatments	12.022	3	4.007	27.63
Residual	0.868	6	0.145	
Total	17.964	15		

The critical value for treatments is $F_{0.05,3,6} = 4.76$. The calculated $F = 27.63$ is greater than the critical value, thus, H_0 is rejected, and it can be concluded that treatments influence hay intake of steers.

16.3.1 SAS Example for Latin Square

The SAS program for a Latin square is shown for the example measuring intake of steers. Recall that the objective of the experiment was to test the effect of four different supplements (*A*, *B*, *C* and *D*) on hay intake of fattening steers. The experiment was defined as a Latin square with four animals in four periods of 20 days. The data are:

Periods	Steers			
	1	2	3	4
1	10.0(*B*)	9.0(*D*)	11.1(*C*)	10.8(*A*)
2	10.2(*C*)	11.3(*A*)	9.5(*D*)	11.4(*B*)
3	8.5(*D*)	11.2(*B*)	12.8(*A*)	11.0(*C*)
4	11.1(*A*)	11.4(*C*)	11.7(*B*)	9.9(*D*)

SAS program:

```
DATA a;
INPUT period steer suppl $ hay @@;
DATALINES;
```

```
1 1 B 10.0        3 1 D  8.5
1 2 D  9.0        3 2 B 11.2
1 3 C 11.1        3 3 A 12.8
1 4 A 10.8        3 4 C 11.0
2 1 C 10.2        4 1 A 11.1
2 2 A 11.3        4 2 C 11.4
2 3 D  9.5        4 3 B 11.7
2 4 B 11.4        4 4 D  9.9
;
PROC GLM;
CLASS period steer suppl;
MODEL hay = period steer suppl;
LSMEANS suppl / STDERR PDIFF ADJUST=TUKEY;
RUN;
```

Explanation: The GLM procedure was used. The CLASS statement defines categorical (class) variables. The MODEL statement defines *hay* as the dependent and *period*, *steer* and *suppl* as independent variables. The LSMEANS statement calculates the treatment (*suppl*) means. The options after the slash request standard errors and test the differences between means by using a Tukey test.

Note: If steer is defined as a random effect, MIXED procedure can be used. The *F* value for the *suppl* will be the same; however, the standard errors of least squares means will be different due to inclusion of steer variance component.

SAS output:

Dependent Variable: hay

Source	DF	Sum of Squares	Mean Square	F Value	Pr > F
Model	9	17.09562500	1.89951389	13.12	0.0027
Error	6	0.86875000	0.14479167		
Corrected Total	15	17.96437500			

R-Square	Coeff Var	Root MSE	hay Mean
0.951640	3.562458	0.380515	10.68125

Source	DF	Type III SS	Mean Square	F Value	Pr > F
period	3	1.48187500	0.49395833	3.41	0.0938
steer	3	3.59187500	1.19729167	8.27	0.0149
suppl	3	12.02187500	4.00729167	27.68	0.0007

Least Squares Means
Adjustment for Multiple Comparisons: Tukey

suppl	hay LSMEAN	Standard Error	Pr > |t|	LSMEAN Number
A	11.5000000	0.1902575	<.0001	1
B	11.0750000	0.1902575	<.0001	2
C	10.9250000	0.1902575	<.0001	3
D	9.2250000	0.1902575	<.0001	4

```
            Least Squares Means for Effect suppl
          t for H0: LSMean(i)=LSMean(j) / Pr > |t|
                  Dependent Variable: hay


   i/j        1            2            3            4
    1                   0.4536       0.2427       0.0006
    2      0.4536                    0.9411       0.0019
    3      0.2427       0.9411                    0.0030
    4      0.0006       0.0019       0.0030
```

Explanation: First is the *ANOVA* table for the *Dependent Variable hay*. The *Source* of variability are *Model*, residual (*Error*) and *Corrected Total*. In the table are listed degrees of freedom (*DF*), *Sum of Squares*, *Mean Square*, calculated *F Value* and *P* value (*Pr > F*). In the next table the explained sources of variability (*MODEL*) are partitioned to *period, steer* and *suppl*. The calculated *F* and *P* values for *suppl* are 27.68 and 0.0007, respectively. At the end of output are least squares means (*LSMEAN*) with their standard errors (*Standard Error*), and then the Tukey tests between all pairs of *suppl* are shown. The *P* values for the tests of differences are given. These tables shows that supplement D yields significantly lower intake than the other three supplements.

16.4   Change-over Design Set as Several Latin Squares

The main disadvantage of a Latin square is that the number of columns, rows and treatments must be equal. If there are many treatments the Latin square becomes impractical. On the other hand, small Latin squares have few degrees of freedom for experimental error, and because of that are imprecise. In general, precision and the power of test can be increased by using more animals in an experiment. Another way of improving an experiment is the use of a change-over design with periods as block effects. Such a design allows testing of a larger number of animals and accounting for the effect of blocks. In a Latin square design greater precision can be achieved if the experiment is designed as a set of several Latin squares. This is also a change-over design with the effect of squares defined as blocks. For example, consider an experiment designed as two Latin squares with three treatments in three periods:

Rows (periods)	Square I Columns (animals)			Square II Columns (animals)		
	1	2	3	4	5	6
1	T1	T3	T2	T1	T2	T3
2	T3	T2	T1	T2	T3	T1
3	T2	T1	T3	T3	T1	T2

The model is:

$$y_{ij(k)m} = \mu + SQ_m + ROW(SQ)_{im} + COL(SQ)_{jm} + \tau_{(k)} + \varepsilon_{ij(k)m}$$

$$i, j, k = 1,..., r; \quad m = 1,..., b$$

where:

$y_{ij(k)m}$ = observation $ij(k)m$
μ = the overall mean
SQ_m = the effect of square m
$ROW(SQ)_{im}$ = the effect of row i within square m
$COL(SQ)_{jm}$ = the effect of column j within square m
$\tau_{(k)}$ = the effect of treatment k
$\varepsilon_{ij(k)m}$ = random error with mean 0 and variance σ^2
r = the number of treatments, and the number of rows and columns within each square
b = the number of squares

The variability and degrees of freedom are partitioned to sources in the following table:

Source	Degrees of freedom
Squares (blocks)	$b - 1$
Rows within squares	$b(r - 1)$
Columns within squares	$b(r - 1)$
Treatments	$r - 1$
Residual	$b(r - 1)(r - 2) + (b - 1)(r - 1)$
Total	$b\,r^2 - 1$

The F statistic for testing treatments is:

$$F = \frac{MS_{TRT}}{MS_{RES}}$$

Example: The aim of this experiment was to test the effect of four different supplements (A, B, C and D) on hay intake of fattening steers. The experiment was designed as two Latin squares, each with four animals in four periods of 20 days. The steers were housed individually. Each period consists of 10 days of adaptation and 10 days of measurements. The data shown in the following table are the means of forage intakes for the 10 day data collection period:

SQUARE I

	Steers			
Periods	1	2	3	4
1	10.0(B)	9.0(D)	11.1(C)	10.8(A)
2	10.2(C)	11.3(A)	9.5(D)	11.4(B)
3	8.5(D)	11.2(B)	12.8(A)	11.0(C)
4	11.1(A)	11.4(C)	11.7(B)	9.9(D)

SQUARE II

Periods	Steers			
	5	6	7	8
1	11.1(C)	11.4(A)	9.6(D)	11.4(B)
2	10.7(B)	9.8(D)	11.6(C)	11.3(A)
3	11.3(A)	11.6(C)	11.9(B)	10.0(D)
4	9.0(D)	13.1(B)	11.6(A)	11.4(C)

The results are shown in the following *ANOVA* table:

Source	SS	df	MS	F
Squares	1.087813	1	1.087813	
Periods within squares	2.056875	6	0.342813	1.74
Steers within squares	5.481875	6	0.913646	4.64
Treatments	23.380938	3	7.631146	38.77
Residual	2.952188	15	0.196813	
Total	34.472188	31		

The critical value for treatments is $F_{0.05,3,15} = 3.29$. The calculated $F = 38.77$ is greater than the critical value, H_0 is rejected, and it can be concluded that treatments influence hay intake of steers.

16.4.1 SAS Example for Several Latin Squares

The SAS program for the example of intake of hay by steers designed as two Latin squares is as follows.

SAS program, *squares and steers defined as fixed:*

```
DATA a;
   INPUT square period steer suppl $ hay @@;
   DATALINES;
1   1 1 B 10.0      2   1 5 C 11.1
1   1 2 D  9.0      2   1 6 A 11.4
1   1 3 C 11.1      2   1 7 D  9.6
1   1 4 A 10.8      2   1 8 B 11.4
1   2 1 C 10.2      2   2 5 B 10.7
1   2 2 A 11.3      2   2 6 D  9.8
1   2 3 D  9.5      2   2 7 C 11.6
1   2 4 B 11.4      2   2 8 A 11.3
1   3 1 D  8.5      2   3 5 A 11.3
1   3 2 B 11.2      2   3 6 C 11.6
1   3 3 A 12.8      2   3 7 B 11.9
```

```
1  3 4 C 11.0        2  3 8 D 10.0
1  4 1 A 11.1        2  4 5 D  9.0
1  4 2 C 11.4        2  4 6 B 13.1
1  4 3 B 11.7        2  4 7 A 11.6
1  4 4 D  9.9        2  4 8 C 11.4
;
PROC GLM DATA = a;
CLASS square period steer suppl;
MODEL hay = square period(square) steer(square) suppl;
LSMEANS suppl  / STDERR PDIFF ADJUST=TUKEY;
RUN;
```

Explanation: The GLM procedure was used. The CLASS statement defines categorical (class) variables. The MODEL statement defines *hay* as the dependent and *square*, *period(square)*, *steer(square)* and *suppl* as independent variables. The LSMEANS statement calculates the treatment (*suppl*) means. The options after the slash request standard errors and test the difference between means by using a Tukey test.

SAS output:

Dependent Variable: hay

Source	DF	Sum of Squares	Mean Square	F Value	Pr > F
Model	16	31.52000000	1.97000000	10.01	<.0001
Error	15	2.95218750	0.19681250		
Corrected Total	31	34.47218750			

R-Square	Coeff Var	Root MSE	hay Mean
0.914360	4.082927	0.443636	10.86563

Source	DF	Type III SS	Mean Square	F Value	Pr > F
square	1	1.08781250	1.08781250	5.53	0.0328
period(square)	6	2.05687500	0.34281250	1.74	0.1793
steer(square)	6	5.48187500	0.91364583	4.64	0.0074
suppl	3	22.89343750	7.63114583	38.77	<.0001

Least Squares Means
Adjustment for Multiple Comparisons: Tukey

suppl	hay LSMEAN	Standard Error	Pr > |t|	LSMEAN Number
A	11.4500000	0.1568489	<.0001	1
B	11.4250000	0.1568489	<.0001	2
C	11.1750000	0.1568489	<.0001	3
D	9.4125000	0.1568489	<.0001	4

```
        Least Squares Means for Effect suppl
       t for H0: LSMean(i)=LSMean(j) / Pr > |t|

             Dependent Variable: hay
 i/j           1            2            3            4
   1                      0.9995       0.6125       <.0001
   2         0.9995                    0.6792       <.0001
   3         0.6125       0.6792                    <.0001
   4         <.0001       <.0001       <.0001
```

Explanation: First is the *ANOVA* table for the *Dependent Variable hay*. The *Sources* of variability are *Model*, residual (*Error*) and *Corrected Total*. In the table are listed degrees of freedom (*DF*), *Sum of Squares*, *Mean Square*, calculated *F Value* and *P* value (*Pr > F*). In the next table the explained sources of variability (*MODEL*) are partitioned to *square*, *period(square)*, *steer(square)* and *suppl*. The calculated *F* and *P* values for *suppl* are 38.77 and <0.0001, respectively. At the end of output the least squares means (*LSMEAN*) of supplements with their standard errors (*Std Err*), and then the Tukey test between all pairs of *suppl* are shown.

SAS program, *squares and steers defined as random:*

```
PROC MIXED DATA=a;
CLASS square period steer suppl;
MODEL hay = suppl;
RANDOM square period(square) steer(square);
LSMEANS suppl / DIFF ADJUST=TUKEY;
RUN;
```

Explanation: The MODEL statement defines *hay* as the dependent and *suppl* as a fixed effect independent variable. The RANDOM statement define *square*, *period(square)* and *steer(square)* as random variables. The LSMEANS statement calculates the treatment means. The options after the slash test the difference between means by using a Tukey test. Note: To account for missing values it is recommended that degrees of freedom and standard errors are corrected by using options DDFM = SATTERTH or DDFM = KENWARDROGER with the MODEL statement.

SAS output:

```
         Covariance Parameter Estimates

         Cov Parm             Estimate
         square               0.001760
         period(square)        0.03650
         steer(square)         0.1792
         Residual              0.1968


     Type 3 Tests of Fixed Effects

                  Num      Den
      Effect       DF       DF     F Value    Pr > F
      suppl         3       15      38.77     <.0001
```

Least Squares Means

Effect	suppl	Estimate	Standard Error	DF	t Value	Pr > |t|
suppl	A	11.4500	0.2290	15	50.00	<.0001
suppl	B	11.4250	0.2290	15	49.89	<.0001
suppl	C	11.1750	0.2290	15	48.80	<.0001
suppl	D	9.4125	0.2290	15	41.10	<.0001

Differences of Least Squares Means

Effect	suppl	_suppl	Estimate	Standard Error	DF	t Value	Pr>|t|	Adjust	Adj P
suppl	A	B	0.02500	0.2218	15	0.11	0.9118	Tu-Kr	0.9995
suppl	A	C	0.2750	0.2218	15	1.24	0.2341	Tu-Kr	0.6125
suppl	A	D	2.0375	0.2218	15	9.19	<.0001	Tu-Kr	<.0001
suppl	B	C	0.2500	0.2218	15	1.13	0.2774	Tu-Kr	0.6792
suppl	B	D	2.0125	0.2218	15	9.07	<.0001	Tu-Kr	<.0001
suppl	C	D	1.7625	0.2218	15	7.95	<.0001	Tu-Kr	<.0001

Explanation: The estimates of variance components for random effects (*Covariance Parameter Estimates*) are given. Next, the *F* test for the fixed effects (*Type 3 Tests of Fixed Effects*) are given. In the table are listed *Effect*, degrees of freedom for the numerator (*Num DF*), degrees of freedom for the denominator (*Den DF*), *F Value* and *P* value (*Pr > F*). Note the F and P values for *suppl* are the same as from the GLM procedure. In the *Least Squares Means* table the least squares means (*Estimates*) together with their *Standard Error* are shown. In the *Differences of Least Squares Means* table the *Estimates* of mean differences with their *Standard Error* and *P* values (*Pr > |t|*) are shown. The difference from the GLM procedure and all effects defined as fixed is in the standard errors of least squares means due to inclusion of variance components of random effects.

Exercise

16.1. The objective of this experiment was to test the effect of ambient temperature on the progesterone concentration of sows. The sows were subjected to different lengths of temperature stress: Treatment 1 = stress for 24 hours, Treatment 2 = stress for 12 hours, Treatment 3 = no stress. The experiment was conducted on nine sows in three chambers to determine the effect of stress. Each sow was treated with all three treatments over three periods. The design is a set of three Latin squares:

Sow	Treatment	Period	Progesterone	Sow	Treatment	Period	Progesterone
1	*TRT1*	1	5.3	6	*TRT3*	1	7.9
1	TRT2	2	6.3	6	TRT1	2	4.7
1	TRT3	3	4.2	6	TRT2	3	6.8
2	TRT2	1	6.6	7	TRT1	1	5.5
2	TRT3	2	5.6	7	TRT2	2	4.6
2	TRT1	3	6.3	7	TRT3	3	3.4
3	TRT3	1	4.3	8	TRT2	1	7.8
3	TRT1	2	7.0	8	TRT3	2	7.0
3	TRT2	3	7.9	8	TRT1	3	7.9
4	TRT1	1	4.2	9	TRT3	1	3.6
4	TRT2	2	5.6	9	TRT1	2	6.5
4	TRT3	3	6.6	9	TRT2	3	5.8
5	TRT2	1	8.1				
5	TRT3	2	7.9				
5	TRT1	3	5.8				

Draw a scheme of the experiment. Test the effects of treatments.

Chapter 17

Factorial Experiments

A factorial experiment has two or more sets of treatments that are analyzed at the same time. Recall that treatments denote particular levels of an independent categorical variable, often called a factor. Therefore, if two or more factors are examined in an experiment, it is a factorial experiment. A characteristic of a factorial experiment is that all combinations of factor levels are tested. The effect of a factor alone is called a main effect. The effect of different factors acting together is called an interaction. The experimental design is completely randomized. Combinations of factors are randomly applied to experimental units. Consider an experiment to test the effect of protein content and type of feed on milk yield of dairy cows. The first factor is protein content and the second is type of feed. Protein content is defined in three levels, and two types of feed are used. Each cow in the experiment receives one of the six protein × feed combinations. This experiment is called a 3 × 2 factorial experiment, because three levels of the first factor and two levels of the second factor are defined. An objective could be to determine if cows' response to different protein levels is different with different feeds. This is the analysis of interaction. The main characteristic of a factorial experiment is the possibility to analyze interactions between factor levels. Further, the factorial experiment is particularly useful when little is known about factors and all combinations have to be analyzed in order to conclude which combination is the best. There can be two, three, or more factors in an experiment. Accordingly, factorial experiments are defined by the number, two, three, etc., of factors in the experiment.

17.1 The Two Factor Factorial Experiment

Consider a factorial experiment with two factors A and B. Factor A has a levels, and factor B has b levels. Let the number of experimental units for each $A \times B$ combination be n. There is a total of nab experimental units divided into ab combinations of A and B. The set of treatments consists of ab possible combinations of factor levels.

The model for a factorial experiment with two factors A and B is:

$$y_{ijk} = \mu + A_i + B_j + (AB)_{ij} + \varepsilon_{ijk} \qquad i = 1,\ldots, a; j = 1,\ldots, b; \ k = 1,\ldots, n$$

where:

y_{ijk} = observation k in level i of factor A and level j of factor B
μ = the overall mean
A_i = the effect of level i of factor A
B_j = the effect of level j of factor B
$(AB)_{ij}$ = the effect of the interaction of level i of factor A with level j of factor B
ε_{ijk} = random error with mean 0 and variance σ^2

356

a = number of levels of factor A; b = number of levels of factor B; n = number of observations for each $A \times B$ combination.

The simplest factorial experiment is a 2×2, an experiment with two factors each with two levels. The principles for this experiment are generally valid for any factorial experiment. Possible combinations of levels are shown in the following table:

	Factor B	
Factor A	B_1	B_2
A_1	$A_1 B_1$	$A_1 B_2$
A_2	$A_2 B_1$	$A_2 B_2$

There are four combinations of factor levels. Using measurements y_{ijk}, the schema of the experiment is:

A_1		A_2	
B_1	B_2	B_1	B_2
y_{111}	y_{121}	y_{211}	y_{221}
y_{112}	y_{122}	y_{212}	y_{222}
...	...	...	...
y_{11n}	y_{12n}	y_{21n}	y_{22n}

The symbol y_{ijk} denotes measurement k of level i of factor A and level j of factor B.

The total sum of squares is partitioned to the sum of squares for factor A, the sum of squares for factor B, the sum of squares for the interaction of $A \times B$ and the residual sum of squares (unexplained sum of squares):

$$SS_{TOT} = SS_A + SS_B + SS_{AB} + SS_{RES}$$

with corresponding degrees of freedom:

$$(abn - 1) = (a - 1) + (b - 1) + (a - 1)(b - 1) + ab(n - 1)$$

The sums of squares are:

$$SS_{TOT} = \sum_i \sum_j \sum_k (y_{ijk} - \bar{y}...)^2$$

$$SS_A = \sum_i \sum_j \sum_k (\bar{y}_i.. - \bar{y}...)^2 = bn \sum_i (\bar{y}_i.. - \bar{y}...)^2$$

$$SS_B = \sum_i \sum_j \sum_k (\bar{y}._j. - \bar{y}...)^2 = an \sum_j (\bar{y}._j. - \bar{y}...)^2$$

$$SS_{AB} = n \sum_i \sum_j (\bar{y}_{ij}. - \bar{y}...)^2 - SS_A - SS_B$$

$$SS_{RES} = \sum_i \sum_j \sum_k (\bar{y}_{ijk} - \bar{y}_{ij}.)^2$$

The sums of squares can be calculated using short-cut computations:

1. Total sum:

$$\sum_i \sum_j \sum_k y_{ijk}$$

2. Correction for the mean:

$$C = \frac{\left(\sum_i \sum_j \sum_k y_{ijk}\right)^2}{abn}$$

3. Total sum of squares:

$$SS_{TOT} = \sum_i \sum_j \sum_k (y_{ijk})^2 - C$$

4. Sum of squares for factor A:

$$SS_A = \sum_i \frac{\left(\sum_j \sum_k y_{ijk}\right)^2}{nb} - C$$

5. Sum of squares for factor B:

$$SS_B = \sum_j \frac{\left(\sum_i \sum_k y_{ijk}\right)^2}{na} - C$$

6. Sum of squares for interaction:

$$SS_{AB} = \sum_i \sum_j \frac{\left(\sum_k y_{ijk}\right)^2}{n} - SS_A - SS_B - C$$

7. Residual sum of squares:

$$SS_{RES} = SS_{TOT} - SS_A - SS_B - SS_{AB}$$

Dividing the sums of squares by their corresponding degrees of freedom yields the mean squares:

Mean square for factor A: $MS_A = \dfrac{SS_A}{a-1}$

Mean square for factor B: $MS_B = \dfrac{SS_B}{b-1}$

Mean square for the $A \times B$ interaction: $MS_{A \times B} = \dfrac{SS_{A \times B}}{(a-1)(b-1)}$

Mean square for residual (experimental error): $MS_{RES} = \dfrac{SS_{RES}}{ab(n-1)}$

The sums of squares, mean squares and degrees of freedom are shown in an *ANOVA* table:

Source	SS	df	MS	F	
A	SS_A	$a-1$	MS_A	MS_A/MS_{RES}	(2)
B	SS_B	$b-1$	MS_B	MS_B/MS_{RES}	(3)
$A \times B$	$SS_{A \times B}$	$(a-1)(b-1)$	$MS_{A \times B}$	$MS_{A \times B}/MS_{RES}$	(1)
Residual	SS_{RES}	$ab(n-1)$	MS_{RES}		
Total	SS_{TOT}	$abn-1$			

In the table, the tests for A, B and $A \times B$ effects are depicted with numbers (2), (3) and (1), respectively:

(1) The F test for the interaction follows the hypotheses:

H_0: $\mu_{ij} = \mu_{i'j'}$ for all i, j, i', j'
H_1: $\mu_{ij} \neq \mu_{i'j'}$ for at least one pair $(ij, i'j')$

The test statistic:

$$F = \frac{MS_{A \times B}}{MS_{RES}}$$

has an F distribution with $(a-1)(b-1)$ and $ab(n-1)$ degrees of freedom if H_0 holds.

(2) The F test for factor A (if there is no interaction) follows the hypotheses:
H_0: $\mu_i = \mu_{i'}$ for each pair i, i'
H_1: $\mu_i \neq \mu_{i'}$ for at least one pair i, i'

The test statistic:

$$F = \frac{MS_A}{MS_{RES}}$$

has an F distribution with $(a-1)$ and $ab(n-1)$ degrees of freedom if H_0 holds.

(3) The F test for factor B (if there is no interaction) follows the hypotheses:
H_0: $\mu_j = \mu_{j'}$ for each pair j, j'
H_1: $\mu_j \neq \mu_{j'}$ for at least one pair j, j'

The test statistic:

$$F = \frac{MS_B}{MS_{RES}}$$

has an F distribution with $(b-1)$ and $ab(n-1)$ degrees of freedom if H_0 holds.

The hypothesis test for interaction must be carried out first, and only if the effect of interaction is not significant the main effects are tested. If the interaction is significant, tests for the main effects are meaningless, because the magnitude of effect of one factor depends on the level of the second factor in which it is being tested, i.e., the effect of factor A is not consistent across the levels of factor B.

Interactions can be shown graphically (Figure 17.1). The vertical axis represents measures and the horizontal axis represents levels of factor A. The connected symbols represent the levels of factor B. If the lines are roughly parallel, this means that there is no interaction. Any difference in slope between the lines indicates a possible interaction, the greater the difference in slope the stronger the interaction.

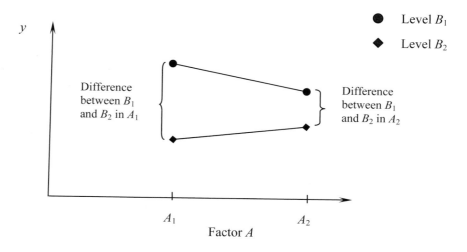

Figure 17.1 Illustration of interaction between two factors A and B

If an interaction exists there are two possible approaches to the problem:

1. Use a two-way model with interaction. The total sum of squares is partitioned to the sum of squares for factor A, the sum of squares for factor B, the sum of squares for interaction and the residual sum of squares:

$$SS_{TOT} = SS_A + SS_B + SS_{AB} + SS_{RES}$$

2. Use a one-way model, the combination of levels of $A \times B$ are treatments. With this procedure, the treatment sum of squares is equal to the summation of the sum of squares for factor A, the sum of squares for factor B, and the sum of squares for interaction:

$$SS_{TRT} = SS_A + SS_B + SS_{AB}$$

The total sum of squares is:

$$SS_{TOT} = SS_{TRT} + SS_{RES}$$

If interaction does not exist, an additive model is more appropriate. The additive model contains only main effects and interaction is not included:

$$y_{ijk} = \mu + A_i + B_j + \varepsilon_{ijk}$$

In the additive model the total sum of squares is partitioned to:

$$SS_{TOT} = SS_A + SS_B + SS_{RES'}$$

The residual sum of squares (SS_{RES}) is equal to the sum of squares for interaction plus the residual sum of squares for the model with interaction:

$$SS_{RES}' = SS_{AB} + SS_{RES}$$

In factorial experiments with three or more factors, there are additional combinations of interactions. For example, in an experiment with three factors A, B and C, it is possible to define the following interactions: $A \times B$, $A \times C$, $B \times C$ and $A \times B \times C$. A problem connected with three-way and more complex interactions is that it is often difficult to explain their practical meaning.

Example: An experiment was conducted to determine the effect of adding two vitamins (I and II) in feed on average daily gain of pigs. Two levels of vitamin I (0 and 4 mg) and two levels of vitamin II (0 and 5 mg) were used. The total sample size was 20 pigs, on which the four combinations of vitamin I and vitamin II were randomly assigned. The following daily gains were measured:

Vitamin I	0 mg		4 mg	
Vitamin II	0 mg	5 mg	0 mg	5 mg
	0.585	0.567	0.473	0.684
	0.536	0.545	0.450	0.702
	0.458	0.589	0.869	0.900
	0.486	0.536	0.473	0.698
	0.536	0.549	0.464	0.693
Sum	2.601	2.786	2.729	3.677
Average	0.520	0.557	0.546	0.735

The sums of squares are calculated:
1. Total sum:

$$\sum_i \sum_j \sum_k y_{ijk} = (0.585 + \ldots\ldots + 0.693) = 11.793$$

2. Correction for the mean:

$$C = \frac{\left(\sum_i \sum_j \sum_k y_{ijk}\right)^2}{abn} = \frac{(11.793)^2}{20} = 6.953742$$

3. Total sum of squares:

$$SS_{TOT} = \sum_i \sum_j \sum_k (y_{ijk})^2 - C = 0.585^2 + 0.536^2 + \ldots + 0.693^2 = 7.275437 -$$
$$6.953742 = 0.32169455$$

4. Sum of squares for vitamin I:

$$SS_{Vit\,I} = \sum_i \frac{\left(\sum_j \sum_k y_{ijk}\right)^2}{nb} - C = \frac{(2.601 + 2.786)^2}{10} + \frac{(2.729 + 3.677)^2}{10} - 6.953742 = 0.05191805$$

5. Sum of squares for vitamin II:

$$SS_{Vit\,II} = \sum_j \frac{\left(\sum_i \sum_k y_{ijk}\right)^2}{na} - C = \frac{(2.601+2.729)^2}{10} + \frac{(2.786+3.677)^2}{10} - 6.953742 = 0.06418445$$

6. Sum of squares for interaction:

$$SS_{Vit\,I \times Vit\,II} = \sum_i \sum_j \frac{\left(\sum_k y_{ijk}\right)^2}{n} - SS_A - SS_B - C =$$

$$\frac{(2.601)^2}{5} + \frac{(2.786)^2}{5} + \frac{(2.729)^2}{5} + \frac{(3.677)^2}{5} - 0.05191805 - 0.06418445 - 6.953742 = 0.02910845$$

7. Residual sum of squares:

$$SS_{RES} = SS_{TOT} - SS_{Vit\,I} - SS_{Vit\,II} - SS_{Vit\,I \times Vit\,II} =$$
$$0.32169455 - 0.05191805 - 0.06418445 - 0.02910845 = 0.17648360$$

The *ANOVA* table is:

Source	SS	df	MS	F
Vitamin I	0.05191805	1	0.05191805	4.71
Vitamin II	0.06418445	1	0.06418445	5.82
Vit I x Vit II	0.02910845	1	0.02910845	2.64
Residual	0.17648360	16	0.01103023	
Total	0.32169455	19		

The critical value for $\alpha = 0.05$ is $F_{0.05,1,16} = 4.49$. The computed F value for the interaction is 2.64. In this case the calculated F value is less than the critical value. The means of the factor level combinations are shown in Figure 17.2. Recall that if lines are roughly parallel, interaction is not present. In this case, the figure suggests that interaction may exist, but five measurements per group does not give sufficient power to detect it. Both main effects are significant indicating additive beneficial effects of Vitamin I and II on gain of pigs.

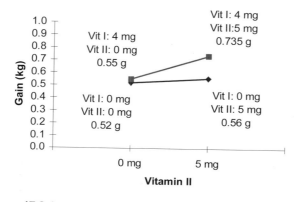

Figure 17.2 Interaction of vitamins I and II

17.1.1 SAS Example for Factorial Experiment

The SAS program for the example of vitamin supplementation is as follows.

SAS program:

```
DATA gain;
INPUT vitI vitII gain @@;
DATALINES;
1 1 0.585      2 1 0.473
1 1 0.536      2 1 0.450
1 1 0.458      2 1 0.869
1 1 0.486      2 1 0.473
1 1 0.536      2 1 0.464
1 2 0.567      2 2 0.684
1 2 0.545      2 2 0.702
1 2 0.589      2 2 0.900
1 2 0.536      2 2 0.698
1 2 0.549      2 2 0.693
;
PROC GLM DATA = gain ;
CLASS vitI vitII;
MODEL gain = vitI vitII vitI*vitII;
LSMEANS vitI*vitII / STDERR PDIFF ADJUST=TUKEY;
RUN;
QUIT;
```

Explanation: The GLM procedure is used. The CLASS statement defines classification (categorical) independent variables. The statement, *MODEL Gain = vitI vitII vitI*vitII* defines the dependent variable *gain*, and independent variables *vitI*, *vitII* and their interaction *vitI*vitII*. The LSMEANS statement calculates means. The options after the slash specify calculation of standard errors and tests of differences between least squares means using a Tukey test.

SAS output:

Dependent Variable: gain

Source	DF	Sum of Squares	Mean Square	F Value	Pr > F
Model	3	0.14521095	0.04840365	4.39	0.0196
Error	16	0.17648360	0.01103023		
Corrected Total	19	0.32169455			

R-Square	Coeff Var	Root MSE	gain Mean
0.451394	17.81139	0.10502	0.58965

Source	DF	Type III SS	Mean Square	F Value	Pr > F
vitI	1	0.05191805	0.05191805	4.71	0.0454
vitII	1	0.06418445	0.06418445	5.82	0.0282
vitI*vitII	1	0.02910845	0.02910845	2.64	0.1238

```
                              Least Squares Means
                      Adjustment for multiple comparisons: Tukey
        vit     vit                      Standard                    LSMEAN
         I      II     gain LSMEAN        Error        Pr > |t|      Number
         1       1      0.52020000      0.04696855      0.0001          1
         1       2      0.55720000      0.04696855      0.0001          2
         2       1      0.54580000      0.04696855      0.0001          3
         2       2      0.73540000      0.04696855      0.0001          4

            Least Squares Means for effect vitI*vitII
        Pr > |t|  for H0: LSMean(i)=LSMean(j), Dependent Variable: gain

             i/j        1          2          3          4
              1                  0.9433     0.9799     0.0238
              2       0.9433                0.9981     0.0701
              3       0.9799     0.9981                0.0506
              4       0.0238     0.0701     0.0506
```

Explanation: First in the GLM output is an *ANOVA* table for the *Dependent Variable gain*. The *Sources* of variability are *Model*, *Error* (residual) and *Corrected Total*. In the table are listed degrees of freedom (*DF*), *Sum of Squares*, *Mean Square*, calculated *F* (*F Value*) and *P* value (*Pr > F*). In the next table the explained sources of variability are partitioned to *vitI*, *vitII* and *vitI*vitII*. For example, for the interaction effect *vitI*vitII* the calculated *F* and *P* values are 2.64 and 0.1238, respectively. At the end of output least squares means (*LSMEAN*) with their standard errors (*Standard Errors*) are given, and then the Tukey test between all pairs of treatment combinations. The *P* values adjusted for multiple comparisons are shown. For example, in row 1 and column 4 the number 0.0238 denotes the *P* value for testing the differences between combinations of 0 mg vitamin I and 0 mg vitamin II, versus 4 mg vitamin I and 5 mg vitamin II.

Exercise

17.1. The objective of this experiment was to determine possible interactions of three protein sources with increasing energy on milk yield in dairy cows. Three types of protein were used: rape seed + soybean, sunflower + soybean and sunflower + rape seed meal, and two energy levels: standard and high. The base diet was the same for all cows. The following average daily milk yields were measured:

Protein Source	Rape seed + soybean		Sunflower + soybean		Rape seed + sunflower	
Energy level	High	Standard	High	Standard	High	Standard
	32	25	30	29	28	25
	29	26	29	28	27	30
	38	25	26	34	32	26
	36	31	34	36	33	27
	30	28	34	32	33	28
	25	23	30	30	37	24
	29	26	32	27	36	22
	32	26	33	29	26	28

Test the effect of interaction between protein source and energy level.

Chapter 18

Hierarchical or Nested Design

In some experiments samples must be chosen in two or more steps. For example, if the objective is to test if corn silage quality varies more between regions than within regions, a random sample of regions must be chosen, and then within each region a sample of farms must be chosen. The farms within region 1 are different than the farms within region 2, etc. Therefore, the first step is to choose regions, and the second step is to choose farms within the regions. This is an example of a hierarchical or nested design. Samples can be chosen in more steps giving, two-, three- or multiple-step hierarchical designs.

18.1 Hierarchical Design with Two Factors

Consider an experiment with two factors. Let factor A have three levels, and factor B three levels within each level of factor A. The levels of B are nested within levels of A, that is, the three levels of B are different for each level of A. Within each level of B some number n random samples are chosen. The schema of this design is:

A		1			2			3	
B	1	2	3	4	5	6	7	8	9
	y_{111}	y_{121}	y_{131}	y_{241}	y_{251}	y_{261}	y_{371}	y_{381}	y_{391}
	y_{112}	y_{122}	y_{132}	y_{242}	y_{252}	y_{262}	y_{372}	y_{382}	y_{392}
	y_{11n}	y_{12n}	y_{13n}	y_{24n}	y_{25n}	y_{26n}	y_{37n}	y_{38n}	y_{39n}

The model for this design is:

$$y_{ijk} = \mu + A_i + B(A)_{ij} + \varepsilon_{ijk} \qquad i = 1,\dots, a; \; j = 1,\dots, b; \; k = 1,\dots, n$$

where:

y_{ijk} = observation k in level i of factor A and level j of factor B
μ = the overall mean
A_i = the effect of level i of factor A
$B(A)_{ij}$ = the effect of level j of factor B within level i of factor A
ε_{ijk} = random error with mean 0 and variance σ^2
a = the number of levels of A; b = the number of levels of B; n = the number of observations per level of B

For example, assume that the levels of factor A are boars of the Landrace breed, and the levels of factor B are sows mated to those boars. The sows are a random sample within the boars. Daily gain was measured on offspring of those boars and sows. The offspring represent random samples within the sows. If any relationship among the sows is ignored, then the sows bred by different boars are independent. Also, the offspring of different sows and boars are independent of each other.

Similarly to other designs, the total sum of squares can be partitioned into the sums of squares for each source of variability. They are the sum of squares for factor A, the sum of squares for factor B within factor A, and the sum of squares within B (the residual sum of squares):

$$SS_{TOT} = SS_A + SS_{B(A)} + SS_{WITHIN\ B}$$

Their corresponding degrees of freedom are:

$$(abn - 1) = (a - 1) + a(b - 1) + ab(n - 1)$$

The sums of squares are:

$$SS_{TOT} = \sum_i \sum_j \sum_k (y_{ijk} - \bar{y}...)^2$$

$$SS_A = \sum_i \sum_j \sum_k (\bar{y}_i.. - \bar{y}...)^2 = bn \sum_i (\bar{y}_i.. - \bar{y}...)^2$$

$$SS_{B(A)} = \sum_i \sum_j \sum_k (\bar{y}_{ij}. - \bar{y}_i..)^2 = n \sum_i \sum_j (\bar{y}_{ij}. - \bar{y}_i..)^2$$

$$SS_{WITHIN\ B} = \sum_i \sum_j \sum_k (\bar{y}_{ijk} - \bar{y}_{ij}.)^2$$

Sums of squares can be calculated by short-cut computations:

1. Total sum:

$$\sum_i \sum_j \sum_k y_{ijk}$$

2. Correction for the mean:

$$C = \frac{\left(\sum_i \sum_j \sum_k y_{ijk} \right)^2}{abn}$$

3. Total sum of squares:

$$SS_{TOT} = \sum_i \sum_j \sum_k (y_{ijk})^2 - C$$

4. Sum of squares for factor A:

$$SS_A = \sum_i \frac{\left(\sum_j \sum_k y_{ijk} \right)^2}{nb} - C$$

5. Sum of squares for factor B within factor A:

$$SS_{B(A)} = \sum_i \sum_j \frac{\left(\sum_k y_{ijk} \right)^2}{n} - SS_A - C$$

6. Sum of squares within factor B (the residual sum of squares):

$$SS_{WITHIN\ B} = SS_{TOT} - SS_A - SS_{B(A)}$$

Mean squares (MS) are obtained by dividing the sums of squares (SS) by their corresponding degrees of freedom (df). The $ANOVA$ table is:

Source	SS	df	MS = SS / df
A	SS_A	$a - 1$	MS_A
B within A	$SS_{B(A)}$	$a(b - 1)$	$MS_{B(A)}$
Within B	$SS_{WITHIN\ B}$	$ab(n - 1)$	$MS_{WITHIN\ B}$
Total	SS_{TOT}	$abn - 1$	

The effect '*Within B*' is an unexplained effect or residual. Expectations of mean squares, $E(MS)$, are defined according whether the effects of A and B are fixed or random:

$E(MS)$	A and B fixed	A fixed and B random	A and B random
$E(MS_A)$	$\sigma^2 + Q(A)$	$\sigma^2 + n\,\sigma^2_B + Q(A)$	$\sigma^2 + n\,\sigma^2_B + nb\,\sigma^2_A$
$E(MS_{B(A)})$	$\sigma^2 + Q(B(A))$	$\sigma^2 + n\,\sigma^2_B$	$\sigma^2 + n\,\sigma^2_B$
$E(MS_{WITHIN\ B})$	σ^2	σ^2	σ^2

where σ^2, σ^2_B and σ^2_A are variance components for error, factor B and factor A, and $Q(A)$ and $Q(B(A))$ are fixed values of squares of factors A and B, respectively.

The experimental error for particular effects depends on whether the effects are fixed or random. Most often B is random. In that case the experimental error term to test the effect of A is the $MS_{B(A)}$, and the experimental error term for the effect of B is the $MS_{WITHIN\ B}$. The F statistic for the effect of A is:

$$F = \frac{MS_A}{MS_{B(A)}}$$

The F statistic for the effect of B is:

$$F = \frac{MS_{B(A)}}{MS_{WITHIN\ B}}$$

Example: The aim of this experiment was to determine effects of boars and sows on variability of birth weight of their offspring. A nested design was used: 4 boars were randomly chosen with 3 sows per boar and 2 piglets per sow. The data, together with sums and sum of squares, are shown in the following table:

Boars	Sows	Piglets	Weight	Total sum	Sums per boar	Sums per sow
1	1	1	1.2			
1	1	2	1.2			2.4
1	2	3	1.2			
1	2	4	1.3			2.5
1	3	5	1.1			
1	3	6	1.2		7.2	2.3
2	4	7	1.2			
2	4	8	1.2			2.4
2	5	9	1.1			
2	5	10	1.2			2.3
2	6	11	1.2			
2	6	12	1.1		7.0	2.3
3	7	13	1.2			
3	7	14	1.2			2.4
3	8	15	1.3			
3	8	16	1.3			2.6
3	9	17	1.2			
3	9	18	1.2		7.4	2.4
4	10	19	1.3			
4	10	20	1.3			2.6
4	11	21	1.4			
4	11	22	1.4			2.8
4	12	23	1.3			
4	12	24	1.3	29.6	8.0	2.6
Sum			29.6	29.6	29.6	29.6
Number			24			
Sum of squares (uncorrected)			36.66		219.6	73.28

a = the number of boars = 4; b = the number of sows per boar = 3; n = the number of piglets per sow = 2

Short-cut computations of sums of squares:

1. Total sum:

$$\sum_i \sum_j \sum_k y_{ijk} = (1.2 + 1.2 + 1.2 + ... + 1.3 + 1.3) = 29.6$$

2. Correction for the mean:

$$C = \frac{\left(\sum_i \sum_j \sum_k y_{ijk}\right)^2}{abn} = \frac{(29.6)^2}{24} = 36.50667$$

abn = 24 = the total number of observations

3. Total sum of squares:

$$SS_{TOT} = \sum_i \sum_j \sum_k (y_{ijk})^2 - C = (1.2)^2 + (1.2)^2 + (1.2)^2 + \ldots\ldots + (1.3)^2 + (1.3)^2 - C$$

$$= 0.15333$$

4. Sum of squares for boars:

$$SS_{BOAR} = \sum_i \frac{\left(\sum_j \sum_k y_{ijk}\right)^2}{nb} - C \frac{1}{6}\left[(7.2)^2 + (7.0)^2 + (7.4)^2 + (8.0)^2\right] - 36.50667 = 0.09333$$

$nb = 6 =$ the number of observations per boar

5. Sum of squares for sows within boars:

$$SS_{SOW(BOAR)} = \sum_i \sum_j \frac{\left(\sum_k y_{ijk}\right)^2}{n} - SS_{BOAR} - C$$

$$= \frac{1}{2}\left[(2.4)^2 + (2.5)^2 + \ldots + (2.8)^2 + (2.6)^2\right] - 0.09333 - 36.50667 = 0.040$$

$n = 2 =$ the number of observations per sow

6. Sum of squares within sows (the residual sum of squares):

$$SS_{PIGLET(SOW)} = SS_{TOT} - SS_{BOAR} - SS_{SOW(BOAR)} = 0.15333 - 0\,09333 - 0.040 = 0.020$$

The *ANOVA* table:

Source	SS	df	MS	F
Boars	0.093	3	0.031	6.22
Sows within boars	0.040	8	0.005	3.00
Piglets within sows	0.020	12	0.002	
Total	0.153	23		

It was assumed that the effects of boars and sows are random. The experimental error term for boars is the mean square for sows within boars, and the experimental error term for sows is the mean square for piglets within sows. The critical value for boars is $F_{0.05,3,8} = 4.07$, and the critical value for sows within boars is $F_{0.05,8,12} = 2.85$. The calculated F values are greater than the critical values and thus the effects of sows and boars are significant. The estimates of variance components are shown in the following table:

Source	E(MS)	Variance components	Percentage of the total variability
Boars	$\sigma^2 + 2\sigma^2_{SOWS} + 6\sigma^2_{BOARS}$	0.004352	56.63
Sows within boars	$\sigma^2 + 2\sigma^2_{SOWS}$	0.001667	21.69
Piglets within sows	σ^2	0.001667	21.69
Total		0.007685	100.00

18.1.1 SAS Example for Hierarchical Design

SAS program for the example with variability of piglets' birth weight is as follows. The use of the NESTED and MIXED procedures are shown.

SAS program *for the NESTED procedure:*

```
DATA pig;
INPUT boar sow piglet birth_wt @@;
DATALINES;
1  1  1   1.2      1  1  2   1.2      1  2  1   1.2
1  2  2   1.3      1  3  1   1.1      1  3  2   1.2
2  1  1   1.2      2  1  2   1.2      2  2  1   1.1
2  2  2   1.2      2  3  1   1.2      2  3  2   1.1
3  1  1   1.2      3  1  2   1.2      3  2  1   1.3
3  2  2   1.3      3  3  1   1.2      3  3  2   1.2
4  1  1   1.3      4  1  2   1.3      4  2  1   1.4
4  2  2   1.4      4  3  1   1.3      4  3  2   1.3
;
PROC NESTED DATA = pig;
CLASS boar sow;
VAR birth_wt;
RUN;
```

Explanation: The NESTED procedure applies *ANOVA* estimation, which estimates mean squares by using degrees of freedom as for a fixed effects model. Variance components are calculated by using definitions of expected mean squares and numbers of observations per boar and sow. The NESTED procedure is appropriate only if there are not additional fixed effects in the model. The CLASS statement defines categorical variables, later variables are defined to be nested in previous ones. The data must be sorted by boar and sow in the order that they are given in the CLASS statement. The VAR statement defines the dependent variable *birth_wt*.

SAS output *of the NESTED procedure:*

```
          Coefficients of Expected Mean Squares

      Source        boar        sow       Error
       boar           6           2          1
       sow            0           2          1
       Error          0           0          1
```

Nested Random Effects Analysis of Variance for Variable birth_wt

Variance Source	DF	Sum of Squares	F Value	Pr>F	Error Term	Mean Square	Variance Component	Percent of Total
Total	23	0.153333				0.006667	0.007685	100.0000
boar	3	0.093333	6.22	0.0174	sow	0.031111	0.004352	56.6365
sow	8	0.040000	3.00	0.0424	Error	0.005000	0.001667	21.6867
Error	12	0.020000				0.001667	0.001667	21.6867

```
          birth_wt Mean                        1.23333333
          Standard Error of birth_wt Mean      0.03600411
```

Explanation: The first table presents the coefficients for estimating mean squares by the *ANOVA* method. Next is the *ANOVA* table for the *Dependent Variable birth_wt*. The *Sources* of variability are *Total*, *boar*, *sow* and *Error* (piglets within sows). In the table are listed degrees of freedom (*DF*), *Sum of Squares*, *F Value* and *P* value (*Pr > F*). Also, the correct *Error Term* is given; to test the effect of *boar* the appropriate error term is *sow*. Also are given, *Mean Squares*, *Variance Components* and each source's percentage of the total variance (*Percent of Total*). The variance components for *boar*, *sow* and *residual* (piglets) are 0.004352, 0.001667 and 0.001667, respectively. The NESTED procedure can be used to estimate variance components if data are missing or unequal numbers of observations per level of class variables (i.e. unbalanced data), but *F* tests are not valid, since sums of squares for unbalanced data do not have *chi*-square distributions. So, is data are unbalanced, no *F* test is given.

SAS program *for the MIXED procedure:*

```
PROC MIXED DATA = pig METHOD=TYPE3;
CLASS boar sow;
MODEL birth_wt =  / DDFM = KENWARDROGER;
RANDOM boar sow(boar)/S;
RUN;
```

Explanation: The MIXED procedure by default uses Restricted Maximum Likelihood (REML) estimation. The ANOVA method can be used by defining *METHOD = TYPE3*. The MIXED procedure is a more general procedure, and is appropriate with additional fixed effects in the model. The CLASS statement defines categorical variables, and the statement, *MODEL birth_wt = ;* denotes that the dependent variable is *birth_wt* and the only fixed effect in the model is the overall mean. The RANDOM statement defines the random effects *boar* and *sow(boar)*. The expression *sow(boar)* denotes that *sow* is nested within *boar*. The S option directs computation of predictions and their standard errors. The *DDFM = KENWARDROGER* option in the MODEL statement is recommended to estimate correct degrees of freedom when the mean square for experimental error or standard errors are linear combination of variances, and generally when there is heterogeneity of variances. Also, it can be used to account for possible missing data or different number of observations per boar or sow. Since there are no fixed effects in this model, the LSMEANS statement is not needed.

SAS output *of the MIXED procedure:*

```
                  Type 3 Analysis of Variance

                 Sum of
Source      DF   Squares   Mean Square      Expected Mean Square
boar        3    0.093333   0.031111    Var(Residual)+2Var(sow(boar))+6Var(boar)
sow(boar)   8    0.040000   0.005000    Var(Residual)+2Var(sow(boar))
Residual    12   0.020000   0.001667    Var(Residual)

                                     Error
Source          Error Term            DF     F Value    Pr > F
boar            MS(sow(boar))          8      6.22      0.0174
sow(boar)       MS(Residual)          12      3.00      0.0424
Residual        .                      .       .          .
```

Covariance Parameter Estimates

Cov Parm	Estimate
boar	0.004352
sow(boar)	0.001667
Residual	0.001667

Solution for Random Effects

Effect	boar	sow	Estimate	Std Err Pred	DF	t Value	Pr > |t|
boar	1		-0.02798	0.04239	3.96	-0.66	0.5456
boar	2		-0.05595	0.04239	3.96	-1.32	0.2579
boar	3		3.26E-15	0.04239	3.96	0.00	1.0000
boar	4		0.08393	0.04239	3.96	1.98	0.1195
sow(boar)	1	1	-0.00357	0.03390	7.87	-0.11	0.9187
sow(boar)	1	2	0.02976	0.03390	7.87	0.88	0.4060
sow(boar)	1	3	-0.03690	0.03390	7.87	-1.09	0.3085
sow(boar)	2	1	0.01508	0.03390	7.87	0.44	0.6684
sow(boar)	2	2	-0.01825	0.03390	7.87	-0.54	0.6051
sow(boar)	2	3	-0.01825	0.03390	7.87	-0.54	0.6051
sow(boar)	3	1	-0.02222	0.03390	7.87	-0.66	0.5308
sow(boar)	3	2	0.04444	0.03390	7.87	1.31	0.2268
sow(boar)	3	3	-0.02222	0.03390	7.87	-0.66	0.5308
sow(boar)	4	1	-0.01151	0.03390	7.87	-0.34	0.7431
sow(boar)	4	2	0.05516	0.03390	7.87	1.63	0.1430
sow(boar)	4	3	-0.01151	0.03390	7.87	-0.34	0.7431

Explanation: When the *type3 ANOVA* method is used, *ANOVA* tables are similar to those from the NESTED procedure, followed by the estimated variance components (*Cov Parm Estimate*). Under the title, *Solution for Random Effects*, the predictions for each *boar* and *sow* (*Estimate*) with their standard errors (*Std Err Pred*), *t* and *P* value (*t Value, Pr > |t|*) are shown. Note degrees of freedom are corrected by the *DDFM = KENWARDROGER* option. For example, for boars DF would be 12 without correction.

The MIXED procedure will use Restricted Maximum Likelihood (REML) estimation by default (i.e. if no method is specified) or if METHOD = REML is specified. The output using REML estimation will contain no ANOVA table outputs. For balanced data, ANOVA (TYPE3) and REML will yield the same estimates. If there are missing values (unbalanced data), REML method is highly recommended over TYPE3, as it yield correct estimates and tests.

Chapter 19

More about Blocking

If the results of an experiment are to be applied to livestock production, then experimental housing should be similar to housing on commercial farms. For example if animals in production are held in pens or paddocks, then the same should be applied in the experiment. It can often be difficult to treat animals individually. Choice of an experimental design can depend on grouping of animals and the way treatments are applied. The effect of blocking on the efficiency of a design was shown in Chapter 15. The precision of experiments can sometimes be enhanced by defining double blocks. For example, if animals to be used in an experiment are from two breeds and have different initial weights, breed can be defined as one block, and groups of initial weights as another. The use of multiple blocking variables can improve the precision of an experiment by removing blocks' contribution to variance.

19.1 Blocking With Pens, Corrals and Paddocks

In planning an experimental design it is necessary to define the experimental unit. If multiple animals are held in cages or pens it may be impossible to treat them individually. If the whole cage or pen is treated together, then the cage or pen is an experimental unit. Similarly, in experiments with a single treatment applied to all animals in each paddock, all animals in a paddock are one experimental unit. Multiple paddocks per treatment represent replications. This is true even when animals can be measured individually. In that case, multiple samples are taken on each experimental unit. Animals represent sample units. It is necessary to define the experimental error and the sample error. The definition of the experimental design and the statistical analysis of the experimental design depends on how the experimental unit is defined. For example, assume a design with two blocks ($b = 2$), two treatments ($a = 2$), and two animals per treatment × block combination ($n = 2$). Denote blocks by I and II, and treatments by T_1 and T_2. If it is possible to treat animals individually, then a possible design is:

Block I	Block II
T_2	T_1
T_1	T_2
T_1	T_1
T_2	T_2

There are four animals per block, and treatments are randomly assigned to them. This is a randomized complete block design with two units per treatment × block combination. The table with sources of variability is:

Source	Degrees of freedom	
Block	$(b - 1) =$	1
Treatment	$(a - 1) =$	1
Block x treatment	$(b - 1)(a - 1) =$	1
Exp. error = Residual	$ab(n - 1) =$	4
Total	$(abn - 1) =$	7

By using this design it is possible to estimate the block × treatment interaction. The experimental error is equal to the residual after accounting for the effects of block, treatment and their interaction.

It is often the case that animals cannot be treated individually. For example, assume again two blocks and two treatments, but two animals are held in each of four cages. The same treatment is applied to both animals in each cage. A possible design can be as follows:

Block I	Block II
T_1	T_2
T_1	T_2
T_2	T_1
T_2	T_1

Two animals are in each cage, there are two cages per block, and the treatments are randomly assigned to the cages within each block. The table with sources of variability is:

Source	Degrees of freedom	
Block	$(b - 1) =$	1
Treatment	$(a - 1) =$	1
Exp. error = Block x treatment	$(b - 1)(a - 1) =$	1
Residual	$ab(n - 1) =$	4
Total	$(abn - 1) =$	7

The experimental error term for testing the effect of treatment is the mean square for the block × treatment interaction because the experimental unit is a cage, and there is one cage per combinations of treatment × block. More generally, if there is more than one cage per treatment × block combinations, the experimental error term would be the mean square for cage within treatment × block.

The statistical model of this design is:

$$y_{ijk} = \mu + \tau_i + \beta_j + \delta_{ij} + \varepsilon_{ijk} \qquad i = 1,..., a; j = 1,..., b; k = 1,..., n$$

where:

y_{ijk} = observation k of treatment i in block j
μ = the overall mean
τ_i = the effect of treatment i
β_j = the effect of block j

δ_{ij} = random error between experimental units within block × treatment interaction with mean 0 and variance σ^2_δ (it corresponds to interaction of treatment × block when there is only one experimental unit per treatment × block combination)

ε_{ijk} = random error within experimental units with mean 0 and variance σ^2

a = the number of treatments, b = the number of blocks, n = the number of observations within experimental unit

The hypotheses of treatment effects are of primary interest:

H_0: $\tau_1 = \tau_2 = ... = \tau_a$, no treatment effects

H_1: $\tau_i \neq \tau_{i'}$, for at least one pair (i, i') a difference exists

To test hypotheses an F statistic can be used which, if H_0 holds, has an F distribution with $(a - 1)$ and $(a - 1)(b - 1)$ degrees of freedom:

$$F = \frac{MS_{TRT}}{MS_{Exp.Error}}$$

where MS_{TRT} is the treatment mean square, and $MS_{Exp.Error}$ is the mean square for error δ.

The *ANOVA* table is:

Source	SS	df	MS	F
Blocks	SS_{BLK}	$b - 1$	MS_{BLK}	$MS_{BLK} / MS_{Exp.Error}$
Treatments	SS_{TRT}	$a - 1$	MS_{TRT}	$MS_{TRT} / MS_{Exp.Error}$
Exp. error	$SS_{Exp.Error}$	$(a - 1)(b - 1)$	$MS_{Exp.Error}$	$MS_{Exp.Error} / MS_{RES}$
Residual	SS_{RES}	$ab(n - 1)$	MS_{RES}	
Total	SS_{TOT}	$abn - 1$		

The expected mean squares are:

$$E(MS_{Exp.Error}) = \sigma^2 + n\,\sigma^2_\delta$$
$$E(MS_{RES}) = \sigma^2$$

When calculating standard errors of the estimated treatment means and the difference between treatment means, the appropriate mean square must also be used. The standard error of the estimated treatment mean is:

$$s_{\bar{y}_{i..}} = \sqrt{\frac{MS_{Exp.Error}}{bn}}$$

Generally, using variance components, the standard error of the estimated mean of treatment i is:

$$s_{\bar{y}_{i..}} = \sqrt{\frac{\sigma^2 + n\sigma^2_\delta}{bn}}$$

The standard error of the estimated difference between means of two treatments i and i' is:

$$s_{\bar{y}_{i..}-\bar{y}_{i'..}} = \sqrt{MS_{Exp.Error}\left(\frac{1}{bn} + \frac{1}{bn}\right)}$$

Example: The effect of four treatments on daily gain of steers was investigated. The steers were grouped into three blocks according to their initial weight. Twenty-four steers were held in 12 pens, two steers per pen. Pen is the experimental unit. The following average daily gains were recorded:

Block I	Block II	Block III
Treatment 1	Treatment 2	Treatment 3
826	871	736
806	881	740
Treatment 3	Treatment 1	Treatment 4
795	827	820
810	800	835
Treatment 4	Treatment 4	Treatment 2
850	860	801
845	840	821
Treatment 2	Treatment 3	Treatment 1
864	729	753
834	709	773

The results are shown in the *ANOVA* table shown below, and conclusions are made as usual comparing the calculated F values with the critical values.

Source	SS	df	MS	F
Block	8025.5833	2	4012.7917	2.98
Treatment	33816.8333	3	11272.2778	8.36
Exp. Error	8087.4167	6	1347.9028	7.67
Residual	2110.0000	12	175.8333	
Total	52039.8333	23		

For the 0.05 level of significance, the critical value $F_{0.05,3,6}$ is 4.76. The calculated F for treatments is 8.36; thus, treatments affect daily gain of steers. There is only one pen per block × treatment combination and thus not sufficient replications in this experiment to test the fixed block × treatment interaction. If there was additional replication, pen (block x treatment) would be used to test the block, treatment and their interaction.

The standard error of an estimated treatment mean is:

$$s_{\bar{y}_i} = \sqrt{\frac{1347.9028}{(3)(2)}} = 14.9883$$

The standard error of the estimated difference between means of two treatments is:

$$s_{\bar{y}_i - \bar{y}_{i'}} = \sqrt{1347.9028\left(\frac{1}{(3)(2)} + \frac{1}{(3)(2)}\right)} = 21.1967$$

19.1.1 SAS Example for Designs with Pens and Paddocks

The SAS program for the example of daily gain of steers is as follows:

SAS program:

```
DATA steer;
INPUT pen block trt $ d_gain @@;
DATALINES;
  1   1  T1  826        1   1  T1  806
  2   1  T2  864        2   1  T2  834
  3   1  T3  795        3   1  T3  810
  4   1  T4  850        4   1  T4  845
  5   2  T1  827        5   2  T1  800
  6   2  T2  871        6   2  T2  881
  7   2  T3  729        7   2  T3  709
  8   2  T4  860        8   2  T4  840
  9   3  T1  753        9   3  T1  773
 10   3  T2  801       10   3  T2  821
 11   3  T3  736       11   3  T3  740
 12   3  T4  820       12   3  T4  835
;
PROC MIXED  DATA = steer;
CLASS block trt;
MODEL d_gain = block trt ;
RANDOM pen(block*trt);
LSMEANS trt / DIFF ADJUST=TUKEY;
RUN;
```

Explanation: The MIXED procedure by default uses Restricted Maximum Likelihood (REML) estimation. The CLASS statement defines categorical (classification) variables. The MODEL statement defines the dependent variable and the independent variables fitted in the model. The RANDOM statement defines random effects *pen(block*trt)*, which will thus be defined as the experimental error for testing treatments. The LSMEANS statement calculates treatment least squares means. The options after the slash specify calculation of standard errors and tests of differences between least squares means using a Tukey test. If there are missing values it is recommended that degrees of freedom are corrected by using DDFM = SATTERTH or DDFM = KENWARDROGER as an option in the MODEL statement (MODEL milk = order trt period / DDFM = SATTERTH;)

SAS output:

```
            Covariance Parameter Estimates (REML)

        Cov Parm            Estimate
        block*trt        586.03472222
        Residual         175.83333333
```

```
            Type 3 Tests of Fixed Effects

   Effect        Num DF   Den DF    F Value   Pr > F
   block            2        6        2.98     0.1264
   trt              3        6        8.36     0.0145
```

```
                  Least Squares Means

                            Standard
   Effect    trt   Estimate    Error    DF     t      Pr > |t|
   trt   T1    797.5000    14.9883     6    53.21     0.0001
   trt   T2    845.3333    14.9883     6    56.40     0.0001
   trt   T3    753.1667    14.9883     6    50.25     0.0001
   trt   T4    841.6667    14.9883     6    56.15     0.0001
```

```
              Differences of Least Squares Means

                            Standard
   Effect trt _trt  Diff     Error    DF   t    Pr>|t|   Adjust.   Adj P
   trt     T1  T2  -47.83   21.1967    6  -2.26  0.0648   Tukey    0.2106
   trt     T1  T3   44.33   21.1967    6   2.09  0.0814   Tukey    0.2561
   trt     T1  T4  -44.17   21.1967    6  -2.08  0.0823   Tukey    0.2585
   trt     T2  T3   92.17   21.1967    6   4.35  0.0048   Tukey    0.0188
   trt     T2  T4    3.67   21.1967    6   0.17  0.8684   Tukey    0.9980
   trt     T3  T4  -88.50   21.1967    6  -4.18  0.0058   Tukey    0.0226
```

Explanation: The MIXED procedure estimates variance components for random effects (*Covariance Parameter Estimates*) and provides *F* tests for fixed effects (*Type 3 Tests of Fixed Effects*). These values will be the same as from the GLM procedure if the data are balanced. If the numbers of observations are not equal, the MIXED procedure must be used. In the *Least Squares Means* table, the means (*Estimate*) with their *Standard Error* are presented. In the *Differences of Least Squares Means* table the differences between pairs of means are shown (*Diff*). The differences are tested using the Tukey-Kramer procedure, which adjusts for the multiple comparison and unequal subgroup size. The correct *P* value is the adjusted *P* value (*Adj P*). For example, the *P* value for testing the difference between treatments 3 and 4 is 0.0226.

19.2   Double Blocking

If two explained sources of variability are known along with treatment, then the experimental units can be grouped into double blocks. For example, animals can be grouped to blocks according to their initial weight and also their sex. Consider a design with three treatments, four blocks according to initial weight, and two blocks according to sex. Thus, there are eight blocks, with four within each sex. There is a total of $3 \times 2 \times 4 = 24$ animals. A possible design is:

Males		Females	
Block I	Block II	Block V	Block VI
T1	T2	T3	T1
T2	T1	T2	T2
T3	T3	T1	T3
Block III	Block IV	Block VII	Block VIII
T1	T2	T2	T3
T3	T1	T1	T2
T2	T3	T3	T1

The number of sexes is $s = 2$, the number of blocks within sex is $b = 4$, and the number of treatments is $a = 3$. The *ANOVA* table is:

Source	Degrees of freedom	
Blocks	$(sb - 1) =$	7
Sex	$(s - 1) =$	1
Blocks within sex	$s(b - 1) =$	6
Treatment	$(a - 1) =$	2
Block x treatment	$(sb - 1)(a - 1) =$	14
Sex x treatment	$(s - 1)(a - 1) =$	2
Residual	$s(b - 1)(a - 1) =$	12
Total	$(abs - 1) =$	23

The effects in the table shifted to the right denote partitions of the effects above them. The effects of all eight blocks are partitioned into the effects of sex and blocks within sex. The interaction of block × treatment is divided into the sex × treatment interaction and residual.

The experimental design and statistical model depend on how sources of variability are defined, as blocks or treatments. If the objective is to test an effect, it is defined as a treatment. If an effect is defined just to reduce unexplained variability, it should be defined as a block. For example, the aim of an experiment is to investigate the effects of three treatments on dairy cows. Groups of cows from each of two breeds were used. The cows were also grouped according to their number of lactations: I, II, III and IV. The number of breeds is $b = 2$, the number of lactations is $m = 4$, and the number of treatments is $a = 3$. Several experimental designs can be defined depending on the objective and possible configurations of animal housing.

Experimental design 1:

The objective is to test the effect of treatment with breed defined as a block. The animals are first divided according to breed into two pens. For each breed there are cows in each of the four lactation numbers. The treatments are randomly assigned within each lactation × breed combination.

Breed A		Breed B	
Lactation I	Lactation II	Lactation I	Lactation II
T1	T2	T3	T1
T2	T1	T2	T2
T3	T3	T1	T3
Lactation III	Lactation IV	Lactation III	Lactation IV
T1	T2	T2	T3
T3	T1	T1	T2
T2	T3	T3	T1

The *ANOVA* table is:

Source	Degrees of freedom	
Breed	$(b - 1) =$	1
Lactation within breed	$b(m - 1) =$	6
Treatment	$(a - 1) =$	2
Breed x treatment	$(b - 1)(a - 1) =$	2
Residual	$b(m - 1)(a - 1) =$	12
Total	$(abm - 1) =$	23

Experimental design 2:

If breed is defined as a 'treatment', then a factorial experiment is defined with $2 \times 3 = 6$ combinations of breed × treatment assigned to a randomized block plan. The lactations are blocks and cows in the same lactation are held in the same pen. This design is appropriate if the objective is to test the effects of the breed and breed × treatment interaction. In the following scheme letters *A* and *B* denote breeds:

Lactation I	Lactation II	Lactation III	Lactation IV
A T1	B T3	A T1	A T2
B T2	A T2	B T3	A T1
B T3	A T1	B T2	B T3
A T2	A T1	A T2	A T3
B T1	B T2	B T1	B T2
A T3	A T3	A T3	B T1

The *ANOVA* table is:

Source	Degrees of freedom	
Lactation	$(m - 1) =$	3
Breed	$(b - 1) =$	1
Treatment	$(a - 1) =$	2
Breed x treatment	$(b - 1)(a - 1) =$	2
Residual	$(m - 1)[(b - 1) + (a - 1) + (b - 1)(a - 1)] =$	15
Total	$(amb - 1) =$	23

Experimental design 3:

The cows are grouped according to lactations into four blocks. Each block is then divided into two pens and one breed is randomly assigned to each, and treatments are randomly assigned within each pen. Thus, there is a total of eight pens. This is a split-plot design which will be explained in detail in the next chapter. Note that two experimental errors are defined, because two types of experimental units exist: breed within lactation and treatment within breed within lactation.

Lactation I		*Lactation II*	
Breed A	*Breed B*	*Breed B*	*Breed A*
T1	T2	T3	T1
T2	T1	T2	T2
T3	T3	T1	T3

Lactation III		*Lactation IV*	
Breed B	*Breed A*	*Breed A*	*Breed B*
T1	T2	T2	T3
T3	T1	T1	T2
T2	T3	T3	T1

The *ANOVA* table is:

Source	Degrees of freedom	
Lactation	$(m - 1) =$	3
Breed	$(b - 1) =$	1
Exp. error I (Lactation x Breed)	$(m - 1)(b - 1) =$	3
Subtotal	$(m - 1) + (b - 1) + (m - 1)(b - 1) =$	7
Treatment	$(a - 1) =$	2
Breed x treatment	$(b - 1)(a - 1) =$	2
Exp. error II	$b(a - 1)(m - 1) =$	12
Total	$(amb - 1) =$	23

The most appropriate experimental design depends on the objective and the housing and grouping configuration. There may be appropriate designs that make use of combinations of double blocking and experimental units defined as pens, paddocks or corrals.

Chapter 20

Split-plot Design

The split-plot design is applicable when the effects of two factors are organized in the following manner. Experimental material is divided into several main units, to which the levels of the first factor are randomly assigned. Each of the main units is then divided into subunits to which the levels of the second factor are randomly assigned. For example, consider an experiment conducted on a meadow in which we wish to investigate the effects of three levels of nitrogen fertilizer and two grass mixtures on green mass yield. The experiment can be designed in such a way that each of several blocks of land is divided into three main plots, and on each plot a level of nitrogen is randomly assigned. Each of the main plots is again divided into two subplots, and on each subplot within main plots one of the two grass mixtures is sown, again randomly. To obtain replicates, everything is repeated on each block. The name split-plot came from this type of application in agricultural experiments. The main units were called plots, and the subunits split-plots. The split-plot design plan can include combinations of completely randomized designs, randomized block designs, or Latin square designs, which can be applied either on the plots or subplots.

The split-plot design is used when one of the factors needs more experiment material than the second factor. For example, in field experiments one of the factors is land tillage or application of fertilizer. Such factors need large experimental units, therefore they are applied on the main plots. The other factor can be for example, different grass species, which can be compared on subplots. As a common rule, if one factor is applied later than the other, then this later factor is assigned to subplots. Also, if from experience we know that larger differences are to be expected from one of the factors, then that factor is assigned to the main plots. If we need more precise analyses of one factor, then that factor is assigned to the subplots.

20.1 Split-Plot Design – Main Plots in Randomized Blocks

One example of a split-plot design has one of the factors applied to main plots in randomized block design. Consider a factor A with four levels (A_1, A_2, A_3 and A_4), and a factor B with two levels (B_1 and B_2). The levels of factor A are applied to main plots in three blocks. This is a randomized block plan. Each of the plots is divided into two subplots and the levels of B are randomly assigned to them.

One possible plan is:

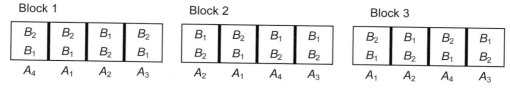

The model for this design is:

$$y_{ijk} = \mu + Block_k + A_i + \delta_{ik} + B_j + (AB)_{ij} + \varepsilon_{ijk} \qquad i = 1,..., a; j = 1,..., b ; k = 1,..., n$$

where:

y_{ijk} = observation k in level i of factor A and level j of factor B

μ = the overall mean

$Block_k$ = the effect of the kth block

A_i = the effect of level i of factor A

B_j = the effect of level j of factor B

$(AB)_{ij}$ = the effect of the ij^{th} interaction of $A \times B$

δ_{ik} = the main plot error (the interaction $Block_k \times A_i$) with mean and variance σ^2_{δ}

ε_{ijk} = the split-plot error with mean 0 and variance σ^2

Also, $\mu_{ij} = \mu + A_i + B_j + (AB)_{ij}$ = the mean of ij^{th} $A \times B$ interaction

n = number of blocks

a = number of levels of factor A

b = number of levels of factor B

It is assumed that main plot and split-plot errors are independent.

The *ANOVA* table for the design with three blocks, four levels of factor A and two levels of factor B:

Source	Degrees of freedom	
Block	$(n-1) =$	2
Factor A	$(a-1) =$	3
Main plot exp. error	$(n-1)(a-1) =$	6
Factor B	$(b-1) =$	1
$A \times B$	$(a-1)(b-1) =$	3
Split-plot exp. error	$a(b-1)(n-1) =$	8
Total	$(abn-1) =$	23

$a = 4$ = number of levels of factor A

$b = 2$ = number of levels of factor B

$n = 3$ = number of blocks

The effects of factors and interactions can be tested by using F tests.

The F statistic for factor A:

$$F = \frac{MS_A}{MS_{Main\ plot\ error}}$$

The main plot experimental error term is the mean square for the *Block* $\times A$ interaction. The F statistic for factor B:

$$F = \frac{MS_B}{MS_{Split\text{-}plot\ error}}$$

The split-plot experimental error term is the residual mean square.

The F statistic for the $A \times B$ interaction:

$$F = \frac{MS_{A \times B}}{MS_{Split\text{-}plot\ error}}$$

Example: An experiment was conducted in order to investigate four different treatments of pasture and two mineral supplements on milk yield. Twenty-four cows were used in the experiment. The experiment was designed as a split-plot, with pasture treatments (factor A) assigned to the main plots and mineral supplements (factor B) assigned to split-plots. The experiment was replicated in three blocks. The following milk yields were measured:

Plot	Block	Pasture	Mineral	Milk (kg)	Plot	Block	Pasture	Mineral	Milk (kg)
1	1	4	2	30	7	2	4	1	34
1	1	4	1	29	7	2	4	2	37
2	1	1	2	27	8	2	3	1	33
2	1	1	1	25	8	2	3	2	32
3	1	2	1	26	9	3	1	2	34
3	1	2	2	28	9	3	1	1	31
4	1	3	2	26	10	3	2	1	30
4	1	3	1	24	10	3	2	2	31
5	2	2	1	32	11	3	4	2	36
5	2	2	2	37	11	3	4	1	38
6	2	1	2	30	12	3	3	1	33
6	2	1	1	31	12	3	3	2	32

The results are shown in the *ANOVA* table.

Source	SS	df	MS	F
Block	212.583	2	106.292	
Pasture treatment	71.167	3	23.722	5.46
Main plot exp. error	26.083	6	4.347	
Mineral supplement	8.167	1	8.167	3.63
Pasture x Mineral	5.833	3	1.944	0.86
Split-plot exp. error	18.000	8	2.250	
Total	341.833	23		

The critical value for the Pasture treatment is $F_{0.05,3,6} = 4.76$. The critical value for the Mineral supplement is $F_{0.05,1,8} = 5.32$. The critical value for the Pasture treatment $\times$ Mineral supplement interaction is $F_{0.05,3,8} = 4.07$. From the table it can be concluded that the effect of the Pasture treatment was significant.

Means that may be of interest include the means of the levels of factor A, the means of the levels of factor B, and the means of the combinations of factors A and B. In a balanced design (i.e. in a design with equal number of observations per level of factors), means are estimated by using arithmetic means. For example, the means of combinations A and B, denoted as μ_{ij} are estimated with $\bar{y}_{ij}.$. The variance of the estimator depends if blocks are defined as fixed or random. For example, if blocks are fixed, the variance of $\bar{y}_{ij}.$ is:

$$Var(\bar{y}_{ij}.) = Var\left(\frac{1}{n}\sum_k y_{ijk}\right) = \frac{1}{n^2}\left[\sum_k Var(y_{ijk})\right] = \frac{n}{n^2}Var(\mu_{ij} + \delta_{ik} + \varepsilon_{ijk})$$

$$= \frac{1}{n}\left(\sigma_\delta^2 + \sigma^2\right)$$

The standard error of the mean of combinations of factors A and B with fixed blocks is:

$$s_{\bar{y}_{ij}.} = \sqrt{\frac{1}{n}\left(\hat{\sigma}_\delta^2 + \hat{\sigma}^2\right)}$$

in which n is the number of blocks. The variances and standard errors of other means and differences between means can be derived similarly. The means, estimators and appropriate standard errors are shown in the following table:

Effects	Means	Estimators	Standard errors	
			Blocks fixed	Blocks random
Interaction $A \times B$	μ_{ij}	$\bar{y}_{ij}.$	$s_{\bar{y}_{ij}.} = \sqrt{\frac{1}{n}\left(\hat{\sigma}_\delta^2 + \hat{\sigma}^2\right)}$	$s_{\bar{y}_{ij}.} = \sqrt{\frac{1}{n}\left(\hat{\sigma}_{block}^2 + \hat{\sigma}_\delta^2 + \hat{\sigma}^2\right)}$
Factor A	$\mu_{i.}$	$\bar{y}_{i}..$	$s_{\bar{y}_{i}..} = \sqrt{\frac{1}{bn}\left(b\hat{\sigma}_\delta^2 + \hat{\sigma}^2\right)}$	$s_{\bar{y}_{i}..} = \sqrt{\frac{1}{bn}\left(b\hat{\sigma}_{block}^2 + b\hat{\sigma}_\delta^2 + \hat{\sigma}^2\right)}$
Factor B	$\mu_{.j}$	$\bar{y}_{.j}.$	$s_{\bar{y}_{.j}.} = \sqrt{\frac{1}{an}\left(\hat{\sigma}_\delta^2 + \hat{\sigma}^2\right)}$	$s_{\bar{y}_{.j}.} = \sqrt{\frac{1}{an}\left(a\hat{\sigma}_{block}^2 + \hat{\sigma}_\delta^2 + \hat{\sigma}^2\right)}$
Differences of factor A	$\mu_{i.} - \mu_{i'.}$	$\bar{y}_{i}.. - \bar{y}_{i'}..$	$s_{\bar{y}_{i}..-\bar{y}_{i'}..} = \sqrt{\frac{2}{bn}\left(b\hat{\sigma}_\delta^2 + \hat{\sigma}^2\right)}$	$s_{\bar{y}_{i}..-\bar{y}_{i'}..} = \sqrt{\frac{2}{bn}\left(b\hat{\sigma}_\delta^2 + \hat{\sigma}^2\right)}$
Differences of factor B	$\mu_{.j} - \mu_{.j'}$	$\bar{y}_{.j}. - \bar{y}_{.j'}.$	$s_{\bar{y}_{.j}.-\bar{y}_{.j'}.} = \sqrt{\frac{2}{an}\left(\hat{\sigma}^2\right)}$	$s_{\bar{y}_{.j}.-\bar{y}_{.j'}.} = \sqrt{\frac{2}{an}\left(\hat{\sigma}^2\right)}$
Differences of factor B within factor A	$\mu_{ij} - \mu_{ij'}$	$\bar{y}_{ij}. - \bar{y}_{ij'}.$	$s_{\bar{y}_{ij}.-\bar{y}_{ij'}.} = \sqrt{\frac{2}{n}\left(\hat{\sigma}^2\right)}$	$s_{\bar{y}_{ij}.-\bar{y}_{ij'}.} = \sqrt{\frac{2}{n}\left(\hat{\sigma}^2\right)}$
Differences of factor A within factor B	$\mu_{ij} - \mu_{i'j}$	$\bar{y}_{ij}. - \bar{y}_{i'j}.$	$s_{\bar{y}_{ij}.-\bar{y}_{i'j}.} = \sqrt{\frac{2}{n}\left(\hat{\sigma}_\delta^2 + \hat{\sigma}^2\right)}$	$s_{\bar{y}_{ij}.-\bar{y}_{i'j}.} = \sqrt{\frac{2}{n}\left(\hat{\sigma}_\delta^2 + \hat{\sigma}^2\right)}$

20.1.1 SAS Example: Main Plots in Randomized Blocks

The SAS program for the example of the effect of four pasture treatments and two mineral supplements on milk production of cows from section 20.1 is as follows. Four pasture treatments were assigned to the main plots in a randomized block design.

SAS program:

```
DATA spltblk;
   INPUT block past min milk @@;
   DATALINES;
   1 4 2 30    1 4 1 29    1 1 2 27    1 1 1 25
   1 2 1 26    1 2 2 28    1 3 2 26    1 3 1 24
   2 2 1 32    2 2 2 37    2 1 2 30    2 1 1 31
   2 4 1 34    2 4 2 37    2 3 1 33    2 3 2 32
   3 1 2 34    3 1 1 31    3 2 1 30    3 2 2 31
   3 4 2 36    3 4 1 38    3 3 1 33    3 3 2 32
;
PROC MIXED DATA = spltblk;
CLASS block past min;
MODEL milk =past min past*min / DDFM = KENWARDROGER;
RANDOM block block*past/;
LSMEANS past min past*min / DIFF ADJUST=TUKEY ;
RUN;
```

Explanation: The MIXED procedure by default uses Restricted Maximum Likelihood (REML) estimation. The CLASS statement defines categorical (classification) variables. The MODEL statement defines the dependent variable and the independent variables fitted in the model. The RANDOM statement defines random effects (*block* and *block*past*). Here, *block*past* will be used as the experimental error for testing *past*. The LSMEANS statement calculates effect means. The options after the slash specify calculation of standard errors and tests of differences between least-squares means using a Tukey test with adjustment for multiple comparisons. The *DDFM = KENWARDROGER* option with the MODEL statement is recommended to estimate correct degrees of freedom when the mean square of experimental error term or standard errors are linear combinations of variances, especially if there are missing values.

SAS output:

```
             Covariance Parameter Estimates

               Cov Parm          Estimate
               block             12.7431
               block*past         1.0486
               Residual           2.2500

          Type 3 Tests of Fixed Effects
```

Effect	Num DF	Den DF	F Value	Pr > F
past	3	6	5.46	0.0377
min	1	8	3.63	0.0932
past*min	3	8	0.86	0.4981

Least Squares Means

Effect	past	min	Estimate	Standard Error	DF	t Value	Pr > |t|
past	1		29.6667	2.2298	2.51	13.30	0.0022
past	2		30.6667	2.2298	2.51	13.75	0.0020
past	3		30.0000	2.2298	2.51	13.45	0.0021
past	4		34.0000	2.2298	2.51	15.25	0.0015
min		1	30.5000	2.1266	2.09	14.34	0.0041
min		2	31.6667	2.1266	2.09	14.89	0.0038
past*min	1	1	29.0000	2.3124	2.9	12.54	0.0013
past*min	1	2	30.3333	2.3124	2.9	13.12	0.0011
past*min	2	1	29.3333	2.3124	2.9	12.69	0.0013
past*min	2	2	32.0000	2.3124	2.9	13.84	0.0010
past*min	3	1	30.0000	2.3124	2.9	12.97	0.0012
past*min	3	2	30.0000	2.3124	2.9	12.97	0.0012
past*min	4	1	33.6667	2.3124	2.9	14.56	0.0008
past*min	4	2	34.3333	2.3124	2.9	14.85	0.0008

Differences of Least Squares Means

Effect	past	min	_past	_min	Estimate	Standard Error	DF	t Value	Pr>|t|	Adjust	Adj P
past	1		2		-1.0000	1.2038	6	-0.83	0.4379	Tuk-Kr	0.8385
past	1		3		-0.3333	1.2038	6	-0.28	0.7911	Tuk-Kr	0.9918
past	1		4		-4.3333	1.2038	6	-3.60	0.0114	Tuk-Kr	0.0427
past	2		3		0.6667	1.2038	6	0.55	0.5997	Tuk-Kr	0.9421
past	2		4		-3.3333	1.2038	6	-2.77	0.0325	Tuk-Kr	0.1135
past	3		4		-4.0000	1.2038	6	-3.32	0.0159	Tuk-Kr	0.0587
min	1		2		-1.1667	0.6124	8	-1.91	0.0932	Tuk-Kr	0.0932
past*min	1	1	1	2	-1.3333	1.2247	8	-1.09	0.3080	Tuk-Kr	0.9425
past*min	1	1	2	1	-0.3333	1.4829	11.5	-0.22	0.8261	Tuk-Kr	1.0000
past*min	1	1	2	2	-3.0000	1.4829	11.5	-2.02	0.0669	Tuk-Kr	0.5207
past*min	1	1	3	1	-1.0000	1.4829	11.5	-0.67	0.5134	Tuk-Kr	0.9955
past*min	1	1	3	2	-1.0000	1.4829	11.5	0.67	0.5134	Tuk-Kr	0.9955
past*min	1	1	4	1	-4.6667	1.4829	11.5	-3.15	0.0088	Tuk-Kr	0.1407
...					...						...
past*min	3	1	4	2	-4.3333	1.4829	11.5	-2.92	0.0133	Tuk-Kr	0.1868
past*min	3	2	4	1	-3.6667	1.4829	11.5	-2.47	0.0301	Tuk-Kr	0.3217
past*min	3	2	4	2	-4.3333	1.4829	11.5	-2.92	0.0133	Tuk-Kr	0.1868
past*min	4	1	4	2	-0.6667	1.2247	8	-0.54	0.6011	Tuk-Kr	0.9988

Explanation: The MIXED procedure estimates variance components for random effects (*Covariance Parameter Estimates*) and provides *F* tests for fixed effects (*Type 3 Tests of Fixed Effects*). In the *Least Squares Means* table, the means (*Estimate*) with their *Standard Error* are presented. In the *Differences of Least Squares Means* table the differences among means are shown (*Estimate*). The differences are tested using the Tukey-Kramer procedure, which adjusts for the multiple comparison and unequal subgroup size. The correct *P* value is the adjusted *P* value (*Adj P*). For example, the *P* value for the difference between levels 3 and 4 for *pasture* is 0.0587. The MIXED procedure calculates appropriate standard errors for the least squares means and differences between them. Note the degrees of freedom (*DF*) are corrected by the Kenward-Roger adjustment for compound experimental error terms.

20.2 Split-plot Design – Main Plots in a Completely Randomized Design

In an alternative split-plot design one of the factors can be assigned to the main plots in completely randomized design. For example, consider a factor A with four levels (A_1, A_2, A_3 and A_4) assigned randomly on 12 plots. This is a completely randomized design. Each level of factor A is repeated three times. Let a second factor denoted B, have two levels (B_1 and B_2). Thus, each of the main plots is divided into two split-plots, on which levels B_1 and B_2 are randomly assigned. One possible scheme of such a design is:

B_2	B_2	B_1	B_2	B_1	B_2	B_1	B_1	B_1	B_2	B_1	B_2
B_1	B_1	B_2	B_1	B_2	B_1	B_2	B_2	B_2	B_1	B_2	B_1
A_4	A_1	A_2	A_3	A_2	A_1	A_4	A_3	A_4	A_3	A_1	A_2

The model is:

$$y_{ijk} = \mu + A_i + \delta_{ik} + B_j + (AB)_{ij} + \varepsilon_{ijk} \qquad i = 1,..., a; j = 1,..., b ; k = 1,..., n$$

where:

y_{ijk} = observation k in level i of factor A and level j of factor B
μ = the overall mean
A_i = the effect of level i of factor A
B_j = the effect of level j of factor B
$(AB)_{ij}$ = the effect of the ij^{th} interaction of $A \times B$
δ_{ik} = the main plot error (the main plots within factor A) with mean 0 and variance σ^2_δ
ε_{ijk} = the split-plot error with mean 0 and the variance σ^2

Also, $\mu_{ij} = \mu + A_i + B_j + (AB)_{ij}$ = the mean of the ij^{th} $A \times B$ interaction

a = number of levels of factor A
b = number of levels of factor B
n = number of replicates

It is assumed that main plot and split-plot errors are independent.

The *ANOVA* table for the design with three replicates, four levels of factor A and two levels of factor B:

Source	Degrees of freedom	
Factor A	$(a - 1) =$	3
Main plot exp. error	$a(n - 1) =$	8
Factor B	$(b - 1) =$	1
A × B	$(a - 1)(b - 1) =$	3
Split-plot exp. error	$a(b - 1)(n - 1) =$	8
Total	$(abn - 1) =$	23

$a = 4$ = number of levels of factor A
$b = 2$ = number of levels of factor B
$n = 3$ = number of replicates (plots) per level of factor A

The F statistic for factor A:

$$F = \frac{MS_A}{MS_{Main\ plot\ error}}$$

The main plot experimental error term is the mean square among plots within factor A.

The F statistic for factor B is:

$$F = \frac{MS_B}{MS_{Split\text{-}plot\ error}}$$

The split-plot experimental error term is the residual mean square.

The F statistic for the $A \times B$ interaction:

$$F = \frac{MS_{A \times B}}{MS_{Split\text{-}plot\ error}}$$

Example: Consider an experiment similar to that in 20.1.1. The effects of four different treatments of pasture and two mineral supplements are tested on milk yield. Twenty-four cows are used. However, this time blocks are not defined. The levels of factor A (pasture treatments) are assigned to the main plots in a completely randomized design.

Plot	Pasture	Mineral	Milk (kg)	Plot	Pasture	Mineral	Milk (kg)
1	4	2	30	7	4	1	34
1	4	1	29	7	4	2	37
2	1	2	27	8	3	1	33
2	1	1	25	8	3	2	32
3	2	1	26	9	1	2	34
3	2	2	28	9	1	1	31
4	3	2	26	10	2	1	30
4	3	1	24	10	2	2	31
5	2	1	32	11	4	2	36
5	2	2	37	11	4	1	38
6	1	2	30	12	3	1	33
6	1	1	31	12	3	2	32

The results are shown in the *ANOVA* table.

Source	SS	df	MS	F
Pasture treatment	71.167	3	23.722	0.80
Main plot exp. error	238.667	8	29.833	
Mineral supplement	8.167	1	8.167	3.63
Pasture × Mineral	5.833	3	1.944	0.86
Split-plot exp. error	18.000	8	2.250	
Total	341.833	23		

The critical value for the Pasture treatment is $F_{0.05,3,8} = 4.07$. The critical value for the Mineral supplement is $F_{0.05,1,8} = 5.32$. The critical value for the Pasture treatment × Mineral supplement interaction is $F_{0.05,3,8} = 4.07$.

 Comparing the two examples of split-plot designs, note that the method of randomizing Pasture treatment has not influenced the test for Mineral supplement; however, using blocks improved the precision of the test for Pasture treatment. Neighboring paddocks tend to be alike, and that is why a split-plot design with randomized blocks is appropriate in this research. Note that the sum of squares for plots within Pasture treatment (main plot experimental error) is equal to the sum of squares for Block plus the sum of squares for Pasture treatment × Block from the example with in section 20.1.1. (238.667 = 212.583 + 26.083).

The means and their estimators and corresponding standard errors for a split-plot design with assignment of treatments to main plots completely at random are shown in the following table:

Effects	Means	Estimators	Standard errors
Interaction A x B	μ_{ij}	$\bar{y}_{ij\cdot}$	$s_{\bar{y}_{ij\cdot}} = \sqrt{\dfrac{1}{n}\left(\hat{\sigma}_\delta^2 + \hat{\sigma}^2\right)}$
Factor A	$\mu_{i\cdot}$	$\bar{y}_{i\cdot\cdot}$	$s_{\bar{y}_{i\cdot\cdot}} = \sqrt{\dfrac{1}{bn}\left(b\hat{\sigma}_\delta^2 + \hat{\sigma}^2\right)}$
Factor B	$\mu_{\cdot j}$	$\bar{y}_{\cdot j\cdot}$	$s_{\bar{y}_{\cdot j\cdot}} = \sqrt{\dfrac{1}{an}\left(\hat{\sigma}_\delta^2 + \hat{\sigma}^2\right)}$
Differences for factor A	$\mu_{i\cdot} - \mu_{i'\cdot}$	$\bar{y}_{i\cdot\cdot} - \bar{y}_{i'\cdot\cdot}$	$s_{\bar{y}_{i\cdot\cdot} - \bar{y}_{i'\cdot\cdot}} = \sqrt{\dfrac{2}{bn}\left(b\hat{\sigma}_\delta^2 + \hat{\sigma}^2\right)}$
Differences for factor B	$\mu_{\cdot j} - \mu_{\cdot j'}$	$\bar{y}_{\cdot j\cdot} - \bar{y}_{\cdot j'\cdot}$	$s_{\bar{y}_{\cdot j\cdot} - \bar{y}_{\cdot j'\cdot}} = \sqrt{\dfrac{2}{an}\left(\hat{\sigma}^2\right)}$
Differences for factor B within factor A	$\mu_{ij} - \mu_{ij'}$	$\bar{y}_{ij\cdot} - \bar{y}_{ij'\cdot}$	$s_{\bar{y}_{ij\cdot} - \bar{y}_{ij'\cdot}} = \sqrt{\dfrac{2}{n}\left(\hat{\sigma}^2\right)}$
Differences for factor A within factor B	$\mu_{ij} - \mu_{i'j}$	$\bar{y}_{ij\cdot} - \bar{y}_{i'j\cdot}$	$s_{\bar{y}_{ij\cdot} - \bar{y}_{i'j\cdot}} = \sqrt{\dfrac{2}{n}\left(\hat{\sigma}_\delta^2 + \hat{\sigma}^2\right)}$

20.2.1 SAS Example: Main Plots in a Completely Randomized Design

The SAS program for the example of the effect of four pasture treatments and two mineral supplements on milk production of cows when pasture treatments were assigned to the main plots as a completely randomized design is as follows.

SAS program:

```
DATA splt;
    INPUT plot past min  milk @@;
    DATALINES;
  1 4 2 30   1 4 1 29   2 1 2 27   2 1 1 25
  3 2 1 26   3 2 2 28   4 3 2 26   4 3 1 24
  5 2 1 32   5 2 2 37   6 1 2 30   6 1 1 31
  7 4 1 34   7 4 2 37   8 3 1 33   8 3 2 32
  9 1 2 34   9 1 1 31  10 2 1 30  10 2 2 31
 11 4 2 36  11 4 1 38  12 3 1 33  12 3 2 32
;
PROC MIXED DATA = splt;
CLASS plot past min;
MODEL milk =past min past*min  / DDFM = KENWARDROGER;
RANDOM plot(past) /;
LSMEANS past min past*min / DIFF ADJUST=TUKEY ;
RUN;
```

Explanation: The MIXED procedure by default uses Restricted Maximum Likelihood (REML) estimation. The CLASS statement defines categorical (classification) variables. Note that *plots* must be defined as class variable to ensure proper testing of pasture treatment effects. The MODEL statement defines the dependent variable and the independent variables fitted in the model. The RANDOM statement defines the random effect, plots within pasture treatments (*plot(past)*), which will thus be defined as the experimental error for testing *past*. The LSMEANS statement calculates effect means. The options after the slash specify calculation of standard errors and tests of differences between least squares means using a Tukey test with adjustment for multiple comparisons. The *DDFM = KENWARDROGER* option is used here to correct degrees of freedom because the mean square of the main plot error and standard errors are linear combinations of variances. This option is especially useful when some values are missing.

SAS output:

Covariance Parameter Estimates

Cov Parm	Estimate
plot(past)	13.7917
Residual	2.2500

Type 3 Tests of Fixed Effects

Effect	Num DF	Den DF	F Value	Pr > F
past	3	8	0.80	0.5302
min	1	8	3.63	0.0932
past*min	3	8	0.86	0.4981

Least Squares Means

Effect	past	min	Estimate	Standard Error	DF	t Value	Pr > |t|
past	1		29.6667	2.2298	8	13.30	<.0001
past	2		30.6667	2.2298	8	13.75	<.0001
past	3		30.0000	2.2298	8	13.45	<.0001
past	4		34.0000	2.2298	8	15.25	<.0001
min		1	30.5000	1.1562	9.2	26.38	<.0001
min		2	31.6667	1.1562	9.2	27.39	<.0001
past*min	1	1	29.0000	2.3124	9.2	12.54	<.0001
past*min	1	2	30.3333	2.3124	9.2	13.12	<.0001
past*min	2	1	29.3333	2.3124	9.2	12.69	<.0001
past*min	2	2	32.0000	2.3124	9.2	13.84	<.0001
past*min	3	1	30.0000	2.3124	9.2	12.97	<.0001
past*min	3	2	30.0000	2.3124	9.2	12.97	<.0001
past*min	4	1	33.6667	2.3124	9.2	14.56	<.0001
past*min	4	2	34.3333	2.3124	9.2	14.85	<.0001

Differences of Least Squares Means

Effect	past	min	_past	_min	Estimate	Stand Error	DF	t	Pr>|t|	Adj	Adj P
past	1		2		-1.0000	3.1535	8	-0.32	0.7593	Tuk-Kr	0.9881
past	1		3		-0.3333	3.1535	8	-0.11	0.9184	Tuk-Kr	0.9995
past	1		4		-4.3333	3.1535	8	-1.37	0.2067	Tuk-Kr	0.5469
past	2		3		0.6667	3.1535	8	0.21	0.8379	Tuk-Kr	0.9964
past	2		4		-3.3333	3.1535	8	-1.06	0.3214	Tuk-Kr	0.7231
past	3		4		-4.0000	3.1535	8	-1.27	0.2403	Tuk-Kr	0.6053
min		1		2	-1.1667	0.6124	8	-1.91	0.0932	Tuk-Kr	0.0932

Explanation: The MIXED procedure estimates variance components for random effects (*Covariance Parameter Estimates*) and provides *F* tests for fixed effects (*Type 3 Tests of Fixed Effects*). In the *Least Squares Means* table, the means (*Estimate*) with their *Standard Error* are presented. In the *Differences of Least Squares Means* table the differences among means are shown (*Estimate*). The differences are tested using the Tukey-Kramer procedure, which adjusts for multiple comparison and unequal subgroup size. The correct *P* value is the adjusted *P* value (*Adj P*). For example, the *P* value for the difference between levels 3 and 4 for *pasture* is 0.6053. The MIXED procedure calculates appropriate standard errors for the least squares means and differences between them. Differences between the *past*min* interaction means are not shown.

Exercise

20.1. The objective of the study was to test effects of grass species and stocking density on the daily gain of Suffolk lambs kept on a pasture. The experiment was designed as a split-plot on three different 1 ha pastures. Each pasture was divided into two plots, one randomly assigned to fescue and the other to rye-grass. Each plot was then split into two split-plots with different numbers of sheep on each (20 and 24). The length of the experiment was 2 weeks. At the end of the experiment the following daily gains were calculated:

Pasture	Grass	Number of sheep	Daily gain (g)
1	fescue	20	290
1	fescue	24	310
1	rye-grass	20	310
1	rye-grass	24	330
2	fescue	20	320
2	fescue	24	350
2	rye-grass	20	380
2	rye-grass	24	400
3	fescue	20	320
3	fescue	24	320
3	rye-grass	20	380
3	rye-grass	24	410

Describe the experimental design. Analyze the data to test the effect of grass species and stocking density on daily gain.

Chapter 21

Analysis of Covariance

Analysis of covariance is a term for a statistical procedure in which variability of a dependent variable is explained by both categorical and continuous independent variables. The continuous variable in the model is called a covariate. Common application of analysis of covariance is to adjust treatment means for a known source of variability that can be explained by a continuous variable. For example, in an experiment designed to test the effects of three diets on yearling weight of animals, different initial weight or different age at the beginning of the experiment will influence animal performance and precision of the experiment. It is necessary to adjust yearling weights for differences in initial weight or initial age. This can be accomplished by defining initial weight or age as a covariate in the model. This will improve the precision of the experiment, since part of the unexplained variability is explained by the covariate and consequently the experimental error is reduced. Another application of analysis covariance includes testing differences of regression slopes among groups. For example, a test to determine if the regression of daily gain on initial weight is different for males than females.

21.1 Completely Randomized Design with a Covariate

In a completely randomized design with a covariate, the analysis of covariance is utilized for correcting treatment means, controlling the experimental error, and increasing precision. The statistical model is:

$$y_{ij} = \beta_0 + \beta_1 x_{ij} + \tau_i + \varepsilon_{ij} \qquad\qquad i = 1,..., a \; ; \; j = 1,..., n$$

where:

y_{ij} = observation j in group i (treatment i)
β_0 = the intercept
β_1 = the regression coefficient
x_{ij} = a continuous independent variable with mean μ_x (covariate)
τ_i = the fixed effect of group or treatment i
ε_{ij} = random error

The overall mean is: $\mu = \beta_0 + \beta_1 \mu_x$
The mean of group or treatment i is: $\mu_i = \beta_0 + \beta_1 \mu_x + \tau_i$
where μ_x is the mean of the covariate x.

The assumptions are:

1. the covariate is fixed and independent of treatments
2. errors are independent of each other
3. usually, errors have a normal distribution with mean 0 and homogeneous variance σ^2

Example: The effect of three diets on daily gain of steers was investigated. The experiment used a completely randomized design. Weight at the beginning of the experiment (initial weight) was recorded, but not used in the assignment of animals to diet. At the end of the experiment the following daily gains were measured:

Diet A		Diet B		Diet C	
Initial weight (kg)	Gain (g/day)	Initial weight (kg)	Gain (g/day)	Initial weight (kg)	Gain (g/day)
350	970	390	990	400	990
400	1000	340	950	320	940
360	980	410	980	330	930
350	980	430	990	390	1000
340	970	390	980	420	1000

To show the efficiency of including the effect of initial weight in the model, the model for the completely randomized design without a covariate is first fitted. The *ANOVA* table is:

Source	SS	df	MS	F
Treatment	173.333	2	86.667	0.16
Residual	6360.000	12	530.000	
Total	6533.333	14		

The critical value for the treatment effect is $F_{0.05,2,12} = 3.89$. Thus, the effect of treatments is not significant. When initial weight is included in the model as a covariate the *ANOVA* table is:

Source	SS	df	MS	F
Initial weight	4441.253	1	4441.253	46.92
Treatment	1050.762	2	525.381032	5.55
Residual	1041.319	11	94.665	
Total	6533.333	14		

The critical value for treatment is $F_{0.05,2,11} = 3.98$. The critical value for the regression of daily gain on initial weight is $F_{0.05,1,11} = 4.84$. Since the calculated F values are 5.55 and 46.92, the effects of both the initial weight and treatment are significant. It appears that the first model was not adequate. By including initial weight in the model, a significant difference between treatments was found.

21.1.1 SAS Example for a Completely Randomized Design with a Covariate

The SAS program for the example of the effect of three diets on daily gain of steers is as follows.

SAS program:

```
DATA gain;
INPUT treatment $ initial gain @@;

DATALINES;
A 350   970   B 390   990   C 400   990
A 400 1000   B 340   950   C 320   940
A 360   980   B 410   980   C 330   930
A 350   980   B 430   990   C 390 1000
A 340   970   B 390   980   C 420 1000
;
PROC GLM DATA = gain ;
CLASS treatment;
MODEL gain = initial treatment  / SOLUTION SS1;
LSMEANS treatment / STDERR PDIFF ADJUST=TUKEY;
RUN;
```

Explanation: The GLM procedure is used. The CLASS statement defines *treatment as* a classification variable. The statement, *MODEL gain = initial treatment* defines *gain* as the dependent variable, and *initial* and *treatment* as independent variables. Since the variable *initial* is not listed in the CLASS statement, the procedure fits it as a continuous variable. The SOLUTION option directs estimates of regression parameters and the SS1 option directs the use of type I sums of squares (sequential sum of squares) which are appropriate for this kind of analysis. Sequential sums of squares remove the effect of the covariate before consideration of effects of treatment. The LSMEANS statement estimates the treatment means adjusted for the effect of the covariate. Options after the slash calculate standard errors and test the difference between means using a Tukey test.

SAS output:

Dependent Variable: gain

Source	DF	Sum of Squares	Mean Square	F Value	Pr > F
Model	3	5492.014652	1830.671551	19.34	0.0001
Error	11	1041.318681	94.665335		
Corrected Total	14	6533.333333			

R-Square	Coeff Var	Root MSE	gain Mean
0.840614	0.996206	9.729611	976.6667

Source	DF	Type I SS	Mean Square	F Value	Pr > F
initial	1	4441.252588	4441.252588	46.92	<.0001
treatment	2	1050.762064	525.381032	5.55	0.0216

			Standard		
Parameter		Estimate	Error	t Value	Pr > |t|
Intercept		747.1648352 B	30.30956710	24.65	<.0001
initial		0.6043956	0.08063337	7.50	<.0001
treatment	A	15.2527473 B	6.22915600	2.45	0.0323
treatment	B	-6.0879121 B	6.36135441	-0.96	0.3591
treatment	C	0.0000000 B	.	.	.

NOTE: The X'X matrix has been found to be singular, and a
generalized inverse was used to solve the normal equations.
Terms whose estimates are followed by the letter 'B' are not
uniquely estimable.

Least Squares Means
Adjustment for Multiple Comparisons: Tukey-Kramer

	gain	Standard		LSMEAN
treatment	LSMEAN	Error	Pr > |t|	Number
A	988.864469	4.509065	<.0001	1
B	967.523810	4.570173	<.0001	2
C	973.611722	4.356524	<.0001	3

Least Squares Means for Effect treatment
t for H0: LSMean(i)=LSMean(j) / Pr > \|t\|

Dependent Variable: gain

i/j	1	2	3
1		0.0213	0.0765
2	0.0213		0.6175
3	0.0765	0.6175	

Explanation: Listed first is an *ANOVA* table for the dependent variable *gain*. The sources of variation are *Model*, residual (*Error*) and *Corrected Total*. In the table are listed degrees of freedom (*DF*), *Sum of Squares*, *Mean Square*, calculated F (*F Value*) and P value (*Pr > F*). In the next table F tests of the effects of the independent variables *initial* and *treatment* are given. It is appropriate to use sequential sums of squares (*Type I SS*) because the variable *initial* is defined in order to adjust data prior to the effect of *treatment*, and *treatment* does not affect *initial*. The F and P values for *treatment* are 5.55 and 0.0216. Thus, the effect of treatment is significant in the sample. The next table presents parameter estimates. The letter '*B*' behind the estimates denotes that the corresponding solution is not unique. Only the slope (*initial*) has a unique solution (0.6043956). Under the title *Least Squares Means*, the means adjusted for differences in initial weight (*LSMEAN*) with their *Standard Errors* are shown. At the end the Tukey test between means of all treatment pairs are given. For example, in column 3 and row 1 the number 0.0765 denotes the P value between treatments 1 and 3. The P values are corrected for multiple comparisons and possible unbalanced data.

21.2 Testing the Difference between Regression Slopes

The difference between regression lines of groups can be tested by defining an interaction between a categorical variable representing the groups and the continuous variable (covariate). The interaction produces a separate regression curve for each group. The model including the group effect and simple linear regression is:

$$y_{ij} = \beta_0 + \tau_i + \beta_1 x_{ij} + \beta_{2i}(\tau^*x)_{ij} + \varepsilon_{ij} \quad i = 1,...,a; \; j = 1,...,n$$

where:

y_{ij} = observation j in group i
τ_i = the effect of group i
β_0, β_1 and β_{2i} = regression parameters
x_{ij} = the value of the continuous independent variable for observation j in group i
$(\tau^*x)_{ij}$ = the group $\times$ covariate interaction
ε_{ij} = random error

The overall mean is: $\mu = \beta_0 + \beta_1 \mu_x$
The mean of group i is: $\mu_i = \beta_0 + \tau_i + \beta_1 \mu_x + \beta_{2i} \mu_x$
The intercept for group i is: $\beta_0 + \tau_i$
The regression coefficient for group i is: $\beta_1 + \beta_{2i}$

The hypotheses are the following:

a) H_0: $\tau_i = 0$ for all i, there is no group effect
   H_1: $\tau_i \neq 0$ for at least one i, there is a group effect

b) H_0: $\beta_1 = 0$, the overall slope is equal to zero, there is no regression
   H_1: $\beta_1 \neq 0$, the overall slope is different from zero, there is a regression

c) H_0: $\beta_{2i} = 0$, the slope in group i is not different than the average slope
   H_1: $\beta_{2i} \neq 0$, the slope in group i is different than the average slope.

The difference between regression lines can also be tested by using a multiple regression. The categorical variable (group) can be defined as a set of binary variables with assigned numerical values of 0 or 1. The value 1 denotes that an observation belongs to some particular group, and 0 denotes that the observation does not belong to that group. Thus, for a number of groups there are $(a - 1)$ new variables that can be used as independent variables in a multiple regression setting. For each group there is a regression coefficient that can be tested against zero to determine if the slope for that group is different than the average slope of all groups. This multiple regression model is equivalent to the model with the group effect as a categorical variable, a covariate and their interaction, and parameter estimates and inferences are in both cases the same.

To show the logic of testing the difference between regression slopes, a simple model with two groups will be shown. Consider a regression of variable y on variable x. The variables are measured on animals that are grouped according to sex. There are two questions of interest:

a) whether females and males have separate regression lines
b) whether there is a difference between regression slopes for males and females.

For this example the multiple regression model is:

$$y_i = \beta_0 + \beta_1 x_{1i} + \beta_2 x_{2i} + \beta_3 x_{1i} x_{2i} + \varepsilon_i$$

where x_{1i} is a continuous variable and x_{2i} is a variable that explains if an animal is a male or female with values $x_{2i} = 1$ if male and 0 if female. The term $x_{1i} x_{2i}$ denotes interaction between x_{1i} and x_{2i}.

Figure 21.1 shows possible models that can explain changes in the dependent variable due to changes in a continuous independent variable.

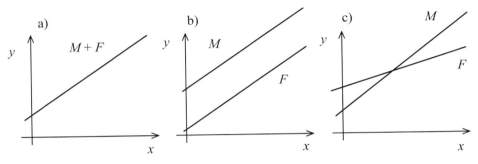

Figure 21.1 Regression models with sex as a categorical independent variable: a) no difference between males (*M*) and females (*F*); b) a difference exists between *M* and *F*, but the slopes are equal; c) a difference exists and slopes are different

There are three possible models.

Model a): No difference between males and females. The expectation of the dependent variable is:

$$E(y_i) = \beta_0 + \beta_1 x_i$$

One model explains changes in y when x is changed.

Model b): A difference exists between males and females, but the slopes are equal. The expectation is:

$$E(y_i) = \beta_0 + \beta_1 x_{1i} + \beta_2 x_{2i}$$

For males (*M*) the model is:

$$E(y_i) = \beta_0 + \beta_1 x_{1i} + \beta_2(1) = (\beta_0 + \beta_2) + \beta_1 x_{1i}$$

For females (*F*) the model is:

$$E(y_i) = \beta_0 + \beta_1 x_{1i} + \beta_2(0) = \beta_0 + \beta_1 x_{1i}$$

The hypotheses H_0: $\beta_2 = 0$ vs. H_1: $\beta_2 \neq 0$ test whether the same line explains the regression for both males and females. If H_0 is true the lines are the same, and if H_1 is true the lines are different but parallel. The difference between males and females is equal to β_2 for any value of x_1.

Model c): A difference between males and females is shown by different regression slopes, indicating interaction between x_{1i} and x_{2i}. The expectation of the dependent variable is:

$$E(y_i) = \beta_0 + \beta_1 x_{1i} + \beta_2 x_{2i} + \beta_3 x_{1i} x_{2i}$$

For males (M) the model is:

$$E(y_i) = (\beta_0 + \beta_2) + (\beta_1 + \beta_3)x_{1i}$$

For females (F) the model is:

$$E(y_i) = \beta_0 + \beta_1 x_{1i} + \beta_2(0) + \beta_3 x_{1i}(0) = \beta_0 + \beta_1 x_{1i}$$

The hypotheses $H_0: \beta_3 = 0$ vs. $H_1: \beta_3 \neq 0$ test whether the slopes are equal. If H_0 is true there is no interaction and the slope is the same for both males and females.

Example: Consider an experiment in which the effect of two treatments on daily gain of steers was investigated. A completely randomized design was used. The following initial weights and daily gains were measured:

Treatment A		Treatment B	
Initial weight (kg)	Gain (g/day)	Initial weight (kg)	Gain (g/day)
340	900	340	920
350	950	360	930
350	980	370	950
360	980	380	930
370	990	390	930
380	1020	410	970
400	1050	430	990

Is there a significant difference in daily gains between the two treatment groups and does the initial weight influence daily gain differently in the two groups?

Figure 21.2 indicates a linear relationship between initial weight and daily gain measured in the experiment. Also, the slopes appear to be different which indicates a possible interaction between treatments and initial weight.

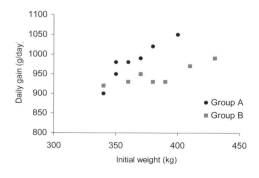

Figure 21.2 Daily gain of two treatment groups of steers dependent on initial weight

The following model can be defined:

$$y_i = \beta_0 + \beta_1 x_{1i} + \beta_2 x_{2i} + \beta_3 x_{1i} x_{2i} + \varepsilon_i \qquad i = 1,...,7$$

where:

y_i = daily gain of steer i

$\beta_0, \beta_1, \beta_2, \beta_3$ = regression parameters

x_{1i} = initial weight of steer i

x_{2i} = assignment to treatment (1 if treatment A, 0 if treatment B)

$x_{1i} x_{2i}$ = interaction of treatment × initial weight

ε_j = random error

The hypotheses are:

$$H_0: \beta_2 = 0 \text{ vs. } H_1: \beta_2 \neq 0$$

If H_0 is true the lines are identical. If H_1 is true the lines are different but parallel.

$$H_0: \beta_3 = 0 \text{ vs. } H_1: \beta_3 \neq 0$$

If H_0 is true there is no interaction and the regression slopes are identical. If H_1 is true the slopes are different. The *ANOVA* table is:

Source	SS	df	MS	F
Model	19485.524	3	6495.175	22.90
Residual	2835.905	10	283.590	
Total	22321.429	13		

The critical value for the model is $F_{0.05,3,10} = 3.71$. The null hypotheses if particular parameters are equal to zero can be tested by using t tests. The parameter estimates with their corresponding standard errors and t tests are shown in the following table:

Parameter	Estimate	Std. error	t value	Critical t
β_0	663.505	86.833	7.641	2.228
β_1	0.737	0.226	3.259	2.228
β_2	−469.338	149.050	−3.149	2.228
β_3	1.424	0.402	3.544	2.228

Note that the absolute value of the calculated t is greater than the critical value for all parameters, thus all parameters are required in the model. There are effects of initial weight, treatment and their interaction on daily gain of steers.

The estimated regression for treatment A is:

$$E(y_i) = (\beta_0 + \beta_2) + (\beta_1 + \beta_3) x_{1i} = (663.505 - 469.338) + (0.737 + 1.424) x_{1i}$$
$$= 194.167 + 2.161 x_{1i}$$

The estimated regression for treatment B is:

$$E(y_i) = \beta_0 + \beta_1 x_{1i} = 663.505 + 0.737 x_{1i}$$

21.2.1 SAS Example for Testing the Difference between Regression Slopes

A SAS program for the example examining the effect of two treatments and initial weight on daily gain of steers is as follows:

SAS program:

```
DATA gain;
INPUT treatment $ initial gain @@;
DATALINES;
A    340    900    A    350    950    A    350    980
A    360    980    A    370    990    A    380    1020
A    400    1050   B    340    920    B    360    930
B    370    950    B    380    930    B    390    930
B    410    970    B    430    990
;
PROC GLM DATA = gain;
CLASS treatment;
MODEL gain = initial treatment treatment*initial / SS1;
LSMEANS treatment / STDERR PDIFF ADJUST = TUKEY;
LSMEANS treatment / STDERR PDIFF ADJUST = TUKEY AT initial=340;
LSMEANS treatment / STDERR PDIFF ADJUST = TUKEY AT initial=400;
RUN;

PROC GLM DATA = gain;
CLASS treatment;
MODEL gain = treatment treatment*initial / NOINT SOLUTION;
RUN;
QUIT;
```

Explanation: The GLM procedure is used. The CLASS statement defines *treatment* as a categorical variable. The statement, *MODEL gain = initial treatment treatment*initial* defines *gain* as the dependent variable, *treatment* as a categorical independent variable, *initial* as a continuous independent variable, and the interaction of *treatment*gain*. A test of interaction effect *treatment*initial* tests whether regressions are different in different treatments. Two GLM procedures are used; the first gives the correct F tests, and the second estimates the regression parameters. There are three LSMEAN statements. The first calculates the mean of *treatments* over all *initial* weights and does Tukey tests between those means. The second calculates the means of the treatments at *initial* weight 340 based on the covariate *initial*, and also tests treatment for that particular *initial* weight. The third does the same for *initial* of 400. The initial weights 340 and 400 were chosen arbitrarily to be near the extremes of the range of initial weights, but any *initial* weight can be chosen.

SAS output:

Dependent Variable: gain

Source	DF	Sum of Squares	Mean Square	F Value	Pr > F
Model	3	19485.52365	6495.17455	22.90	0.0001
Error	10	2835.90493	283.59049		
Corrected Total	13	22321.42857			

R-Square	Coeff Var	Root MSE	gain Mean
0.872951	1.747680	16.84015	963.5714

Source	DF	Type I SS	Mean Square	F Value	Pr > F
initial	1	5750.54735	5750.54735	20.28	0.0011
treatment	1	10173.11966	10173.11966	35.87	0.0001
initial*treatment	1	3561.85664	3561.85664	12.56	0.0053

Least Squares Means
Adjustment for Multiple Comparisons: Tukey-Kramer

treatment	gain LSMEAN	Standard Error	H0:LSMean1= LSMean2 Pr > |t|
A	1001.49603	7.07264	<.0001
B	938.86966	6.70257	

Least Squares Means at initial=340
Adjustment for Multiple Comparisons: Tukey-Kramer

treatment	gain LSMEAN	Standard Error	H0:LSMean1=LSMean2 Pr > |t|
A	928.944444	10.274176	0.3613
B	914.123711	11.596819	

Least Squares Means at initial=400
Adjustment for Multiple Comparisons: Tukey-Kramer

treatment	gain LSMEAN	Standard Error	H0:LSMean1=LSMean2 Pr > |t|
A	1058.61111	13.46042	<.0001
B	958.35052	7.45310	

Parameter		Estimate	Standard Error	t Value	Pr > |t|
treatment	A	194.1666667	121.1437493	1.60	0.1401
	B	663.5051546	86.8331663	7.64	0.0001
initial*treat	A	2.1611111	0.3320921	6.51	0.0001
initial*treat	B	0.7371134	0.2261929	3.26	0.0086

Explanation: First is an *ANOVA* table for the dependent variable *gain*. The sources of variation are *Model*, residual (*Error*) and *Corrected Total*. In the table are listed degrees of freedom (*DF*), *Sum of Squares*, *Mean Square*, calculated *F* (*F Value*) and *P* value (*Pr > F*). In the next table *F* tests for *initial*, *treatment* and *initial*treatment* interaction are given. The *Type I SS* (sequential sum of squares) was used. The *F* test for treatment tests the effects of *treatments* adjusted on *initial* weight (by using Type 1 sums of squares) but is not appropriate if there is a significant interaction. If the interaction is present, the effects of *initial* are different for each treatment group. Next are tables with least square means and test between treatments. The first shows the treatment means at average initial weight (thus corrected on initial weight). Since there is an interaction between initial weight and treatments, we my wish to test the difference between treatments at a particular initial weight. The means and tests for two initial weight are shown: 340 and 400 kg. At 340 kg the difference between treatments is not significant, but at 400 kg it is.

The solutions given in the last table are the final part of the output of the second GLM procedure. These are the regression parameter estimates for each group. The estimated regression for treatment A is:

$gain = 194.1666667 + 2.161111\ initial$

The estimated regression for treatment B is:

$gain = 663.5051546 + 0.7371134\ initial$

Alternatively, the MIXED procedure can be used which gives the same results, but the output is somehow nicer and easier to comprehend.

SAS program:

```
PROC MIXED  DATA = gain;
CLASS treatment;
MODEL gain = initial treatment  treatment*initial / HTYPE=1 ;
LSMEANS treatment / DIFF ADJUST = TUKEY;
LSMEANS treatment / DIFF ADJUST = TUKEY AT initial=340;
LSMEANS treatment / DIFF ADJUST = TUKEY AT initial=400;
RUN;
```

Explanation: The syntax is similar to the GLM procedure, except sequential analysis of effects in the model is defined with *HTYPE=1*.

The results may be more meaningful if the independent variable is centered on its mean. This can be done by subtracting the mean from the values of the independent variable. For example, centering *initial* would give the same F test, but the regression equations would be calculated with the intercept at the mean *initial*. The mean of initial is 373.57.

From the second GLM program:

Parameter		Estimate	Standard Error	t Value	Pr > |t|
treatment	A	1001.492944	7.07243142	141.61	<.0001
treatment	B	938.868608	6.70267176	140.07	<.0001
initial*treatment	A	2.161111	0.33209214	6.51	<.0001
initial*treatment	B	0.737113	0.22619291	3.26	0.0086

The value for the treatments are the intercepts for the corresponding regression equations. The regression equation for treatment A is:

$gain = 1001.492944 + 2.161111\ (initial - 373.57).$

The estimated regression for treatment B is:

$gain = 938.868608 + 0.7371134\ (initial - 373.57)$

Chapter 22

Repeated Measures

Experimental units are often measured more than once if precision of single measurements is not adequate or if changes are expected over time. Variability among measurements on the same experimental unit can be homogeneous, but may alternatively be expected to change through time. Typical examples are milk yield during lactation, hormone concentrations in blood, or growth measurements over some period. In a repeated measures design the effect of a treatment is tested on experimental units that have been measured repeatedly over time. An experimental unit measured repeatedly is often called a subject. Note that 'change-over' designs can be considered repeated measured designs, but they differ in that two or more treatments are assigned to each subject. Here we will consider repeated measurements on an experimental unit receiving the same treatment over time.

The problem posed by repeated measurements on the same subject is that the measurements are usually correlated. For example, if a particular cow has high milk yield in the third month of lactation, it is likely that she will also have high yield in the fourth month, regardless of treatment. Measurements on the same subject are not independent, and thus it may be necessary to define an appropriate covariance structure for such measurements. Since the experimental unit is the animal (the subject), and not a single measurement on the animal, it is consequently necessary to define the appropriate experimental error for testing hypotheses. There may also be a treatment × time interaction, that is, the effect of particular treatments may be different at different time points.

Models for analyzing repeated measures can have the effects of period (time) defined as a categorical or a continuous independent variable. They can also include homogeneous or heterogeneous variances and covariances by defining appropriate variance-covariance structures or variance-covariance functions.

22.1 Homogeneous Variances and Covariances among Repeated Measures

The simplest model for describing repeated measures defines equal variance of and covariance between repeated measures, regardless of distance in time or space. The effects of periods can accounted for in the model by defining periods as values of a categorical independent variable. For example, consider an experiment with a treatments and b animals assigned to each treatment with each animal measured n times in n periods. The model is:

$$y_{ijk} = \mu + \tau_i + \delta_{ij} + t_k + (\tau^*t)_{ik} + \varepsilon_{ijk} \qquad i = 1,...,a; \ j = 1,...,b; \ k = 1,...,n$$

where:

y_{ijk} = observation ijk

μ = the overall mean

τ_i = the effect of treatment i

t_k = the effect of period k

$(\tau^*t)_{ik}$ = the effect of interaction between treatment i and period k

δ_{ij} = random error with mean 0 and variance σ^2_δ, the variance between animals (subjects) within treatment, equal to the covariance between repeated measurements within animals

ε_{ijk} = random error with mean 0 and variance σ^2, the variance between measurements within animals

Also, a = the number of treatments; b = the number of subjects (animals); and n = the number of periods

The mean of treatment i in period k is: $\mu_{ik} = \mu + \tau_i + t_k + (\tau^*t)_{ik}$

The variance between observations is:

$$Var(y_{ijk}) = Var(\delta_{ij} + \varepsilon_{ijk}) = \sigma^2_\delta + \sigma^2$$

The covariance between observations within animal is:

$$Cov(y_{ijk}, y_{ijk'}) = Var(\delta_{ij}) = \sigma^2_\delta$$

It is assumed that covariances between measures on different subjects are zero.

An equivalent model with a variance-covariance structure between subjects included in the error term (ε'_{ijk}) can be expressed as:

$$y_{ijk} = \mu + \tau_i + t_k + (\tau^*t)_{ik} + \varepsilon'_{ijk} \qquad i = 1,..., a; \; j = 1,..., b; \; k = 1,..., n$$

The equivalent model has one error term (ε'_{ijk}), but this error term is of a structure containing both variability between and within subjects. For example, a structure for four measurements on one subject shown in matrix form is:

$$\begin{bmatrix} \sigma^2 + \sigma^2_\delta & \sigma^2_\delta & \sigma^2_\delta & \sigma^2_\delta \\ \sigma^2_\delta & \sigma^2 + \sigma^2_\delta & \sigma^2_\delta & \sigma^2_\delta \\ \sigma^2_\delta & \sigma^2_\delta & \sigma^2 + \sigma^2_\delta & \sigma^2_\delta \\ \sigma^2_\delta & \sigma^2_\delta & \sigma^2_\delta & \sigma^2 + \sigma^2_\delta \end{bmatrix}$$

in which:

σ^2 = variance within subjects

σ^2_δ = covariance between measurements within subjects = variance between subjects

This variance-covariance structure is called *compound symmetry*, because it is diagonally symmetric and it is a compound of two variances.

Example: The effect of three treatments ($a = 3$) on milk fat yield of dairy cows was investigated. Fat yield was measured weekly for 6 weeks ($n = 6$). There were four cows per treatment ($b = 4$), and a total of 12 cows in the experiment ($ab = 12$). A table with the sources of variability, degrees of freedom and appropriate experimental errors defined is shown below:

Source	Degrees of freedom	
Treatment	$(a - 1) =$	2
Exp. error for treatments (Cow within treatment)	$a(b - 1) =$	9
Weeks	$(n - 1) =$	5
Treatment x weeks	$(a - 1)(n - 1) =$	10
Exp. error	$a(b - 1)(n - 1) =$	45
Total	$(abn - 1) =$	71

The experimental error for testing the effect of treatment is cow within treatment.

If changes of a dependent variable over time can be explained with a regression function, periods can be defined as values of a continuous variable. Note that the variance structure between measurements can still be defined. When period is a continuous variable and a linear change is assumed, the model is:

$$y_{ijk} = \mu + \tau_i + \delta_{ij} + \beta_1 (t_k) + \beta_{2i}(\tau^*t)_{ik} + \varepsilon_{ijk} \qquad i = 1,..., a;\ j = 1,..., b;\ k = 1,..., n$$

where:

y_{ijk} = observation ijk
μ = the overall mean
τ_i = the effect of treatment i
δ_{ij} = random error with mean 0 and variance σ^2_δ, the variance between animals within treatment
β_1 = regression coefficient of observations on periods t_k
β_{2i} = regression coefficient of observations on the treatment × period interaction $(\tau^*t)_{ik}$
ε_{ijk} = random error with mean 0 and variance σ^2, the variance between measurement within animals

Also, a = the number of treatments; b = the number of subjects (animals); n = the number of periods

Example: A table with sources of variability, degrees of freedom, and appropriate experimental error terms defined for the example with three treatments, four animals per treatment, and six weekly measurements per animal with week described as a continuous variable is:

Source	Degrees of freedom	
Treatment	$(a - 1) =$	2
Exp. error for treatments (Cow within treatment)	$a(b - 1) =$	9
Weeks	$1 =$	1
Treatment x weeks	$(a - 1)\, 1 =$	2
Exp. error	$ab(n - 1) - a =$	57
Total	$(abn - 1) =$	71

22.1.1 SAS Example for Homogeneous Variances and Covariances

The SAS programs for repeated measurements of variables with homogeneous variances and covariances will be shown for the following example. The aim of this experiment was to test the difference between two treatments on gain of kids. A sample of 18 kids was chosen, nine animals for each treatment. One kid in treatment 1 was removed from the experiment due to illness. Kids were placed on the experiment at age 8 weeks. Weekly gain was measured at ages 9, 10, 11 and 12 weeks. Two approaches will be shown: a) using week as a categorical variable; and b) using week as a continuous variable. The measurements are shown in the following table:

	Week	Kid 1	Kid 2	Kid 3	Kid 4	Kid 5	Kid 6	Kid 7	Kid 8
Treatment 1	9	1.2	1.2	1.3	1.1	1.2	1.1	1.1	1.3
	10	1.0	1.1	1.4	1.1	1.3	1.1	1.2	1.3
	11	1.1	1.4	1.4	1.2	1.2	1.1	1.3	1.3
	12	1.3	1.5	1.6	1.3	1.3	1.2	1.5	1.4

	Week	Kid 9	Kid 10	Kid 11	Kid 12	Kid 13	Kid 14	Kid 15	Kid 16	Kid 17
Treatment 2	9	1.2	1.3	1.5	1.4	1.2	1.0	1.4	1.1	1.2
	10	1.5	1.2	1.7	1.5	1.2	1.1	1.8	1.3	1.5
	11	1.9	1.4	1.6	1.7	1.4	1.4	2.1	1.4	1.7
	12	2.1	1.7	1.7	1.8	1.6	1.5	2.1	1.8	1.9

SAS program *with week defined as a categorical variable:*

```
DATA reps;
INPUT kid week trt gain @@;
DATALINES;
 1  9 1 1.2     1 10 1 1.0     1 11 1 1.1     1 12 1 1.3
 2  9 1 1.2     2 10 1 1.1     2 11 1 1.4     2 12 1 1.5
 3  9 1 1.3     3 10 1 1.4     3 11 1 1.4     3 12 1 1.6
 4  9 1 1.1     4 10 1 1.1     4 11 1 1.2     4 12 1 1.3
 5  9 1 1.2     5 10 1 1.3     5 11 1 1.2     5 12 1 1.3
 6  9 1 1.1     6 10 1 1.1     6 11 1 1.1     6 12 1 1.2
 7  9 1 1.1     7 10 1 1.2     7 11 1 1.3     7 12 1 1.5
 8  9 1 1.3     8 10 1 1.3     8 11 1 1.3     8 12 1 1.4
 9  9 2 1.2     9 10 2 1.5     9 11 2 1.9     9 12 2 2.1
10  9 2 1.3    10 10 2 1.2    10 11 2 1.4    10 12 2 1.7
11  9 2 1.5    11 10 2 1.7    11 11 2 1.6    11 12 2 1.7
12  9 2 1.4    12 10 2 1.5    12 11 2 1.7    12 12 2 1.8
13  9 2 1.2    13 10 2 1.2    13 11 2 1.4    13 12 2 1.6
14  9 2 1.0    14 10 2 1.1    14 11 2 1.4    14 12 2 1.5
15  9 2 1.4    15 10 2 1.8    15 11 2 2.1    15 12 2 2.1
16  9 2 1.1    16 10 2 1.3    16 11 2 1.4    16 12 2 1.8
17  9 2 1.2    17 10 2 1.5    17 11 2 1.7    17 12 2 1.9
;
```

```
PROC MIXED DATA = reps;
CLASS kid trt week;
MODEL gain = trt week trt*week / DDFM = KENWARDROGER;
REPEATED / TYPE=CS SUB=kid(trt) ;
LSMEANS trt / DIFF ADJUST = TUKEY;
LSMEANS trt*week / DIFF ADJUST = TUKEY;
RUN;
```

Explanation: The MIXED procedure was used. The CLASS statement defines categorical variables. The MODEL statement defines the dependent variable *gain*, and independent variables *trt*, *week* and *trt*week* interaction. It is recommended that the Kenward-Roger option be used to estimate degrees of freedom (option *DDFM = KENWARDROGER* with the MODEL statement) when the subjects mean square and standard errors are linear combination of variances. This option is also useful when some values are missing. The REPEATED statement defines the variance structure for repeated measurements. The subject (*SUB = kid(trt)*) defines the variable on which repeated measurements were taken. The type of variance-covariance structure is compound symmetry (*TYPE = CS*). The LSMEANS statement calculates treatment means.

An excerpt of the most important information from the SAS output is shown below.

SAS output:

Covariance Parameter Estimates

Cov Parm	Subject	Estimate
CS	kid(trt)	0.02083
Residual		0.01116

Fit Statistics

-2 Res Log Likelihood	-50.3
AIC (smaller is better)	-46.3
AICC (smaller is better)	-46.1
BIC (smaller is better)	-44.6

Null Model Likelihood Ratio Test

DF	Chi-Square	Pr > ChiSq
1	31.13	<.0001

Type 3 Tests of Fixed Effects

Effect	Num DF	Den DF	F Value	Pr > F
trt	1	15	13.25	0.0024
week	3	45	40.09	<.0001
trt*week	3	45	9.20	<.0001

Least Squares Means

Effect	trt	week	Estimate	Standard Error	DF	t Value	Pr>|t|
trt	1		1.2531	0.05434	15	23.06	<.0001
trt	2		1.5250	0.05123	15	29.77	<.0001
trt*week	1	9	1.1875	0.06324	26.4	18.78	<.0001
trt*week	1	10	1.1875	0.06324	26.4	18.78	<.0001
trt*week	1	11	1.2500	0.06324	26.4	19.77	<.0001
trt*week	1	12	1.3875	0.06324	26.4	21.94	<.0001
trt*week	2	9	1.2556	0.05962	26.4	21.06	<.0001
trt*week	2	10	1.4222	0.05962	26.4	23.85	<.0001
trt*week	2	11	1.6222	0.05962	26.4	27.21	<.0001
trt*week	2	12	1.8000	0.05962	26.4	30.19	<.0001

Differences of Least Squares Means

Effect	trt	week	_trt	_week	Estim.	Stand Error	DF	t Value	Pr>|t|	Adj.	Adj P
trt	1		2		-0.2719	0.07468	15	-3.64	0.0024	Tuk-Kr	0.0024
trt*week	1	9	2	9	-0.06806	0.08691	26.4	-0.78	0.4406	Tuk-Kr	0.9932
trt*week	1	10	2	10	-0.2347	0.08691	26.4	-2.70	0.0119	Tuk-Kr	0.1490
trt*week	1	11	2	11	-0.3722	0.08691	26.4	-4.28	0.0002	Tuk-Kr	0.0023
trt*week	1	12	2	12	-0.4125	0.08691	26.4	-4.75	<.0001	Tuk-Kr	0.0005

Explanation: The table *Covariance Parameter Estimates* gives estimates of the following: *CS* = the variance between subjects, and *Residual* = the variance within subject. The *Fit Statistics* are criteria describing model fit. The *Null Model Likelihood Ratio Test* tests significance and appropriateness of the model. The *Type 3 Tests of Fixed Effects* tests the *Effect* in the model. Degrees of freedom for the effects are *Num DF*, degrees of freedom for the experimental error terms are *Den DF*, and *P* values are (*Pr > F*). The *P* values for fixed effects in the model are each smaller than 0.05 indicating that each effect is significant. Note the different denominator degrees of freedom (*Den DF*) indicate that appropriate experimental error terms were used for testing particular effects. In the table *Least Squares Means*, estimated means (*Estimate*) with the corresponding *Standard Error*s are shown for each treatments week combination and the overall treatment effects. The table *Differences of Least Squares Means* shows the difference between treatments (*Estimate*), the standard error of the difference (*Stand Error*) and *P* value (*Pr > |t|*). For the *trt*week* interaction only the difference of *trt* means within weeks are shown. Note the degrees of freedom (*DF*) corrected with the Kenward-Roger adjustment. *P*-values were adjusted for multiple comparisons by using the Tukey-Kramer procedure (*Adj P*).

SAS program *with week defined as a continuous variable:*

```
PROC MIXED DATA=reps;
CLASS kid trt;
MODEL gain = trt week trt*week
            / HTYPE=1 DDFM = KENWARDROGER;
REPEATED / TYPE=CS SUB=kid(trt) ;
LSMEANS trt / DIFF ADJUST = TUKEY;
LSMEANS trt / DIFF ADJUST = TUKEY AT week = 9;
LSMEANS trt / DIFF ADJUST = TUKEY AT week = 10;
```

```
LSMEANS trt / DIFF ADJUST = TUKEY AT week = 11;
LSMEANS trt / DIFF ADJUST = TUKEY AT week = 12;
RUN;
```

Explanation: The MIXED procedure was used. Note, that the variable *week* is not listed in the CLASS statement and the procedure fits it as a continuous variable. The option *HTYPE* = *1* in the MODEL statement tests the effects sequentially. The five LSMEANS statements direct calculation of the treatment means. The first calculates the mean of *treatments* over all *weeks* and tests the differences between those means. The second through fifth calculate the means of the treatments at *weeks* 9 to 12, respectively, and also test for treatment effects at each of those *weeks*. The Kenward-Roger correction for degrees of freedom is used because the standard errors are linear combinations of variances that cannot be estimated by a single mean square.

SAS output:

Covariance Parameter Estimates

Cov Parm	Subject	Estimate
CS	kid(trt)	0.02085
Residual		0.01106

Fit Statistics

-2 Res Log Likelihood	-59.9
AIC (smaller is better)	-55.9
AICC (smaller is better)	-55.7
BIC (smaller is better)	-54.3

Null Model Likelihood Ratio Test

DF	Chi-Square	Pr > ChiSq
1	32.98	<.0001

Type 1 Tests of Fixed Effects

Effect	Num DF	Den DF	F Value	Pr > F
trt	1	15	13.25	0.0024
week	1	49	126.38	<.0001
week*trt	1	49	26.25	<.0001

Least Squares Means

Effect	trt	week	Estimate	Standard Error	DF	t Value	Pr > |t|
trt	1	10.50	1.2531	0.05434	15	23.06	<.0001
trt	2	10.50	1.5250	0.05123	15	29.77	<.0001
trt	1	9.00	1.1537	0.05979	21.7	19.30	<.0001
trt	2	9.00	1.2500	0.05637	21.7	22.18	<.0001
trt	1	10.00	1.2200	0.05497	15.7	22.19	<.0001
trt	2	10.00	1.4333	0.05183	15.7	27.66	<.0001
trt	1	11.00	1.2863	0.05497	15.7	23.40	<.0001
trt	2	11.00	1.6167	0.05183	15.7	31.19	<.0001
trt	1	12.00	1.3525	0.05979	21.7	22.62	<.0001
trt	2	12.00	1.8000	0.05637	21.7	31.93	<.0001

Differences of Least Squares Means

Effect	trt	_trt	week	Estimate	Standard Error	DF	t Value	Pr>|t|	Adjust	Adj P
trt	1	2	10.50	-0.2719	0.07468	15	-3.64	0.0024	Tuk-Kr	0.0024
trt	1	2	9.00	-0.09625	0.08217	21.7	-1.17	0.2542	Tuk-Kr	0.2597
trt	1	2	10.00	-0.2133	0.07555	15.7	-2.82	0.0124	Tuk-Kr	0.0128
trt	1	2	11.00	-0.3304	0.07555	15.7	-4.37	0.0005	Tuk-Kr	0.0005
trt	1	2	12.00	-0.4475	0.08217	21.7	-5.45	<.0001	Tuk-Kr	<.0001

Explanation: The output is similar to that from the model in which week was fitted as a categorical variable. The table *Type 1 Tests of Fixed Effects* shows that there is a *week*trt* interaction indicating the effect of each treatment over time is different. In other words, the slopes of the regression of gain on week differ between treatments. The *F* test for *trt* tests the difference between averages of treatments across all *week*s, and since interaction exists, the results should be interpreted for the individual weeks. (Note that a Type 3 test for *trt* would give the test of difference at the intercept, i.e. at week 0, which is meaningless). The least squares means are listed for the average week (10.5) and for weeks 9 to 12 for each treatment. Also the tests of difference between treatments are given overall (average week 10.5) and for individual weeks. Note degrees of freedom are corrected using the Kenward-Rogers correction.

The regression coefficient estimates for each treatment can be obtained as follows:

```
PROC MIXED DATA=reps;
CLASS kid trt;
MODEL gain = trt trt*week
      / NOINT SOLUTION DDFM = KENWARDROGER;
REPEATED / TYPE=CS SUB=kid(trt) ;
RUN;
```

22.2 Heterogeneous Variances and Covariances among Repeated Measures

Variances and covariances (or correlations) are not always constant between measurements. Variance and covariance can be modeled with a variety of variance-covariance structures. The most general model, called an unstructured model, defines different variances for each period and different covariances between periods, but assumes that covariance between measurements on different animals is zero. An example of an unstructured variance-covariance matrix for four measures on each subject is:

$$\begin{bmatrix} \sigma_1^2 & \sigma_{12} & \sigma_{13} & \sigma_{14} \\ \sigma_{21} & \sigma_2^2 & \sigma_{23} & \sigma_{24} \\ \sigma_{31} & \sigma_{32} & \sigma_3^2 & \sigma_{34} \\ \sigma_{41} & \sigma_{42} & \sigma_{43} & \sigma_4^2 \end{bmatrix}$$

in which:

σ_i^2 = variance of measures in period i

σ_{ij} = covariance within subjects between measures in periods i and j

Another model is called an autoregressive model. It assumes that with greater distance between periods, correlations are smaller. The correlation is ρ^t, for which t is the number of periods between measurements. An example of the variance-covariance matrix of the autoregressive structure for four measurements on each subject is:

$$\sigma^2 \begin{bmatrix} 1 & \rho & \rho^2 & \rho^3 \\ \rho & 1 & \rho & \rho^2 \\ \rho^2 & \rho & 1 & \rho \\ \rho^3 & \rho^2 & \rho & 1 \end{bmatrix}$$

in which:

σ^2 = variance of measures

ρ^t = correlation between measurements taken t periods apart ($t = 0, 1, 2, 3$) within subjects

Another variance-covariance structure is the Toeplitz structure, in which correlations between measurements also depend on the number of periods. Measurements taken one period apart have the same covariance, for example $\sigma_{12} = \sigma_{23}$, measurements two periods apart have the same covariance but it is different from the first, for example $\sigma_{13} = \sigma_{24} \neq \sigma_{12} = \sigma_{23}$. An example of the variance-covariance matrix for a Toeplitz structure for four measurements on each subject is:

$$\begin{bmatrix} \sigma^2 & \sigma_1 & \sigma_2 & \sigma_3 \\ \sigma_1 & \sigma^2 & \sigma_1 & \sigma_2 \\ \sigma_2 & \sigma_1 & \sigma^2 & \sigma_1 \\ \sigma_3 & \sigma_2 & \sigma_1 & \sigma^2 \end{bmatrix}$$

in which:

σ^2 = variance of measures

$\sigma_1, \sigma_2, \sigma_3$ = covariances between measurements within subjects

The power model variance-covariance structure defines covariances between measures on the same animal with a power function. The covariance of two measurements taken at time points t_i and t_j depends on the distance between those two points:

$$\sigma^2 \rho^{|t_i - t_j|}$$

This variance-covariance structure can be used when periods between measurements are not equal. An example of the covariance matrix of the power variance-covariance structure for four measurements taken at time points t_1, t_2, t_3, and t_4 on one animal is:

$$\sigma^2 \begin{bmatrix} 1 & \rho^{|t_1-t_2|} & \rho^{|t_1-t_3|} & \rho^{|t_1-t_4|} \\ \rho^{|t_2-t_1|} & 1 & \rho^{|t_2-t_3|} & \rho^{|t_2-t_4|} \\ \rho^{|t_3-t_1|} & \rho^{|t_3-t_2|} & 1 & \rho^{|t_3-t_4|} \\ \rho^{|t_4-t_1|} & \rho^{|t_4-t_2|} & \rho^{|t_4-t_3|} & 1 \end{bmatrix}$$

in which:

σ^2 = variance of measures

$\rho^{|t_i-t_j|}$ = the correlation between measurements on the same animal taken at time points t_i and t_j

Note that the power model can be equivalently expressed in terms of an exponential function and is also known as the exponential variance-covariance structure model. If we define:

$$exp(-\theta) = \rho$$

in which θ is a parameter, then:

$$\sigma^2 \, exp\left(-\theta^{|t_i-t_j|}\right)$$

which is an exponential model, but equivalent to the power model defined previously.

22.2.1 SAS Examples for Heterogeneous Variances and Covariances

SAS programs for repeated measurement with heterogeneous variances and covariances will be shown by analyzing data from the example examining the effects of two treatments on weekly gain of kids. Data were collected on a sample of 17 kids, eight and nine animals for treatments one and two, respectively. Weekly gain was measured four times in 4 weeks. The use of unstructured, autoregressive and Toeplitz and power variance-covariance structures will be shown. Week will be defined as a categorical variable.

SAS program:

```
PROC MIXED DATA = reps;
CLASS kid trt week;
MODEL gain = trt week trt*week / DDFM = KENWARDROGER;
REPEATED / TYPE=UN SUB=kid(trt) R RCORR;
LSMEANS trt trt*week / DIFF ADJUST = TUKEY;
RUN;
```

Explanation: The MIXED procedure is used. The CLASS statement defines categorical variables. The MODEL statement defines the dependent and independent variables. The dependent variable is *gain*, and the independent variables are *trt*, *week* and *trt*week* interaction. The REPEATED statement defines the variance structure for repeated measurements. The subject statement (*SUB* = *kid(trt)*) defines *kid* nested within treatment (*trt*) as the variable on which measures are repeated, and the type of variance-covariance structure is defined as unstructured by *TYPE* = *UN* in examples below [for autoregressive *TYPE* = *AR(1)*, for Toeplitz: *TYPE* = *TOEP*, and for power *TYPE* = *SP(POW)(week)*]. When using the power structure, the internal distance between time is not required to be equal, but the time variable must be specified along with SP(POW) option. The R and RCORR options direct output of the variance-covariance and correlation matrices. The Kenward-Rogers estimation of degrees of freedom is recommended for repeated measures analysis because the mean square of the subjects and the standard errors consist of linear combinations of variances, and because of possible heterogeneity of variances and covariances. This option also accounts for missing values.

SAS output *for the unstructured model:*

```
          Covariance Parameter Estimates

    Cov Parm    Subject        Estimate
    UN(1,1)     kid(trt)       0.01673
    UN(2,1)     kid(trt)       0.01851
    UN(2,2)     kid(trt)       0.03895
    UN(3,1)     kid(trt)       0.01226
    UN(3,2)     kid(trt)       0.03137
    UN(3,3)     kid(trt)       0.04104
    UN(4,1)     kid(trt)       0.00792
    UN(4,2)     kid(trt)       0.02325
    UN(4,3)     kid(trt)       0.03167
    UN(4,4)     kid(trt)       0.03125


              Fit Statistics

    -2 Res Log Likelihood           -72.2
    AIC (smaller is better)         -52.2
    AICC (smaller is better)        -47.7
    BIC (smaller is better)         -43.9


        Estimated R Matrix for kid(trt) 1 1

 Row       Col1        Col2        Col3        Col4
  1      0.01673     0.01851     0.01226     0.007917
  2      0.01851     0.03895     0.03137     0.02325
  3      0.01226     0.03137     0.04104     0.03167
  4      0.007917    0.02325     0.03167     0.03125
```

```
              Estimated R Correlation Matrix for kid(trt) 1 1

   Row         Col1           Col2           Col3           Col4
    1         1.0000         0.7250         0.4679         0.3462
    2         0.7250         1.0000         0.7846         0.6664
    3         0.4679         0.7846         1.0000         0.8843
    4         0.3462         0.6664         0.8843         1.0000
```

Explanation: Only the variance-covariance estimates, fit statistics and variance-covariance and correlation matrices are shown. There are 10 parameters in this model. The $UN(i, j)$ denotes covariance between at time points i and j. For example, $UN(1,1) = 0.01673$ denotes the variance of measurements taken in period 1 (week 9), and $UN(3,1) = 0.01226$ denotes the covariance between measures within animals taken in periods 1 and 3 (weeks 9 and 11). This can be more clearly seen from *Estimated R Matrix for kid(trt) 1 1*, which presents the variance-covariance estimates in matrix form.

$$\begin{bmatrix} 0.01673 & 0.01851 & 0.01226 & 0.00792 \\ 0.01851 & 0.03895 & 0.03137 & 0.02325 \\ 0.01226 & 0.03137 & 0.04104 & 0.03167 \\ 0.00792 & 0.02325 & 0.03167 & 0.03125 \end{bmatrix}$$

The last table (*Estimated R Correlation Matrix for kid(trt) 1 1*) shows the correlations between measurements taken at different time points. Correlations are higher for measurements closer in time.

SAS output *for the autoregressive structure:*

```
                 Covariance Parameter Estimates

      Cov Parm         Subject            Estimate
      AR(1)            kid(trt)             0.7491
      Residual                             0.02888

                      Fit Statistics

       -2 Res Log Likelihood              -62.4
       AIC (smaller is better)            -58.4
       AICC (smaller is better)           -58.2
       BIC (smaller is better)            -56.7

             Estimated R Matrix for kid(trt) 1 1

   Row         Col1           Col2           Col3           Col4
    1         0.02888        0.02163        0.01620        0.01214
    2         0.02163        0.02888        0.02163        0.01620
    3         0.01620        0.02163        0.02888        0.02163
    4         0.01214        0.01620        0.02163        0.02888
```

```
          Estimated R Correlation Matrix for kid(trt) 1 1

  Row          Col1          Col2          Col3          Col4
   1          1.0000        0.7491        0.5612        0.4204
   2          0.7491        1.0000        0.7491        0.5612
   3          0.5612        0.7491        1.0000        0.7491
   4          0.4204        0.5612        0.7491        1.0000
```

Explanation: Only the variance-covariance estimates, fit statistics and variance-covariance and correlation matrices between measures within subjects are shown. There are two parameters in this model. The variance of measures is denoted by *Residual*, and the correlation between adjacent measurements (one period apart) on the same animal by *AR(1)*. The variance-covariance estimates are presented in the table titled *Estimated R Matrix for kid(trt) 1 1* and are calculated as follows:

$$0.02888 \begin{bmatrix} 1 & 0.7491 & 0.7491^2 & 0.7491^3 \\ 0.7491 & 1 & 0.7491 & 0.7491^2 \\ 0.7491^2 & 0.7491 & 1 & 0.7491 \\ 0.7491^3 & 0.7491^2 & 0.7491 & 1 \end{bmatrix} = \begin{bmatrix} 0.028888 & 0.021634 & 0.016206 & 0.012140 \\ 0.021634 & 0.028888 & 0.021634 & 0.016206 \\ 0.016206 & 0.021634 & 0.028888 & 0.021634 \\ 0.012140 & 0.016206 & 0.021634 & 0.028888 \end{bmatrix}$$

The corresponding correlation matrix is titled *Estimated R Correlation Matrix for kid(trt) 1 1*.

SAS output *for the Toeplitz structure:*

```
              Covariance Parameter Estimates

  Cov Parm       Subject              Estimate
  TOEP(2)        kid(trt)             0.02062
  TOEP(3)        kid(trt)             0.01127
  TOEP(4)        kid(trt)            -0.00015
  Residual                           0.02849


                  Fit Statistics

   -2 Res Log Likelihood              -64.3
   AIC (smaller is better)            -56.3
   AICC (smaller is better)           -55.6
   BIC (smaller is better)            -53.0


          Estimated R Matrix for kid(trt) 1 1

  Row          Col1          Col2          Col3          Col4
   1          0.02849       0.02062       0.01127      -0.00015
   2          0.02062       0.02849       0.02062       0.01127
   3          0.01127       0.02062       0.02849       0.02062
   4         -0.00015       0.01127       0.02062       0.02849
```

```
          Estimated R Correlation Matrix for kid(trt) 1 1

      Row          Col1          Col2          Col3          Col4
       1          1.0000        0.7237        0.3956       -0.00524
       2          0.7237        1.0000        0.7237        0.3956
       3          0.3956        0.7237        1.0000        0.7237
       4         -0.00524       0.3956        0.7237        1.0000
```

Explanation: Only the variance-covariance estimates, fit statistics and variance-covariance and correlation matrices between measures within subjects are shown. There are four parameters in this model. The *TOEP(2)*, *TOEP(3)* and *TOEP(4)* denote covariances between measures on the same subject (*kid*) one, two and three periods apart, respectively. The variance of measures is denoted by *Residual*. The variance-covariance structure for one subject is titled *Estimated R Matrix for kid(trt) 1 1*. The corresponding correlation matrix is titled *Estimated R Correlation Matrix for kid(trt) 1 1*.

SAS output *for the power* (*exponential*) *structure:*

```
              Covariance Parameter Estimates

        Cov Parm        Subject        Estimate
        SP(POW)         kid(trt)        0.7491
        Residual                        0.02888


                   Fit Statistics

        -2 Res Log Likelihood             -62.4
        AIC (smaller is better)           -58.4
        AICC (smaller is better)          -58.2
        BIC (smaller is better)           -56.7

            Estimated R Matrix for kid(trt) 1 1

   Row          Col1          Col2          Col3          Col4
    1          0.02887       0.02163       0.01620       0.01214
    2          0.02163       0.02887       0.02163       0.01620
    3          0.01620       0.02163       0.02887       0.02163
    4          0.01214       0.01620       0.02163       0.02887


        Estimated R Correlation Matrix for kid(trt) 1 1

   Row          Col1          Col2          Col3          Col4
    1          1.0000        0.7491        0.5611        0.4203
    2          0.7491        1.0000        0.7491        0.5611
    3          0.5611        0.7491        1.0000        0.7491
    4          0.4203        0.5611        0.7491        1.0000
```

Explanation: Only the variance-covariance estimates, fit statistics and variance-covariance and correlation matrices between measures within subjects are shown. There are two parameters in this model. The variance of measures is denoted by *Residual*. The covariance between two measurements at two weeks, t_i and t_j, is:

$$\sigma^2 \rho^{|t_i - t_j|}$$

The estimate of σ^2 is *Residual* = 0.02888 and of ρ is *SP(POW)* = 0.7491.

The variance-covariance structure for one subject is:

$$\sigma^2 \begin{bmatrix} 1 & \rho^{|9-10|} & \rho^{|9-11|} & \rho^{|9-12|} \\ \rho^{|10-9|} & 1 & \rho^{|10-11|} & \rho^{|10-12|} \\ \rho^{|11-9|} & \rho^{|11-10|} & 1 & \rho^{|11-12|} \\ \rho^{|12-9|} & \rho^{|12-10|} & \rho^{|12-11|} & 1 \end{bmatrix} = \begin{bmatrix} 0.02887 & 0.02163 & 0.01620 & 0.01214 \\ 0.02163 & 0.02887 & 0.02163 & 0.01620 \\ 0.01620 & 0.02163 & 0.02887 & 0.02163 \\ 0.01214 & 0.01620 & 0.02163 & 0.02887 \end{bmatrix}$$

This is shown in the output under the title *Estimated R Matrix for kid(trt) 1 1*.

Note: The covariances are expected to be positive, and if the estimates are negative they are multiplied by $\cos(\pi|t_i - t_j|)$ to account for the negative values.

The corresponding correlation matrix is titled *Estimated R Correlation Matrix for kid(trt) 1 1*.

Note that when time units are equally spaced the power structure is equivalent to autoregressive.

SAS gives several criteria for evaluating model fit including Akaike information criteria (*AIC*) and the Schwarz Bayesian information criteria (*BIC*). These calculations are based on the log likelihood (or log restricted likelihood) for the model. They depend on method of estimation, number of observations and number of parameters. In SAS the better model will have a smaller *AIC* and *BIC* value. In the following table the values of –2 restricted log likelihood and *AIC* for the variance-covariance structure models are listed for this example:

Model	–2 Res Log Likelihood	AIC
Unstructured (UN)	–72.2	–52.2
Compound symmetry (CS)	–59.9	–55.9
Autoregressive [AR(1)]	–62.4	–58.4
Toeplitz (TOEP)	–64.3	–56.3
Power (SP(POW)(week)	–62.4	–58.4

These criteria indicate that two models appear to be better than the others, the autoregressive and power models. Their values are –58.4, smaller than the other models. Note that AIC is computed in SAS as –2 times residual log likelihood plus twice the number of variance-covariance parameters. For example, for the unstructured model the numbers of parameters is 10 and the –2 Res loglikelihood is –72.2, and the AIC = –72.2 + 20= –52.2.

22.3 Random Coefficient Regression

Another approach for analyzing repeated measures for situations with heterogeneous variance and covariance is random coefficient regression. The assumption is that each subject has its own regression defined over time, thus the regression coefficients are assumed to be a random sample from some population. The main advantage of a random coefficient regression model is that the time or distance between measures need not be equal, and the number of observations per subject can be different. This gives more flexibility than other mopdels of variance-covariance structure. For example, a simple linear random regression model is:

$$y_{ij} = b_{0i} + b_{1i}t_{ij} + \varepsilon_{ij} \qquad\qquad i = 1,\ldots, \text{number of subjects}$$

where:

y_{ij} = dependent variable

t_{ij} = independent variable

b_{0i}, b_{1i} = regression coefficients with means β_{0i}, β_{1i}, and variance covariance matrix

$$\begin{bmatrix} \sigma_{b_0}^2 & \sigma_{b_0 b_1} \\ \sigma_{b_0 b_1} & \sigma_{b_1}^2 \end{bmatrix}$$

ε_{ij} = random error

Alternatively, the random coefficient regression model can be expressed as:

$$y_{ij} = \beta_0 + \beta_1 t_{ij} + b_{0i} + b_{1i}t_{ij} + \varepsilon_{ij}$$

$\beta_0 + \beta_1 t_{ij}$ represent the fixed component and $b_{0i} + b_{1i}t_{ij} + \varepsilon_{ij}$ represent the random component. The means of b_{0i} and b_{1i} are zero, and the variance-covariance matrix is:

$$\begin{bmatrix} \sigma_{b_0}^2 & \sigma_{b_0 b_1} \\ \sigma_{b_0 b_1} & \sigma_{b_1}^2 \end{bmatrix}$$

The effect of subject i at time point t_{ij} is $b_{0i} + b_{1i}t_{ij}$.

An important characteristic of random coefficient regression is that a covariance function can be defined which describes the variance-covariance structure between repeated measures in time. The covariance function that describes covariance between measures j and j' on the same subject is:

$$\sigma_{t_j t_{j'}} = \begin{bmatrix} 1 & t_j \end{bmatrix} \begin{bmatrix} \sigma_{b_0}^2 & \sigma_{b_0 b_1} \\ \sigma_{b_0 b_1} & \sigma_{b_1}^2 \end{bmatrix} \begin{bmatrix} 1 \\ t_{j'} \end{bmatrix}$$

It is possible to estimate the covariance within subject between measures at any two time points t_j and $t_{j'}$, and variance between subjects at any time t_j. If the common error variance within subjects is denoted by σ^2, the variance of an observation taken at time t_j is:

$$\sigma^2 + \sigma_{t_j t_j}$$

If measures are taken at the same ages for all subjects, say at ages t_1, t_2,..., t_k, then the variance-covariance structure that describes covariance between measures for one subject is:

$$\mathbf{R} = \begin{bmatrix} 1 & t_1 \\ 1 & t_2 \\ \cdots & \cdots \\ 1 & t_k \end{bmatrix} \begin{bmatrix} \sigma_{b_0}^2 & \sigma_{b_0 b_1} \\ \sigma_{b_0 b_1} & \sigma_{b_1}^2 \end{bmatrix} \begin{bmatrix} 1 & 1 & \cdots & 1 \\ t_1 & t_2 & \cdots & t_k \end{bmatrix} + \begin{bmatrix} \sigma^2 & 0 & 0 & 0 \\ 0 & \sigma^2 & 0 & 0 \\ \cdots & \cdots & \cdots & \cdots \\ 0 & 0 & 0 & \sigma^2 \end{bmatrix}_{k \times k}$$

For example, the variance-covariance structure for four measures per subject taken at times t_1, t_2, t_3 and t_4 is:

$$\begin{bmatrix} \sigma^2 + \sigma_{t_1 t_1} & \sigma_{t_1 t_2} & \sigma_{t_1 t_3} & \sigma_{t_1 t_4} \\ \sigma_{t_1 t_2} & \sigma^2 + \sigma_{t_2 t_2} & \sigma_{t_2 t_3} & \sigma_{t_2 t_4} \\ \sigma_{t_1 t_3} & \sigma_{t_2 t_3} & \sigma^2 + \sigma_{t_3 t_3} & \sigma_{t_3 t_4} \\ \sigma_{t_1 t_4} & \sigma_{t_2 t_4} & \sigma_{t_3 t_4} & \sigma^2 + \sigma_{t_4 t_4} \end{bmatrix}$$

Note again that covariance between measures between different subjects at different times is equal to zero.

More complex models can include separate variances and covariances both between and within subjects for each treatment group. These will be shown using SAS examples.

22.3.1 SAS Examples for Random Coefficient Regression

The SAS programs for random coefficient regression will be shown by analysis of the example examining the effects of two treatments on weekly gain of kids.

Homogeneous Variance-Covariance Parameters across Treatments

SAS program:

```
PROC MIXED DATA = reps;
CLASS kid trt;
MODEL gain = trt week trt*week / DDFM = KENWARDROGER;
RANDOM int week / TYPE=UN  SUB = kid(trt);
LSMEANS trt / DIFF ADJUST = TUKEY;
LSMEANS trt / DIFF ADJUST = TUKEY AT week = 9;
LSMEANS trt / DIFF ADJUST = TUKEY AT week = 10;
LSMEANS trt / DIFF ADJUST = TUKEY AT week = 11;
LSMEANS trt / DIFF ADJUST = TUKEY AT week = 12;
RUN;
```

Explanation: The MIXED procedure was used. The CLASS statement defines categorical variables. The MODEL statement defines the dependent and independent variables. The dependent variable is *gain*, and independent variables are *trt*, *week* and *trt*week* interaction. The RANDOM statement defines the regression coefficients (*int* and *week* for intercept and slope) as random variables. The variance-covariance structure is designated unstructured (*TYPE = UN*), and the subject is *SUB = kid(trt)*. The LSMEANS statements direct calculation of treatment means. The first calculates the mean of *treatments* over all *weeks* and tests differences between those means. The subsequent statements calculate the means of the treatments at *weeks* 9 to 12 and also test the effects of treatment for each of those *weeks*. The Kenward-Rogers correction for degrees of freedom is used because the standard errors are linear combinations of variances. This option is also useful when some values are missing and when there is heterogeneity of variances and covariances.

SAS output:

```
                Covariance Parameter Estimates

        Cov Parm        Subject         Estimate

        UN(1,1)         kid(trt)         0.234200
        UN(2,1)         kid(trt)        -0.023230
        UN(2,2)         kid(trt)         0.002499
        Residual                         0.007235
```

Explanation: Only the variance-covariance estimates from the SAS output are shown. The covariance matrix of the regression coefficients is:

$$\begin{bmatrix} \hat{\sigma}^2_{b_0} & \hat{\sigma}_{b_0 b_1} \\ \hat{\sigma}_{b_0 b_1} & \hat{\sigma}^2_{b_1} \end{bmatrix} = \begin{bmatrix} 0.234200 & -0.023230 \\ -0.023230 & 0.002499 \end{bmatrix}$$

The variance of measures within animals or the estimate of the error variance denoted here as *Residual* is:

$$\hat{\sigma}^2 = 0.007235$$

The covariance function between measures on the same animal is:

$$\hat{\sigma}_{t_j t_{j'}} = \begin{bmatrix} 1 & t_j \end{bmatrix} \begin{bmatrix} 0.234200 & -0.023230 \\ -0.023230 & 0.002499 \end{bmatrix} \begin{bmatrix} 1 \\ t_{j'} \end{bmatrix}$$

For example, the variance between animals at the age of 9 weeks is:

$$\hat{\sigma}_{t_9 t_9} = \begin{bmatrix} 1 & 9 \end{bmatrix} \begin{bmatrix} 0.234200 & -0.023230 \\ -0.023230 & 0.002499 \end{bmatrix} \begin{bmatrix} 1 \\ 9 \end{bmatrix} = 0.018479$$

The variance of measures at the age of 9 weeks is the sum of the residual plus the variance between animals:

$$\hat{\sigma}^2 + \hat{\sigma}_{t_9 t_9} = 0.007235 + 0.018479 = 0.025714$$

The covariance between measures within animal at 9 and 10 weeks is:

$$\hat{\sigma}_{t_9 t_{10}} = \begin{bmatrix} 1 & 9 \end{bmatrix} \begin{bmatrix} 0.234200 & -0.023230 \\ -0.023230 & 0.002499 \end{bmatrix} \begin{bmatrix} 1 \\ 10 \end{bmatrix} = 0.017740$$

If measures are taken at the same ages for all animals, as is the case here, then the variance-covariance structure for one animal is:

$$\hat{\mathbf{R}} = \begin{bmatrix} 1 & t_1 \\ 1 & t_2 \\ 1 & t_3 \\ 1 & t_4 \end{bmatrix} \begin{bmatrix} \hat{\sigma}_{b_0}^2 & \hat{\sigma}_{b_0 b_1} \\ \hat{\sigma}_{b_0 b_1} & \hat{\sigma}_{b_1}^2 \end{bmatrix} \begin{bmatrix} 1 & 1 & 1 & 1 \\ t_1 & t_2 & t_3 & t_4 \end{bmatrix} + \begin{bmatrix} \hat{\sigma}^2 & 0 & 0 & 0 \\ 0 & \hat{\sigma}^2 & 0 & 0 \\ 0 & 0 & \hat{\sigma}^2 & 0 \\ 0 & 0 & 0 & \hat{\sigma}^2 \end{bmatrix}$$

$$= \begin{bmatrix} 1 & 9 \\ 1 & 10 \\ 1 & 11 \\ 1 & 12 \end{bmatrix} \begin{bmatrix} 0.234200 & -0.023230 \\ -0.023230 & 0.002499 \end{bmatrix} \begin{bmatrix} 1 & 1 & 1 & 1 \\ 9 & 10 & 11 & 12 \end{bmatrix}$$

$$+ \begin{bmatrix} 0.007235 & 0 & 0 & 0 \\ 0 & 0.007235 & 0 & 0 \\ 0 & 0 & 0.007235 & 0 \\ 0 & 0 & 0 & 0.007235 \end{bmatrix}$$

$$= \begin{bmatrix} 0.025714 & 0.017740 & 0.017001 & 0.016262 \\ 0.017740 & 0.026735 & 0.021260 & 0.023020 \\ 0.017001 & 0.021260 & 0.032754 & 0.029778 \\ 0.016262 & 0.023020 & 0.029778 & 0.043771 \end{bmatrix}$$

Heterogeneous Variance-Covariance Parameters across Treatments

Between group heterogeneous random coefficient regressions can be estimated by using a RANDOM and/or REPEATED statement with the GROUP option.

Defining different variance-covariance parameters for each treatment and having a common error variance, the SAS program is:

```
PROC MIXED DATA=reps;
CLASS kid trt;
MODEL gain = trt week trt*week / DDFM = KENWARDROGER;
RANDOM int week / TYPE=UN  SUB = kid(trt)  GROUP = trt;
LSMEANS trt / DIFF ADJUST = TUKEY;
LSMEANS trt / DIFF ADJUST = TUKEY AT week = 9;
LSMEANS trt / DIFF ADJUST = TUKEY AT week = 10;
LSMEANS trt / DIFF ADJUST = TUKEY AT week = 11;
LSMEANS trt / DIFF ADJUST = TUKEY AT week = 12;
RUN;
```

Explanation: The option *GROUP* = *trt* is added to the RANDOM statement to direct computations of separate random regressions for each treatment.

SAS output:

```
                Covariance Parameter Estimates

Cov Parm        Subject       Group       Estimate
UN(1,1)         kid(trt)      trt 1        0.015500
UN(2,1)         kid(trt)      trt 1       -0.002490
UN(2,2)         kid(trt)      trt 1        0.000408
UN(1,1)         kid(trt)      trt 2        0.425500
UN(2,1)         kid(trt)      trt 2       -0.041380
UN(2,2)         kid(trt)      trt 2        0.004328
Residual                                   0.007235
```

Explanation: There are seven parameters in this model. The $UN(i, j)$ and *Group* denote the variance-covariance structure of regression coefficients within *trt* 1 and 2, respectively. There is just one *Residual*, indicating that homogeneous error variance across treatments is assumed. The covariance matrix of regression coefficients within *trt* 1 is:

$$\begin{bmatrix} \hat{\sigma}^2_{b_0} & \hat{\sigma}_{b_0 b_1} \\ \hat{\sigma}_{b_0 b_1} & \hat{\sigma}^2_{b_1} \end{bmatrix} = \begin{bmatrix} 0.015500 & -0.002490 \\ -0.002490 & 0.000408 \end{bmatrix}$$

The covariance matrix of regression coefficients within *trt* 2 is:

$$\begin{bmatrix} \hat{\sigma}^2_{b_0} & \hat{\sigma}_{b_0 b_1} \\ \hat{\sigma}_{b_0 b_1} & \hat{\sigma}^2_{b_1} \end{bmatrix} = \begin{bmatrix} 0.425500 & -0.041380 \\ -0.041380 & 0.004328 \end{bmatrix}$$

The common error variance is:

$$\hat{\sigma}^2 = 0.007235$$

The variance-covariance structure for one animal within *trt* 1 is:

$$\hat{\mathbf{R}} = \begin{bmatrix} 1 & 9 \\ 1 & 10 \\ 1 & 11 \\ 1 & 12 \end{bmatrix} \begin{bmatrix} 0.015500 & -0.002490 \\ -0.002490 & 0.000408 \end{bmatrix} \begin{bmatrix} 1 & 1 & 1 & 1 \\ 9 & 10 & 11 & 12 \end{bmatrix}$$

$$+ \begin{bmatrix} 0.007235 & 0 & 0 & 0 \\ 0 & 0.007235 & 0 & 0 \\ 0 & 0 & 0.007235 & 0 \\ 0 & 0 & 0 & 0.007235 \end{bmatrix}$$

$$= \begin{bmatrix} 0.010963 & 0.004910 & 0.006092 & 0.007274 \\ 0.004910 & 0.013735 & 0.008090 & 0.009680 \\ 0.006092 & 0.008090 & 0.017323 & 0.012086 \\ 0.007274 & 0.009680 & 0.012086 & 0.021727 \end{bmatrix}$$

The variance-covariance structure for one animal within *trt* 2 is:

$$\hat{R} = \begin{bmatrix} 1 & 9 \\ 1 & 10 \\ 1 & 11 \\ 1 & 12 \end{bmatrix} \begin{bmatrix} 0.425500 & -0.041380 \\ -0.041380 & 0.004328 \end{bmatrix} \begin{bmatrix} 1 & 1 & 1 & 1 \\ 9 & 10 & 11 & 12 \end{bmatrix}$$

$$+ \begin{bmatrix} 0.007235 & 0 & 0 & 0 \\ 0 & 0.007235 & 0 & 0 \\ 0 & 0 & 0.007235 & 0 \\ 0 & 0 & 0 & 0.007235 \end{bmatrix}$$

$$= \begin{bmatrix} 0.038463 & 0.02880 & 0.026372 & 0.023944 \\ 0.028800 & 0.037935 & 0.032600 & 0.034500 \\ 0.026372 & 0.032600 & 0.046063 & 0.045056 \\ 0.023944 & 0.034500 & 0.045056 & 0.062847 \end{bmatrix}$$

Defining separate variance-covariance parameters for each treatment, and also separate error variances for each treatment, the SAS program is:

```
PROC MIXED DATA=reps;
CLASS kid trt;
MODEL gain = trt week trt*week / DDFM = KENWARDROGER;
RANDOM int week / TYPE=UN  SUB=kid(trt)  GROUP = trt;
REPEATED / SUB = kid(trt)  GROUP = trt;
LSMEANS trt / DIFF ADJUST = TUKEY;
LSMEANS trt / DIFF ADJUST = TUKEY AT week = 9;
LSMEANS trt / DIFF ADJUST = TUKEY AT week = 10;
LSMEANS trt / DIFF ADJUST = TUKEY AT week = 11;
LSMEANS trt / DIFF ADJUST = TUKEY AT week = 12;
RUN;
```

Explanation: A REPEATED statement with the option *GROUP = trt* is added to the program to direct computations of separate error variances for each treatment.

SAS output:

Covariance Parameter Estimates

Cov Parm	Subject	Group	Estimate
UN(1,1)	kid(trt)	trt 1	0.041660
UN(2,1)	kid(trt)	trt 1	-0.004950
UN(2,2)	kid(trt)	trt 1	0.000643
UN(1,1)	kid(trt)	trt 2	0.402300
UN(2,1)	kid(trt)	trt 2	-0.039190
UN(2,2)	kid(trt)	trt 2	0.004119
Residual	kid(trt)	trt 1	0.006063
Residual	kid(trt)	trt 2	0.008278

Fit Statistics

-2 Res Log Likelihood	-73.0
AIC (smaller is better)	-57.0
AICC (smaller is better)	-54.4
BIC (smaller is better)	-50.4

Explanation: There are eight parameters in this model. The $UN(i, j)$ and *Group* denote the variance-covariance structure of regression coefficients within *trt* 1 and 2, respectively. There are also two *Residual* variances, indicating heterogeneous error variances between treatments are assumed.

The variance-covariance matrix of regression coefficients within *treatment* 1 is:

$$\begin{bmatrix} \hat{\sigma}^2_{b_0} & \hat{\sigma}_{b_0 b_1} \\ \hat{\sigma}_{b_0 b_1} & \hat{\sigma}^2_{b_1} \end{bmatrix} = \begin{bmatrix} 0.041660 & -0.004950 \\ -0.004950 & 0.000643 \end{bmatrix}$$

The variance-covariance matrix of regression coefficients within *treatment* 2 is:

$$\begin{bmatrix} \hat{\sigma}^2_{b_0} & \hat{\sigma}_{b_0 b_1} \\ \hat{\sigma}_{b_0 b_1} & \hat{\sigma}^2_{b_1} \end{bmatrix} = \begin{bmatrix} 0.402300 & -0.039190 \\ -0.039190 & 0.004119 \end{bmatrix}$$

The error variance within *treatment* 1: $\hat{\sigma}^2_1 = 0.006063$

The error variance within *treatment* 2: $\hat{\sigma}^2_2 = 0.008278$

The variance-covariance structure for one animal within *treatment* 1 is:

$$\hat{\mathbf{R}} = \begin{bmatrix} 1 & 9 \\ 1 & 10 \\ 1 & 11 \\ 1 & 12 \end{bmatrix} \begin{bmatrix} 0.041660 & -0.004950 \\ -0.004950 & 0.000643 \end{bmatrix} \begin{bmatrix} 1 & 1 & 1 & 1 \\ 9 & 10 & 11 & 12 \end{bmatrix}$$

$$+ \begin{bmatrix} 0.006063 & 0 & 0 & 0 \\ 0 & 0.006063 & 0 & 0 \\ 0 & 0 & 0.006063 & 0 \\ 0 & 0 & 0 & 0.006063 \end{bmatrix}$$

$$= \begin{bmatrix} 0.010706 & 0.005480 & 0.006317 & 0.007154 \\ 0.005480 & 0.073023 & 0.008440 & 0.009920 \\ 0.006317 & 0.008440 & 0.016626 & 0.012686 \\ 0.007154 & 0.009920 & 0.012686 & 0.021515 \end{bmatrix}$$

The variance-covariance structure for an animal within *treatment* 2 is:

$$\hat{R} = \begin{bmatrix} 1 & 9 \\ 1 & 10 \\ 1 & 11 \\ 1 & 12 \end{bmatrix} \begin{bmatrix} 0.402300 & -0.039190 \\ -0.039190 & 0.004119 \end{bmatrix} \begin{bmatrix} 1 & 1 & 1 & 1 \\ 9 & 10 & 11 & 12 \end{bmatrix}$$

$$+ \begin{bmatrix} 0.008278 & 0 & 0 & 0 \\ 0 & 0.008278 & 0 & 0 \\ 0 & 0 & 0.008278 & 0 \\ 0 & 0 & 0 & 0.008278 \end{bmatrix}$$

$$= \begin{bmatrix} 0.038797 & 0.028400 & 0.026281 & 0.024162 \\ 0.028400 & 0.038678 & 0.032400 & 0.034400 \\ 0.026281 & 0.032400 & 0.046797 & 0.044638 \\ 0.024162 & 0.034400 & 0.044638 & 0.063154 \end{bmatrix}$$

22.4 Accounting for Baseline Measurements

In comparing treatments using repeated measurements, animals (subjects) are often measured before any treatment is applied in an experiment. This pretreatment measurement, known as a baseline measurement, gives an internal control for each subject, and there may be no need to have a control group.

The baseline measurements are not subject to the same treatment as other, later measurements. Including baseline measurements in the set of the repeated measures can cause numerical and conceptual problems, especially if variance-covariance structure models with a reduced set of parameters such as Toeplitz or autoregressive are used. It is usually better to define the baseline measurements as a separate variable to be used as a covariate to adjust differences between subjects existing prior to treatment.

Example: The data from the example with two treatments affecting gain of kids are used, but the measurement at the 9th week is defined as a baseline measurement and denoted *base_m*.

SAS program:

```
DATA reps;
INPUT kid base_m week trt gain @@;
DATALINES;
 1 1.2 10 1 1.0      1 1.2 11 1 1.1      1 1.2 12 1 1.3
 2 1.2 10 1 1.1      2 1.2 11 1 1.4      2 1.2 12 1 1.5
 3 1.3 10 1 1.4      3 1.3 11 1 1.4      3 1.3 12 1 1.6
 4 1.1 10 1 1.1      4 1.1 11 1 1.2      4 1.1 12 1 1.3
 5 1.2 10 1 1.3      5 1.2 11 1 1.2      5 1.2 12 1 1.3
 6 1.1 10 1 1.1      6 1.1 11 1 1.1      6 1.1 12 1 1.2
 7 1.1 10 1 1.2      7 1.1 11 1 1.3      7 1.1 12 1 1.5
```

```
 8 1.3 10 1 1.3      8 1.3 11 1 1.3      8 1.3 12 1 1.4
 9 1.2 10 2 1.5      9 1.2 11 2 1.9      9 1.2 12 2 2.1
10 1.3 10 2 1.2     10 1.3 11 2 1.4     10 1.3 12 2 1.7
11 1.5 10 2 1.7     11 1.5 11 2 1.6     11 1.5 12 2 1.7
12 1.4 10 2 1.5     12 1.4 11 2 1.7     12 1.4 12 2 1.8
13 1.2 10 2 1.2     13 1.2 11 2 1.4     13 1.2 12 2 1.6
14 1.0 10 2 1.1     14 1.0 11 2 1.4     14 1.0 12 2 1.5
15 1.4 10 2 1.8     15 1.4 11 2 2.1     15 1.4 12 2 2.1
16 1.1 10 2 1.3     16 1.1 11 2 1.4     16 1.1 12 2 1.8
17 1.2 10 2 1.5     17 1.2 11 2 1.7     17 1.2 12 2 1.9
;

PROC MIXED DATA=reps;
CLASS kid trt week;
MODEL gain = base_m trt week trt*week / DDFM = KENWARDROGER;
REPEATED / TYPE=AR(1) SUB=kid(trt) ;
LSMEANS trt trt*week / DIFF ADJUST = TUKEY;
RUN;
```

Explanation: The MIXED procedure was used. The CLASS statement defines categorical variables. The MODEL statement defines the dependent variable *gain*, and the independent variables *base_m*, *trt*, *week* and *trt*week* interaction. The Kenward-Roger method is used to estimate correct degrees of freedom (*DFFM=KENWARDROGER*). The REPEATED statement defines the variance-covariance structure for repeated measurements. The subject (*SUB = kid (trt)*) defines the variable on which repeated measurements were taken. The type of variance-covariance structure is autoregressive (*TYPE = AR(1)*). The LSMEANS statement directs calculations of treatment means.

SAS output:

```
Covariance Parameter Estimates

    Cov Parm     Subject      Estimate
    AR(1)        kid(trt)       0.7639
    Residual                   0.02739
```

```
      Type 3 Tests of Fixed Effects

Effect        Num DF    Den DF    F Value    Pr > F
base_m          1         14        7.13     0.0183
trt             1         14       14.69     0.0018
week            2         30       31.20     <.0001
trt*week        2         30        3.63     0.0388
```

Least Squares Means

Effect	trt	week	Estimate	Standard Error	DF	t Value	Pr > |t|
trt	1		1.3032	0.05295	15.1	24.61	<.0001
trt	2		1.5897	0.04981	15.1	31.91	<.0001
trt*week	1	10	1.2157	0.05944	21.8	20.45	<.0001
trt*week	1	11	1.2782	0.05944	21.8	21.50	<.0001
trt*week	1	12	1.4157	0.05944	21.8	23.82	<.0001
trt*week	2	10	1.3971	0.05594	21.8	24.97	<.0001
trt*week	2	11	1.5971	0.05594	21.8	28.55	<.0001
trt*week	2	12	1.7749	0.05594	21.8	31.73	<.0001

Differences of Least Squares Means

Effect	trt	week	_trt	_week	Estimate	Stand Error	DF	t Value	Pr>t|	Adjust	AdjP
trt	1		2		-0.2865	0.07402	15.1	-3.87	0.0015	Tuk-Kr	0.0015
trt*week	1	10	2	10	-0.1814	0.08281	21.6	-2.19	0.0396	Tuk-Kr	0.2713
trt*week	1	11	2	11	-0.3189	0.08281	21.6	-3.85	0.0009	Tuk-Kr	0.0068
trt*week	1	12	2	12	-0.3592	0.08281	21.6	-4.34	0.0003	Tuk-Kr	0.0019

Explanation: The table *Covariance Parameter Estimates* gives the following estimates: $AR(1) = 0.7639$, the correlation between successive measurements within subjects, *Residual* = 0.02739, the estimate of variance of measures (between subjects). The correlation between measurements taken t periods apart is then 0.7639. The table *Type 3 Tests of Fixed Effects* shows that all the effects in the model are significant (P values are all smaller than 0.05), including the baseline measures (*base_m*) which justifies including it as a covariate to adjust measurements taken later in the experiment. Note the different degrees of freedom in the *Least Squares Means*, and *Differences of Least Squares Means* tables. These were calculated by using the Kenward-Rogers correction. For the *trt*week* interaction only the differences between treatment within each time point (weeks 10, 11 and 12) are shown.

22.5 Missing Data in Repeated Measures Analysis

A discussion about missing values is presented in chapter 14. An example of an analysis of data with missing values in a repeated measures design will be given using the MIXED procedure of SAS.

Example: The data on two treatments affecting gain of kids is again used, except that two measurements of kid 1 (at weeks 10 and 12) and one measurement of kid 17 (at week 12) are missing. A model with an autoregressive covariance structure will be shown.

SAS program:

```
DATA repsmiss;
INPUT kid week trt gain @@;
DATALINES;
 1  9 1 1.2      1 10 1  .      1 11 1 1.1      1 12 1  .
 2  9 1 1.2      2 10 1 1.1      2 11 1 1.4      2 12 1 1.5
```

```
  3   9  1  1.3      3  10  1  1.4      3  11  1  1.4      3  12  1  1.6
  4   9  1  1.1      4  10  1  1.1      4  11  1  1.2      4  12  1  1.3
  5   9  1  1.2      5  10  1  1.3      5  11  1  1.2      5  12  1  1.3
  6   9  1  1.1      6  10  1  1.1      6  11  1  1.1      6  12  1  1.2
  7   9  1  1.1      7  10  1  1.2      7  11  1  1.3      7  12  1  1.5
  8   9  1  1.3      8  10  1  1.3      8  11  1  1.3      8  12  1  1.4
  9   9  2  1.2      9  10  2  1.5      9  11  2  1.9      9  12  2  2.1
 10   9  2  1.3     10  10  2  1.2     10  11  2  1.4     10  12  2  1.7
 11   9  2  1.5     11  10  2  1.7     11  11  2  1.6     11  12  2  1.7
 12   9  2  1.4     12  10  2  1.5     12  11  2  1.7     12  12  2  1.8
 13   9  2  1.2     13  10  2  1.2     13  11  2  1.4     13  12  2  1.6
 14   9  2  1.0     14  10  2  1.1     14  11  2  1.4     14  12  2  1.5
 15   9  2  1.4     15  10  2  1.8     15  11  2  2.1     15  12  2  2.1
 16   9  2  1.1     16  10  2  1.3     16  11  2  1.4     16  12  2  1.8
 17   9  2  1.2     17  10  2  1.5     17  11  2  1.7     17  12  2  .
;
PROC MIXED DATA=repsmiss;
CLASS kid trt week;
MODEL gain = trt week trt*week / DDFM = KENWARDROGER;
REPEATED / TYPE=AR(1) SUB=kid(trt);
LSMEANS trt trt*week / DIFF ADJUST = TUKEY;
RUN;
```

Explanation: The MIXED procedure is used. The CLASS statement defines categorical variables. The MODEL statement defines the dependent variable *gain*, and the independent variables *trt*, *week* and *trt*week* interaction. The REPEATED statement defines the variance-covariance structure for repeated measurements. The subject statement (*SUB = kid(trt)*) defines *kid* as the variable on which repeated measures are taken, and the type of variance-covariance structure is defined as autoregressive, *TYPE = AR(1)*. The Kenward-Roger estimation of degrees of freedom is recommended because the mean square of subjects and standard errors are linear combination of variances, and it also accounts for missing values.

SAS output:

```
            Covariance Parameter Estimates

        Cov Parm      Subject       Estimate
        AR(1)         kid(trt)       0.7481
        Residual                     0.02916
```

```
           Type 3 Tests of Fixed Effects

     Effect       Num DF   Den DF   F Value   Pr > F
     trt             1      17.8     14.54    0.0013
     week            3      43.4     20.52    <.0001
     trt*week        3      43.4      5.29    0.0034
```

Least Squares Means

Effect	trt	week	Estimate	Standard Error	DF	t Value	Pr > |t|
trt	1		1.2566	0.05101	17.9	24.63	<.0001
trt	2		1.5237	0.04799	17.8	31.75	<.0001
trt*week	1	9	1.1875	0.06038	29.2	19.67	<.0001
trt*week	1	10	1.2049	0.06164	31.2	19.55	<.0001
trt*week	1	11	1.2500	0.06038	29.2	20.70	<.0001
trt*week	1	12	1.3840	0.06237	31.6	22.19	<.0001
trt*week	2	9	1.2556	0.05692	29.2	22.06	<.0001
trt*week	2	10	1.4222	0.05692	29.2	24.99	<.0001
trt*week	2	11	1.6222	0.05692	29.2	28.50	<.0001
trt*week	2	12	1.7948	0.05857	31.3	30.64	<.0001

Differences of Least Squares Means

Effect	trt	week	_trt	_week	Estimate	Standard Error	DF	t Value	Pr>|t|	Adjust	Adj P
trt	1		2		-0.2671	0.07004	17.8	-3.81	0.0013	Tuk-Kr	0.0013
trt*week	1	9	2	9	-0.0680	0.08298	29.2	-0.82	0.4188	Tuk-Kr	0.9910
trt*week	1	10	2	10	-0.2174	0.08391	30.3	-2.59	0.0146	Tuk-Kr	0.1872
trt*week	1	11	2	11	-0.3722	0.08298	29.2	-4.49	0.0001	Tuk-Kr	0.0013
trt*week	1	12	2	12	-0.4108	0.08557	31.5	-4.80	<.0001	Tuk-Kr	0.0005

Explanation: Note the *Type 3 Tests of Fixed Effects*, *Least Squares Means*, and *Differences of Least Squares Means* tables. The degrees of freedom (*DF*) are computed by Kenward-Roger method. If this method had not been applied, the degrees of freedom would have been 15 and 42, for *trt* and *trt*week*, respectively. The use of the Kenward-Roger correction reduces the probability of type I error.

Chapter 23

Analysis of Numerical Treatment Levels

In biological research there is often more than one measurement of the dependent variable for each of several numerically ordered levels of the independent variable (Figure 23.1). For example, the goal of an experiment might be to evaluate the effect of protein content in a ration on daily gain of animals. Protein level is the independent variable, and daily gain is the dependent variable. For each level of protein several animals are measured. It may not be enough just to determine if there is a significant difference among levels, but it may be of interest to find the optimum protein content by fitting a curve over protein level. This problem can be approached by using regression or by using orthogonal polynomial contrasts. A problem with regression is that it may be difficult to conclude which regression model is most appropriate. Because of replications for each level of the independent variable it may be difficult to determine if simple linear regression is enough to explain the phenomena, or if perhaps a quadratic regression is more appropriate. Testing the appropriateness of a model can be done by Lack of Fit analysis. Similarly, linear, quadratic and other contrasts can be tested in order to make conclusions about linearity or nonlinearity of the phenomena.

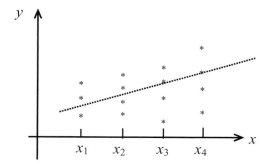

Figure 23.1 Several measurements per level of independent variable

23.1 Lack of Fit

Consider more than one measurement of the dependent variable y on each level of independent variable x. Let y_{ij} depict the jth measurement of level i of x. There are m levels of x, that is, $i = 1, 2,..., m$. The number of measurements for a level i is n_i and $\sum_i n_i = N$ is the total number of measurements. An example with four levels of x is shown in Figure 23.1. From the graph it is difficult to conclude if simple linear regression or quadratic regression is more appropriate for explaining changes in y resulting from changing the level of x. Lack of Fit analysis provides information to aid in determining which model is more appropriate.

First, assume the model of simple linear regression:

$$y_{ij} = \beta_0 + \beta_1 x_i + \varepsilon_{ij}$$

Let $\bar{y}_i$ denote the mean and $\hat{y}_i$ denote the estimated value for level i. If the model is correct, $\hat{y}_i$ will not differ significantly from $\bar{y}_i$.

Thus,

> if $\bar{y}_i \approx \hat{y}_i$ (for all i) then the model is correct,
>
> if $\bar{y}_i \neq \hat{y}_i$ (for some i) then the model is not correct.

The test is based on the fact that the residual sum of squares can be partitioned to a 'pure error' sum of squares and a lack of fit sum of squares:

$$SS_{RES} = SS_{PE} + SS_{LOF}$$

with degrees of freedom:

$$(n-1) = \sum_i (n_i - 1) + (m - p)$$

where p = the number of parameters in the model.

Sums of squares are:

$$SS_{RES} = \sum_i \sum_j \left(y_{ij} - \hat{y}_i\right)^2$$

$$SS_{PE} = \sum_i \sum_j \left(y_{ij} - \bar{y}_i\right)^2$$

$$SS_{LOF} = \sum_i n_i \left(\bar{y}_i - \hat{y}_i\right)^2$$

where

$$\bar{y}_i = \tfrac{1}{n_i} \sum_j y_{ij} = \text{mean for level } i$$

$$\hat{y}_i = \text{estimated value for level } i$$

The mean square for pure error is:

$$MS_{PE} = \frac{SS_{PE}}{\sum (n_i - 1)}$$

The expectation of the MS_{PE} is $E(MS_{PE}) = \sigma^2$, which means that MS_{PE} estimates the variance regardless if model is correct or not. The mean square for lack of fit is:

$$MS_{LOF} = \frac{SS_{LOF}}{m - p}$$

If the model is correct then $E(MS_{LOF}) = \sigma^2$, which means that the mean square for lack of fit estimates the variance only if the regression is linear. The null hypothesis states that the model is correct, that is, a change in x causes a linear change in y:

$$H_0: E(y) = \beta_0 + \beta_1 x_i$$

The alternative hypothesis states that linear model is not correct. For testing the hypotheses one can apply an F statistic:

$$F = \frac{MS_{LOF}}{MS_{PE}}$$

If the impact of lack of fit is significant, the relationship between variables is not adequately described by a simple linear regression. These results are shown in an *ANOVA* table:

Source	SS	df	MS	F
Regression	SS_{REG}	1	$MS_{REG} = SS_{REG} / 1$	$F = MS_{REG} / MS_{RES}$
Residual	SS_{RES}	$n - 2$	$MS_{RES} = SS_{RES} / (n - 2)$	
Lack of fit	SS_{LOF}	$m - 2$	$MS_{LOF} = SS_{LOF} / (m - 2)$	$F = MS_{LOF} / MS_{PE}$
Pure error	SS_{PE}	$n - m$	$MS_{PE} = SS_{PE} / (n - m)$	
Total	SS_{TOT}	$n - 1$		

Example: The objective of this experiment was to analyze the effect of protein level in a pig ration on feed conversion. The experiments started with pigs at an approximate weight of 39 kg and finished at 60 kg. There were five litters with five pigs randomly chosen from each litter. One of the five protein levels (10, 12, 14, 16, and 18%) was randomly assigned to each pig from each litter. The following data were obtained:

Litter	Protein level				
	10%	12%	14%	16%	18%
I	4.61	4.35	4.21	4.02	4.16
II	4.12	3.84	3.54	3.45	3.28
III	4.25	3.93	3.47	3.24	3.59
IV	3.67	3.37	3.19	3.55	3.92
V	4.01	3.98	3.42	3.34	3.57

In the model, litter was defined as a block, and protein level was defined as a regressor. The model is:

$$y_{ij} = \mu + L_j + \beta_1 x_i + \varepsilon_{ij}$$

where:

y_{ij} = feed conversion of pig i in litter j
μ = overall mean
L_j = the effect of litter j
β_1 = regression parameter
x_i = protein level i
ε_{ij} = random error

The number of protein levels is $m = 5$, the total number of pigs is $n = 25$, and the number of litters (blocks) is $b = 5$. Results are presented in the following *ANOVA* table:

Source	SS	df	MS	F
Litter	1.6738	5 − 1 = 4	0.4184	9.42
Regression	0.7565	1	0.7565	11.71
Residual	1.2273	25 − (5 − 1) − 2 = 19	0.0646	
Lack of fit	0.5169	5 − 2 = 3	0.1723	3.88
Pure error	0.7105	25 − 5 − (5 − 1) = 16	0.0444	
Total	3.6575	25 − 1 = 24		

The critical value of $F_{0.05,1,19}$ is 4.38. The calculated F for regression is 11.71. Thus, protein level has a significant linear impact on feed conversion. The calculated F for lack of fit is 3.88, and the critical value of $F_{0.05,3,16}$ is 3.24. This indicates that a linear regression model is not adequate in describing the relationship. The change in feed conversion as protein level increases is not strictly linear. The next step is to try a quadratic model and test the correctness of fit of that model.

23.1.1 SAS Example for Lack of Fit

The example considering the effect of protein level on feed conversion will be used as an illustration of Lack of Fit analysis using SAS.

			Protein level		
Litter	10%	12%	14%	16%	18%
I	4.61	4.35	4.21	4.02	4.16
II	4.12	3.84	3.54	3.45	3.28
III	4.25	3.93	3.47	3.24	3.59
IV	3.67	3.37	3.19	3.55	3.92
V	4.01	3.98	3.42	3.34	3.57

SAS program:

```
DATA a;
INPUT litter $ prot conv @@;
prot1=prot;
DATALINES;
  I 10 4.61      I 12 4.35      I 14 4.21      I 16 4.02      I  18 4.16
 II 10 4.12     II 12 3.84     II 14 3.54     II 16 3.45     II  18 3.28
III 10 4.25    III 12 3.93    III 14 3.47    III 16 3.24    III  18 3.59
 IV 10 3.67     IV 12 3.37     IV 14 3.19     IV 16 3.55     IV  18 3.92
  V 10 4.01      V 12 3.98      V 14 3.42      V 16 3.34      V  18 3.57

;
*the following procedure computes lack of fit for linear regression;
PROC GLM;
CLASS litter prot;
MODEL conv = litter prot1 prot /SS1;
RUN;
```

```
*the following procedure computes lack of fit for quadratic regression;
PROC GLM;
CLASS litter prot;
MODEL conv = litter prot1 prot1*prot1 prot / SS1;
RUN;
```

Explanation: The first procedure tests if a linear regression model adequately describes the relationship between protein level and feed conversion. The CLASS statement defines class (categorical) independent variables. The MODEL statement *conv = litter prot1 prot* defines *conv* as the dependent variable, and *litter*, *prot* and *prot1* as independent variables. The variables *prot1* and *prot* are numerically identical (see DATA step), but the program treats them differently. The variable *prot1* is not in the CLASS statement and the program uses it as a continuous (regressor) variable. Defining the same variable as both class and continuous tests lack of fit. The SS1 option computes sequential sums of squares. The second GLM procedure tests if the quadratic model adequately describes the relationship. The MODEL statement *conv = litter prot1 prot1*prot1 prot / SS1*, defines effects in the model. The variable *prot1*prot1* defines a quadratic effect of protein level.

SAS output:

Dependent Variable: conv

Source	DF	Sum of Squares	Mean Square	F Value	Pr > F
Model	8	2.94708800	0.36838600	8.30	0.0002
Error	16	0.71045600	0.04440350		
Corrected Total	24	3.65754400			

Source	DF	Type I SS	Mean Square	F Value	Pr > F
litter	4	1.67378400	0.41844600	9.42	0.0004
prot1	1	0.75645000	0.75645000	17.04	0.0008
prot	3	0.51685400	0.17228467	3.88	0.0293

Dependent Variable: conv

Source	DF	Sum of Squares	Mean Square	F Value	Pr > F
Model	8	2.947088	0.36838600	8.30	0.0002
Error	16	0.710456	0.04440350		
Corrected Total	24	3.657544			

Source	DF	Type I SS	Mean Square	F Value	Pr > F
litter	4	1.673784	0.41844600	9.42	0.0004
prot1	1	0.756450	0.75645000	17.04	0.0008
prot1*prot1	1	0.452813	0.45281286	10.20	0.0057
prot	2	0.064041	0.03202057	0.72	0.5013

Explanation: The first GLM procedure tests to determine if linear regression is adequate. First is the *ANOVA* table for *conv* as a *Dependent Variable*. The *Source(s)* of variability are

Model, *Error* (pure error) and *Corrected Total*. In the table are presented degrees of freedom (*DF*), *Sum of Squares*, *Mean Square*, calculated *F* (*F value*) and *P* value (*Pr > F*). In the next table the sum of squares for *MODEL from the first table* is partitioned to *litter*, *prot1* and *prot*. The variable *prot* defined as a class variable tests the effect of lack of fit. The calculated *F* and *P* values are 3.88 and 0.0293, respectively. Thus, the effect of lack of fit is significant. That means that the model of linear regression does not adequately describe the relationship. The check if the quadratic model is correct is shown by the second GLM procedure. Analogously to the first procedure, the effect *prot* in the last table tests the lack of fit effect. That effect not being significant (*P* value is 0.5013) indicates that the quadratic regression is appropriate in describing the effect of protein level on feed conversion.

23.2 Polynomial Orthogonal Contrasts

Linear, quadratic, and higher order effects of treatment levels can be tested by using polynomial orthogonal contrasts. The treatment sum of squares can be partitioned into orthogonal polynomial contrasts, and each tested by *F* test. In the following table the contrast coefficients are shown for two to five treatment levels:

No. of treatment levels	Degree of polynom	Coefficient (c)					$\sum_i c_i^2$
2	linear	-1	$+1$				2
3	linear	-1	0	$+1$			2
	quadratic	$+1$	-2	$+1$			6
4	linear	-3	-1	$+1$	$+3$		20
	quadratic	$+1$	-1	-1	$+1$		4
	cubic	-1	$+3$	-3	$+1$		20
5	linear	-2	-1	0	$+1$	$+2$	10
	quadratic	$+2$	-1	-2	-1	$+2$	14
	cubic	-1	$+2$	0	-2	$+1$	10
	quartic	$+1$	-4	$+6$	-4	$+1$	70

For example, if a model has three treatment levels, the treatment sum of squares can be partitioned into two orthogonal polynomial contrasts: linear and quadratic. These two contrasts explain linear and quadratic effects of the independent variable (treatments) on the dependent variable. The quadratic component is equivalent to lack of fit sum of squares for linearity. The significance of each of the components can be tested with an *F* test. Each *F* value is a ratio of contrast mean square to the residual (experimental error) mean square. The null hypothesis is that the particular regression coefficient is equal to zero. If only the linear effect is significant, we can conclude that the changes in values of the dependent variable are linear with respect to the independent variable. If the quadratic component is significant, we can conclude that the changes are not linear but parabolic. Polynomials of higher degree can also be tested by using similar reasoning.

Example: Continuing with the example of the effects of protein levels on feed conversion, recall that litters were used as blocks in a randomized block design. Since five protein levels were defined, the treatment sum of squares can be partitioned into four polynomial orthogonal contrasts. Recall the data:

Litter	Protein level				
	10%	12%	14%	16%	18%
I	4.61	4.35	4.21	4.02	4.16
II	4.12	3.84	3.54	3.45	3.28
III	4.25	3.93	3.47	3.24	3.59
IV	3.67	3.37	3.19	3.55	3.92
V	4.01	3.98	3.42	3.34	3.57

ANOVA table:

Source	SS	df	MS	F
Litter	1.6738	4	0.4184	9.42
Protein level	1.2733	4	0.3183	7.17
Linear contrast	0.7565	1	0.7565	17.04
Quadratic contrast	0.4528	1	0.4528	10.25
Cubic contrast	0.0512	1	0.0512	1.15
Quartic contrast	0.0128	1	0.0128	0.29
Residual	0.7105	16	0.0444	
Total	3.6575	24		

The effect of level of protein is significant. Further, only linear and quadratic contrasts are significant, the others are not. This leads to the conclusion that changes in feed conversion can be explained by a quadratic regression on protein levels. Note that the treatment sum of squares is equal to the sum of contrasts sums of squares:

$$1.2733 = 0.7565 + 0.4528 + 0.0512 + 0.0128$$

The residual term here is equal to the pure error from the lack of fit analysis. The coefficients of the quadratic function are estimated by quadratic regression of feed conversion on protein levels. The following function results:

$$y = 8.4043 - 0.6245x + 0.0201x^2$$

The protein level for minimum feed conversion is determined by taking the first derivative of the quadratic function, setting it equal to zero, and solving. The solution of that equation is an optimum.

The first derivative of y is:

$$y' = -0.6245 + 2(0.0201)x = 0$$

Then $x = 15.5$, and the optimum level of protein is 15.5%.

23.2.1 SAS Example for Polynomial Contrasts

The SAS program for calculation of polynomial contrasts for the example with feed conversion is:

SAS program:

```
DATA a;
INPUT litter $ prot conv @@;
DATALINES;
  I 10 4.61      I 12 4.35      I 14 4.21      I 16 4.02      I  18  4.16
 II 10 4.12     II 12 3.84     II 14 3.54     II 16 3.45     II  18  3.28
III 10 4.25    III 12 3.93    III 14 3.47    III 16 3.24    III  18  3.59
 IV 10 3.67     IV 12 3.37     IV 14 3.19     IV 16 3.55     IV  18  3.92
  V 10 4.01      V 12 3.98      V 14 3.42      V 16 3.34      V  18  3.57

;
*the following procedure computes the contrasts;
PROC GLM DATA = a ;
CLASS litter prot;
MODEL conv = litter prot;
CONTRAST 'linear' prot -2 -1  0 +1 +2;
CONTRAST 'quad'  prot +2 -1 -2 -1 +2;
CONTRAST 'cub'   prot -1 +2  0 -2 +1;
CONTRAST 'quart' prot +1 -4 +6 -4 +1;
LSMEANS prot / stderr;
RUN;

*the following procedure computes regression coefficients;
PROC GLM;
MODEL conv= prot prot*prot /SOLUTION;
RUN;
```

Explanation: The first GLM procedure tests the significance of litter and protein. The CLASS statement defines class (categorical) variables. The statement, *MODEL conv = litter prot*, denotes *conv* as the dependent variable, and *litter* and *prot* as independent variables. The CONTRAST statement defines contrasts. Each contrast requires a distinctive CONTRAST statement. Words between quotation marks, i.e. *'lin'*, *'quad'*, *'cub'* and *'quart'*, label contrasts as they will be shown in the output. The word *prot* specifies the variable for which the contrast is calculated, and is followed by the contrast coefficients. The second GLM procedure estimates the quadratic regression.

SAS output:

Dependent Variable: conv

Source	DF	Sum of Squares	Mean Square	F Value	Pr > F
Model	8	2.94708800	0.36838600	8.30	0.0002
Error	16	0.71045600	0.04440350		
Corrected Total	24	3.65754400			

Source	DF	Type III SS	Mean Square	F Value	Pr > F
litter	4	1.67378400	0.41844600	9.42	0.0004
prot	4	1.27330400	0.31832600	7.17	0.0017

Contrast	DF	Contrast SS	Mean Square	F Value	Pr > F
linear	1	0.75645000	0.75645000	17.04	0.0008
quad	1	0.45281286	0.45281286	10.20	0.0057
cub	1	0.05120000	0.05120000	1.15	0.2988
quart	1	0.01284114	0.01284114	0.29	0.5981

Least Squares Means

prot	conv LSMEAN	Standard Error	Pr > |t|
10	4.13200000	0.09423747	<.0001
12	3.89400000	0.09423747	<.0001
14	3.56600000	0.09423747	<.0001
16	3.52000000	0.09423747	<.0001
18	3.70400000	0.09423747	<.0001

Dependent Variable: conv

Source	DF	Sum of Squares	Mean Square	F Value	Pr > F
Model	2	1.20926286	0.60463143	5.43	0.0121
Error	22	2.44828114	0.11128551		
Corrected Total	24	3.65754400			

Source	DF	Type I SS	Mean Square	F Value	Pr > F
prot1	1	0.75645000	0.75645000	6.80	0.0161
prot1*prot1	1	0.45281286	0.45281286	4.07	0.0560

Parameter	Estimate	Standard Error	t Value	Pr > |t|
Intercept	8.404342857	1.90403882	4.41	0.0002
prot1	-0.624500000	0.28010049	-2.23	0.0363
prot1*prot1	0.020107143	0.00996805	2.02	0.0560

Explanation: First is an *ANOVA* table for the *Dependent Variable conv*. The *Sources* of variability are *Model*, *Error* (residual) and *Corrected Total*. In the table are shown degrees of freedom (*DF*), *Sum of Squares*, *Mean Square*, calculated *F* (*F value*) and *P* values (*Pr > F*). In the next table the explained source of variability (*MODEL*) is partitioned into *litter* and *prot*. For *prot* the calculated *F* and *P* values are 7.17 and 0.0017, respectively. There exists an effect of protein level. Next, the contrasts are shown. Both the *linear* and *quad* contrasts are significant. The last table of the first GLM procedure shows the least squares means (*conv LSMEAN*) together with *Standard Errors*. The second GLM procedure estimates quadratic regression coefficients. An *ANOVA* table and the parameter estimates are shown. Thus, the quadratic function is:

$Conversion = 8.40434 - 0.6245 \ (protein) + 0.0201 \ (protein^2)$.

Chapter 24

Discrete Dependent Variables

Up to now we have emphasized analysis of continuous dependent variables; however, dependent variables can be discrete or categorical as well. For example, we could evaluate the effect of housing systems on calf survival with survival coded as living = 1 or dead = 0. Another example is an experiment in which the objective is to test the effect of a treatment on botanical content of pastures. The dependent variable can be defined as the number of plants per unit area, and is often described as a count variable. In these examples the dependent variables are not continuous, and classical regression or analysis of variance may not be appropriate because assumptions such as homogeneity of variance and linearity are often not satisfied. Further, these variables do not have normal distributions and F or t tests are not valid. In chapter 6 an analysis of proportions using the normal approximation and a test of difference between an observed and theoretical frequency were shown using a *chi*-square test. In this chapter generalized linear models will be shown for analysis of binary and other discrete dependent variables.

Generalized linear models are models in which independent variables explain a function of the mean of a dependent variable. This is in contrast to classical linear models in which the independent variables explain the dependent variable or its mean directly. Which function is applicable depends on the distribution of the dependent variable.

To introduce a generalized linear model, denote $\mu = E(y)$ as the expectation or mean of a dependent variable y, and $\mathbf{x}\boldsymbol{\beta}$ as a linear combination of the vector of independent variables $\mathbf{x}$ and the corresponding vector of parameters $\boldsymbol{\beta}$. For example, for two independent continuous variables x_1 and x_2:

$$\mathbf{x} = \begin{bmatrix} 1 & x_1 & x_2 \end{bmatrix} \quad \boldsymbol{\beta} = \begin{bmatrix} \beta_0 \\ \beta_1 \\ \beta_2 \end{bmatrix} \quad \text{and}$$

$$\mathbf{x}\boldsymbol{\beta} = \begin{bmatrix} 1 & x_1 & x_2 \end{bmatrix} \begin{bmatrix} \beta_0 \\ \beta_1 \\ \beta_2 \end{bmatrix} = \beta_0 + \beta_1 x_1 + \beta_1 x_2$$

The generalized linear model in matrix notation is:

$$\eta = g(\mu) = \mathbf{x}\boldsymbol{\beta}$$

in which $\eta = g(\mu)$ is a function of the mean of the dependent variable known as a link function. It follows that the mean is:

$$\mu = g^{-1}(\eta)$$

in which g^{-1} = an inverse 'link' function, that is, a function that transforms $\mathbf{x}\boldsymbol{\beta}$ back to the mean. Observations of variable y can be expressed as:

$$y = \mu + \varepsilon$$

in which ε is an error that can have a distribution other than normal. If the independent variables are fixed, it is assumed that the error variance is equal to the variance of the dependent variable, that is:

$$Var(y) = Var(\varepsilon)$$

The model can also account for heterogeneity of variance by defining the variance to depend on the mean. The variance can be expressed as:

$$Var(y) = V(\mu)\phi^2$$

in which $V(\mu)$ is a function of the mean that contributes to the $Var(y)$, $V(\mu)$ is called the variance function, and ϕ^2 is a dispersion parameter.

Example: For a normal distribution with mean μ, variance σ^2, and a link function $\eta = g(\mu) = 1$; the variance function $V(\mu) = 1$, and the dispersion parameter $\phi^2 = \sigma^2$.

24.1 Logit Models, Logistic Regression

The influence of independent variables on a binary dependent variable can be explained by using a generalized linear model and a logit link function. These models are often called logit models. Recall that a binary variable can have only two outcomes, for example Yes and No, or 0 and 1. The probability distribution of a binary variable y has a Bernoulli distribution:

$$p(y) = p^y q^{1-y} \qquad\qquad y = 0, 1$$

The probabilities of outcomes are:

$$P(y_i = 1) = p$$
$$P(y_i = 0) = q = 1 - p$$

The expectation and variance of a binary variable are:

$$E(y) = \mu = p \qquad \text{and} \qquad Var(y) = \sigma^2 = pq$$

The binomial distribution is a distribution of y successes from a total of n trials:

$$p(y) = \binom{n}{y} p^y q^{n-y} \qquad\qquad y = 0, 1, 2, \ldots, n$$

in which p = the probability of success in a single trial, and $q = 1 - p$ = the probability of failure.

The expectation and variance of a binomial variable are:

$$E(y) = \mu = np \qquad \text{and} \qquad Var(y) = \sigma^2 = npq$$

For $n = 1$, a binomial variable is identical to a binary variable.

It is often practical to express binomial data as binomial proportions. A binomial proportion is the value of a binomial variable y divided by the total number of trials n. The mean and variance of binomial proportions are:

$$E(y/n) = \mu = p \qquad \text{and} \qquad Var(y/n) = \sigma^2 = pq/n$$

Given the mean is $\mu = p$, the model that explains changes in the mean of a binary variable or in the binomial proportion is:

$$\eta_i = g(\mu_i) = g(p_i) = \mathbf{x}_i\boldsymbol{\beta}$$

As a link function a logit function, g, can be used:

$$\eta_i = logit(p_i) = log[p_i /(1 - p_i)]$$

An inverse link function that transforms the logit value back to a proportion is the logistic function:

$$p_i = \frac{e^{\eta_i}}{1 + e^{\eta_i}}$$

A model which uses logit and logistic functions is called a logit or logistic model. For continuous independent variables, the corresponding model is a logistic regression model.

$$\eta_i = log[p_i /(1 - p_i)] = \beta_0 + \beta_1 x_{1i} + \beta_2 x_{2i} + \dots + \beta_{p-1} x_{(p-1)i}$$

where:

$x_{1i}, x_{2i}, \dots, x_{(p-1)i}$ = independent variables

$\beta_0, \beta_1, \beta_2, \dots, \beta_{p-1}$ = regression parameters

A simple logistic regression is a logistic regression with only one independent continuous variable:

$$\eta_i = log[p_i /(1 - p_i)] = \beta_0 + \beta_1 x_i$$

Independent variables can also be categorical. For example, a one-way logit model can be defined as follows:

$$\eta_i = log[p_i /(1 - p_i)] = m + \tau_i$$

where:

m = the overall mean of the proportion on the logarithmic scale

τ_i = the effect of group i

Defining the logit function assures that estimates or predicted values of the dependent variable are always between 0 and 1. Errors in the model have a Bernoulli distribution or a binomial distribution divided by n. A variance function is also defined:

$$V(\mu) = V(p) = p\,(1 - p) = pq$$

where $q = 1 - p$

Thus, the variance of binomial proportions y/n is:

$$Var(y / n) = pq / n = \tfrac{1}{n} V(p)\phi^2$$

The variance function $V(p)$ must be divided by n because a proportion is a binomial variable divided by n. It follows that the dispersion parameter is:

$$\phi^2 = 1$$

A property of logistic regression is that the variance of *y/n* is a function of *p*. The model takes into account variance heterogeneity by defining a variance function. The mean and variance depend on the parameter *p*. Thus, if the independent variables influence the parameter *p*, they will also influence the mean and variance.

24.1.1 Testing Hypotheses

Recall that for a linear regression the expression:

$$\chi^2 = \frac{SS_{RES\_REDUCED} - SS_{RES\_FULL}}{\hat{\sigma}^2}$$

is utilized to test if particular parameters are needed in a model (section 9.3). Here, SS_{RES} are residual sums of squares. That expression is equal to:

$$\chi^2 = -2log\frac{L(reduced\_model)}{L(full\_model)} = 2\left[-logL(reduced\_model) + logL(full\_model)\right]$$

where *L* and *logL* are values of the likelihood function and log likelihood function. This expression has a *chi*-square distribution with degrees of freedom equal to the difference in numbers of parameters.

The same holds for generalized linear models. A measure of deviation between the estimated and observed values for generalized linear models is called the deviance. The deviance is analogous to the SS_{RES} for linear models, that is, deviance for linear models is SS_{RES}. For the logistic model the deviance is:

$$D = 2\sum_i \left\{ y_i\, log\left(\frac{y_j}{n_i\hat{p}_i}\right) + (n_i - y_i)log\left(\frac{n_i - y_i}{n_i - n_i\hat{p}_i}\right) \right\}$$

$$i = 1,\ldots, \text{number of observations}$$

where:

y_i = number of successes from a total of n_i trials for observation *i*
$\hat{p}_i$ = the estimated probability of success for observation *i*

The distribution of the difference between the full and reduced model deviances is an approximate *chi*-square distribution with degrees of freedom equal to the difference in numbers of parameters.

Example: Consider a simple logistic model to explain changes in a binomial proportion *p* due to changes in an independent variable *x*:

$$log[p_i/(1 - p_i)] = \beta_0 + \beta_1 x_i$$

The null hypothesis is:

$$H_0: \beta_1 = 0$$

The reduced model is:

$$log[p_i/(1 - p_i)] = \beta_0$$

Let $D\left(\hat{\beta}_0 + \hat{\beta}_1 x\right)$ denote a deviance for the full model, and $D\left(\hat{\beta}_0\right)$ a deviance for the reduced model. For large samples the difference:

$$\chi^2 = D\left(\hat{\beta}_0\right) - D\left(\hat{\beta}_0 + \hat{\beta}_1 x\right)$$

has an approximate *chi*-square distribution with $(2 - 1) = 1$ degree of freedom. If the calculated difference is greater than the critical value χ^2_α, H_0 is rejected.

Sometimes binomial proportion data show variance that differs from the theoretical variance pq/n. In that case the dispersion parameter ϕ^2 differs from one and usually is denoted the extra-dispersion parameter. The variance is:

$$Var(y/n) = (pq/n)\phi^2 = \frac{1}{n}V(\mu)\phi^2$$

The parameter ϕ^2 can be estimated from the data with the deviance (D) divided by the degrees of freedom (df):

$$\hat{\phi}^2 = \frac{D}{df}$$

The degrees of freedom are defined similarly to the residual degrees of freedom in a linear model. For example in regression they are equal to the number of observations minus the number of regression parameters. The value $\phi^2 = 1$ indicates that the variance is consistent with the assumed distribution, $\phi^2 < 1$ indicates under-dispersion, and $\phi^2 > 1$ indicates over-dispersion from the assumed distribution. If the extra-dispersion parameter ϕ^2 is different than 1, the test must be adjusted by dividing the deviances by ϕ^2/n. The estimates do not depend on the parameter ϕ^2 and they need not be adjusted.

Example: Is there an effect of age at first calving on incidence of mastitis in cows? On a sample of 21 cows the incidence of mastitis and age at first calving (in months) were recorded:

Age	19	20	20	20	21	21	21	22	22	22	23
Mastitis	1	1	0	1	0	1	1	1	1	0	1
Age	26	27	27	27	27	29	30	30	31	32	
Mastitis	1	0	1	0	0	1	0	0	0	0	

A logit model was assumed:

$$log[p_i/(1 - p_i)] = \beta_0 + \beta_1 x_i$$

where:

p_i = the proportion with mastitis for observation i
x_i = age at first calving for observation i
β_0, β_1 = regression parameters

The following estimates were obtained:

$$\hat{\beta}_0 = 6.7439 \quad \hat{\beta}_1 = -0.2701$$

The deviances for the full and reduced models are:

$$D(\hat{\beta}_0 + \hat{\beta}_1 x) = 23.8416$$
$$D(\hat{\beta}_0) = 29.0645$$
$$\chi^2 = D(\hat{\beta}_0) - D(\hat{\beta}_0 + \hat{\beta}_1 x) = 5.2229$$

The critical value is $\chi^2_{0.05,1} = 3.841$. Since the calculated difference is greater than the critical value, H_0 is rejected, and we can conclude that age at first calving influences incidence of mastitis.

The estimated curve describing incidence of mastitis can be seen in Figure 24.1. To estimate the proportion for a particular age x_i, a logistic function is used. For example, the estimate for the age $x_i = 22$ is:

$$\hat{\mu}_{x=22} = \hat{p}_{x=22} = \frac{e^{(\hat{\beta}_0 + \hat{\beta}_1 x_i)}}{1 + e^{(\hat{\beta}_0 + \hat{\beta}_1 x_i)}} = \frac{e^{6.7439 - (0.2701)(22)}}{1 + e^{6.7439 - (0.2701)(22)}} = 0.6904$$

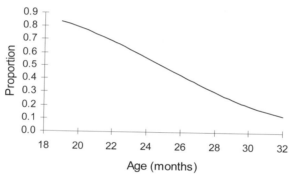

Figure 24.1 Logistic curve of proportion with mastitis as affected by age at first calving

Logistic regression is also applicable when the independent variables are categorical. Recall that the effects of categorical variables can be analyzed through a regression model by assigning codes, usually 0 and 1, to the observations of a particular group or treatment. The code 1 denotes that the observation belongs to the group, 0 denotes that it does not belong to the group.

Example: Are the proportions of cows with mastitis significantly different among three farms? The total number of cows and the number of cows with mastitis are shown in the following table:

Farm	Total no. of cows	No. of cows with mastitis
A	96	36
B	132	29
C	72	10

The model is:

$$\eta_i = log[p_i/(1 - p_i)] = m + \tau_i \qquad\qquad i = A, B, C$$

where:

p_i = the proportion with mastitis on farm i
m = the overall mean of the proportion on the logarithmic scale
τ_i = the effect of farm i

As shown for linear models with categorical independent variables, there are no unique solutions for $\hat{m}$ and the $\hat{\tau}_i$. For example, one set of the solutions is obtained by setting one of the $\hat{\tau}_i$ to zero:

$$\hat{m} = -1.8245$$
$$\hat{\tau}_A = 1.3137$$
$$\hat{\tau}_B = 0.5571$$
$$\hat{\tau}_C = 0.000$$

The estimate of the proportion for farm A is:

$$\hat{\mu}_A = \hat{p}_A = \frac{e^{(\hat{m}+\hat{\tau}_A)}}{1+e^{(\hat{m}+\hat{\tau}_A)}} = \frac{e^{(-1.8245+1.3437)}}{1+e^{-1.8245+1.3437}} = 0.3750$$

The estimate of the proportion for farm B is:

$$\hat{\mu}_B = \hat{p}_B = \frac{e^{(\hat{m}+\hat{\tau}_B)}}{1+e^{(\hat{m}+\hat{\tau}_B)}} = \frac{e^{(-1.8245+0.5571)}}{1+e^{(-1.8245+0.5571)}} = 0.2197$$

The estimate of the proportion for farm C is:

$$\hat{\mu}_C = \hat{p}_C = \frac{e^{(\hat{m}+\hat{\tau}_C)}}{1+e^{(\hat{m}+\hat{\tau}_C)}} = \frac{e^{(-1.8245)}}{1+e^{(-1.8245)}} = 0.1389$$

The deviances for the full and reduced models are:

$$D(\hat{m} + \hat{\tau}_i) = 0$$
$$D(\hat{m}) = 13.3550$$

The value of *chi*-square statistic is:

$$\chi^2 = D(\hat{m}) - D(\hat{m} + \hat{\tau}_i) = 13.3550$$

For $(3 - 1) = 2$ degrees of freedom, the critical value $\chi^2_{0.05} = 5.991$. The difference in incidence of mastitis among the three farms is significant at the 5% level.

Another approach to solving this example is to define an equivalent model in the form of a logistic regression:

$$log[p_i/(1 - p_i)] = \beta_0 + \beta_1 x_{1i} + \beta_2 x_{2i}$$

where:

p_i = the proportion with mastitis on farm i
x_{1i} = an independent variable with the values 1 if observation i is on farm A or 0 if observation i is not on farm A

x_{2i} = an independent variable with the value 1 if observation i is on farm B or 0 if observation i is not on farm B

$\beta_0, \beta_1, \beta_2$ = regression parameters

The following parameter estimates were obtained:

$$\hat{\beta}_0 = -1.8245$$

$$\hat{\beta}_1 = 1.3137$$

$$\hat{\beta}_2 = 0.5571$$

The estimate of the incidence of mastitis for farm A is:

$$\mu_{x_1=1, x_2=0} = \frac{e^{(\hat{\beta}_0 + \hat{\beta}_1 x_{1i} + \hat{\beta}_2 x_{2i})}}{1 + e^{(\hat{\beta}_0 + \hat{\beta}_1 x_{1i} + \hat{\beta}_2 x_{2i})}} = \frac{e^{-1.8245+(1.3137)(1)+(0.5571)(0)}}{1 + e^{-1.8245+(1.3137)(1)+(0.5571)(0)}} = 0.3750$$

The estimate of the incidence of mastitis for farm B is:

$$\mu_{x_1=0, x_2=1} = \frac{e^{(\hat{\beta}_0 + \hat{\beta}_1 x_{1i} + \hat{\beta}_2 x_{2i})}}{1 + e^{(\hat{\beta}_0 + \hat{\beta}_1 x_{1i} + \hat{\beta}_2 x_{2i})}} = \frac{e^{-1.8245+(1.3137)(0)+(0.5571)(1)}}{1 + e^{-1.8245+(1.3137)(0)+(0.5571)(1)}} = 0.2197$$

The estimate of the incidence of mastitis for farm C is:

$$\mu_{x_1=0, x_2=0} = \frac{e^{(\hat{\beta}_0 + \hat{\beta}_1 x_{1i} + \hat{\beta}_2 x_{2i})}}{1 + e^{(\hat{\beta}_0 + \hat{\beta}_1 x_{1i} + \hat{\beta}_2 x_{2i})}} = \frac{e^{-1.8245+(1.3137)(0)+(0.5571)(0)}}{1 + e^{-1.8245+(1.3137)(0)+(0.5571)(0)}} = 0.1389$$

The deviance for the full model is equal to zero because the data are completely described by the model:

$$D\left(\hat{\beta}_0 + \hat{\beta}_1 x_1 + \hat{\beta}_2 x_2\right) = 0$$

The deviance for the reduced model is:

$$D\left(\hat{\beta}_0\right) = 13.3550$$

The difference between deviances is:

$$\chi^2 = D\left(\hat{\beta}_0\right) - D\left(\hat{\beta}_0 + \hat{\beta}_1 x_1 + \hat{\beta}_2 x_2\right) = 13.3550$$

The critical value for $(3 - 1) = 2$ degrees of freedom is $\chi^2_{0.05} = 5.991$. The calculated χ^2 is greater than the critical value and the differences between farms are significant. Note the same estimates and calculated χ^2 value were obtained when data were analyzed as a one-way model with a categorical independent variable.

24.1.2 SAS Examples for Logistic Models

The SAS program for the example examining the effect of age on incidence of mastitis is the following. Recall the data:

Age	19	20	20	20	21	21	21	22	22	22	23
Mastitis	1	1	0	1	0	1	1	1	1	0	1
Age	26	27	27	27	27	29	30	30	31	32
Mastitis	1	0	1	0	0	1	0	0	0	0

SAS program:

```
DATA a;
INPUT age mastitis @@;
DATALINES;
19 1    20 1    20 0    20 1    21 0    21 1    21 1
22 1    22 1    22 0    23 1    26 1    27 0    27 1
27 0    27 0    29 1    30 0    30 0    31 0    32 0
;

PROC GENMOD DATA = a;
MODEL mastitis = age / DIST = BIN
                       LINK = LOGIT
                       TYPE1
                       TYPE3;
RUN;
```

Explanation: The GENMOD procedure is used. The statement, *MODEL mastitis = age* denotes *mastitis* as the dependent variable and *age* as the independent variable. The options *DIST = BIN* defines a binomial distribution, and *LINK = LOGIT* denotes that the model is a logit model, that is, the 'link' function is a logit. The TYPE1 and TYPE3 commands direct calculation of sequential and partial tests using the deviances for the full and reduced models.

SAS output:

Criteria For Assessing Goodness Of Fit

Criterion	DF	Value	Value/DF
Deviance	19	23.8416	1.2548
Scaled Deviance	19	23.8416	1.2548
Pearson Chi-Square	19	20.4851	1.0782
Scaled Pearson X2	19	20.4851	1.0782
Log Likelihood		-11.9208	

Analysis Of Parameter Estimates

Parameter	DF	Estimate	Standard Error	Wald 95% Confidence Limits		Chi-Square	Pr>ChiSq
Intercept	1	6.7439	3.2640	0.3466	13.1412	4.27	0.0388
age	1	-0.2701	0.1315	-0.5278	-0.0124	4.22	0.0399
Scale	0	1.0000	0.0000	1.0000	1.0000		

NOTE: The scale parameter was held fixed.

LR Statistics For Type 1 Analysis

Source	Deviance	DF	Chi-Square	Pr > ChiSQ
INTERCEPT	29.0645	0	.	.
AGE	23.8416	1	5.2230	0.0223

LR Statistics For Type 3 Analysis

Source	DF	Chi-Square	Pr > ChiSq
age	1	5.22	0.0223

Explanation: The first table shows measures of the correctness of the model. Several criteria are shown (*Criterion*), along with the degrees of freedom (*DF*), a *Value* and the value divided by degrees of freedom (*Value/DF*). The *Deviance* is 23.8416. The extra-dispersion parameter (*Scale*) is 1, and thus the *Scaled Deviance* is equal to the *Deviance*. The *Pearson Chi-Square* and *Log Likelihood* are also shown. The next table presents the parameter estimates (*Analysis of Parameter Estimates*). The parameter estimates are $b_0 = 6.7439$ and $b_1 = -0.2701$. Below the table is a note that the extra-dispersion parameter (*Scale*) is held fixed (=1) for every value of the *x* variable (*NOTE: The scale parameter was held fixed*). At the end of the output the *Type1* and *Type3* tests of significance of regression are shown: *Source* of variability, *Deviance*, degrees of freedom (*DF*), *Chi-Square* and *P* value (*Pr > ChiSQ*). The deviance for β_0 (*INTERCEPT*), that is for the reduced model, is 29.0645. The deviance for β_1 (*AGE*), that is for the full model, is 23.8416. The *Chi-Square* value (5.2230) is the difference between the deviances. Since the *P* value = 0.0223, H_0 is rejected, indicating an effect of age on development of mastitis.

Note: The same results and similar output can be obtained from the LOGISTIC procedure, which does logistic regression by default. The SAS program is:

```
PROC LOGISTIC DATA=a ORDER=DATA;
MODEL mastitis = age / ;
RUN;
```

The GENMOD (or LOGISTIC) procedure can also be used to analyze the data expressed as proportions. The SAS program is:

```
DATA a;
INPUT age mastitis n @@;
DATALINES;
19  1  1    20  2  3    21  2  3
22  2  3    23  1  1    26  1  1
27  1  4    29  1  1    30  0  2
31  0  1    32  0  1
;
PROC GENMOD DATA=a;
MODEL mastitis/n = age / DIST = BIN
                        LINK = LOGIT
                        TYPE1
                        TYPE3
                        PREDICTED;
RUN;
```

Explanation: The variables defined with the statement INPUT are *age*, the number of cows with mastitis for a particular age (*mastitis*), and the total number of cows for the respective age (*n*). In the MODEL statement, the dependent variable is expressed as a proportion *mastitis/n*. The options are as before with the addition of the PREDICTED option, which produces output of the estimated proportions for each observed *age* as follows:

Observation Statistics

Obs	mastitis	n	Pred	Xbeta	Std	HessWgt
1	1	1	0.8337473	1.6124213	0.8824404	0.1386127
2	2	3	0.7928749	1.3423427	0.7775911	0.4926728
3	2	3	0.7450272	1.0722641	0.6820283	0.569885
4	2	3	0.6904418	0.8021854	0.6002043	0.6411958
5	1	1	0.6299744	0.5321068	0.5384196	0.2331067
6	1	1	0.4309125	-0.278129	0.535027	0.2452269
7	1	4	0.3662803	-0.548208	0.5951266	0.9284762
8	1	1	0.2519263	-1.088365	0.770534	0.1884594
9	0	2	0.2044934	-1.358444	0.8748417	0.3253516
10	0	1	0.1640329	-1.628522	0.9856678	0.1371261
11	0	1	0.1302669	-1.898601	1.1010459	0.1132974

The table *Observation Statistics* shows for each age the predicted proportions (*Pred*), estimate of $\hat{\beta}_0 + \hat{\beta}_0$ *age* (*Xbeta*), standard error (*Std*), and diagonal element of the weight matrix used in computing the Hessian matrix (matrix of the second derivatives of the likelihood function), which is needed for iterative estimation of parameters (*HessWgt*).

The SAS program for the example examining the incidence of mastitis in cows on three farms, which uses a logit model with categorical independent variables, is as follows. Recall the data:

Farm	Total no. of cows	No. of cows with mastitis
A	96	36
B	132	29
C	72	10

SAS program:

```
DATA a;
INPUT n y farm $;
DATALINES;
 96   36   A
132   29   B
 72   10   C
;
PROC GENMOD DATA=a;
CLASS farm;
MODEL y/n = farm / DIST = BIN
                   LINK = LOGIT
                   TYPE1
                   TYPE3
                   PREDICTED;
LSMEANS farm / DIFF CL;
RUN;
```

Explanation: The GENMOD procedure is used. The CLASS statement defines *farm* as a classification variable. The statement, MODEL *y/n* = *farm*, defines the dependent variable as a binomial proportion, with *y* = the number of cows with mastitis and *n* = the total number of cows on the particular farm. The independent variable is *farm*. The *DIST* = *BIN* option defines a binomial distribution, and *LINK* = *LOGIT* denotes a logit model. The TYPE1 and TYPE3 options direct calculation of sequential and partial tests using deviances for the full and reduced models. The PREDICTED option produces output including predicted proportions for each farm. The LSMEANS statement directs output of parameter estimates for each farm.

SAS output:

Criteria For Assessing Goodness Of Fit

Criterion	DF	Value	Value/DF
Deviance	0	0.0000	.
Scaled Deviance	0	0.0000	.
Pearson Chi-Square	0	0.0000	.
Scaled Pearson X2	0	0.0000	.
Log Likelihood		-162.0230	

Analysis Of Parameter Estimates

Parameter	DF	Estimate	Standard Error	Wald 95% Confidence Limits		Chi-Square	Pr>ChiSq
Intercept	1	-1.8245	0.3408	-2.4925	-1.1566	28.67	<.0001
farm A	1	1.3137	0.4007	0.5283	2.0991	10.75	0.0010
farm B	1	0.5571	0.4004	-0.2277	1.3419	1.94	0.1641
farm C	0	0.0000	0.0000	0.0000	0.0000	.	.
Scale	0	1.0000	0.0000	1.0000	1.0000		

NOTE: The scale parameter was held fixed.

LR Statistics For Type 1 Analysis

Source	Deviance	DF	Chi-Square	Pr > ChiSq
Intercept	13.3550			
farm	0.0000	2	13.36	0.0013

LR Statistics For Type 3 Analysis

Source	DF	Chi-Square	Pr > ChiSq
farm	2	13.36	0.0013

Least Squares Means

Effect	farm	Estimate	Standard Error	DF	Chi-Square	Pr > ChiSq	Alpha
farm	A	-0.5108	0.2108	1	5.87	0.0154	0.05
farm	B	-1.2674	0.2102	1	36.35	<.0001	0.05
farm	C	-1.8245	0.3408	1	28.67	<.0001	0.05

Least Squares Means

Effect	farm	Confidence Limits	
farm	A	-0.9240	-0.0976
farm	B	-1.6795	-0.8554
farm	C	-2.4925	-1.1566

Differences of Least Squares Means

Effect	farm	_farm	Estimate	Standard Error	DF	Chi-Square	Pr>ChiSq	Alpha
farm	A	B	0.7566	0.2977	1	6.46	0.0110	0.05
farm	A	C	1.3137	0.4007	1	10.75	0.0010	0.05
farm	B	C	0.5571	0.4004	1	1.94	0.1641	0.05

Differences of Least Squares Means

Effect	farm	_farm	Confidence Limits	
farm	A	B	0.1731	1.3401
farm	A	C	0.5283	2.0991
farm	B	C	-0.2277	1.3419

Observation Statistics

Observation	y	n	Pred	Xbeta	Std	HessWgt
1	36	96	0.375	-0.510826	0.2108185	22.5
2	29	132	0.219697	-1.267433	0.2102177	22.628788
3	10	72	0.1388889	-1.824549	0.3407771	8.6111111

Explanation: The first table presents statistics describing the correctness of the model. Several criteria are shown (*Criterion*), along with degrees of freedom (*DF*), *Value* and value divided by degrees of freedom (*Value/DF*). The *Deviance* = 0, since the model exactly describes the data (a saturated model). The next table presents parameter estimates (*Analysis Of Parameter Estimates*). For a model with categorical independent variables SAS defines an equivalent regression model. The estimates for *Intercept, farm A, farm B*, and *farm C* are equivalent to the solution from the one-way model when the estimate of *farm C* is set to zero. Thus, the parameter estimates are $\hat{\beta}_0 = -1.8245$ (*Intercept*), $\hat{\beta}_1 = 1.3137$ (*farm A*), and $\hat{\beta}_2 = 0.5571$ (*farm B*) for the regression model $log[p_i/(1 - p_i)] = \beta_0 + \beta_1 x_{1i} + \beta_2 x_{2i}$ or analogously $\hat{m} = -1.8245$ (*Intercept*), $\hat{\tau}_A = 1.3137$ (*farm A*), $\hat{\tau}_B = 0.5571$ (*farm B*), and $\hat{\tau}_C = 0.000$ (*farm C*) for the one-way model $log[p_i/(1 - p_i)] = m + \tau_i$ (see example in section 24.1.1 for model definition). The extra-dispersion parameter (*Scale*) is 1, and the *Scaled Deviance* is equal to *Deviance*. (*NOTE: The scale parameter was held fixed.*) Next, the *Type1* and *Type3* tests of significance of the regression parameters are shown. Listed are: *Source* of variability, *Deviance*, degrees of freedom (*DF*), *Chi-Square* and *P* value (*Pr>ChiSq*). The deviance for the reduced model$_0$ (*INTERCEPT*) is 13.3550. The deviance for the full model (*farm*) is 0. The *Chi-Square* value (13.36) is for the difference between the deviances. Since the *P* value = 0.0013, the H_0 is rejected, these data show evidence for an effect of farm on mastitis. The next table shows *Least Squares Means* and corresponding analyses in logit values: *Estimate, Standard Errors*, degrees of freedom (*DF*), *ChiSquare, P* value (*Pr>ChiSq*), and confidence level (*Alpha*) for *Confidence Limits*. The next table presents the *Difference of Least Squares Means*. This output is useful to test which farms are significantly different from others. From the last table (*Observation Statistics*) note the predicted proportions (*Pred*) for each farm.

24.2 Diagnostics Test – ROC Curve

Diagnostic tests can be used to identify the health or nutritional status of an animal and the reaction of an animal to treatment. In their simplest form, diagnostic tests can give positive or negative predictions of status. Statistical analyses of the results of diagnostic tests can specify the required level of treatment that can yield a response or the dosage of drug that maximizes a positive outcome.

When using a diagnostic test, we seek to be as sure as possible that the diagnosis is correct. Rarely will there be a test that distinguishes positive from negative results or affected from healthy individuals without error. Often there is some overlap or error in both directions, some affected animals will be identified as healthy and some healthy animals will be identified as affected. This happens because individual animals express different symptoms of a condition, or may react differently to a test or drug. We seek to minimize these types of error. For example, a pregnancy test for cattle could be used to predict which cows are open and should be culled versus those that are predicted to be pregnant and should be kept in the herd. A false positive test result (predicting a cow to be pregnant when she is open) could result in a cow being kept in the herd when she should be culled, and a false negative test result (predicting a cow to be open when she is truly pregnant) could result in keeping an open cow that should be culled (Figure 24.2).

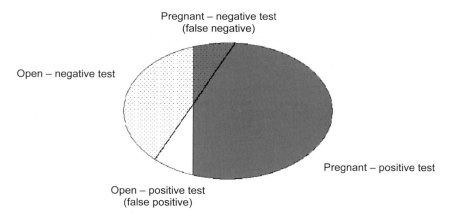

Figure 24.2 Diagram of possible diagnostic test results and true outcomes

Consider a diagnostic test in its simplest form, one which can yield two results: positive and negative prediction of an illness. Some of the test results will be erroneous, some of the healthy animals will be falsely classified as affected, and some animals that are affected will be falsely classified as healthy. We can categorize the combination of test results and unknown but true disease status as in the following table:

Test result		True disease status	
		Affected (+)	Healthy (−)
	Positive (+)	True positive (TP)	False positive (FP)
	Negative (−)	False negative (FN)	True negative (TN)

It may be convenient to express the counts as percentages or relative numbers. Probabilities usually associated with the diagnostic test are:

- *Positive predictive value*: probability that an animal is affected when the test is positive.
- *Negative predictive value*: probability that an animal is not affected when the test is negative.
- *Sensitivity*: probability of a positive test result when an animal is affected.
- *Specificity*: probability of a negative test result when an animal is not affected.
- *Prevalence*: probability that an animal in the population is affected.

These probabilities are summarized in the following table:

Test results		True status		
		Positive (+)	Negative (−)	
	Positive (+)	True positive (TP)	False positive (FP)	Positive predictive value $\dfrac{TP}{TP + FP}$
	Negative (−)	False negative (FN)	True negative (TN)	Negative predictive value $\dfrac{TN}{TN + FN}$
		Sensitivity (True positive rate) $\dfrac{TP}{TP + FN}$	False positive rate $\dfrac{FP}{TN + FP}$	
		False negative rate $\dfrac{FN}{TP + FN}$	Specificity (True negative rate) $\dfrac{TN}{TN + FP}$	Prevalence $\dfrac{TP + FN}{TN + FN + FP + TN}$

Another useful expression is the *positive likelihood ratio*, which denotes how many times more likely than not an animal with a positive test is affected and is expressed as:

$$\text{Positive likelihood ratio} = \frac{\textit{True positive rate}}{\textit{False positive rate}} = \frac{\textit{Sensitivity}}{1 - \textit{Specificity}}$$

Example: A study was conducted to validate a pregnancy test. Out of a total of 180 animals, 20 had a positive test. However, the number of pregnant animals was 30, and the number of the animals found positive which were pregnant was 16.

The number of animals and proportions (estimated probabilities) are in the following table:

		True status		
		Positive (+) Pregnant $n = 30$	Negative (−) Open $n = 150$	
Test results	Positive (+) $n = 20$	True positive (*TP*) 16	False positive (*FP*) 4	Positive predictive value $\dfrac{TP}{TP + FP} = \dfrac{16}{16 + 4}$ = 0.80
	Negative (−) $n = 160$	False negative (*FN*) 14	True negative (*TN*) 146	Negative predictive value $\dfrac{TN}{TN + FN} = \dfrac{146}{146 + 14}$ = 0.912
		Sensitivity (True positive rate) $\dfrac{TP}{TP + FN} = \dfrac{16}{16 + 14} =$ 0.533	False positive rate $\dfrac{FP}{TN + FP} = \dfrac{4}{146 + 4}$ = 0.027	
		False negative rate $\dfrac{FN}{TP + FN} = \dfrac{14}{16 + 14}$ = 0.467	Specificity (True negative rate) $\dfrac{TN}{TN + FP} = \dfrac{146}{146 + 4}$ = 0.973	Prevalence $\dfrac{TP + FN}{TP + FN + FP + TN}$ $= \dfrac{14 + 16}{14 + 16 + 4 + 146}$ = 0.167

From the table we can conclude:
- The probability that the pregnancy test was positive given that the animal was pregnant is 0.533 (sensitivity or true positive rate).
- The probability that the pregnancy test was negative given that the animal was open is 0.973 (specificity or true negative rate).
- The probability that an animal was pregnant given a positive test is 0.80 (positive predictive value).
- The probability that an animal was open given a negative test is 0.912 (negative predictive value).

Sensitivity and specificity depend only on the diagnostics test, because they are probabilities of test results given true status. Predictive values depend on the probabilities of the status in the population and on the diagnostics test. Once the probability from the diagnostics test is established, they depend on the true status in the population. If prevalence is low (low occurrence of the condition), the predictive value of a positive test will be low. To validate a diagnostic test, it is very important that the test sample has similar frequencies of incidence in the population in which the test will be applied. For example, testing a sample that is 50% affected and 50% healthy may yield misleading results, if the true frequency in the population is much lower. However, sensitivity and specificity depend only on the test, and thus a sample with equal numbers of positives and negatives is good for evaluating the quality of a diagnostic test.

Example: For a diagnostic test with a known sensitivity of 0.9 and specificity of 0.8, we can calculate the positive and negative predictive values for the given sensitivity and specificity when prevalence of disease in the population is: a) 0.25, b) 0.1, and c) 0.01 by using Bayes theorem:

a) For prevalence $= P(+\ true) = 0.25$:

Sensitivity $= P(+test\ |\ +true) = 0.9$, the probability that a test is positive given true positive status

Specificity $= P(-test\ |\ -true) = 0.8$, the probability that a test is negative given true negative status

Recall that positive predictive value is the probability that an animal is positive (ill) given a positive test:

$$+PRED = P(+\ true\ |\ +test).$$

Using Bayes formula:

$$P(+\ true\ |+test) = \frac{P(+\ true \cap +test)}{P(+test)} = \frac{P(+true)P(+test\ |+true)}{P(-true)P(+test\ |-true) + P(+true)P(+test\ |+true)}$$

We need:

$P(-true) = 1 - P(+true) = 1 - 0.25 = 0.75$ (the probability that animal in the population is healthy)

$P(+test\ |\ -true) = 1 - P(-test\ |\ -true) = 1 - 0.8 = 0.2$, $(1 - $ specificity$=$ false positive rate)

Thus, the positive predictive value is:

$$P(+\ true\ |+test) = \frac{(0.25)(0.9)}{(0.75)(0.2) + (0.25)(0.9)} = 0.6$$

Similarly, the negative predictive value (the probability that an animal is negative (healthy) given a negative test):

$$-PRED = P(-\ true\ |-test).$$

Using Bayes formula:

$$P(-\ true\ |-test) = \frac{P(-\ true \cap -test)}{P(-test)} = \frac{P(-true)P(-test\ |-true)}{P(+true)P(-test\ |+true) + P(-true)P(-test\ |-true)}$$

We need:

$P(-true) = 1 - P(+true) = 1 - 0.25 = 0.75$ (the probability that an animal in the population is healthy)

$P(-test\ |\ +true) = 1 - P(+test\ |\ +true) = 1 - 0.9 = 0.1$, $(1 - $ sensitivity $=$ false negative rate)

Thus, the negative predictive value is:

$$P(-\,true\,|\,-test) = \frac{(0.75)(0.8)}{(0.25)(0.1) + (0.75)(0.8)} = 0.96$$

b) For prevalence = $P(+\,true) = 0.1$:

$$P(-true) = 1 - P(+true) = 1 - 0.1 = 0.9$$

Positive predictive value:

$$P(+\,true\,|\,+test) = \frac{(0.1)(0.9)}{(0.9)(0.2) + (0.1)(0.9)} = 0.333$$

Negative predictive value:

$$P(-\,true\,|\,-test) = \frac{(0.9)(0.8)}{(0.1)(0.1) + (0.9)(0.8)} = 0.986$$

c) For prevalence = $P(+\,true) = 0.01$:

$$P(-true) = 1 - P(+true) = 1 - 0.01 = 0.9$$

Positive predictive value:

$$P(+\,true\,|\,+test) = \frac{(0.01)(0.9)}{(0.99)(0.2) + (0.01)(0.9)} = 0.043$$

Negative predictive value:

$$P(-\,true\,|\,-test) = \frac{(0.99)(0.8)}{(0.01)(0.1) + (0.99)(0.8)} = 0.999$$

This example shows that predictive value depends on prevalence (the proportion of animals in a population which are ill). In statistical terms we say that the diagnostic test depends on prior information. If the proportion of animals which are affected is low (and consequently the proportion of animals which are healthy is high), the positive predictive value will be low, and the negative predictive value will be high. For a given number of animals tested, and known sensitivity and specificity of the diagnostic test, we can easily calculate the expected numbers of true and false positives and negatives.

Consider further a sample of 1,000 animals, the sensitivity and specificity as shown above and prevalence of 0.1, the number of true and false positives and negatives are:

		True status		
		Positive (+) $n = 100$	Negative (−) $n = 900$	
Test results	Positive (+) $n = 270$	True positive (*TP*) 90	False positive (*FP*) 180	Positive predictive value $\dfrac{TP}{TP + FP} = 0.333$
	Negative (−) $n = 730$	False negative (*FN*) 10	True negative (*TN*) 720	Negative predictive value $\dfrac{TN}{TN + FN} = 0.986$
		Sensitivity (True positive rate) $\dfrac{TP}{TP + FN} = 0.9$	False positive rate $\dfrac{FP}{TN + FP} = 0.2$	
		False negative rate $\dfrac{FN}{TP + FN} = 0.1$	Specificity (True negative rate) $\dfrac{TN}{TN + FP} = 0.8$	Prevalence $\dfrac{TP + FN}{TP + FN + FP + TN}$ $= 0.1$

From the table we can see that of 1,000 animals, 100 (positive) are affected and 900 (negative) are not. Given a sensitivity of 0.90, the number of animals with a true positive diagnosis is $0.90 \times 100 = 90$. Given 100 positive animals of which 90 are correctly diagnosed, $100 - 90 = 10$ have a false negative diagnosis. Given a specificity of 0.80, the number of animals with a true negative diagnosis is $0.80 \times 900 = 720$. Given 900 negative animals of which 720 are correctly diagnosed, $900 - 720 = 180$ have a false positive diagnosis. In total, 270 tested positive, of which only 90 were true positive and 180 false positive. The positive predictive value is the number of true positives divided by the overall number positives, $90 / 270 = 0.33$.

In the previous discussion and example, the diagnostic tests were assumed to yield only a positive or negative result, that is, there was a true distinction between what was positive and what was negative; however, a diagnostics test or symptoms can show a continuous character. For example, observed swelling can range from slightly visible to large, or a visual diagnostic test can be scored from 1 to 10. Measures can also be on a true continuous scale, for example blood sugar level in mg per ml. The question then becomes what 'cutoff' point best distinguishes between negative and positive animals. To illustrate this, we can draw an analogous table, with diagnostic test having more than two levels. This will be shown in an example.

Example: For a sample of 480 animals with known status, the results of a diagnostic test scored from 1 to 5, were reported as the numbers of animals with each score as follows:

		True status	
		Positive (+) $n = 120$	Negative (−) $n = 360$
Test results (scores 1 − 5)	Score 5 $n = 52$	50	2
	Score 4 $n = 48$	40	8
	Score 3 $n = 52$	22	30
	Score 2 $n = 125$	5	120
	Score 1 $n = 203$	3	200

A higher score is clearly related to a greater probability that an animal is positive (has disease, affected). The next step is to choose a cutoff point that will distinguish between affected and healthy animals. It should be set to give maximal true positive and minimal false positive results. To determine that, sensitivity, specificity, predicted positive, and predicted negative values will be calculated for each score based on the cumulative observations. For example, consider a cutoff point of 2 and greater for the diagnostic to be positive, and a score 1 for the diagnostic test to be negative:

		True status		
		Positive (+) $n = 120$	Negative (−) $n = 360$	
Test results	Positive (+) $n = 277$	True positive (*TP*) 117	False positive (*FP*) 160	Positive predictive value $$\frac{TP}{TP+FP} = \frac{117}{117+160}$$ $= 0.422$
	Negative (−) $n = 203$	False negative (*FN*) 3	True negative (*TN*) 200	Negative predictive value $$\frac{TN}{TN+FN} = \frac{200}{200+3}$$ $= 0.985$
		Sensitivity (True positive rate) $$\frac{TP}{TP+FN} = \frac{117}{117+3} =$$ 0.975	False positive rate $\frac{FP}{TN+FP} = \frac{160}{200+160}$ $= 0.444$	
		False negative rate $$\frac{FN}{TP+FN} = \frac{3}{117+3} =$$ 0. 025	Specificity (True negative rate) $$\frac{TN}{TN+FP} = \frac{200}{200+160} =$$ 0.556	Prevalence $\frac{TP+FN}{TP+FN+FP+TN}$ $$= \frac{117+3}{117+3+160+200}$$ $= 0.250$

It is apparent that because the cutoff point was low, the sensitivity (true positive rate) is very high (0.975); however, the false positive rate is also high (0.444). This means that by using this cutoff point, 44.4% of healthy animals will be identified as being affected.

Similarly, sensitivity, specificity and other values can be calculated for other possible cutoff points. The possible cutoff points are: 1 and greater, 2 and greater, 3 and greater, 4 and greater, and 5. Note that 1 and greater means that all animals are stated affected, and more than 5 that all animals are stated healthy, regardless of test results.

A summary for all scores defined as cutoff points is presented in the following table:

Negative score	Positive score	True positive	False negative	True negative	False positive	Prevalence
1–5	None	0	120	360	0	0.25
1–4	5	50	70	358	2	0.25
1–3	4–5	90	30	350	10	0.25
1–2	3–5	112	8	320	40	0.25
1	2–5	117	3	200	160	0.25
None	1–5	120	0	0	360	0.25

Negative score	Positive score	Sensitivity	False positive rate	Specificity	Positive predictive value	Negative predictive value
1–5	None	0.000	0.000	1.000	–	0.250
1–4	5	0.417	0.006	0.994	0.962	0.836
1–3	4–5	0.750	0.028	0.972	0.900	0.921
1–2	3–5	0.933	0.111	0.889	0.737	0.976
1	2–5	0.975	0.444	0.556	0.422	0.985
None	1–5	1.000	1.000	0.000	0.250	–

It is apparent that if a lower cutoff is used as the criterion to predict affected animals, the sensitivity is higher, but the false positive rate is also higher (specificity is lower). This means that the probability of identifying affected animals is high, but since the threshold is low, many healthy animals will also be diagnosed as affected. A reasonable cutoff point would be between scores 2 and 3, or between 3 and 4, depending which is more costly, to misclassify an affected animal as healthy, or a healthy animal as affected.

The adequacy of the overall diagnostic test and the trade off between specificity and sensitivity can be shown graphically by constructing a receiver operating characteristic (ROC) curve. Such analyses were first done in signal detection theory, specifically in the quality of radar signals during World War II. It can be used whenever there is a binary response as a 'signal'.

A ROC curve is a plot of sensitivity (true positive rate) versus false positive rate (1 – specificity) as a function of the change in cutoff point. The diagnostics test is better if the curve goes up steeply, closer to the left top corner. A straight line from bottom left to top right corner (from coordinate 0,0 to coordinate 1,1) implies no diagnostic value, since the sensitivity (true rate) and false positive rate are always the same. The area under the curve is a good indicator of the quality of diagnostic test. If the total area within all four corners is defined as 1, then the closer the area under the curve is to 1, the better the diagnostic test. Two tests can be compared by comparing their ROC curves.

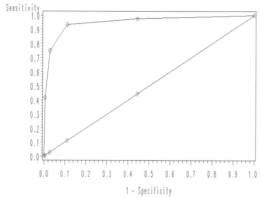

Figure 24.3 Receiver operating characteristic (ROC) curve

Figure 24.3 illustrates an ROC curve from the previous example. The curve increases rapidly and then plateaus towards the coordinate 1,1. The straight line represents a hypothesized diagnostic test that is meaningless, since each animal tested will have a 50% chance to be ill or healthy, regardless of the result of the test. The difference between lines can be used in formal testing of hypotheses. The null and alternative hypotheses are:

$$H_0: A = 0.5$$
$$H_1: A > 0.5$$

where A is the area under the ROC curve.

For a large sample and under H_0, the test statistic:

$$z = \frac{(A - 0.5)}{s_A}$$

has a standard normal distribution. Here, s_A is the standard error of the area under the ROC curve A, which can be calculated by using the following formula:

$$s_A = \sqrt{\frac{1}{n_p n_n}\left[A(1-A) + (n_p - 1)\left(\frac{A(1-A)^2}{2-A}\right) + (n_n - 1)\left(\frac{A^2(1-A)}{1+A}\right)\right]}$$

where n_p and n_n are the number of positive and negative animals, respectively.

Example: From the previous example, the area under the ROC curve was calculated to be 0.951. To test the hypothesis that this is different than 0.5, recall that the number of positive and negative animals was $n_p = 120$, and $n_n = 360$, respectively.

The null and alternative hypotheses are:

$$H_0: A = 0.5$$
$$H_1: A > 0.5$$

The standard error is:

$$s_A = \sqrt{\frac{1}{n_p n_n}\left[A(1-A) + (n_p - 1)\left(\frac{A(1-A)^2}{2-A}\right) + (n_n - 1)\left(\frac{A^2(1-A)}{1+A}\right)\right]} =$$

$$\sqrt{\frac{1}{(120)(360)}\left[0.951(1-0.951) + (120-1)\left(\frac{0.951(1-0.951)^2}{2-0.951}\right) + (360-1)\left(\frac{0.951^2(1-0.951)}{1+0.951}\right)\right]} = 0.014$$

The value of the z statistics is:

$$z = \frac{(0.951 - 0.5)}{0.014} = 32.23$$

For the significance level $\alpha = 0.05$, the critical value is $z_\alpha = 1.65$. Since the calculated value is greater than the critical value, H_0 is rejected.

To compare two diagnostics procedures the standard normal distribution can be also used. The null and alternative hypotheses are:

$$H_0: A_1 - A_2 = 0$$
$$H_1: A_1 - A_2 \neq 0$$

where A_1 and A_2 are the areas under two ROC curves.

For a large sample and under H_0, the test statistic:

$$z = \frac{(A_1 - A_2) - 0}{s_{A_1 - A_2}}$$

has a standard normal distribution. Here, $s_{A_1 - A_2}$ is the standard error of the difference between two areas under the ROC. If the samples for calculating the two areas are independent, the standard error of the difference can be calculated as:

$$s_{A_1 - A_2} = \sqrt{s_{A_1}^2 + s_{A_2}^2}$$

where s_{A_1} and s_{A_2} are the standard errors of areas A_1 and A_2, respectively.

24.2.1 SAS Example for Diagnostic Tests

The SAS program for the example of a diagnostic test with 5 scores is shown below. These data were collected to validate a diagnostic test scores from 1 to 5. The number of animals for each score and their true positive and negative status is given in the following table:

		True status	
		Positive (+) $n = 120$	Negative (−) $n = 360$
Test results (scores 1–5)	Score 5 $n = 52$	50	2
	Score 4 $n = 48$	40	8
	Score 3 $n = 52$	22	30
	Score 2 $n = 125$	5	120
	Score 1 $n = 203$	3	200

We will test the association between the true health status and the diagnostic test results and draw a ROC curve.

SAS program:

```
DATA a;
INPUT diagn disease n;
DATALINES;
1  1      3
2  1      5
3  1     22
4  1     40
5  1     50
1  0    200
2  0    120
3  0     30
4  0      8
5  0      2
;
PROC LOGISTIC DATA = a ORDER=DATA ;
FREQ n;
MODEL disease = diagn / LINK=LOGIT OUTROC = rocdata;
RUN;
QUIT;

DATA roc0;
_sensit_=0;
_1mspec_=0;
DATA rocdata1;
SET roc0 rocdata;
sensitivity=_pos_ /(_pos_ + _falneg_);
false_pos_rate=_falpos_ /(_neg_ + _falpos_);
specificity=_neg_ /(_neg_ + _falpos_);
pred_neg=_neg_ /(_neg_ + _falneg_);
pred_pos=_pos_ /(_pos_ + _falpos_);
prevalence = (_pos_ + _falneg_) / (_pos_ + _neg_ + _falpos_ + _falneg_);
RUN;
```

```
PROC PRINT DATA = rocdata1;
RUN;

SYMBOL1 i=JOIN l=1 v=DIAMOND;
SYMBOL2 i=JOIN l=2 v=NONE;
PROC GPLOT DATA=rocdata1;
 PLOT ( _sensit_ _1mspec_ ) * _1mspec_ / OVERLAY VAXIS=0 to 1 BY .1 ;
RUN;
QUIT;
```

Explanation: The data were entered as 1 being positive and 0 being negative for the results of the diagnostic test (variable *diagn*), and 1 being truly affected and 0 being truly healthy (variable *disease*). The LOGISTIC procedure was used to evaluate the quality of the diagnostic test, that is, how well it describes true status. The *FREQ n* statement defines the variable *n* as count variable (how many animals were observed for the particular status and diagnosis). The statement, *MODEL disease = diagn* defines *disease* as the dependent variable (true status) and *diagn* as the independent variable (result of diagnostic test). The LINK function for the logistic regression is LOGIT (also defined by default). The calculated sensitivity and specificity needed for the ROC curve are output to the file named *rocdata*. SAS specifies the variable names *_sensit_* for sensitivity, *_1mspec_* for false positive rate (1 – specificity), *_pos_* for positive value, *_neg_* for negative value, *_falpos_* for false positive value, and *_falneg_* for false negative value; however, SAS does not give the initial values *_sensit_* = 0 and *_1mspec_* = 0, which are required to plot the ROC curve. This can be done by using the data steps *DATA roc0* and *DATA rocdata1*, and these steps can also be used to calculate the false positive rate (*false_pos_rate*), positive predictive value (*pred_pos*), negative predictive value (*pred_neg*), and prevalence. The PRINT procedure prints those values. The GPLOT procedure can be used to draw the ROC curve from *DATA = rocdata1*, with options specifying the appearance of the curve.

SAS output:

Response Profile

Ordered Value	disease	Total Frequency
1	1	120
2	0	360

Probability modeled is disease=1.

Model Fit Statistics

Criterion	Intercept Only	Intercept and Covariates
AIC	541.842	211.664
SC	546.016	220.011
-2 Log L	539.842	207.664

Testing Global Null Hypothesis: BETA=0

Test	Chi-Square	DF	Pr > ChiSq
Likelihood Ratio	332.1781	1	<.0001
Score	298.7513	1	<.0001
Wald	117.7367	1	<.0001

Analysis of Maximum Likelihood Estimates

Parameter	DF	Estimate	Standard Error	Wald Chi-Square	Pr > ChiSq
Intercept	1	-6.8058	0.5940	131.2577	<.0001
diagn	1	2.0930	0.1929	117.7367	<.0001

Odds Ratio Estimates

Effect	Point Estimate	95% Wald Confidence Limits	
diagn	8.109	5.556	11.835

Association of Predicted Probabilities and Observed Responses

Percent Concordant	92.5	Somers' D	0.902
Percent Discordant	2.3	Gamma	0.952
Percent Tied	5.3	Tau-a	0.339
Pairs	43200	c	0.951

Obs	_sensit_	_1mspec_	_PROB_	_POS_	_NEG_	_FALPOS_	_FALNEG_	sensitivity
1	0.00000	0.00000	.	.	.	.	.	.
2	0.41667	0.00556	0.97490	50	358	2	70	0.41667
3	0.75000	0.02778	0.82725	90	350	10	30	0.75000
4	0.93333	0.11111	0.37128	112	320	40	8	0.93333
5	0.97500	0.44444	0.06788	117	200	160	3	0.97500
6	1.00000	1.00000	0.00890	120	0	360	0	1.00000

Obs	false_pos_rate	specificity	pred_neg	pred_pos	prevalence
1	.	.	.	.	.
2	0.00556	0.99444	0.83645	0.96154	0.25
3	0.02778	0.97222	0.92105	0.90000	0.25
4	0.11111	0.88889	0.97561	0.73684	0.25
5	0.44444	0.55556	0.98522	0.42238	0.25
6	1.00000	0.00000	.	0.25000	0.25

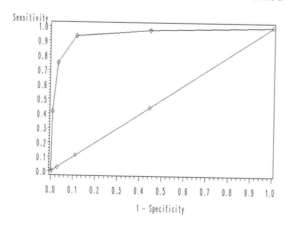

Explanation: An excerpt of the SAS output is given with relevant statistics required for analysis. The first table presents information of the dependent variable (*disease*), including its total frequency, and indicates the value 1 used to model probability. The next table is *Fit Statistics*, with Akaike (*AIC*), Schwarz (*SC*) and –2 loglikelihood (*–2Log L*) values. The values are given for the null model (*Intercept Only*, with no effects in the mode other than the mean) and for the full model (*Intercept and Covariates*, all effects in the model). Reduced values of criteria denote that the effects are needed in the model. The next table *Testing Global Null Hypothesis: BETA=0*, presents significance tests of the association between the true health status and the diagnostic tests results by using Likelihood ratio, Score and Wald statistics which all use a *chi*-square distribution for the model. For example, for the likelihood ratio test the calculated *Chi-Square* is 332.1781 which gives a *P* value (*Pr > ChiSq*) of less than 0.0001. The table *Analysis of Maximum Likelihood Estimates* presents *Estimates* with their *Standard Errors* and hypothesis tests for the particular effects in the model. Next are *Odds Ratio Estimates* with their *Confidence Limits*. The odds ratio is the ratio of positive vs. negative. The next table presents measures of association between observed and predicted value, the strength of the predictive ability of the model. The theory behind these values is beyond the scope of this book. Readers are kindly referred to SAS or other relevant literature for further information. Here the *c* value is of interest because it denotes the area under ROC curve (0.951).

The last table is from the PRINT procedure presenting sensitivity (*_sensit_* and *sensitivity*), false positive rate (*_lmspec_* and *false_pos_rate*), the positive value (*_POS_*), negative value (*_NEG_*), false positive value (*_FALPOS_*), false negative value (*_FALNEG_*), *specificity*, negative predictive value (*pred_neg*), positive predictive value (*pred_pos*), and *prevalence*. The figure in the output is the ROC curve. For the comparison, the line from 0,0 to 1,1 is also presented and denotes the null hypothesis (H_0) indicating no association between the diagnostic test and the true values. From the plot it is evident that an association exists (area under the curve is 0.951).

24.3 Probit Model

A standard normal variable can be transformed to a binary variable by defining the following: for all values less than some value η, the value 1 is assigned; for all values greater than η, the value 0 is assigned (Figure 24.4). The proportion of values equal to 1 and 0 is determined from the area under the normal distribution by using the cumulative normal distribution.

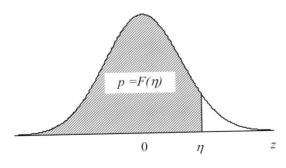

Figure 24.4 Connection between a binomial and a normal variable

Thus, although a binary variable or binomial proportion is being considered, a probability can be estimated by using the cumulative standard normal distribution. The effects of independent variables on the probability of success can be estimated. The inverse cumulative normal distribution is called a probit function, and consequently such models are called probit models. Probit models can be applied to proportions of more than two categories as well.

The inverse link function is the cumulative normal distribution and the mean is:

$$\mu = p = F(\eta) = \int_{-\infty}^{\eta} \frac{1}{\sqrt{2\pi}} e^{-0.5z^2} dz$$

where z is a standard normal variable with mean 0 and variance 1. The link function is called the probit link:

$$\eta = F^{-1}(\mu)$$

The effects of independent variables on η are defined as:

$$\eta_i = F^{-1}(\mu_i) = \mathbf{x}_i\boldsymbol{\beta}$$

For example, for regression:

$$\eta_i = F^{-1}(\mu_i) = \beta_0 + \beta_1 x_{1i} + \beta_2 x_{2i} + ... + \beta_{p-1} x_{(p-1)i}$$

where:

$x_{1i}, x_{2i},..., x_{(p-1)i}$ = independent variables

$\beta_0, \beta_1, \beta_2,..., \beta_{p-1}$ = regression parameters

The estimation of parameters and tests of hypotheses follow a similar approach as shown for logistic regression.

Example: Using a probit model, test the difference in proportions of cows with mastitis among three farms:

Farm	Total no. of cows	No. of cows with mastitis
A	96	36
B	132	29
C	72	10

The model is:

$$\eta = F^{-1}(p) = m + \tau_i$$

where:

m = the overall mean of the proportion on the probit scale

τ_i = the effect of farm i

A set of solutions obtained by setting one $\hat{\tau}_i$ to zero is:

$$\hat{m} = -1.0853$$
$$\hat{\tau}_A = 0.7667$$
$$\hat{\tau}_B = 0.3121$$
$$\hat{\tau}_C = 0.000$$

The estimate of the proportion for farm A is:

$$\hat{\mu}_A = \hat{p}_A = F^{-1}(\hat{m} + \hat{\tau}_A) = F^{-1}(-1.0853 + 0.7667) = F^{-1}(-0.3186) = 0.3750$$

The estimate of the proportion for farm B is:

$$\hat{\mu}_B = \hat{p}_B = F^{-1}(\hat{m} + \hat{\tau}_B) = F^{-1}(-1.0853 + 0.7667) = F^{-1}(-0.7732) = 0.2197$$

The estimate of the proportion for farm C is:

$$\hat{\mu}_C = \hat{p}_C = F^{-1}(\hat{m} + \hat{\tau}_C) = F^{-1}(-1.0853) = 0.1389$$

The deviances for the full and reduced models are:

$$D(\hat{m} + \hat{\tau}_i) = 0$$
$$D(\hat{m}) = 13.3550$$

The value of the *chi*-square statistic is:

$$\chi^2 = D(\hat{m}) - D(\hat{m} + \hat{\tau}_i) = 13.3550$$

For $(3 - 1) = 2$ degrees of freedom, the critical value $\chi^2_{0.05} = 5.991$. The difference in incidence of mastitis among the three farms is significant at the 5% level.

24.3.1 SAS Example for a Probit Model

The SAS program using a probit model for analyzing data from the example comparing incidence of mastitis on three farms is the following. Recall the data:

Farm	Total no. of cows	No. of cows with mastitis
A	96	36
B	132	29
C	72	10

SAS program:

```
DATA aa;
INPUT n y farm$;
DATALINES;
 96   36   A
132   29   B
 72   10   C
;
```

```
PROC GENMOD DATA = aa;
   CLASS farm;
   MODEL y/n = farm / DIST = BIN
            LINK = PROBIT
            TYPE1
            TYPE3
            PREDICTED;
   LSMEANS farm /DIFF CL;
RUN;
```

Explanation: The GENMOD procedure is used. The CLASS statement defines *farm* as a classification variable. The statement, *MODEL y/n = farm*, defines the dependent variable as a binomial proportion, y = the number of cows with mastitis, and n = the total number of cows on that particular farm. The independent variable is *farm*. The *DIST = BIN* option defines the distribution as binomial, and *LINK = PROBIT* denotes a probit model. The TYPE1 and TYPE3 statements direct calculation of sequential and partial tests using deviances from the full and reduced models. The PREDICTED statement requests predicted proportions for each farm. The LSMEANS statement produces parameter estimates for each farm.

SAS output:

Analysis Of Parameter Estimates

Parameter		DF	Estimate	Standard Error	Wald 95% Confidence Limits		Chi-Square	Pr>ChiSq
Intercept		1	-1.0853	0.1841	-1.4462	-0.7245	34.75	<.0001
farm	A	1	0.7667	0.2256	0.3246	1.2088	11.55	0.0007
farm	B	1	0.3121	0.2208	-0.1206	0.7448	2.00	0.1574
farm	C	0	0.0000	0.0000	0.0000	0.0000	.	.
Scale		0	1.0000	0.0000	1.0000	1.0000		

NOTE: The scale parameter was held fixed.

LR Statistics For Type 1 Analysis

Source	Deviance	DF	Chi-Square	Pr > ChiSq
Intercept	13.3550			
farm	0.0000	2	13.36	0.0013

LR Statistics For Type 3 Analysis

Source	DF	Chi-Square	Pr > ChiSq
farm	2	13.36	0.0013

Least Squares Means

Effect	farm	Estimate	Standard Error	DF	Chi-Square	Pr > ChiSq	Alpha
farm	A	-0.3186	0.1303	1	5.98	0.0145	0.05
farm	B	-0.7732	0.1218	1	40.30	<.0001	0.05
farm	C	-1.0853	0.1841	1	34.75	<.0001	0.05

Least Squares Means

Effect	farm	Confidence Limits	
farm	A	-0.5740	-0.0632
farm	B	-1.0120	-0.5345
farm	C	-1.4462	-0.7245

Differences of Least Squares Means

Effect	farm	_farm	Estimate	Standard Error	DF	Chi-Square	Pr>ChiSq	Alpha
farm	A	B	0.4546	0.1784	1	6.49	0.0108	0.05
farm	A	C	0.7667	0.2256	1	11.55	0.0007	0.05
farm	B	C	0.3121	0.2208	1	2.00	0.1574	0.05

Differences of Least Squares Means

Effect	farm	_farm	Confidence Limits	
farm	A	B	0.1050	0.8042
farm	A	C	0.3246	1.2088
farm	B	C	-0.1206	0.7448

Observation Statistics

Observation	y	n	Pred	Xbeta	Std	HessWgt
1	36	96	0.375	-0.318639	0.1303038	58.895987
2	29	132	0.219697	-0.773217	0.1218067	67.399613
3	10	72	0.1388889	-1.085325	0.1841074	29.502401

Explanation: The first table presents parameter estimates (*Analysis Of Parameter Estimates*). The parameter estimates are $\hat{m} = -1.0853$ (*Intercept*), $\hat{\tau}_A = 0.7667$ (*farm A*), and $\hat{\tau}_B = 0.3121$ (*farm B*), and $\hat{\tau}_C = 0.000$ (*farm C*).The extra-dispersion parameter (*Scale*) is 1, and the *Scaled Deviance* is equal to *Deviance*. (*NOTE: The scale parameter was held fixed.*) Next, the *Type1* and *Type3* tests of significance of the regression parameters are shown. Listed are: *Source* of variability, *Deviance*, degrees of freedom (*DF*), *Chi-Square* and *P* value (*Pr>ChiSq*). The deviance for the reduced model (*INTERCEPT*) is 13.3550. The deviance for the full model (*farm*) is 0. The *Chi-Square* value 13.36 is the difference of the deviances. Since the *P* value = 0.0013, H_0 is rejected, these data show evidence of an effect of farm on incidence of mastitis. The next table shows the *Least Squares Means* and corresponding analyses in probit values: *Estimate*, *Standard Errors*, degrees of freedom (*DF*), *Chi-Square*, *P* value (*Pr>ChiSq*), confidence level (*Alpha*) and *Confidence Limits*. In the next table are presented the *Difference of Least Squares Means*. This output is useful to test which farms are significantly different. The last table (*Observation Statistics*) shows predicted proportions of animals with mastitis (*Pred*) for each farm.

24.4 Log-linear Models

When a dependent variable is the number of units in some area or volume, classical linear regression is often not appropriate to test the effects of independent variables. A count variable usually does not have a normal distribution and the variance is often not homogeneous. To analyze such problems a log-linear model and Poisson distribution can be used.

The log-linear model is a generalized linear model with a logarithm function as a link function:

$$\eta = log(\mu)$$

The inverse link function is an exponential function. The mean is:

$$\mu = E(y) = e^{\eta}$$

Recall the Poisson distribution and its probability function:

$$p(y) = \frac{e^{-\lambda}\lambda^{y}}{y!}$$

where λ is the mean number of successes in a given time, volume or area, and e is the base of the natural logarithm ($e = 2.71828$).

A characteristic of a Poisson variable is that the expectation and variance are equal to the parameter λ:

$$\mu = Var(y) = \lambda$$

The log-linear model for a Poisson variable is:

$$log(\mu_i) = log(\lambda_i) = \mathbf{x}_i\boldsymbol{\beta}$$

where $\mathbf{x}_i\boldsymbol{\beta}$ is a linear combination of the vector of independent variables $\mathbf{x}_i$ and the corresponding vector of parameters $\boldsymbol{\beta}$. The independent variables can be continuous or categorical.

The variance function is:

$$V(\mu) = V(\lambda) = \lambda$$

Since the dispersion parameter is equal to one ($\phi^2 = 1$), the variance of a Poisson variable is equal to variance function:

$$Var(y) = V(\mu) = V(\lambda) = \lambda$$

The Poisson log-linear model takes into account heterogeneity of variance by defining that the variance depends on the mean. The mean can be expressed as an exponential function:

$$\mu_i = \lambda_i = e^{\mathbf{x}_i\boldsymbol{\beta}}$$

Similarly to logit models, the measure of the difference between the observed and estimated values is the deviance. The deviance for a Poisson variable is:

$$D = 2\sum_i\left\{y_i log\left(\frac{y_i}{\hat{\mu}_i}\right) + (y_i - \hat{\mu}_i)\right\}$$

To test if a particular parameter is needed in the model a *chi*-square distribution can be used. The difference of deviances between the full and reduced models has an approximate *chi*-square distribution with degrees of freedom equal to the difference in the number of parameters of the full and reduced models.

Similarly to binomial proportions, count data can sometimes have variance which differs from the theoretical variance $Var(y) = \lambda$. In these situations the dispersion parameter ϕ^2 differs from one and is often called an extra-dispersion parameter. The variance is:

$$Var(y) = V(\mu)\phi^2$$

The parameter ϕ^2 can be estimated from the data as the deviance (D) divided by degrees of freedom (df):

$$\hat{\phi}^2 = \frac{D}{df}$$

The degrees of freedom are defined similarly as the residual degrees of freedom in a linear model. The value $\phi^2 = 1$ indicates that the variance is consistent with the assumed distribution, $\phi^2 < 1$ indicates under-dispersion, and $\phi^2 > 1$ indicates over-dispersion from the assumed distribution. If the extra-dispersion parameter ϕ^2 is different than 1, the test must be adjusted by dividing the deviances by ϕ^2. Estimates do not depend on the parameter ϕ^2 and they need not be adjusted.

Example: The objective of this experiment was to test the difference of somatic cell counts in milk between the first, second and third lactations of dairy cows. Six cows each from the first, second and third lactations were randomly chosen from each of three farms. The somatic cell counts in thousand are shown in the following table:

Lactation	Farm		
	A	B	C
1	50 200	40 35	180 90
2	250 500	150 45	210 100
3	150 200	60 120	80 150

A log-linear model is assumed:

$$log(\lambda_{ij}) = m + \tau_i + \gamma_j$$

where:

λ_{ij} = the expected number of somatic cells
m = the overall mean on the logarithmic scale
τ_i = the effect of farm i, $i = A, B, C$
γ_j = the effect of lactation j, $j = 1, 2, 3$

Similarly as shown for linear models with categorical independent variables, there are no unique solutions for $\hat{m}$, $\hat{\tau}_i$ and $\hat{\gamma}_j$. For example, one set of solutions can be obtained by setting one of the $\hat{\tau}_i$ and one of the $\hat{\gamma}_j$ to zero:

$$\hat{m} = 4.7701$$
$$\hat{\tau}_A = 0.5108$$
$$\hat{\tau}_B = -0.5878$$
$$\hat{\tau}_C = 0.0000$$
$$\hat{\gamma}_1 = -0.2448$$
$$\hat{\gamma}_2 = 0.5016$$
$$\hat{\gamma}_3 = 0.0000$$

The estimates of the means are unique. For example, the estimate of the mean number of cells for the first lactation on farm B is:

$$\hat{\lambda}(\tau_B, \gamma_1) = e^{\hat{m}+\tau_B+\gamma_1} = e^{4.7701-0.5878-0.2448} = 51.290$$

The deviance for the full model is:

$$D(\hat{m} + \hat{\tau}_i + \hat{\gamma}_j) = 471.2148$$

The estimate of the extra-dispersion parameter ϕ^2 is:

$$\phi^2 = \frac{D(\hat{m} + \hat{\tau}_i + \hat{\gamma}_j)}{df} = \frac{471.2148}{13} = 36.2473$$

The degrees of freedom are defined similarly to the residual degrees of freedom for a linear model. Here degrees of freedom are defined as for the residual mean square in the two-way analysis of variance in an ordinary linear model: $df = n - a - b + 1$, where $n = 18 =$ the number of observations; $a = 3 =$ the number of farms; and $b = 3 =$ the number of lactations, yielding $df = 13$.

The effects of farm and lactation are tested using the differences of deviances adjusted for the estimate of the over-dispersion parameter. To test the effect of lactation the adjusted deviance of the full model is:

$$D^*(\hat{m} + \hat{\tau}_i + \hat{\gamma}_j) = \frac{D(\hat{m} + \hat{\tau}_i + \hat{\gamma}_j)}{\phi^2} = \frac{471.2148}{36.2473} = 13.0$$

The adjusted deviance of the reduced model is:

$$D^*(\hat{m} + \hat{\tau}_i) = \frac{D(\hat{m} + \hat{\tau}_i)}{\phi^2} = \frac{733.2884}{36.2473} = 20.23$$

The difference between the adjusted deviances is:

$$\chi^2 = D^*(\hat{m} + \hat{\tau}_i) - D^*(\hat{m} + \hat{\tau}_i + \hat{\gamma}_j) = 20.23 - 13.0 = 7.23$$

The critical value for 2 degrees of freedom is $\chi^2_{0.05,1} = 5.991$. The calculated difference is greater than the critical value; thus there is a significant effect of lactation on the somatic cell count.

24.4.1 SAS Example for a Log-linear Model

The SAS program for the example examining the effect of lactation on somatic cell count is as follows. The somatic cell counts are in thousands.

SAS program:

```
DATA cow;
   INPUT cow farm lact SCC@@;
   DATALINES;
  1  1 1    50     2 1 1 200      3 1 2 250
  4  1 2 500       5 1 3 150      6 1 3 200
  7  2 1    40     8 2 1   35     9 2 2 150
 10  2 2    45    11 2 3   60    12 2 3 120
 13  3 1 180      14 3 1   90    15 3 2 210
 16  3 2 100      17 3 3   80    18 3 3 150

   ;

PROC GENMOD DATA = cow;
CLASS  farm lact;
MODEL SCC = farm lact / DIST = POISSON
                        LINK = LOG
                        TYPE1
                        TYPE3
                        DSCALE PREDICTED;
LSMEANS farm lact / DIFF CL;
RUN;
```

Explanation: The GENMOD procedure is used. The CLASS statement defines *farm* and *lact* as categorical variables. The statement, *MODEL SCC = farm lact*, defines somatic cell count as the dependent variable, and *farm* and *lact* as independent variables. The *DIST = POISSON* option defines the distribution as Poisson, and *LINK = LOG* denotes that the link function is logarithmic. The TYPE1 and TYPE3 statements calculate sequential and partial tests using the deviances for the full and reduced models. The DSCALE option estimates the over-dispersion parameter. The PREDICTED statement produces an output of predicted proportions for each farm. The LSMEANS statement gives the parameter estimates for each farm.

SAS output:

Criteria For Assessing Goodness Of Fit

Criterion	DF	Value	Value/DF
Deviance	13	471.2148	36.2473
Scaled Deviance	13	13.0000	1.0000
Pearson Chi-Square	13	465.0043	35.7696
Scaled Pearson X2	13	12.8287	0.9868
Log Likelihood		296.5439	

Analysis Of Parameter Estimates

Parameter		DF	Estimate	Standard Error	Wald 95% Confidence Limits		Chi-Square	Pr>ChiSq
Intercept		1	4.7701	0.2803	4.2208	5.3194	289.65	<.0001
farm	1	1	0.5108	0.2676	-0.0136	1.0353	3.64	0.0563
farm	2	1	-0.5878	0.3540	-1.2816	0.1060	2.76	0.0968
farm	3	0	0.0000	0.0000	0.0000	0.0000	.	.
lact	1	1	-0.2448	0.3296	-0.8907	0.4012	0.55	0.4577
lact	2	1	0.5016	0.2767	-0.0408	1.0439	3.29	0.0699
lact	3	0	0.0000	0.0000	0.0000	0.0000	.	.
Scale		0	6.0206	0.0000	6.0206	6.0206		

NOTE: The scale parameter was estimated by the square root of DEVIANCE/DOF.

LR Statistics For Type 1 Analysis

Source	Deviance	Num DF	Den DF	F Value	Pr>F	Chi-Square	Pr > ChiSq
Intercept	1210.4938						
farm	733.2884	2	13	6.58	0.0106	13.17	0.0014
lact	471.2148	2	13	3.62	0.0564	7.23	0.0269

LR Statistics For Type 3 Analysis

Source	Num DF	Den DF	F Value	Pr > F	Chi-Square	Pr > ChiSq
farm	2	13	6.58	0.0106	13.17	0.0014
lact	2	13	3.62	0.0564	7.23	0.0269

Least Squares Means

Effect	farm	lact	Estim	Stand Error	DF	Chi-Square	Pr>ChiSq	Alpha	Conf Limits	
farm	1		5.3665	0.1680	1	1019.8	<.0001	0.05	5.0372	5.6959
farm	2		4.2679	0.2862	1	222.30	<.0001	0.05	3.7069	4.8290
farm	3		4.8557	0.2148	1	511.02	<.0001	0.05	4.4347	5.2767
lact		1	4.4997	0.2529	1	316.67	<.0001	0.05	4.0041	4.9953
lact		2	5.2460	0.1786	1	862.72	<.0001	0.05	4.8960	5.5961
lact		3	4.7444	0.2252	1	443.88	<.0001	0.05	4.3031	5.1858

Differences of Least Squares Means

Effect	farm	lact	_farm	_lact	Estimate	Stand Error	DF	Chi-Square	Pr>ChiSq	Alpha
farm	1		2		1.0986	0.3277	1	11.24	0.0008	0.05
farm	1		3		0.5108	0.2676	1	3.64	0.0563	0.05
farm	2		3		-0.5878	0.3540	1	2.76	0.0968	0.05
lact		1		2	-0.7463	0.2997	1	6.20	0.0128	0.05
lact		1		3	-0.2448	0.3296	1	0.55	0.4577	0.05
lact		2		3	0.5016	0.2767	1	3.29	0.0699	0.05

Differences of Least Squares Means

Effect	farm	lact	_farm	_lact	Confidence Limits	
farm	1		2		0.4563	1.7409
farm	1		3		-0.0136	1.0353
farm	2		3		-1.2816	0.1060
lact		1		2	-1.3337	-0.1590
lact		1		3	-0.8907	0.4012
lact		2		3	-0.0408	1.0439

Observation Statistics

Observation	SCC	Pred	Xbeta	Std	HessWgt
1	50	153.87931	5.0361686	0.2718121	4.2452636
2	200	153.87931	5.0361686	0.2718121	4.2452636
3	250	324.56897	5.782498	0.2045588	8.9542954
4	500	324.56897	5.782498	0.2045588	8.9542954
5	150	196.55172	5.2809256	0.246284	5.4225215
6	200	196.55172	5.2809256	0.246284	5.4225215
7	40	51.293104	3.9375563	0.3571855	1.4150879
8	35	51.293104	3.9375563	0.3571855	1.4150879
9	150	108.18966	4.6838858	0.3091019	2.9847651
10	45	108.18966	4.6838858	0.3091019	2.9847651
11	60	65.517241	4.1823133	0.3381649	1.8075072
12	120	65.517241	4.1823133	0.3381649	1.8075072
13	180	92.327586	4.525343	0.302955	2.5471582
14	90	92.327586	4.525343	0.302955	2.5471582
15	210	194.74138	5.2716724	0.2444263	5.3725773
16	100	194.74138	5.2716724	0.2444263	5.3725773
17	80	117.93103	4.7701	0.2802779	3.2535129
18	150	117.93103	4.7701	0.2802779	3.2535129

Explanation: The first table shows statistics describing correctness of the model. Several criteria are shown (*Criterion*), along with degrees of freedom (*DF*), *Value* and value divided by degrees of freedom (*Value/DF*). The *Deviance* = 471.214, and the scaled deviance on the extra-dispersion parameter is *Scaled Deviance* = 13.0. The next table presents parameter estimates (*Analysis of Parameter Estimates*). The extra-dispersion parameter (*Scale* = 6.0206) is expressed here as the square root of the deviance divided by the degrees of freedom. Next, the *Type1* and *Type3* tests of significance of regression parameters are shown including: *Source* of variability, *Deviance*, degrees of freedom (*DF*), *Chi-Square* and *P* value (*Pr>ChiSq*). The values of *Chi-Square* are for testing the difference of deviances corrected on the parameter of dispersion. There are significant effects of farms and lactations on somatic cell count, the *P* values (*Pr >ChiSq*) are 0.0014 and 0.0269. SAS also calculates *F* tests for farms and lactation by calculating the *F* values as the difference of deviances divided by their corresponding degrees of freedom. In the table are degrees of freedom for the numerator and denominator (*Num DF* and *Den DF*), *F Value* and *P* value (*Pr > F*). The next table shows *Least Squares Means* and corresponding analyses in logit values: *Estimate*, *Standard Errors*, degrees of freedom (*DF*), *Chi-Square*, *P* value (*Pr>ChiSq*), confidence level (*Alpha*) and *Confidence Limits*. The next table presents *Difference of Least Squares Means*. This output is useful to determine significant differences among farms and lactations. For example, there is a significant difference between the first and second lactations with a *P* value (*Pr > ChiSq*) = 0.0128. The last table (*Observation Statistics*) shows predicted proportions (*Pred*) for each combination of farm and lactation among other statistics. For example the estimated number of somatic cells in the first lactation on farm 2 is equal to 51.293104.

Solutions of Exercises

1.1. Mean = 26.625; Variance = 3.625; Standard deviation = 1.9039; Coefficient of variation = 7.15%; Median = 26; Mode = 26

1.2. Variance = 22.6207

1.3. The number of observations = 46; Mean = 20.0869; Variance = 12.6145; Standard deviation = 3.5517; Coefficient of variation = 17.68%

1.4. The number of observations = 17; Mean = 28.00; Variance = 31.3750; Standard deviation = 5.6013; Coefficient of variation = 20.0%

2.1. a) $^2/_3$; b) $^1/_3$; c) $^5/_{12}$; d) $^{11}/_{12}$; e) $^7/_{12}$

3.1. a) 0.10292; b) 0.38278

3.2. Ordinate = 0.22988

3.3. a) 0.5 b) 0.025921; c) 0.10133; d) 184.524; e) 211.664

3.4. a) 52; b) 10; c) 67; d) 16.9; e) 300; f) 360

3.5. a) 0.36944; b) 0.63055; c) 0.88604; d) 4.30235; e) 4.48133

5.1. (26.0161; 27.2339)

5.2. (19.0322; 21.1417)

5.3. (25.1200572; 30.8799)

6.1. $z = -1.7678$; P value = 0.0833

6.2. $t = -2.0202$, degrees of freedom = 16; P value = 0.0605

6.3. $t = -6.504$

6.4. Chi-square = 21.049; P value = 0.0008

6.5. Chi-square = 7.50; P value = 0.0062

6.6. $z = 2.582$

6.7. $z = 3.015$

7.1. $b_0 = 25.4286$; $b_1 = 8.5714$; $F = 12.384$; P value = 0.0079; $R^2 = 0.6075$

7.2. $b_0 = 1.2959$; $b_1 = 0.334014$; $F = 8.318$; P value = 0.0279; $R^2 = 0.5809$

7.3. a) *the origin between years 1985 and 1986; b) $b_0 = 93.917$; $b_1 = -1.470$; c) expected number of horses in 2002 year is 74.803

8.1. $r = 0.97982$, P value < 0.001

8.2. $r = 0.65$; $t = 3.084$; P value = 0.0081

11.1. $MS_{TRT} = 41.68889$; $MS_{RES} = 9.461$; $F = 4.41$; P value = 0.0137

11.2. $MS_{TRT} = 28.1575$; $MS_{RES} = 3.2742$; $F = 8.60$; P value = 0.0082

12.1. $\sigma^2 + 20\,\sigma^2_\tau = 1050.5$; $\sigma^2 = 210$; intraclass correlation = 0.8334

15.1. $MS_{TRT} = 26.6667$; $MS_{BLOCK} = 3.125$; $MS_{RES} = 1.7917$; F for treatment = 14.88; P value = 0.0002

16.1.

Source	df	SS	MS	F	P value
QUAD	2	1.81555556	0.90777778	0.42	0.6658
SOW(QUAD)	6	22.21111111	3.70185185	1.73	0.2120
PERIOD(QUAD)	6	2.31777778	0.38629630	0.18	0.9759
TRT	2	4.74000000	2.37000000	1.11	0.3681

17.1.

Source	df	SS	MS	F	*P* value
PROT	2	41.37500000	20.68750000	1.95	0.1544
ENERG	1	154.08333333	154.08333333	14.55	0.0004
PROT*ENERG	2	61.79166667	30.89583333	2.92	0.0651
Residual	42	444.75000000	10.58928571		

20.1.

Source	df num	df den	F	P value
Grass	1	2	9.82	0.0924
Density	1	4	73.36	0.0033
Grass x Density	1	4	0.11	0.7617

Appendix A: Vectors and Matrices

A matrix is a collection of elements that are organized in rows and columns according to some criteria. Examples of two matrices, $\mathbf{A}$ and $\mathbf{B}$, follow:

$$\mathbf{A} = \begin{bmatrix} a_{11} & a_{12} \\ a_{21} & a_{22} \\ a_{31} & a_{32} \end{bmatrix}_{3\times2} = \begin{bmatrix} 1 & 3 \\ 1 & 1 \\ 2 & -1 \end{bmatrix}_{3\times2} \qquad \mathbf{B} = \begin{bmatrix} b_{11} & b_{12} \\ b_{21} & b_{22} \\ b_{31} & b_{32} \end{bmatrix}_{3\times2} = \begin{bmatrix} 2 & 1 \\ 1 & 3 \\ 1 & 2 \end{bmatrix}_{3\times2}$$

The symbols a_{11}, a_{12}, etc., denote the row and column position of the element. An element a_{ij}, is in the i-th row and j-th column.

A matrix defined with only one column or one row is called a vector. For example, a vector $\mathbf{b}$ is:

$$\mathbf{b} = \begin{bmatrix} 1 \\ 2 \end{bmatrix}_{2\times1}$$

Types and Properties of Matrices

A *square matrix* is a matrix that has equal numbers of rows and columns. The symmetric matrix is a square matrix with $a_{ij} = a_{ji}$. For example, the matrix $\mathbf{C}$ is a symmetric matrix because the element in the second row and first column is equal to the element in the first row and second column:

$$\mathbf{C} = \begin{bmatrix} 2 & 1 \\ 1 & 3 \end{bmatrix}_{2\times2}$$

A *diagonal matrix* is a square matrix with $a_{ij} = 0$ for each $i \neq j$.

$$\mathbf{D} = \begin{bmatrix} 2 & 0 \\ 0 & 4 \end{bmatrix}_{2\times2}$$

An *identity matrix* is a diagonal matrix with $a_{ii} = 1$.

$$\mathbf{I}_2 = \begin{bmatrix} 1 & 0 \\ 0 & 1 \end{bmatrix}, \quad \mathbf{I}_3 = \begin{bmatrix} 1 & 0 & 0 \\ 0 & 1 & 0 \\ 0 & 0 & 1 \end{bmatrix}$$

A *null matrix* is a matrix with all elements equal to zero. A *null vector* is a vector with all elements equal to zero.

$$\mathbf{0} = \begin{bmatrix} 0 & 0 \\ 0 & 0 \end{bmatrix}, \quad \mathbf{0} = \begin{bmatrix} 0 \\ 0 \\ 0 \end{bmatrix}$$

A matrix with all elements equal to 1 is usually denoted with **J**. A vector with all elements equal to 1, is usually denoted with **1**.

$$\mathbf{J} = \begin{bmatrix} 1 & 1 \\ 1 & 1 \end{bmatrix}, \quad \mathbf{1} = \begin{bmatrix} 1 \\ 1 \\ 1 \end{bmatrix}$$

The transpose matrix of a matrix **A**, denoted by **A'**, is obtained by interchanging columns and rows of the matrix **A**. For example, if:

$$\mathbf{A} = \begin{bmatrix} 1 & 3 \\ 1 & 1 \\ 2 & -1 \end{bmatrix}_{3 \times 2} \quad \text{then} \quad \mathbf{A'} = \begin{bmatrix} 1 & 1 & 2 \\ 3 & 1 & -1 \end{bmatrix}_{2 \times 3}$$

The *rank* of a matrix is the number of linearly independent columns or rows. Columns (rows) are linearly dependent if some columns (rows) can be expressed as linear combinations of some other columns (rows). The rank determined by columns is equal to the rank determined by rows.

Example: The matrix $\begin{bmatrix} 1 & -2 & 3 \\ 3 & 1 & 2 \\ 5 & 4 & 1 \end{bmatrix}$ has a rank of two because the number of linearly

independent columns is two. Any column can be presented as the linear combination of the other two columns, that is, only two columns are needed to give the same information as all three columns. For example, the first column is the sum of the second and third columns:

$$\begin{bmatrix} 1 \\ 3 \\ 5 \end{bmatrix} = \begin{bmatrix} -2 \\ 1 \\ 4 \end{bmatrix} + \begin{bmatrix} 3 \\ 2 \\ 1 \end{bmatrix}$$

Also, there are only two independent rows. For example the first row can be expressed as the second row multiplied by 2 minus the third row:

$$\begin{bmatrix} 1 & -2 & 3 \end{bmatrix} = 2\begin{bmatrix} 3 & 1 & 2 \end{bmatrix} - \begin{bmatrix} 5 & 4 & 1 \end{bmatrix}$$

Thus, the rank of the matrix equals two.

Matrix and Vector Operations

A matrix is not only a collection of numbers, but numerical operations are also defined on matrices. *Addition of matrices* is defined such that corresponding elements are added:

$$\mathbf{A} + \mathbf{B} = \begin{bmatrix} a_{11} + b_{11} & a_{12} + b_{12} \\ a_{21} + b_{21} & a_{22} + b_{22} \\ a_{31} + b_{31} & a_{32} + b_{32} \end{bmatrix}$$

For example, if:

$$A = \begin{bmatrix} 1 & 3 \\ 1 & 1 \\ 2 & -1 \end{bmatrix}_{3\times2} \quad \text{and} \quad B = \begin{bmatrix} 2 & 1 \\ 1 & 3 \\ 1 & 2 \end{bmatrix}_{3\times2}, \quad \text{then} \quad A + B = \begin{bmatrix} 1+2 & 3+1 \\ 1+1 & 1+3 \\ 2+1 & -1+2 \end{bmatrix} = \begin{bmatrix} 3 & 4 \\ 2 & 4 \\ 3 & 1 \end{bmatrix}_{3\times2}$$

Matrix multiplication with a number is defined such that each matrix element is multiplied by that number. For example, if:

$$A = \begin{bmatrix} 1 & 3 \\ 1 & 1 \\ 2 & -1 \end{bmatrix}_{3\times2} \quad \text{then} \quad 2A = \begin{bmatrix} 2 & 6 \\ 2 & 2 \\ 4 & -2 \end{bmatrix}_{3\times2}$$

The multiplication of two matrices is possible only if the number of columns of the first (left) matrix is equal to the number of rows of the second (right) matrix. Generally, if a matrix A has dimension $r \times c$, and a matrix B has dimension $c \times s$, then the product AB is a matrix with dimension $r \times s$ and its element in the i-th row and j-th column is defined as:

$$\sum_{k=1}^{c} a_{ik} b_{kj}$$

Example: Let $A = \begin{bmatrix} a_{11} & a_{12} \\ a_{21} & a_{22} \\ a_{31} & a_{32} \end{bmatrix}_{3\times2} = \begin{bmatrix} 1 & 3 \\ 1 & 1 \\ 2 & -1 \end{bmatrix}_{3\times2}$ and $C = \begin{bmatrix} c_{11} & c_{12} \\ c_{21} & c_{22} \end{bmatrix}_{2\times2} = \begin{bmatrix} 2 & 1 \\ 1 & 2 \end{bmatrix}_{2\times2}$

Calculate AC.

$$AC = \begin{bmatrix} a_{11}*c_{11}+a_{12}*c_{21} & a_{11}*c_{21}+a_{12}*c_{22} \\ a_{21}*c_{11}+a_{22}*c_{21} & a_{21}*c_{21}+a_{22}*c_{22} \\ a_{31}*c_{11}+a_{32}*c_{21} & a_{31}*c_{21}+a_{32}*c_{22} \end{bmatrix} = \begin{bmatrix} 1*2+3*1 & 1*2+3*2 \\ 1*2+1*1 & 1*1+1*2 \\ 2*2-1*1 & 2*1-1*2 \end{bmatrix} = \begin{bmatrix} 5 & 7 \\ 3 & 3 \\ 3 & 0 \end{bmatrix}_{3\times2}$$

Example 2:

Let $b = \begin{bmatrix} 1 \\ 2 \end{bmatrix}_{2\times1}$. Calculate Ab.

$$Ab = \begin{bmatrix} 1*1+3*2 \\ 1*1+1*2 \\ 2*2-1*1 \end{bmatrix}_{3\times1} = \begin{bmatrix} 7 \\ 3 \\ 0 \end{bmatrix}_{3\times1}$$

The product of the transpose of a vector and the vector itself is known as a quadratic form and denotes the sum of squares of the vector elements. If y is a vector:

$$y = \begin{bmatrix} y_1 \\ y_2 \\ ... \\ y_n \end{bmatrix}_{n\times1}$$

The quadratic form is:

$$\mathbf{y'y} = \begin{bmatrix} y_1 & y_2 & \cdots & y_n \end{bmatrix} \begin{bmatrix} y_1 \\ y_2 \\ \cdots \\ y_n \end{bmatrix} = \sum_i y_i^2$$

A *trace* of a matrix is the sum of the diagonal elements of the matrix. For example:

$$\mathbf{D} = \begin{bmatrix} 2 & 4 & 2 \\ 1 & 5 & 4 \\ 3 & 4 & 11 \end{bmatrix}$$

then the trace is $tr(\mathbf{D}) = 2 + 5 + 11 = 18$

The inverse of some square matrix $\mathbf{C}$ is a matrix $\mathbf{C}^{-1}$ such that $\mathbf{C}^{-1}\mathbf{C} = \mathbf{I}$ and $\mathbf{CC}^{-1} = \mathbf{I}$, that is, the product of a matrix with its inverse is equal to the identity matrix. A matrix has an inverse if its rows and columns are linearly independent.

A *generalized inverse* of some matrix $\mathbf{C}$ is the matrix $\mathbf{C}^-$ such that $\mathbf{CC}^-\mathbf{C} = \mathbf{C}$. Any matrix, even a nonsquare matrix with linearly dependent rows or columns, has a generalized inverse. Generally, $\mathbf{CC}^-$ or $\mathbf{C}^-\mathbf{C}$ is not equal to identity matrix $\mathbf{I}$, unless $\mathbf{C}^- = \mathbf{C}^{-1}$.

A *system of linear equations* can be expressed and solved using matrices. For example, the system of equations with two unknowns:

$$\begin{array}{ccccc} 2a_1 & + & a_2 & = & 5 \\ a_1 & - & a_2 & = & 1 \end{array}$$

$$\mathbf{a} = \begin{bmatrix} a_1 \\ a_2 \end{bmatrix} \qquad \mathbf{X} = \begin{bmatrix} 2 & 1 \\ 1 & -1 \end{bmatrix} \qquad \mathbf{y} = \begin{bmatrix} 5 \\ 1 \end{bmatrix}$$

$\mathbf{Xa} = \mathbf{y}$ by multiplication of the left and right sides with $\mathbf{X}^{-1}$
$\mathbf{X}^{-1}\mathbf{Xa} = \mathbf{X}^{-1}\mathbf{y}$ because $\mathbf{X}^{-1}\mathbf{X} = \mathbf{I}$
$\mathbf{a} = \mathbf{X}^{-1}\mathbf{y}$

$$\begin{bmatrix} a_1 \\ a_2 \end{bmatrix} = \begin{bmatrix} 2 & 1 \\ 1 & -1 \end{bmatrix}^{-1} \begin{bmatrix} 5 \\ 1 \end{bmatrix} = \begin{bmatrix} 1/3 & 1/3 \\ 1/3 & -2/3 \end{bmatrix} \begin{bmatrix} 5 \\ 1 \end{bmatrix} = \begin{bmatrix} 2 \\ 1 \end{bmatrix}$$

Normal equations are defined as:

$$\mathbf{X'Xa} = \mathbf{X'y}$$

Multiplying both sides with $(\mathbf{X'X})^{-1}$ the solution of $\mathbf{a}$ is:

$$\mathbf{a} = (\mathbf{X'X})^{-1}\mathbf{X'y}$$

The normal equations are useful for solving a system of equations when the number of equations is greater than the number of unknowns.

Appendix B: Statistical Tables

Area under the Standard Normal Curve, $z > z_\alpha$

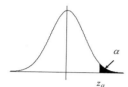

z_α	0.00	0.01	0.02	0.03	0.04	0.05	0.06	0.07	0.08	0.09
0.0	0.5000	0.4960	0.4920	0.4880	0.4840	0.4801	0.4761	0.4721	0.4681	0.4641
0.1	0.4602	0.4562	0.4522	0.4483	0.4443	0.4404	0.4364	0.4325	0.4286	0.4247
0.2	0.4207	0.4168	0.4129	0.4090	0.4052	0.4013	0.3974	0.3936	0.3897	0.3859
0.3	0.3821	0.3783	0.3745	0.3707	0.3669	0.3632	0.3594	0.3557	0.3520	0.3483
0.4	0.3446	0.3409	0.3372	0.3336	0.3300	0.3264	0.3228	0.3192	0.3156	0.3121
0.5	0.3085	0.3050	0.3015	0.2981	0.2946	0.2912	0.2877	0.2843	0.2810	0.2776
0.6	0.2743	0.2709	0.2676	0.2643	0.2611	0.2578	0.2546	0.2514	0.2483	0.2451
0.7	0.2420	0.2389	0.2358	0.2327	0.2296	0.2266	0.2236	0.2206	0.2177	0.2148
0.8	0.2119	0.2090	0.2061	0.2033	0.2005	0.1977	0.1949	0.1922	0.1894	0.1867
0.9	0.1841	0.1814	0.1788	0.1762	0.1736	0.1711	0.1685	0.1660	0.1635	0.1611
1.0	0.1587	0.1562	0.1539	0.1515	0.1492	0.1469	0.1446	0.1423	0.1401	0.1379
1.1	0.1357	0.1335	0.1314	0.1292	0.1271	0.1251	0.1230	0.1210	0.1190	0.1170
1.2	0.1151	0.1131	0.1112	0.1093	0.1075	0.1056	0.1038	0.1020	0.1003	0.0985
1.3	0.0968	0.0951	0.0934	0.0918	0.0901	0.0885	0.0869	0.0853	0.0838	0.0823
1.4	0.0808	0.0793	0.0778	0.0764	0.0749	0.0735	0.0721	0.0708	0.0694	0.0681
1.5	0.0668	0.0655	0.0643	0.0630	0.0618	0.0606	0.0594	0.0582	0.0571	0.0559
1.6	0.0548	0.0537	0.0526	0.0516	0.0505	0.0495	0.0485	0.0475	0.0465	0.0455
1.7	0.0446	0.0436	0.0427	0.0418	0.0409	0.0401	0.0392	0.0384	0.0375	0.0367
1.8	0.0359	0.0351	0.0344	0.0336	0.0329	0.0322	0.0314	0.0307	0.0301	0.0294
1.9	0.0287	0.0281	0.0274	0.0268	0.0262	0.0256	0.0250	0.0244	0.0239	0.0233
2.0	0.0228	0.0222	0.0217	0.0212	0.0207	0.0202	0.0197	0.0192	0.0188	0.0183
2.1	0.0179	0.0174	0.0170	0.0166	0.0162	0.0158	0.0154	0.0150	0.0146	0.0143
2.2	0.0139	0.0136	0.0132	0.0129	0.0125	0.0122	0.0119	0.0116	0.0113	0.0110
2.3	0.0107	0.0104	0.0102	0.0099	0.0096	0.0094	0.0091	0.0089	0.0087	0.0084
2.4	0.0082	0.0080	0.0078	0.0075	0.0073	0.0071	0.0069	0.0068	0.0066	0.0064
2.5	0.0062	0.0060	0.0059	0.0057	0.0055	0.0054	0.0052	0.0051	0.0049	0.0048
2.6	0.0047	0.0045	0.0044	0.0043	0.0041	0.0040	0.0039	0.0038	0.0037	0.0036
2.7	0.0035	0.0034	0.0033	0.0032	0.0031	0.0030	0.0029	0.0028	0.0027	0.0026
2.8	0.0026	0.0025	0.0024	0.0023	0.0023	0.0022	0.0021	0.0021	0.0020	0.0019
2.9	0.0019	0.0018	0.0018	0.0017	0.0016	0.0016	0.0015	0.0015	0.0014	0.0014
3.0	0.0013	0.0013	0.0013	0.0012	0.0012	0.0011	0.0011	0.0011	0.0010	0.0010
3.1	0.0010	0.0009	0.0009	0.0009	0.0008	0.0008	0.0008	0.0008	0.0007	0.0007
3.2	0.0007	0.0007	0.0006	0.0006	0.0006	0.0006	0.0006	0.0005	0.0005	0.0005
3.3	0.0005	0.0005	0.0005	0.0004	0.0004	0.0004	0.0004	0.0004	0.0004	0.0003
3.4	0.0003	0.0003	0.0003	0.0003	0.0003	0.0003	0.0003	0.0003	0.0003	0.0002

Critical Values of Student t Distributions, $t > t_\alpha$

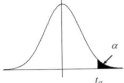

Degrees of freedom	$t_{0.1}$	$t_{0.05}$	$t_{0.025}$	$t_{0.01}$	$t_{0.005}$	$t_{0.001}$
1	3.078	6.314	12.706	31.821	63.656	318.289
2	1.886	2.920	4.303	6.965	9.925	22.328
3	1.638	2.353	3.182	4.541	5.841	10.214
4	1.533	2.132	2.776	3.747	4.604	7.173
5	1.476	2.015	2.571	3.365	4.032	5.894
6	1.440	1.943	2.447	3.143	3.707	5.208
7	1.415	1.895	2.365	2.998	3.499	4.785
8	1.397	1.860	2.306	2.896	3.355	4.501
9	1.383	1.833	2.262	2.821	3.250	4.297
10	1.372	1.812	2.228	2.764	3.169	4.144
11	1.363	1.796	2.201	2.718	3.106	4.025
12	1.356	1.782	2.179	2.681	3.055	3.930
13	1.350	1.771	2.160	2.650	3.012	3.852
14	1.345	1.761	2.145	2.624	2.977	3.787
15	1.341	1.753	2.131	2.602	2.947	3.733
16	1.337	1.746	2.120	2.583	2.921	3.686
17	1.333	1.740	2.110	2.567	2.898	3.646
18	1.330	1.734	2.101	2.552	2.878	3.610
19	1.328	1.729	2.093	2.539	2.861	3.579
20	1.325	1.725	2.086	2.528	2.845	3.552

Critical Values of Student t Distributions, $t > t_\alpha$(cont...)

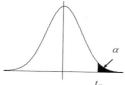

Degrees of freedom	$t_{0.1}$	$t_{0.05}$	$t_{0.025}$	$t_{0.01}$	$t_{0.005}$	$t_{0.001}$
21	1.323	1.721	2.080	2.518	2.831	3.527
22	1.321	1.717	2.074	2.508	2.819	3.505
23	1.319	1.714	2.069	2.500	2.807	3.485
24	1.318	1.711	2.064	2.492	2.797	3.467
25	1.316	1.708	2.060	2.485	2.787	3.450
26	1.315	1.706	2.056	2.479	2.779	3.435
27	1.314	1.703	2.052	2.473	2.771	3.421
28	1.313	1.701	2.048	2.467	2.763	3.408
29	1.311	1.699	2.045	2.462	2.756	3.396
30	1.310	1.697	2.042	2.457	2.750	3.385
40	1.303	1.684	2.021	2.423	2.704	3.307
50	1.299	1.676	2.009	2.403	2.678	3.261
60	1.296	1.671	2.000	2.390	2.660	3.232
120	1.289	1.658	1.980	2.358	2.617	3.160
∞	1.282	1.645	1.960	2.326	2.576	3.090

Critical Values of *Chi*-square Distributions, $\chi^2 > \chi^2_\alpha$

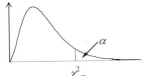

Degrees of freedom	$\chi^2_{0.1}$	$\chi^2_{0.05}$	$\chi^2_{0.025}$	$\chi^2_{0.01}$	$\chi^2_{0.005}$	$\chi^2_{0.001}$
1	2.706	3.841	5.024	6.635	7.879	10.827
2	4.605	5.991	7.378	9.210	10.597	13.815
3	6.251	7.815	9.348	11.345	12.838	16.266
4	7.779	9.488	11.143	13.277	14.860	18.466
5	9.236	11.070	12.832	15.086	16.750	20.515
6	10.645	12.592	14.449	16.812	18.548	22.457
7	12.017	14.067	16.013	18.475	20.278	24.321
8	13.362	15.507	17.535	20.090	21.955	26.124
9	14.684	16.919	19.023	21.666	23.589	27.877
10	15.987	18.307	20.483	23.209	25.188	29.588
11	17.275	19.675	21.920	24.725	26.757	31.264
12	18.549	21.026	23.337	26.217	28.300	32.909
13	19.812	22.362	24.736	27.688	29.819	34.527
14	21.064	23.685	26.119	29.141	31.319	36.124
15	22.307	24.996	27.488	30.578	32.801	37.698
16	23.542	26.296	28.845	32.000	34.267	39.252
17	24.769	27.587	30.191	33.409	35.718	40.791
18	25.989	28.869	31.526	34.805	37.156	42.312
19	27.204	30.144	32.852	36.191	38.582	43.819
20	28.412	31.410	34.170	37.566	39.997	45.314

Critical Values of *Chi*-square Distributions, $\chi^2 > \chi^2{}_\alpha$(cont...)

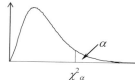

Degrees of freedom	$\chi^2{}_{0.1}$	$\chi^2{}_{0.05}$	$\chi^2{}_{0.025}$	$\chi^2{}_{0.01}$	$\chi^2{}_{0.005}$	$\chi^2{}_{0.001}$
21	29.615	32.671	35.479	38.932	41.401	46.796
22	30.813	33.924	36.781	40.289	42.796	48.268
23	32.007	35.172	38.076	41.638	44.181	49.728
24	33.196	36.415	39.364	42.980	45.558	51.179
25	34.382	37.652	40.646	44.314	46.928	52.619
26	35.563	38.885	41.923	45.642	48.290	54.051
27	36.741	40.113	43.195	46.963	49.645	55.475
28	37.916	41.337	44.461	48.278	50.994	56.892
29	39.087	42.557	45.722	49.588	52.335	58.301
30	40.256	43.773	46.979	50.892	53.672	59.702
40	51.805	55.758	59.342	63.691	66.766	73.403
50	63.167	67.505	71.420	76.154	79.490	86.660
60	74.397	79.082	83.298	88.379	91.952	99.608
70	85.527	90.531	95.023	100.425	104.215	112.317
80	96.578	101.879	106.629	112.329	116.321	124.839
90	107.565	113.145	118.136	124.116	128.299	137.208
100	118.498	124.342	129.561	135.807	140.170	149.449

Critical Values of *F* Distributions, $F > F_\alpha$, $\alpha = 0.05$

F_α

		Numerator degrees of freedom							
		1	2	3	4	5	6	7	8
Denominator degrees of freedom	1	161.45	199.50	215.71	224.58	230.16	233.99	236.77	238.88
	2	18.51	19.00	19.16	19.25	19.30	19.33	19.35	19.37
	3	10.13	9.55	9.28	9.12	9.01	8.94	8.89	8.85
	4	7.71	6.94	6.59	6.39	6.26	6.16	6.09	6.04
	5	6.61	5.79	5.41	5.19	5.05	4.95	4.88	4.82
	6	5.99	5.14	4.76	4.53	4.39	4.28	4.21	4.15
	7	5.59	4.74	4.35	4.12	3.97	3.87	3.79	3.73
	8	5.32	4.46	4.07	3.84	3.69	3.58	3.50	3.44
	9	5.12	4.26	3.86	3.63	3.48	3.37	3.29	3.23
	10	4.96	4.10	3.71	3.48	3.33	3.22	3.14	3.07
	11	4.84	3.98	3.59	3.36	3.20	3.09	3.01	2.95
	12	4.75	3.89	3.49	3.26	3.11	3.00	2.91	2.85
	13	4.67	3.81	3.41	3.18	3.03	2.92	2.83	2.77
	14	4.60	3.74	3.34	3.11	2.96	2.85	2.76	2.70
	15	4.54	3.68	3.29	3.06	2.90	2.79	2.71	2.64
	16	4.49	3.63	3.24	3.01	2.85	2.74	2.66	2.59
	17	4.45	3.59	3.20	2.96	2.81	2.70	2.61	2.55
	18	4.41	3.55	3.16	2.93	2.77	2.66	2.58	2.51
	19	4.38	3.52	3.13	2.90	2.74	2.63	2.54	2.48
	20	4.35	3.49	3.10	2.87	2.71	2.60	2.51	2.45
	21	4.32	3.47	3.07	2.84	2.68	2.57	2.49	2.42
	22	4.30	3.44	3.05	2.82	2.66	2.55	2.46	2.40
	23	4.28	3.42	3.03	2.80	2.64	2.53	2.44	2.37
	24	4.26	3.40	3.01	2.78	2.62	2.51	2.42	2.36
	25	4.24	3.39	2.99	2.76	2.60	2.49	2.40	2.34
	26	4.23	3.37	2.98	2.74	2.59	2.47	2.39	2.32
	27	4.21	3.35	2.96	2.73	2.57	2.46	2.37	2.31
	28	4.20	3.34	2.95	2.71	2.56	2.45	2.36	2.29
	29	4.18	3.33	2.93	2.70	2.55	2.43	2.35	2.28
	30	4.17	3.32	2.92	2.69	2.53	2.42	2.33	2.27
	40	4.08	3.23	2.84	2.61	2.45	2.34	2.25	2.18
	50	4.03	3.18	2.79	2.56	2.40	2.29	2.20	2.13
	60	4.00	3.15	2.76	2.53	2.37	2.25	2.17	2.10
	70	3.98	3.13	2.74	2.50	2.35	2.23	2.14	2.07
	80	3.96	3.11	2.72	2.49	2.33	2.21	2.13	2.06
	90	3.95	3.10	2.71	2.47	2.32	2.20	2.11	2.04
	100	3.94	3.09	2.70	2.46	2.31	2.19	2.10	2.03
	120	3.92	3.07	2.68	2.45	2.29	2.18	2.09	2.02

Critical Values of *F* Distributions, $F > F_\alpha$, $\alpha = 0.05$ (cont...)

	Numerator degrees of freedom								
	9	10	12	15	20	24	30	60	120
1	240.54	241.88	243.90	245.95	248.02	249.05	250.10	252.20	253.25
2	19.38	19.40	19.41	19.43	19.45	19.45	19.46	19.48	19.49
3	8.81	8.79	8.74	8.70	8.66	8.64	8.62	8.57	8.55
4	6.00	5.96	5.91	5.86	5.80	5.77	5.75	5.69	5.66
5	4.77	4.74	4.68	4.62	4.56	4.53	4.50	4.43	4.40
6	4.10	4.06	4.00	3.94	3.87	3.84	3.81	3.74	3.70
7	3.68	3.64	3.57	3.51	3.44	3.41	3.38	3.30	3.27
8	3.39	3.35	3.28	3.22	3.15	3.12	3.08	3.01	2.97
9	3.18	3.14	3.07	3.01	2.94	2.90	2.86	2.79	2.75
10	3.02	2.98	2.91	2.85	2.77	2.74	2.70	2.62	2.58
11	2.90	2.85	2.79	2.72	2.65	2.61	2.57	2.49	2.45
12	2.80	2.75	2.69	2.62	2.54	2.51	2.47	2.38	2.34
13	2.71	2.67	2.60	2.53	2.46	2.42	2.38	2.30	2.25
14	2.65	2.60	2.53	2.46	2.39	2.35	2.31	2.22	2.18
15	2.59	2.54	2.48	2.40	2.33	2.29	2.25	2.16	2.11
16	2.54	2.49	2.42	2.35	2.28	2.24	2.19	2.11	2.06
17	2.49	2.45	2.38	2.31	2.23	2.19	2.15	2.06	2.01
18	2.46	2.41	2.34	2.27	2.19	2.15	2.11	2.02	1.97
19	2.42	2.38	2.31	2.23	2.16	2.11	2.07	1.98	1.93
20	2.39	2.35	2.28	2.20	2.12	2.08	2.04	1.95	1.90
21	2.37	2.32	2.25	2.18	2.10	2.05	2.01	1.92	1.87
22	2.34	2.30	2.23	2.15	2.07	2.03	1.98	1.89	1.84
23	2.32	2.27	2.20	2.13	2.05	2.01	1.96	1.86	1.81
24	2.30	2.25	2.18	2.11	2.03	1.98	1.94	1.84	1.79
25	2.28	2.24	2.16	2.09	2.01	1.96	1.92	1.82	1.77
26	2.27	2.22	2.15	2.07	1.99	1.95	1.90	1.80	1.75
27	2.25	2.20	2.13	2.06	1.97	1.93	1.88	1.79	1.73
28	2.24	2.19	2.12	2.04	1.96	1.91	1.87	1.77	1.71
29	2.22	2.18	2.10	2.03	1.94	1.90	1.85	1.75	1.70
30	2.21	2.16	2.09	2.01	1.93	1.89	1.84	1.74	1.68
40	2.12	2.08	2.00	1.92	1.84	1.79	1.74	1.64	1.58
50	2.07	2.03	1.95	1.87	1.78	1.74	1.69	1.58	1.51
60	2.04	1.99	1.92	1.84	1.75	1.70	1.65	1.53	1.47
70	2.02	1.97	1.89	1.81	1.72	1.67	1.62	1.50	1.44
80	2.00	1.95	1.88	1.79	1.70	1.65	1.60	1.48	1.41
90	1.99	1.94	1.86	1.78	1.69	1.64	1.59	1.46	1.39
100	1.97	1.93	1.85	1.77	1.68	1.63	1.57	1.45	1.38
120	1.96	1.91	1.83	1.75	1.66	1.61	1.55	1.43	1.35

Denominator degrees of freedom

Critical Value of *F* Distributions, $F > F_\alpha$, $\alpha = 0.01$

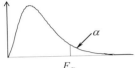

F_α

	Numerator degrees of freedom							
	1	2	3	4	5	6	7	8
1	4052.18	4999.34	5403.53	5624.26	5763.96	5858.95	5928.33	5980.95
2	98.50	99.00	99.16	99.25	99.30	99.33	99.36	99.38
3	34.12	30.82	29.46	28.71	28.24	27.91	27.67	27.49
4	21.20	18.00	16.69	15.98	15.52	15.21	14.98	14.80
5	16.26	13.27	12.06	11.39	10.97	10.67	10.46	10.29
6	13.75	10.92	9.78	9.15	8.75	8.47	8.26	8.10
7	12.25	9.55	8.45	7.85	7.46	7.19	6.99	6.84
8	11.26	8.65	7.59	7.01	6.63	6.37	6.18	6.03
9	10.56	8.02	6.99	6.42	6.06	5.80	5.61	5.47
10	10.04	7.56	6.55	5.99	5.64	5.39	5.20	5.06
11	9.65	7.21	6.22	5.67	5.32	5.07	4.89	4.74
12	9.33	6.93	5.95	5.41	5.06	4.82	4.64	4.50
13	9.07	6.70	5.74	5.21	4.86	4.62	4.44	4.30
14	8.86	6.51	5.56	5.04	4.69	4.46	4.28	4.14
15	8.68	6.36	5.42	4.89	4.56	4.32	4.14	4.00
16	8.53	6.23	5.29	4.77	4.44	4.20	4.03	3.89
17	8.40	6.11	5.19	4.67	4.34	4.10	3.93	3.79
18	8.29	6.01	5.09	4.58	4.25	4.01	3.84	3.71
19	8.18	5.93	5.01	4.50	4.17	3.94	3.77	3.63
20	8.10	5.85	4.94	4.43	4.10	3.87	3.70	3.56
21	8.02	5.78	4.87	4.37	4.04	3.81	3.64	3.51
22	7.95	5.72	4.82	4.31	3.99	3.76	3.59	3.45
23	7.88	5.66	4.76	4.26	3.94	3.71	3.54	3.41
24	7.82	5.61	4.72	4.22	3.90	3.67	3.50	3.36
25	7.77	5.57	4.68	4.18	3.85	3.63	3.46	3.32
26	7.72	5.53	4.64	4.14	3.82	3.59	3.42	3.29
27	7.68	5.49	4.60	4.11	3.78	3.56	3.39	3.26
28	7.64	5.45	4.57	4.07	3.75	3.53	3.36	3.23
29	7.60	5.42	4.54	4.04	3.73	3.50	3.33	3.20
30	7.56	5.39	4.51	4.02	3.70	3.47	3.30	3.17
40	7.31	5.18	4.31	3.83	3.51	3.29	3.12	2.99
50	7.17	5.06	4.20	3.72	3.41	3.19	3.02	2.89
60	7.08	4.98	4.13	3.65	3.34	3.12	2.95	2.82
70	7.01	4.92	4.07	3.60	3.29	3.07	2.91	2.78
80	6.96	4.88	4.04	3.56	3.26	3.04	2.87	2.74
90	6.93	4.85	4.01	3.53	3.23	3.01	2.84	2.72
100	6.90	4.82	3.98	3.51	3.21	2.99	2.82	2.69
120	6.85	4.79	3.95	3.48	3.17	2.96	2.79	2.66

Denominator degrees of freedom

Critical Values of *F* Distributions, $F > F_\alpha$, $\alpha = 0.01$ (cont…)

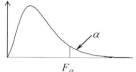

F_α

	Numerator degrees of freedom								
	9	10	12	15	20	24	30	60	120
1	6022.40	6055.93	6106.68	6156.97	6208.66	6234.27	6260.35	6312.97	6339.51
2	99.39	99.40	99.42	99.43	99.45	99.46	99.47	99.48	99.49
3	27.34	27.23	27.05	26.87	26.69	26.60	26.50	26.32	26.22
4	14.66	14.55	14.37	14.20	14.02	13.93	13.84	13.65	13.56
5	10.16	10.05	9.89	9.72	9.55	9.47	9.38	9.20	9.11
6	7.98	7.87	7.72	7.56	7.40	7.31	7.23	7.06	6.97
7	6.72	6.62	6.47	6.31	6.16	6.07	5.99	5.82	5.74
8	5.91	5.81	5.67	5.52	5.36	5.28	5.20	5.03	4.95
9	5.35	5.26	5.11	4.96	4.81	4.73	4.65	4.48	4.40
10	4.94	4.85	4.71	4.56	4.41	4.33	4.25	4.08	4.00
11	4.63	4.54	4.40	4.25	4.10	4.02	3.94	3.78	3.69
12	4.39	4.30	4.16	4.01	3.86	3.78	3.70	3.54	3.45
13	4.19	4.10	3.96	3.82	3.66	3.59	3.51	3.34	3.25
14	4.03	3.94	3.80	3.66	3.51	3.43	3.35	3.18	3.09
15	3.89	3.80	3.67	3.52	3.37	3.29	3.21	3.05	2.96
16	3.78	3.69	3.55	3.41	3.26	3.18	3.10	2.93	2.84
17	3.68	3.59	3.46	3.31	3.16	3.08	3.00	2.83	2.75
18	3.60	3.51	3.37	3.23	3.08	3.00	2.92	2.75	2.66
19	3.52	3.43	3.30	3.15	3.00	2.92	2.84	2.67	2.58
20	3.46	3.37	3.23	3.09	2.94	2.86	2.78	2.61	2.52
21	3.40	3.31	3.17	3.03	2.88	2.80	2.72	2.55	2.46
22	3.35	3.26	3.12	2.98	2.83	2.75	2.67	2.50	2.40
23	3.30	3.21	3.07	2.93	2.78	2.70	2.62	2.45	2.35
24	3.26	3.17	3.03	2.89	2.74	2.66	2.58	2.40	2.31
25	3.22	3.13	2.99	2.85	2.70	2.62	2.54	2.36	2.27
26	3.18	3.09	2.96	2.81	2.66	2.58	2.50	2.33	2.23
27	3.15	3.06	2.93	2.78	2.63	2.55	2.47	2.29	2.20
28	3.12	3.03	2.90	2.75	2.60	2.52	2.44	2.26	2.17
29	3.09	3.00	2.87	2.73	2.57	2.49	2.41	2.23	2.14
30	3.07	2.98	2.84	2.70	2.55	2.47	2.39	2.21	2.11
40	2.89	2.80	2.66	2.52	2.37	2.29	2.20	2.02	1.92
50	2.78	2.70	2.56	2.42	2.27	2.18	2.10	1.91	1.80
60	2.72	2.63	2.50	2.35	2.20	2.12	2.03	1.84	1.73
70	2.67	2.59	2.45	2.31	2.15	2.07	1.98	1.78	1.67
80	2.64	2.55	2.42	2.27	2.12	2.03	1.94	1.75	1.63
90	2.61	2.52	2.39	2.24	2.09	2.00	1.92	1.72	1.60
100	2.59	2.50	2.37	2.22	2.07	1.98	1.89	1.69	1.57
120	2.56	2.47	2.34	2.19	2.03	1.95	1.86	1.66	1.53

Denominator degrees of freedom

Critical Values of the Studentized Range, $q(a,v)$

a = number of groups
df = degrees of freedom for the experimental error
α = 0.05

df						Number of groups (a)										
	2	3	4	5	6	7	8	9	10	11	12	13	14	15	16
1	18.00	27.00	32.80	37.20	40.50	43.10	45.40	47.30	49.10	50.60	51.90	53.20	54.30	55.40	56.30
2	6.09	8.33	9.80	10.89	11.73	12.43	13.03	13.54	13.99	14.39	14.75	15.08	15.38	15.65	15.91
3	4.50	5.91	6.83	7.51	8.04	8.47	8.85	9.18	9.46	9.72	9.95	10.16	10.35	10.52	10.69
4	3.93	5.04	5.76	6.29	6.71	7.06	7.35	7.60	7.83	8.03	8.21	8.37	8.52	8.67	8.80
5	3.64	4.60	5.22	5.67	6.03	6.33	6.58	6.80	6.99	7.17	7.32	7.47	7.60	7.72	7.83
6	3.46	4.34	4.90	5.31	5.63	5.89	6.12	6.32	6.49	6.65	6.79	6.92	7.04	7.14	7.24
7	3.34	4.16	4.68	5.06	5.35	5.59	5.80	5.99	6.15	6.29	6.42	6.54	6.65	6.75	6.84
8	3.26	4.04	4.53	4.89	5.17	5.40	5.60	5.77	5.92	6.05	6.18	6.29	6.39	6.48	6.57
9	3.20	3.95	4.42	4.76	5.02	5.24	5.43	5.60	5.74	5.87	5.98	6.09	6.19	6.28	6.36
10	3.15	3.88	4.33	4.66	4.91	5.12	5.30	5.46	5.60	5.72	5.83	5.93	6.03	6.12	6.20
11	3.11	3.82	4.26	4.58	4.82	5.03	5.20	5.35	5.49	5.61	5.71	5.81	5.90	5.98	6.06
12	3.08	3.77	4.20	4.51	4.75	4.95	5.12	5.27	5.40	5.51	5.61	5.71	5.80	5.88	5.95
13	3.06	3.73	4.15	4.46	4.69	4.88	5.05	5.19	5.32	5.43	5.53	5.63	5.71	5.79	5.86
14	3.03	3.70	4.11	4.41	4.64	4.83	4.99	5.13	5.25	5.36	5.46	5.56	5.64	5.72	5.79
15	3.01	3.67	4.08	4.37	4.59	4.78	4.94	5.08	5.20	5.31	5.40	5.49	5.57	5.65	5.72
16	3.00	3.65	4.05	4.34	4.56	4.74	4.90	5.03	5.15	5.26	5.35	5.44	5.52	5.59	5.66
17	2.98	3.62	4.02	4.31	4.52	4.70	4.86	4.99	5.11	5.21	5.31	5.39	5.47	5.55	5.61
18	2.97	3.61	4.00	4.28	4.49	4.67	4.83	4.96	5.07	5.17	5.27	5.35	5.43	5.50	5.57
19	2.96	3.59	3.98	4.26	4.47	4.64	4.79	4.92	5.04	5.14	5.23	5.32	5.39	5.46	5.53
20	2.95	3.58	3.96	4.24	4.45	4.62	4.77	4.90	5.01	5.11	5.20	5.28	5.36	5.43	5.50
24	2.92	3.53	3.90	4.17	4.37	4.54	4.68	4.81	4.92	5.01	5.10	5.18	5.25	5.32	5.38
30	2.89	3.48	3.84	4.11	4.30	4.46	4.60	4.72	4.83	4.92	5.00	5.08	5.15	5.21	5.27
40	2.86	3.44	3.79	4.04	4.23	4.39	4.52	4.63	4.74	4.82	4.90	4.98	5.05	5.11	5.17
60	2.83	3.40	3.74	3.98	4.16	4.31	4.44	4.55	4.65	4.73	4.81	4.88	4.94	5.00	5.06
120	2.80	3.36	3.69	3.92	4.10	4.24	4.36	4.47	4.56	4.64	4.71	4.78	4.84	4.90	4.95
∞	2.77	3.32	3.63	3.86	4.03	4.17	4.29	4.39	4.47	4.55	4.62	4.68	4.74	4.80	4.84

Bibliography

Allen, M.P. (1977) Understanding Regression Analysis. Plenum Press, New York and London.

Allison, P. D. (1999) Logistic Regression Using the SAS System: Theory and Application. SAS Institute Inc., Cary, North Carolina.

Box, G. E. P. (1978) Statistics for Experimenters: An Introduction to Design, Data Analysis, and Model Building. John Wiley & Sons, New York.

Bozdogan, H. (1987) Model selection and Akaike's information criterion (AIC); the general theory and its analytical extensions. Psychometrika 52:345-370.

Breslow, N. R. and Clayton, D. G. (1993) Approximate inference in generalized linear mixed models. Journal of the American Statistical Association 88:9-25.

Cameron, A. Colin and Pravin K. Trevedi (1998) Regression Analysis of Count Data. Cambridge University Press, Cambridge.

Casella, G. and Berger, R. L. (1990) Statistical Inference. Wadsworth & Brooks / Cole, Belmont, California.

Chatterjee, S., Hadi, A. and Price, B. (2000) Regression Analysis by Example. John Wiley & Sons, New York.

Chen, C. (2002) Robust Regression and Outlier Detection with the ROBUSTREG Procedure. *Proceedings of the Twenty-seventh Annual SAS Users Group International Conference*, SAS Institute Inc., Cary, North Carolina.

Christensen, R., Pearson, L. M. and Johnson, W. (1992) Case-deletion diagnostics for mixed models. Technometrics 34:38-45.

Clarke, G. M. (1994) Statistics and Experimental Design: An Introduction for Biologists and Biochemists. Third Edition. Oxford University Press, New York.

Cochran, W. G. and Cox, G. M. (1992) Experimental Designs, Second Edition. John Wiley & Sons, New York.

Cody, R. P. and Smith, J. K. (2005) Applied Statistics and the SAS Programming Language, Fifth Edition. Prentice Hall, Upper Saddle River, New Jersey.

Cohen, J. and Cohen, P. (1983) Applied Multiple Regression / Correlation Analysis for the Behavioral Sciences, Second Edition. Lawrence Erlbaum Associates, New Jersey.

Collins, C. A. and Seeney, F. M. (1999) Statistical Experiment Design and Interpretation: An Introduction with Agricultural Examples. John Wiley & Sons, New York.

Cox, D. R. (1958) Planning of Experiments. John Wiley & Sons, New York.

Crowder, M. J. and Hand, D. J. (1990) Analysis of Repeated Measures. Chapman & Hall, London.

Daniel, W. W. (1990) Applied Nonparametric Statistics, Second Edition. PWS-Kent Publishing Company, Boston, Massachusetts.

Der, G. and Everitt, B. (2001) A Handbook of Statistical Analyses Using SAS, Second Edition. Chapman & Hall / CRC Press, London.

Diggle, P. J. (1988) An approach to the analysis of repeated measurements. Biometrics 44:959-971.

Diggle, P.J., Liang, K.Y. and Zeger, S.L. (1994) Analysis of Longitudinal Data. Clarendon Press, Oxford.

Draper, N.R. and Smith, H. (1981) Applied Regression Analysis. Second Edition. John Wiley & Sons, New York.

Elliott, R. J. (1995) Learning SAS in the Computer Lab. Duxbury Press, Boston, Massachusetts.

Fitzmaurice, G. M., Laird, N. M. and Ware, J. H. (2004) Applied longitudinal analysis. John Wiley & Sons, Inc., New Jersey.

Fox, J. (1997) Applied Regression Analysis, Linear Models, and Related Methods. Sage Publications, Thousand Oaks, California.

Freund, R. J. and Wilson W. J. (1998) Regression Analysis: Statistical Modeling of a Response Variable. Academic Press, New York.

Freund, R. and Littell, R. (2000) SAS System for Regression, Third Edition. SAS Institute Inc., Cary, North Carolina.

Geoff Der, Everott, G. B. and Everitt, B. (1996) A Handbook of Statistical Analysis Using SAS. CRC Press, London.

Gianola, D. and Hammond, K. (eds) (1990) Advances in Statistical Methods for Genetic Improvement of Livestock. Springer-Verlag, New York.

Hamilton, L.C. (1992) Regression with Graphics: A Second Course in Applied statistics. Wadsworth, Belmont, California.

Hartley, H. O. and Rao, J. N. K. (1967) Maximum likelihood estimation for the mixed analysis of variance model. Biometrika 54:93-108.

Harville, D. A. (1977) Maximum likelihood approaches to variance component estimation and to related problems. Journal of the American Statistical Association 72:320-340.

Harville, D. A. (1997) Matrix Algebra from a Statistician's Perspective. Springer-Verlag, New York.

Heath, D. (1995) Introduction to Experimental Design and Statistics for Biology. UCL Press, London.

Henderson, C. R. (1963) Selection index and expected genetic advance. In: Statistical Genetics and Plant Breeding (W. D Hanson and H. F. Robinson, eds). National Academy of Sciences and National Research Council Publication No 982, pp. 141-146.

Henderson, C. R. (1984) Application of Linear Models in Animal Breeding. University of Guelph, Guelph.

Hinkelmann, K. (1984) Experimental Design, Statistical Models, and Genetic Statistics. Marcel Dekker, New York.

Hinkelmann, K. (1994) Design and Analysis of Experiments: Introduction to Experimental Design. John Wiley & Sons, New York.

Hoshmand, R. (1994) Experimental Research Design and Analysis: A Practical Approach for the Agricultural and Natural Sciences. CRC Press, London.

Iman, R. L. (1994) A Data-Based Approach to Statistics. Duxbury Press, Belmont, California.

Jennrich, R. I. and Schluchter, M. D. (1986) Unbalanced repeated measures model with structured covariance matrices. Biometrics 42:805-820.

Johnson, R. A., Wichern, D. A. and Wichern, D. W. (1998) Applied Multivariate Statistical Analysis. Prentice Hall, Upper Saddle River, New Jersey.

Kenward, M. G. and Roger, J. H. (1997) Small sample inference for fixed effects from Restricted Maximum Likelihood. Biometrics 53:983-997.

Kuehl, R.O. (1999) Design of Experiments: Statistical Principles of Research Design and Analysis. Duxbury Press, New York.

Laird, N. M. and Ware, J. H. (1982) Random effect models for longitudinal data. Biometrics 38:963-974.

LaMotte, L. R. (1973) Quadratic estimation of variance components. Biometrics 29:311-330.

Lindsey, J. K. (1993) Models for Repeated Measurements. Clarendon Press, Oxford.

Lindsey, J. K. (1995) Modeling Frequency and Count Data. Oxford University Press, New York.

Lindsey, J. K. (1997) Applying Generalized Linear Models. Springer-Verlag, New York.

Littel, R. C., Freund, R. J. and Spector, P.C. (1991) SAS® System for Linear Models, Third Edition. Sas Institute Inc., Cary, North Carolina.

Littel, R. C., Miliken, G. A., Stroup, W. W. Wolfinger, R. D. and Schabenberger, O. (2006) SAS® for Mixed Models, Second edition. SAS Institute Inc., Cary, North Carolina.

Littell, R., Stroup, W. and Freund, R. (2002) SAS® System for Linear Models, Fourth Edition. SAS Institute, Cary, North Carolina.

Little, T. M. and Hills, F. J. (1978) Agricultural Experimentation. John Wiley & Sons, New York.

Long, J. S. (1997) Regression Models for Categorical and Limited Dependent Variables. Sage Publications, Thousand Oaks, California.

Louis, T. A. (1988) General methods for analysing repeated measurements. Statistics in Medicine 7:29-45.

McClave, J. T. and Dietrich II, F. H. (1987) Statistics, Third Edition. Duxbury Press, Boston, Massachusetts.

McCullagh, P. and Nelder, J. A. (1989) Generalized Linear Models, Second Edition. Chapman and Hall, New York.

McCulloch, C. E. and Searle, S. R. (2001) Generalized, Linear and Mixed Models. John Wiley & Sons, New York.

McLean, R. A., Sanders, W. L. and Stroup, W. W. (1991) A unified approach to mixed model linear models. The American Statistician 45:54-64.

McNeil, K., Newman, I. and Kelly, F. J. (1996) Testing Research Hypotheses with the General Linear Model. Southern Illinois University Press, Carbondale, Illinois.

Mead, R. (1988) The Design of Experiments. Cambridge University Press, New York.

Mendenhall, W. and Sincich, T. (1988) Statistics for the Engineering and Computer Sciences. Dellen Publishing Company, San Francisco, California.

Milliken, G. A. and Johnson, D. E. (1994) Analysis of Messy Data, Volume 1: Designed Experiments. Chapman Press, New York.

Montgomery, D. C. (2000) Design and Analysis of Experiments, Fifth Edition. John Wiley & Sons, New York.

Mood, A. M., Graybill, F. A. and Boes, D. C. (1974) Introduction to the Theory of Statistics. McGraw-Hill, New York.

Morris, T. R. (1999) Experimental Design and Analysis in Animal Science. CAB International, Wallingford, Oxon, UK.

Mrode, R. A. (1996) Linear Models for the Prediction of Animal Breeding Values. CAB International, Wallingford, Oxon, UK.

Myers, R. H. (1990) Classical and Modern Regression with Applications. PWS-KENT Publishing Company, Boston, Massachusetts.

Nelder, J. A. and Weddenburn, R. W. M. (1972) Generalised linear models. Journal of the Royal Statistical Society A 135:370-384.

Neter, J., Wasserman, W. and Kutner, M. H. (1996) Applied Linear Statistical Models. Fourth Edition. McGraw-Hill Higher Education, New York.

Patterson, H. D. and Thompson, R. (1974) Recovery of inter-block information when block sizes are unequal. Biometrika 58:545-554.

Pollard, J. H. (1977) A Handbook of Numerical and Statistical Techniques. Cambridge University Press, Cambridge.

Rao, C. R. (1972) Estimation of variance and covariance components in linear models. Journal of the American Statistical Association 67:112-115.

Rutter, C. M. and Elashoff, R. M. (1994) Analysis of longitudinal data: random coefficient regression modeling. Statistics in Medicine 13:1211-1231.

SAS® User's Guide: Statistics. (1995) SAS Institute Inc., Cary, North Carolina.

SAS®. SAS/STAT 9.1 User's Guide. (2004) SAS Institute Inc., Cary, North Carolina.

Schefler, W. C. (1969) Statistics for the Biological Sciences. Addison-Wesley Publishing, Reading, Massachusetts.

Searle, S. R. (1971) Linear Models. John Wiley & Sons, New York.

Searle, S. R. (1982) Matrix Algebra Useful for Statisticians. John Wiley & Sons, New York.

Searle, S. R., Casella, G. and McCulloh, C. E. (1992) Variance Components. John Wiley & Sons, New York.

Silobrčić, V. (1989) Kako sastaviti i objaviti znanstveno djelo. JUMENA, Zagreb.

Silverman, B. W. (1992) Density Estimation for Statistics and Data Analysis. Chapman & Hall, London.

Snedecor, G. W. and Cochran, W. G. (1989) Statistical Methods. Eighth Edition. Iowa State University Press, Ames, Iowa.

Sokal, R.R. and Rohlf, F.J. (1995) Biometry. Third Edition. W.H. Freeman and Company, New York.

Stuart, A. and Ord, K. (1991) Kendall's Advanced Theory of Statistics. Volume 2. Classical Inference and Relationship. Edward Arnold.

Tanner, M. (1993) Tools for Statistical Inference. Springer-Verlag, New York.

Weber, D. and Skillings, J. H. (1999) A First Course in the Design of Experiments: A Linear Models Approach. CRC Press, London.

Verbeke, G. and Molenberghs, G. (2000) Linear mixed models for longitudinal data. Springer Verlag, NewYork.

Winer, B. J. (1971) Statistical Principles in Experimental Design, Second edition. McGraw-Hill Inc, New York.

Wolfinger, R. D. (1996) Heterogeneous variance covariance structure for repeated measures. Journal of Agricultural, Biological and Environmental Statistics 1:205-230.

Zeger, S. L., Liang, K. Y. and Albert, P. S. (1988) Models for longitudinal data: a generalized estimating equation approach. Biometrics 44:1049-1060.

Zelterman, D. (2002) Advanced Log-Linear Models Using SAS. SAS Institute Inc., Cary, North Carolina.

Zolman, J. F. (1993) Biostatistics: Experimental Design and Statistical Inference. Oxford University Press, New York.

Subject Index